Advanced Learning Analytics Methods

Mohammed Saqr • Sonsoles López-Pernas
Editors

Advanced Learning Analytics Methods

AI, Precision and Complexity

Springer

Editors
Mohammed Saqr
University of Eastern Finland
Joensuu, Finland

Sonsoles López-Pernas
University of Eastern Finland
Joensuu, Finland

ISBN 978-3-031-95364-4 ISBN 978-3-031-95365-1 (eBook)
https://doi.org/10.1007/978-3-031-95365-1

This work was supported by University of Eastern Finland.

This Springer imprint is published by the registered company Springer Nature Switzerland AG
The registered company address is: Gewerbestrasse 11, 6330 Cham, Switzerland

To our families, friends, and loved ones.

A special dedication from Mohammed to his daughters, the exquisite roses, Carmen and Layla.

Preface

Learning analytics (LA) has grown over the years, both in methods and practices. As an applied field, LA embraced a wide range of research methodologies and traditions, including those from computer science, education, statistics, and psychology. While such growth has produced a vast body of research and brought significant progress, the past few years have raised several critical questions: What has LA achieved? What is the current status of the field, and where are we heading? And, more importantly, what is LA as a field?

The meteoric rise of artificial intelligence (AI), and generative AI in particular, has sent shockwaves through many disciplines, and LA is no exception. As such, the field has further expanded to encompass the new technology as it has always done. In doing so, LA has grown even larger and extended to new areas, raising more questions about what LA is after all. In fact, the borders between LA, AI, educational data mining, and quantitative methods in general have always been unclear.

Notwithstanding the LA identity debate, this book addresses several novel issues in analytics and quantitative methods in education from a wider methodological perspective. This book covers a range of topics, including LA, AI, precision education, and complex systems. While we do not claim to define the direction of the field, our aim is to guide researchers in exploring some of these barely trodden territories. We refrain from emphasizing or asserting that any topic lies within the definitive borders of LA. Instead, this book presents these advanced topics of significant interest to educational researchers, without placing undue focus on taxonomy or rigid classifications. Such borders have always been—at least to us and many others—more artificial than real. In fact, almost every methodology we covered in the previous LA methods book came years before LA included them under its veneer. As such, this book you are reading comprises four overarching methodological themes that reflect its title, "Advanced LA." Still, we make no stress on LA in its narrow definition, and in fact, many chapters and discussions address novel advances and methodologies without being restrictive to LA.

The first theme of this book addresses AI and machine learning (ML), which have become increasingly important nowadays. Including AI and ML topics fills a gap in our previous book and introduces several current topics that are of interest to all researchers and practitioners. We also made sure to include introductory chapters that present these topics in considerable detail to the readers so they can get acquainted with the basic knowledge required to read the dense methodological chapters. This enhancement over the previous book makes it easier for early researchers to have a swift introduction.

This book then introduces the "talk of the town": large language models (LLMs) and natural language processing (NLP) topics. The section introduces the foundational concepts of LLMs and their potential applications in education in an introductory chapter that discusses the main concepts. Then, several tutorials cover areas like using LLMs to explain predictive models, code textual data, and offer insights to researchers. Relevant to this theme is the use of NLP to understand students' written text, which is presented with several advanced techniques.

The third theme of this book covers complex systems. While complex systems have become an important part of several disciplines and brought several breakthroughs in modeling intractable problems, the concept has so far been hardly visible in educational research. Therefore, we introduced several chapters that cover complex systems methods and different modeling techniques. As is the case with other sections, we start with an introduction to the topic of complex systems written in clear language to acquaint the readers with the topic.

More importantly, we included three chapters to introduce transition network analysis (TNA). TNA fills an important gap in modeling the temporal unfolding of the learning process across time from a complex system perspective. What makes TNA unique is that it introduces much-needed rigor. Further, TNA is crafted specifically for educational data, unlike other process-oriented methods, which are adopted from other fields with no emphasis on rigor or validity of inferences.

The last section of this book addresses precision education and, more specifically, person-centered and person-specific methods. While such topics are gaining traction and attention in medicine, psychology, and other fields, they remain faintly visible in education research in general and LA in particular. This section of the book introduces a chapter about the general principles and then moves to demonstrate several idiographic methods.

Of course we have not covered every methodological topic nor have we written every tutorial we hoped for. Still, we think that the two books offer a comprehensive reference to interested researchers. In doing so, we believe that researchers have a guide to help them tread swiftly through the complexities of methods in analytics, data mining, and education research at large.

Supplementary Material Code: https://github.com/lamethods/code2
Data: https://github.com/lamethods/data2

Joensuu, Finland
2025

Mohammed Saqr
Sonsoles López-Pernas

Acknowledgments We are thankful for the great work everyone has done to make this book happen. The authors have spent valuable time compiling these chapters and bringing their expertise to the wider audience of learning analytics. The authors from other fields did a great job and took the effort to navigate concepts they had little contact with to present their methods to a completely different audience than what they are used to. We are also grateful to the reviewers who offered valuable feedback to refine and improve the chapters. A special thanks goes to the associate editors who worked with us to process this large number of chapters and handle the peer review process.

We would like to acknowledge the generous funding by the Academy of Finland (Research Council of Finland) for the project TOPEILA, Decision Number 350560, which was received by Mohammed Saqr, and CRETIC, Decision Number 360746, which was received by Sonsoles López-Pernas.

We would like to thank the generous funding from the University of Eastern Finland and, in particular, Markku Tukianen and Kari Lehtinen, who made this book open access for everyone.

Lastly, we would like to acknowledge the use of large language models in formatting some code snippets for better presentation, commenting on certain parts of the code to enhance readability, or generating some of the illustrative figures. Additionally, some limited passages were reviewed for grammar and spelling to ensure clarity. Notwithstanding these stylistic usage, the entire book is entirely the result of our own work and all writing completed by the authors.

Competing Interests The authors have no conflicts of interest to declare that are relevant to the content of this book.

About This Book

This open-access peer-reviewed book, *Advanced Learning Analytics: AI, Precision, and Complexity*, offers a comprehensive methodological guide to advanced learning analytics (LA), artificial intelligence (AI), machine learning (ML), complex systems, and precision education. This book is well-suited for new and experienced researchers to help them explore the innovative methodologies that are reshaping education nowadays. The text is divided into four main sections, each with a distinct methodological theme. The first section introduces the increasingly important theme of AI and ML in educational contexts. It covers a range of topics, including explainable AI (XAI), predictive analytics, and classification methods. This section is a foundational resource for researchers who wish to explore how AI and ML can be applied in real-world educational settings and provides both introductory and advanced chapters to ensure accessibility for readers at all levels. Closely related to this theme is the hot topic of large language models and natural language processing which receive deep coverage with their potential applications in education in several tutorials covering areas like using LLMs to explain predictive models, code text, and offer insights to researchers. The third section of this book turns to complex dynamic systems—an area that has seen significant advancements across disciplines but remains relatively underexplored in educational research. The chapters in this section introduce methods such as transition network analysis (TNA), which offers a robust approach to modeling the temporal aspects of the learning process. The third section focuses on precision education, which emphasizes personalized, person-centered approaches to teaching and learning. This part explores idiographic analysis techniques, which center on individual students' learning trajectories, and introduces emerging methods that prioritize the specificity and individuality of educational experiences. This book offers a series of tutorials and guides with open code and datasets that provide practical insights into the use of advanced data analysis techniques, from automated machine learning (AutoML) to network modeling and temporal data analysis. These practical guides are designed so that researchers and practitioners can implement these methods in their own work, thereby enhancing the overall impact of learning analytics on educational research.

Contents

Editors and Contributors

Editors

Mohammed Saqr University of Eastern Finland, Kuopio, Finland

Sonsoles López-Pernas University of Eastern Finland, Kuopio, Finland

Associate Editors

Sami Heikkinen LAB University of Applied Sciences, Lahti, Finland

Rwitajit Majumdar Kumamoto University, Kumamoto, Japan

Ahmed Tlili Beijing Normal University, Beijing, China

Contributors

Rémi Altamore Université de Montpellier, Montpellier, France

Okan Bulut University of Alberta, Edmonton, AB, Canada

Daryn Dever University of Florida, Gainesville, FL, USA

Christophe Gernigon University of Montpellier, Montpellier, France

Jiesi Guo Australian Catholic University, Brisbane, QLD, Australia

Feifei Han Australian Catholic University, Brisbane, QLD, Australia

Sami Heikkinen LAB University of Applied Sciences, Lahti, Finland

Hibiki Ito University of Helsinki, Helsinki, Finland

Avi Kaplan Temple University, Philadelphia, PA, USA

Halil Kayaduman University of Eastern Finland, Kuopio, Finland

Tian Li University of Queensland, Brisbane, QLD, Australia

Rwitajit Majumdar Kumamoto University, Kumamoto, Japan

Gwen Marchand University of Nevada, Reno, NV, USA

Sonsoles López-Pernas University of Eastern Finland, Kuopio, Finland

Kamila Misiejuk FernUniversität in Hagen, Hagen, Germany

Eduardo Oliveira University of Melbourne, Parkville, VIC, Australia

Clément Roume Université de Haute-Alsace, Mulhouse, France

Mohammed Saqr University of Eastern Finland, Kuopio, Finland

Yige Song University of Melbourne, Parkville, VIC, Australia

Anne Teboul Université de Montpellier, Montpellier, France

Santtu Tikka University of Jyväskylä, Jyväskylä, Finland

Ahmed Tlili Beijing Normal University, Beijing, China

Leonie V. D. E. Vogelsmeier Tilburg University, Tilburg, The Netherlands

Tarid Wongvorachan University of Alberta, Edmonton, AB, Canada

Jinran Wu Australian Catholic University, Brisbane, QLD, Australia

Reviewers

Rémi Altamore University of Montpellier, Montpellier, France

Okan Bulut University of Alberta, Edmonton, AB, Canada

Shaun Cavestany University of Eastern Finland, Kuopio, Finland

Javier Conde Universidad Politécnica de Madrid, Madrid, Spain

Tudor Cristea Eindhoven University of Technology, Eindhoven, The Netherlands

Mouna Denden Université Polytechnique Hauts-de-France, Valenciennes, France

Melis Dülger Utrecht University, Utrecht, The Netherlands

Ramy Elmoazen University of Eastern Finland, Kuopio, Finland

Hamza Hammami University of Tunis El Manar, Tunis, Tunisia

Sami Heikkinen LAB University of Applied Sciences, Lahti, Finland

Halil Kayaduman Inonu University, Malatya, Turkey

Ville Kivimäki Aalto University, Espoo, Finland

Qinyi Liu University of Bergen, Bergen, Norway

Nidia Guadalupe López Flores Reykjavik University, Reykjavik, Iceland

Sonsoles López-Pernas University of Eastern Finland, Kuopio, Finland

Joonas Merikko Annie Advisor, Helsinki, Finland

Kamila Misiejuk FernUniversität in Hagen, Hagen, Germany

Ramkumar Rajendran IIT Bombay, Mumbai, India

Daevesh Singh IIT Bombay, Mumbai, India

Jaani Väisänen LAB University of Applied Sciences, Lahti, Finland

Pavani Vemuri KU Leuven, Leuven, Belgium

Esteban Villalobos Université Toulouse III, Toulouse, France

Megan Wiedbusch University of Central Florida, Orlando, FL, USA

Tarid Wongvorachan University of Alberta, Edmonton, AB, Canada

Editorial Support

Ibrahim Belayachi Université de Technologie de Compiègne, Compiègne, France

Jules Delannoy Université de Technologie de Compiègne, Compiègne, France

Cyprien Dujardin Université de Technologie de Compiègne, Compiègne, France

Unpacking Learning in the Age of AI: Bridging AI, Complexity, and Precision Education

Sonsoles López-Pernas, Sami Heikkinen, Ahmed Tlili, Rwitajit Majumdar, and Mohammed Saqr

Abstract This chapter serves as an introduction to the book *Advanced Learning Analytics Methods: AI, Precision, and Complexity*. In a previous volume—*Learning Analytics Methods and Tutorials: A Practical Guide Using R*—, we introduced the foundational methods of learning analytics, required to bring researchers up to speed with the state of the art. Since then, recent advances in artificial intelligence and complexity science have permeated the field of learning analytics and education research as a whole. As a response, this new book aims to provide researchers with the resources needed to address the methodological expansion required to meet the demands of these advancements. In this new book, we introduce cutting-edge methods at the intersection of learning analytics, artificial intelligence and complexity science, with the hope that this initiative will drive significant advancements in the field.

Keywords Learning analytics · Idiographic analysis · Complex systems · Explainable artificial intelligence

1 Introduction

The birth of learning analytics as a field marked a vast surge of data-intensive studies and exponential growth of methods and applications. However, there was

S. López-Pernas (✉) · M. Saqr
University of Eastern Finland, Joensuu, Finland
e-mail: sonsoles.lopez@uef.fi

S. Heikkinen
LAB University of Applied Sciences, Lahti, Finland

A. Tlili
Beijing Normal University, Beijing, China

R. Majumdar
Kumamoto University, Kumamoto, Japan

M. Saqr, S. López-Pernas (eds.), *Advanced Learning Analytics Methods*,
https://doi.org/10.1007/978-3-031-95365-1_1

a lack of reliable methodological guidance for researchers and practitioners that remained unfulfilled for more than a decade. As a response, we released the first methodological book in 2024 [1]. The book ***Learning Analytics Methods and Tutorials: A Practical Guide Using R*** offered a much-needed resource for both newcomers and experts in the learning analytics field. The book started by introducing the R programming language to help beginners get up to speed with their R programming skills [2–4]. Most importantly, the book covered the foundational methods in learning analytics, such as sequence analysis, process mining, predictive modeling, Markov models, and social network analysis [5]. While those methods constitute the bedrock of learning analytics research and much of the advanced techniques, several aspects still require coverage. During the last few years—and especially the previous two—the field has witnessed remarkable changes, especially those related to artificial intelligence (AI) and the emergence of generative AI [9, 10]. Taken together, there is still a gap in methods and a need to catch up with the fast-advancing field and the recent surge in AI.

To fill this gap, we set out on a journey to create a new book: ***Advanced Learning Analytics Methods: AI, Precision and Complexity***. This book aims to expand on foundational methods and integrate emerging innovations such as AI-driven insights, complexity science, and idiographic approaches, providing a comprehensive toolset to map the learning process. The book builds on the foundation laid by its predecessor, taking a significant step forward both methodologically —covering AI methods and their application in education—and theoretically—exploring idiographic analysis and complexity science. Given the novelty of several topics, the book included a mixture of foundational chapters that offered a comprehensive introduction to these novel topics in thematic sections, i.e., a chapter on complexity, two others on AI, and one on idiographic analysis.

2 Thematic Overview of the Book

The book is structured in four sections covering four main themes. This thematic categorization—albeit imperfect due to the unavoidable overlap—helps align the chapters within a common narrative to facilitate the book organization. Below, we describe each of these sections as well as the included chapters and give an overview of what these chapters discuss, how they are related together, and which parts of the methodological story they tell.

2.1 Section I: Artificial Intelligence

The first section of the book covers the synergies between AI, learning analytics, and education.

Chapter 2 [11] kicks off this section with an overview of AI applications in education, including adaptive learning systems, profiling, and predictive modeling. Moreover, it introduces the main concepts and techniques related to explainable AI (XAI) and how AI can be designed to ensure trustworthiness and equity in AI-driven educational environments. The chapter also provides readers with an understanding of AI in education and XAI's potentials and serves as a foundation for the remaining chapters presented in this section.

Another focus of this section is predictive modeling with AI, one of the central applications of AI in education [12, 13]. While our previous book covered this topic [14], it was approached from a statistical lens without introducing more recent developments in machine learning (ML). This book section provides a more modern approach to predictive modeling, focusing on methods and workflows to perform ML regression and classification tasks with R. The ability to predict student performance has been a main motivator for adopting learning analytics, and one of the central applications of AI in education. Predictive modeling uses historical data to forecast outcomes such as student grades, providing timely support to at-risk students. Chapter 3 [15] presents ML predictive methods to forecast continuous variables like grades. The methods discussed include Random Forest, K-Nearest Neighbor, Linear Regression, Neural Networks, and Support Vector Machines. The chapter presents two approaches: the traditional predictive modeling workflow and a more contemporary approach using the tidymodels framework. Chapter 4 [16] moves on to classify students based on academic performance levels, such as identifying low achievers or potential dropouts. The chapter begins by illustrating the process with a Random Forest classifier, relying on engagement indicators to distinguish between high and low achievers. It covers the entire modeling workflow, including data preparation, model training, and evaluation with performance metrics, and goes beyond the classic ML prediction algorithms. Chapter 5 [17] introduces advanced regularization techniques—Least Absolute Shrinkage and Selection Operator (LASSO), Smoothly Clipped Absolute Deviation (SCAD), and Minimax Concave Penalty (MCP)—to construct predictive models. These methods are useful when dealing with a large number of potential predictors, enabling the selection of the most relevant variables to create effective and interpretable models. As such, Chap. 5 is particularly useful in scenarios with complex and high-dimensional data.

The focus is then shifted to XAI as a solution to address some of the issues that arise from applying AI in education [18]. A key barrier to AI adoption is the "opaque" nature of many ML models, which can —and actually does— hinder trust, fairness, and accountability. To address this issue, explainability methods are essential for making predictions—such as identifying at-risk students, grading essays, or detecting plagiarism—that are more transparent and interpretable. Chapter 6 [19] introduces the DALEX R package as a tool for explaining ML models in educational settings and demonstrates how to interpret regression models (e.g., predicting student grades) and classification models (e.g., distinguishing high from low achievers) using variable importance measures, partial dependence plots, and accumulated local effects. Even though the aforementioned global explanation methods can offer aggregate insights into a model's overall behavior, local explanations are needed to uncover the reasons behind specific outcomes, such as why a particular student was identified as at risk of dropping out. Chapter 7 [20] places the focus on the importance of local explanations in educational settings and explores three widely used techniques: Break Down plots, SHAP (SHapley Additive exPlanations), and LIME (Local Interpretable Model-agnostic Explanations).

The aforementioned techniques are essential to understand why a certain decision has been made, e.g., why a student has been flagged as "at-risk", why an essay has been marked as "excellent", or why an assignment has been labeled as "plagiarized".

2.2 Section II: Large Language Models and Natural Language Processing

Moving on from predictive modeling, the second section of the book focuses on the transformative role of another aspect of AI—natural language processing (NLP) and large language models (LLMs). NLP offers new opportunities to analyze and interpret textual data, such as student essays, discussion forums, and feedback, which provide deeper insights into learning processes and outcomes. The first chapter [21] introduces LLMs as the foundation of modern advancements in education. LLMs enable a wide range of applications, including automated feedback, question generation, sentiment analysis, and multilingual support, making them invaluable in diverse educational contexts. This chapter explains the foundational mechanisms behind LLMs, such as the transformer architecture. It also highlights the varied applications of LLMs in education, many of which will be explored in detail in later tutorial chapters. The next chapter [22] stresses the importance of utilizing textual data in education to enhance learning analytics systems and support informed pedagogical decisions. It introduces foundational concepts in NLP, including lexical and syntax analysis, corpus creation, and the significance of text preprocessing techniques like tokenization, stemming, lemmatization, and stop word removal, with practical demonstrations. The chapter offers a tutorial on the main NLP applications, including Term Frequency-Inverse Document Frequency (TF-IDF) for keyword identification, Latent Dirichlet Allocation (LDA) for topic modeling, and text summarization and demonstrates how these techniques reveal themes and insights in educational texts. The chapter concludes by showcasing sentiment analysis to evaluate student feedback, providing actionable insights into student perceptions and attitudes.

Chapter 10 [23] addresses the challenges of coding text data in qualitative research, a process often prone to errors and labor-intensive efforts. The chapter explores using LLMs to automate text classification, leveraging word embeddings and BERT-like models through the R package text. The chapter demonstrates an ML pipeline that combines word embeddings with supervised algorithms to achieve accurate text coding. A case study on collaborative problem-solving illustrates how a small sample of manually coded data can train a model to classify the remaining data. The tutorial also includes evaluating coding accuracy using standard machine learning metrics and measures of inter-rater reliability. Finally, the chapter applies learning analytics techniques to compare insights from human-coded and machine-coded data and shows the potential of LLMs to streamline and enhance qualitative research processes.

To conclude this section, Chapter 11 [24] investigates the integration of LLMs into the explainable AI pipeline. The chapter demonstrates how LLMs can automate the transformation of explanations—such as feature importance, partial dependence profiles, and local explanations—into natural language narratives that are accessible to non-technical stakeholders. This work advances the previous chapters on predictive ML, XAI, and learning analytics and emphasizes transparency, inclusivity, and fairness in educational technologies. In doing so, it offers a case study of using LLMs to push the boundaries of traditional predictive analytics and bring nuance to these models.

2.3 Section III: Complex Dynamic Systems

Complex dynamic systems as an analytical framework have lately gained traction in educational research by offering a set of methods that uncover the temporal and relational patterns within the learning process [18, 25–27]. This complex dynamic systems lens has been used to analyze complex learning-related constructs such as engagement [28, 29], self-regulation [26, 30, 31], and language learning [32]. The first chapter in this section [33] addresses the limitations of traditional methods in capturing the dynamic and evolving nature of learning processes by introducing complex systems theory for educational research. It outlines the key characteristics of complex systems, including non-linear relationships, emergent properties, and feedback mechanisms, to explain how educational phenomena develop over time. Methodological approaches such as network analysis and recurrence quantification analysis are described as tools to study relationships and patterns in learning. The chapter concludes by discussing data collection methods that align with a complex systems perspective, providing a foundation for studying learning processes through this innovative lens.

Chapter 13 [34] explores the role of network psychometrics as a complement to traditional factor analysis in education research. It goes one step further from the psychological network chapter in the previous book [35]. While factor analysis simplifies relationships by modeling latent variables like cognitive ability, engagement, and motivation, the network approach examines dynamic, unrestricted associations among observed variables. Building on concepts introduced in our previous book [35], this chapter presents advanced applications of psychological networks in learning analytics. The chapter introduces unique variable analysis to identify the main contributors within complex systems, reducing redundancy and simplifying datasets. Additionally, the chapter demonstrates methods for assessing system integrity through exploratory graph analysis (EGA), hierarchical EGA, and dynamic EGA, which enables tracking changes in system structures over time. Finally, the chapter highlights how psychological networks derived from EGA can be transformed into factor analytic formats, illustrating how these two approaches complement each other in studying the complex structures of learning processes.

Another method suitable for analyzing complex systems is Recurrence Quantification Analysis (RQA), a technique used to identify and quantify the repetition of patterns within nonlinear dynamical systems [36, 37] described in Chapter 14 [38]. RQA examines how system behaviors transition between repetitive and novel sequences over time, analyzing revisits to prior states or the evolution into new behaviors across various time lags. Applicable to both categorical and continuous data, RQA supports the analysis of single or multiple time series across diverse contexts. Although less commonly used in education, RQA has been applied to study collaborative learning, self-regulated learning, and literacy processes, using data sources such as eye tracking, log files, physiology, and verbalizations. This chapter reviews RQA methodologies, highlights its applications in educational research, and discusses how it can quantify learning dynamics, distinguish functional from dysfunctional learning behaviors, and enhance collaborative learning.

Chapter 15 [39] introduces Transition Network Analysis (TNA), a new method for capturing the relational dynamics of temporal processes by modeling transitions between events as a weighted directed network. TNA leverages the potentials of network analysis and process mining, offering insights through graph, node, and edge-level metrics [7, 40], and also provides tools for detecting recurring patterns such as dyads and triads, and for identifying communities [41]. More importantly, TNA incorporates statistical validation methods—e.g., bootstrapping, permutation, and case-dropping techniques—to ensure the robustness of research findings. TNA also enables the inclusion of covariates to explain emergent patterns or examine subgroup differences, enhancing its utility for hypothesis testing and theory development. The chapter includes a practical tutorial using the tna R package and demonstrates these features through a case study on group regulation dynamics [42]. Chapter 16 [43] builds on these foundations and presents Frequency-Based Transition Network Analysis (FTNA) as an alternative approach for modeling the relational dynamics and transitions between states or events. While TNA leverages probabilistic assumptions inherent in Markov models [6], FTNA focuses on summarizing, describing, and visually analyzing observed data without such constraints. The section concluded with Chapter 17 [44], which explores clustering within TNA to uncover heterogeneity in learners' behavioral patterns. The focus is on using Mixture Markov Models (MMM) [6] to identify latent subgroups with distinct transition probabilities, incorporating covariates to explain the clusters. Through the tna R package [42], the chapter demonstrates how to compare and contrast unique transition dynamics in each cluster and examine centrality measures, communities, and cliques. Additionally, it provides examples of alternative clustering methods, such as distance-based grouping [8], and extends the discussion to other types of transition networks, including frequency-based approaches.

2.4 Section IV: Idiographic Analysis

The last section of the book zooms in on the individual learner through idiographic and individualized approaches that emphasize within-person dynamics [29, 45]. The section explores methods and frameworks designed to capture the unique, dynamic processes of individual learners and provides deeper insights into their cognitive, emotional, and behavioral patterns. The section closes the book by exploring methodologies that prioritize the variability inherent in human learning processes [46], offering practical tools for researchers and educators seeking to tailor interventions to each individual.

An introductory chapter [47] presents the two primary research paradigms in educational research: between-person (nomothetic) and within-person (idiographic) approaches. The between-person approach examines variations across individuals, while the within-person approach focuses on variations within the same individual over time. These paradigms align only if ergodicity holds—when individuals are similar and remain unchanged over time—conditions that are rarely met in educational contexts. The chapter emphasizes the need for within-person analysis and zooms in on idiographic analysis, which focuses on a single individual (N=1). The chapter also outlines the key characteristics, data sources, and methods associated with both approaches and sets the stage for personalized analyses in educational research.

Based on these principles, Chapter 19 [48] examines three analytical approaches—variable-centered, person-centered, and person-specific—that differ in their balance between generalizability and individualization [46]. The variable-centered approach seeks generalizable insights by analyzing relationships across a population, often at the expense of capturing individual differences. The person-centered approach identifies distinct subgroups within the population, highlighting shared patterns among similar individuals. The person-specific, or idiographic, approach focuses entirely on the individual, providing detailed insights into intra-individual variability and processes over time. Through a practical tutorial in R, the chapter demonstrates these methods using dynamic networks and Experience Sampling Method (ESM) data on students' task value and happiness.

Chapter 20 [49] proposes network analysis methods, specifically Graphical Vector Autoregression (graphicalVAR) and Unified Structural Equation Modeling (uSEM), to model learning processes as complex, dynamic systems [30]. These methods allow researchers to examine both temporal and contemporaneous relationships among variables within individual learners over time. The chapter begins by conceptualizing learning as a networked system, reviewing relevant literature, and discussing the advantages of probabilistic network models in educational contexts. A step-by-step tutorial in R is included, guiding readers through the estimation of idiographic models, visualization of dynamic relationships, and interpretation of results. The tools presented to analyze multivariate time-series data lay the foundation for personalized interventions in educational settings.

Chapter 21 [50] presents Whittle's approximation of the Maximum Likelihood Estimator (MLE) and highlights its ability to detect long-term memory processes within relatively short time series. The chapter explains the foundational principles of Whittle's MLE and demonstrates its application in R to analyze motivational dynamics, specifically the interplay between approach and avoidance behaviors in academic settings. Finally, the chapter explores the theoretical and practical implications of identifying long-memory processes in education, shedding light on how these insights can inform research and practice in understanding sustained learning and behavior patterns.

The last chapter of this section [51] presents a tutorial on using AutoML to streamline and scale predictive modeling in education, with a focus on idiographic analysis. AutoML simplifies the ML pipeline, enabling the creation of individually optimized models tailored to each student's unique data. The tutorial demonstrates how to use AutoML to fit and manage multiple datasets, automating the process of developing personalized models. Additionally, it showcases explainable AI techniques to interpret key predictors, providing transparent insights into the variables that drive individual outcomes [52]. This complete pipeline highlights the potential of AutoML to deliver automated, real-time, and trustworthy insights for idiographic educational research and practice. As such, this chapter combines ML, explainable AI, idiographic analysis and complex systems.

3 A Call to Innovate

The convergence of AI, complex systems theory, and idiographic analysis opens new paths for expanding the boundaries of education research in general and learning analytics in particular. This section discusses some of these potentials, aligning with Clow's learning analytics lifecycle [53] (Fig. 1).

First, regarding **learners**, the chapters of this book have discussed several important synergy points where innovation is ready to take place. The AI technologies discussed in Chapter 2 [11], and especially those powered by LLMs (discussed in Chapter 8 [21]), offer unprecedented opportunities for personalized learning, tailoring learning experiences to each specific learner. As a result, adopting an idiographic lens [47] to treat each individual learner as a single unit of analysis is needed to better understand individual learning trajectories enabled by such personalization. This lens allows to identify patterns in learners' behaviors, preferences, and progress, which are often concealed by aggregate analyses [48]. Moreover, learning processes usually involve multiple interacting constructs, such as motivation, engagement, and prior knowledge, which influence each other and evolve over time. Viewing these interactions through the lens of complexity allows us to recognize emergent patterns that are not apparent when constructs are studied in isolation [33]. Combining both approaches (idiographic and complexity) in light of recent innovations in AI in education is a first step towards studying personalized learning [49].

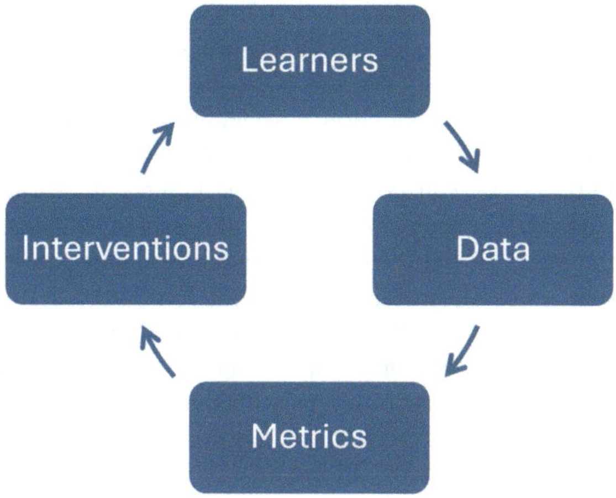

Fig. 1 Learning analytics lifecycle by Clow [53]

When it comes to **data**, the second stage of the learning analytics lifecycle, the focus shifts to the collection, integration, and preparation of diverse data sources to support meaningful analyses. A longstanding problem in learning analytics is balancing the tradeoff between gathering meaningful, high-quality data and automating its analysis to enable timely interventions. While multimodal data—such as behavioral logs, textual inputs, and even physiological signals—can provide a deeper understanding of learning processes, the complexity of such data often requires sophisticated preprocessing and modeling techniques and/or manual coding or inspection, which can delay actionable insights.

To address this challenge, recent advancements in AI, particularly LLMs, have demonstrated significant potential in automating the interpretation of unstructured data [21]. LLMs can extract meaningful patterns from text and synthesize multimodal information. Some chapters of the book deal with the problem of automatically synthesizing or coding textual information [22, 23]. However, it should be reiterated that automated analysis must be accompanied by careful consideration of data quality, representativeness, and fairness. Poorly integrated, unbalanced, or incomplete data can lead to biased models and misguided interventions, which risks compromising the learner-centered approach central to fair and effective learning analytics. In this regard, several chapters of the book cover XAI [11, 19, 20, 24], a central tool to audit AI decisions and detect potential biases in ML models.

The **metrics** phase of the learning analytics lifecycle deals with the selection, development, and application of methods and metrics that are capable of capturing, analyzing, and interpreting the complexity of educational data. As a methodological tutorial-style book, we make our most impactful contributions in this phase. While our previous book focused on the foundational methods of learning analytics, this new volume contributes state-of-the-art techniques and tools for advancing the metrics phase of the learning analytics lifecycle. The most important contribution is a novel method—TNA—, presented for the first time at LAK 2025 [40], which combines Markov models and network analysis, offering a strong statistical framework to model temporal processes [39, 44].

In addition to TNA, this book sheds light on other new methods that have yet to gain widespread adoption in the learning analytics community. EGA aims to identify latent constructs and their relationships within psychological and educational data [34]. EGA's can detect hidden structures in learner interactions and behaviors, particularly when coupled with idiographic analysis. Similarly, RQA [38] enables the study of the recurrence, complexity, and predictability of learning processes over time. RQA offers insights into how learning trajectories evolve by examining patterns of stability and change within temporal sequences, making it an invaluable method for both idiographic and complex systems research. Its emphasis on temporal recurrences aligns with the growing demand for methods strategies that can capture the emergent and non-linear nature of learning.

Lastly, the **intervention** phase crowns the learning analytics lifecycle, where the insights gained from data and metrics are applied to create actionable strategies that support learners and educators. This phase tests the practical value of the analytics process to ensure that the methods and models lead to meaningful interventions. For instance, learning analytics enhanced learning platforms like LEAF [54] considered learning logs to automatically conduct statistical modeling to extract the effectiveness of teaching-learning cases [55] and extract students' learning patterns [56] to aim for precision education. One promising direction is the development of hybrid human-AI systems that combine the computational power and pattern recognition capabilities of AI with the contextual understanding, ethical judgment, and adaptability of human decision-makers. These systems could dynamically adapt to individual learners' needs while maintaining human oversight and interpreting complex learning processes. For instance, AI could analyze real-time learning patterns using the advanced methods described in this book —from transition networks to recurrence quantification—while educators use their expertise to contextualize these insights—perhaps with the help of XAI techniques—and make interventions [13]. This hybrid approach could help bridge the gap between sophisticated analytics and practical pedagogical applications, especially in areas where purely automated systems might miss important contextual or emotional factors.

Furthermore, as learning environments become increasingly digitized and interconnected, we need innovative methods that can handle multimodal data streams while respecting individual learning trajectories. This calls for developing integrated frameworks that can simultaneously analyze different types of data—from clickstream logs to natural language interactions to physiological signals—while maintaining both the precision of idiographic analysis and the scalability of AI-driven approaches, such as AutoML and learning platforms such as LA-ReflecT [57]. Such frameworks could help us better understand the complex interplay between cognitive, emotional, and social aspects of learning, leading to more holistic and effective educational interventions.

4 How to Use this Book

To support your learning and provide hands-on experience with the concepts discussed in this book, we have created a companion code repository for each chapter. This repository includes all the code from the step-by-step tutorials, demonstrating the implementation of various learning analytics and AI methods covered in the chapters. Additionally, the code provides guidance on generating visualizations, graphs, and plots to help interpret and communicate findings effectively. You can access the companion code repository at:
https://github.com/lamethods/code2
Moreover, the data used to demonstrate the methods in most of the chapters have been made available in the following repository:
https://github.com/lamethods/data2
Similar to the previous book, the text and code of all the chapters in the book will be available online at: https://lamethods.org/.

References

1. Saqr M, López-Pernas S (2024) Learning analytics methods and tutorials, 2024th edn. Springer International Publishing, Cham
2. Kopra J, Tikka S, Heinäniemi M, López-Pernas S, Saqr M (2024) An r approach to data cleaning and wrangling for education research. In: Learning analytics methods and tutorials: A practical guide using R. Springer Nature Switzerland, Cham, pp 95–119
3. López-Pernas S, Misiejuk K, Tikka S, Saqr M, Kopra J, Heinäniemi M (2024) Data visualization and reporting in LA with R. In: Saqr M, López-Pernas S (eds) Learning analytics methods and tutorials: a practical guide using R. Springer, Berlin
4. Tikka S, Kopra J, Heinäniemi M, López-Pernas S, Saqr M (2024) Basic statistics with R. In: Saqr M, López-Pernas S (eds) Learning analytics methods and tutorials: a practical guide using R. Springer, Berlin
5. López-Pernas S, Saqr M (2024) The why, how and when of process mining in learning analytics: A guided tutorial in R. In: Saqr M, López-Pernas S (eds) Learning analytics methods and tutorials: a practical guide using R. Springer, Berlin
6. Helske J, Helske S, Saqr M, López-Pernas S, Murphy K (2024) A modern approach to transition analysis and process mining with markov models in education. In: Learning analytics methods and tutorials: A practical guide using R. Springer Nature Switzerland, Cham, pp 381–427
7. Saqr M, López-Pernas S, Conde-González MÁ, Hernández-García Á (2024) Social network analysis: a primer, a guide and a tutorial in R. In: Learning analytics methods and tutorials: A practical guide using R. Springer Nature Switzerland, Cham, pp 491–518
8. Saqr M, López-Pernas S, Helske S, Durand M, Murphy K, Studer M, Ritschard G (2024) Sequence analysis in education: principles, technique, and tutorial with R. In: Learning analytics methods and tutorials. Springer Nature Switzerland, Cham, pp 321–354
9. Cukurova M (2024) The interplay of learning, analytics and artificial intelligence in education: a vision for hybrid intelligence. Br J Educ Technol J Council Educ Technol. https://doi.org/10.1111/bjet.13514
10. Saqr M, Schreuder MJ, López-Pernas S (2024) Why educational research needs a complex system revolution that embraces individual differences, heterogeneity, and uncertainty. In: Saqr

M, López-Pernas S (eds) Learning analytics methods and tutorials: a practical guide using R. Springer Nature Switzerland, Cham, pp 723–734
11. López-Pernas S, Oliveira E, Song Y, Saqr M (2025) AI, explainable AI and evaluative AI: an introduction to informed data-driven decision-making in education. In: Saqr M, López-Pernas S (eds) Advanced learning analytics methods: AI, precision and complexity. Springer Nature Switzerland, Cham
12. Bond M, Khosravi H, De Laat M, Bergdahl N, Negrea V, Oxley E, Pham P, Chong SW, Siemens G (2024) A meta systematic review of artificial intelligence in higher education: a call for increased ethics, collaboration, and rigour. Int J Educ Technol Higher Educ 21. https://doi.org/10.1186/s41239-023-00436-z
13. Mustafa MY, Tlili A, Lampropoulos G, Huang R, Jandrić P, Zhao J, Salha S, Xu L, Panda S, Kinshuk, López-Pernas S, Saqr M (2024) A systematic review of literature reviews on artificial intelligence in education (AIED): a roadmap to a future research agenda. Smart Learn Environ 11:1–33
14. Jovanovic J, López-Pernas S, Saqr M (2024) Predictive modelling in learning analytics: a machine learning approach in R. In: Learning analytics methods and tutorials. Springer Nature Switzerland, Cham, pp 197–229
15. Saqr M, Misiejuk K, Tikka S, López-Pernas S (2025) Artificial intelligence: Using machine learning to predict students' performance. In: Saqr M, López-Pernas S (eds) Advanced learning analytics methods: AI, precision and complexity. Springer Nature Switzerland, Cham
16. Saqr M, Misiejuk K, Tikka S, López-Pernas S (2025) Artificial intelligence: using machine learning to classify students and predict low achievers. In: Saqr M, López-Pernas S (eds) Advanced learning analytics methods: AI, precision and complexity. Springer Nature Switzerland, Cham
17. Li T, Han F, Guo J, Wu J (2025) Comparative analysis of regularization methods for predicting student certification in online courses. In: Saqr M, López-Pernas S (eds) Advanced learning analytics methods: AI, precision and complexity. Springer Nature Switzerland, Cham
18. Saqr M, López-Pernas S (2024) Why explainable AI may not be enough: predictions and mispredictions in decision making in education. Smart Learn Environ 11. https://doi.org/10.1186/s40561-024-00343-4
19. Saqr M, López-Pernas S (2025) Explainable artificial intelligence in education: A tutorial for identifying the variables that matters. In: Saqr M, López-Pernas S (eds) Advanced learning analytics methods: AI, precision and complexity. Springer Nature Switzerland, Cham
20. Saqr M, López-Pernas S (2025) Individualized explainable artificial intelligence: a tutorial for identifying local and individual predictions. In: Saqr M, López-Pernas S (eds) Advanced learning analytics methods: AI, precision and complexity. Springer Nature Switzerland, Cham
21. Oliveira E, Song Y, Saqr M, López-Pernas S (2025) An introduction to large language models in education. In: Saqr M, López-Pernas S (eds) Advanced learning analytics methods: AI, precision and complexity. Springer Nature Switzerland, Cham
22. Wongvorachan T, Bulut O (2025) The use of natural language processing in learning analytics. In: Saqr M, López-Pernas S (eds) Advanced learning analytics methods: AI, precision and complexity. Springer Nature Switzerland, Cham
23. López-Pernas S, Misiejuk K, Saqr M (2025) Using BERT-like language models for automated discourse coding: a primer and tutorial. In: Saqr M, López-Pernas S (eds) Advanced learning analytics methods: AI, precision and complexity. Springer Nature Switzerland, Cham
24. López-Pernas S, Song Y, Oliveira E, Saqr M (2025) LLMs for explainable artificial intelligence: Automating natural language explanations of predictive analytics models. In: Saqr M, López-Pernas S (eds) Advanced learning analytics methods: AI, precision and complexity. Springer Nature Switzerland, Cham
25. Malmberg J, Saqr M, Järvenoja H, Haataja E, Pijeira-Díaz HJ, Järvelä S (2022) Modeling the complex interplay between monitoring events for regulated learning with psychological networks. In: Giannakos M, Spikol D, Di Mitri D, Sharma K, Ochoa X, Hammad R (eds) The multimodal learning analytics handbook. Springer International Publishing, Cham, pp 79–104
26. Malmberg J, Saqr M, Järvenoja H, Järvelä S (2022) How the monitoring events of individual students are associated with phases of regulation: a network analysis approach. J Learn Anal 9:77–92

27. Marchand GC, Hilpert JC (2017) Complex dynamic systems approaches to educational research. J Exp Educ 85:173–174
28. López-Pernas S, Saqr M (2024) How the dynamics of engagement explain the momentum of achievement and the inertia of disengagement: a complex systems theory approach. Comput Human Behav 153:108126
29. Saqr M (2024) Group-level analysis of engagement poorly reflects individual students' processes: why we need idiographic learning analytics. Comput Human Behav 150:107991
30. Saqr M, López-Pernas S (2024) Mapping the self in self-regulation using complex dynamic systems approach. Br J Edu Technol J Council Educ Technol. https://doi.org/10.1111/bjet.13452
31. Törmänen T, Järvenoja H, Saqr M, Malmberg J, Järvelä S (2023) Affective states and regulation of learning during socio-emotional interactions in secondary school collaborative groups. Br J Educ Psychol 93(Suppl 1):48–70
32. Saqr M, Viberg O, Peeters W (2021) Using psychological networks to reveal the interplay between foreign language students' self-regulated learning tactics. In: STELLA2020, pp 1–12
33. Saqr M, Dever D, López-Pernas S, Gernigon C, Marchand G, Kaplan A (2025) Complex dynamic systems in education: beyond the static, the linear and the causal reductionism. In: Saqr M, López-Pernas S (eds) Advanced learning analytics methods: AI, precision and complexity. Springer Nature Switzerland, Cham
34. Wongvorachan T, Bulut O (2025) The advanced applications of psychological networks with exploratory graph analysis. In: Saqr M, López-Pernas S (eds) Advanced learning analytics methods: AI, precision and complexity. Springer Nature Switzerland, Cham
35. Saqr M, Beck E, López-Pernas S (2024) Psychological networks: A modern approach to analysis of learning and complex learning processes. In: Saqr M, López-Pernas S (eds) Learning analytics methods and tutorials: a practical guide using R. Springer Nature Switzerland, Cham, pp 639–671
36. Webber C, Zbilut J (2004) Recurrence quantification analysis of nonlinear dynamical systems. In: Riley MA, Van Orden G (eds) Tutorials in contemporary nonlinear methods for the behavioral sciences, pp 26–94. http://www.nsf.gov/sbe/bcs/pac/nmbs/nmbs.pdf
37. Webber J Charles L., Marwan N, Facchini A, Giuliani A (2009) Simpler methods do it better: Success of recurrence quantification analysis as a general purpose data analysis tool. Phys Lett A 373:3753–3756
38. Dever D (2025) Detecting nonlinear patterns in education research: A tutorial on recurrence quantification analysis. In: Saqr M, López-Pernas S (eds) Advanced learning analytics methods: AI, precision and complexity. Springer Nature Switzerland, Cham
39. Saqr M, López-Pernas S, Tikka S (2025) Mapping the relational dynamics with transition network analysis: a primer and tutorial. In: Saqr M, López-Pernas S (eds) Advanced learning analytics methods: AI, precision and complexity. Springer Nature Switzerland, Cham
40. Saqr M, López-Pernas S, Törmänen T, Kaliisa R, Misiejuk K, Tikka S (2025) Transition network analysis: a novel framework for modeling, visualizing, and identifying the temporal patterns of learners and learning processes. In: Proceedings of learning analytics & knowledge (LAK '25). ACM, New York, NY, USA
41. Hernández-García À, Cuenca-Enrique C, Traxler A, López-Pernas S, Conde-González MÁ, Saqr M (2024) Community detection in learning networks using R. In: Learning analytics methods and tutorials: A practical guide using R. Springer Nature Switzerland, Cham, pp 519–540
42. López-Pernas S, Tikka S, Saqr M (2024). Tna: Transition network analysis (TNA). The R Foundation. https://doi.org/10.32614/cran.package.tna
43. Saqr M, López-Pernas S, Tikka S (2025) Capturing the breadth and dynamics of the temporal processes with frequency transition network analysis: a primer and tutorial. In: Saqr M, López-Pernas S (eds) Advanced learning analytics methods: AI, precision and complexity. Springer Nature Switzerland, Cham
44. López-Pernas S, Tikka S, Saqr M (2025) Mining patterns and clusters with transition network analysis: A heterogeneity approach. In: Saqr M, López-Pernas S (eds) Advanced learning analytics methods: AI, precision and complexity. Springer Nature Switzerland, Cham

45. Saqr M, López-Pernas S (2021) Idiographic learning analytics: A definition and a case study. In: Proceedings of the 2021 international conference on advanced learning technologies (ICALT). https://doi.org/10.1109/ICALT52272.2021.00056
46. Saqr M, Vogelsmeier LVDE, López-Pernas S (2024) Capturing where the learning process takes place: a person-specific and person-centered primer. Learn Indiv Differ 113:102492
47. Saqr M, Ito H, López-Pernas S (2025) A comprehensive introduction to idiographic and within-person analytics. In: Saqr M, López-Pernas S (eds) Advanced learning analytics methods: AI, precision and complexity. Springer Nature Switzerland, Cham
48. López-Pernas S, Kayaduman H, Vogelsmeier LVDE, Saqr M (2025) The three levels of analysis: Variable-centered, person-centered and person-specific analysis in education. In: Saqr M, López-Pernas S (eds) Advanced learning analytics methods: AI, precision and complexity. Springer Nature Switzerland, Cham
49. Saqr M, Dever D, López-Pernas S (2025) Idiographic networks: a tutorial on graphical vector autoregression and unified structural equation modeling. In: Saqr M, López-Pernas S (eds) Advanced learning analytics methods: AI, precision and complexity. Springer Nature Switzerland, Cham
50. Altamore R, Roume C, Teboul A, Gernigon C (2025) Detecting long-memory psychological processes in academic settings using whittle's maximum likelihood estimator: An application with R. In: Saqr M, López-Pernas S (eds) Advanced learning analytics methods: AI, precision and complexity. Springer Nature Switzerland, Cham
51. Saqr M, Tlili A, López-Pernas S (2025) Automating individualized machine learning and AI prediction using AutoML: the case of idiographic predictions. In: Saqr M, López-Pernas S (eds) Advanced learning analytics methods: AI, precision and complexity. Springer Nature Switzerland, Cham
52. Saqr M, Cheng R, López-Pernas S, Beck ED (2024) Idiographic artificial intelligence to explain students' self-regulation: toward precision education. Learn Indiv Differ 114:102499
53. Clow D (2012) The learning analytics cycle. In: Proceedings of the 2nd international conference on learning analytics and knowledge. ACM, New York, NY, USA. https://doi.org/10.1145/2330601.2330636
54. Ogata H, Majumdar R, Flanagan B (2023) Learning in the digital age: power of shared learning logs to support sustainable educational practices. IEICE Trans Inf Syst E106.D:101–109
55. Kuromiya H, Majumdar R, Ogata H (2020) Fostering evidence-based education with learning analytics: capturing teaching-learning cases from log data. Educ Technol Soc 23:14–29
56. Yang CCY, Chen IYL, Ogata H (2021) Toward precision education: educational data mining and learning analytics for identifying students' learning patterns with ebook systems. J Educ Technol Soc 24:152–163
57. Majumdar R, Prasad P, Kadam K, Gatare K, Warriem JM (2023) LA-ReflecT: a platform facilitating micro-learning and its multimodal learning analytics. In: Lecture notes in computer science. Springer Nature Switzerland, Cham, pp 731–735

Part I
Artificial Intelligence

AI, Explainable AI and Evaluative AI: Informed Data-Driven Decision-Making in Education

Sonsoles López-Pernas, Eduardo Oliveira, Yige Song, and Mohammed Saqr

Abstract Artificial Intelligence (AI) has had an immense impact in education through its ability to facilitate personalized learning experiences. However, the increasing adoption of AI systems in education has raised concerns regarding transparency and fairness. Explainable AI (XAI) emerges as a solution to address these issues, offering methods to make AI models interpretable and their decisions understandable to learners, educators, and other stakeholders. This chapter provides an overview of AI applications in education, including adaptive learning systems, profiling, and predictive modeling. Moreover, it provides an introduction to the main XAI concepts and techniques designed to ensure trustworthiness and equity in AI-driven educational environments. Enabling users to scrutinize and refine AI decisions is a necessary step in aligning technological advancements with the ethical and practical needs of education. The chapter aims to provide readers with an understanding of AI in education and XAI's potential, serving as a foundation for the tutorials presented in the book. The chapter also discusses evaluative AI as a new perspective on XAI for decision-making in which evidence is provided against and in favor of human-made hypotheses, rather than providing explanations for the single most-likely AI outcome.

Keywords Artificial intelligence · Explainable AI · Education · Learning analytics

S. López-Pernas (✉) · M. Saqr
University of Eastern Finland, Joensuu, Finland
e-mail: sonsoles.lopez@uef.fi

E. Oliveira · Y. Song
University of Melbourne, Parkville, VIC, Australia

M. Saqr, S. López-Pernas (eds.), *Advanced Learning Analytics Methods*,
https://doi.org/10.1007/978-3-031-95365-1_2

1 Introduction

Artificial Intelligence (AI) has a long and rich history in education. In fact, AI in education predates several other technologies and disciplines (e.g., online learning). Among the first examples of AI applications in education is the creation of the PLATO system at the University of Illinois around the 1960s which allowed for interactive learning and automated feedback [1]. While PLATO was limited in scope, it laid the foundation for future developments to follow and encouraged other researchers to explore the nascent technology in education. Most importantly, interest started to grow in developing intelligent tutoring systems (ITS) and AI powered educational solutions [1].

However, it was not until the early 1980s when the field of AI in education saw its major milestone with the launch of the first International Conference on Artificial Intelligence in Education (AIED) which was held in Montreal [2]. AIED conference brought together researchers from both AI and educational psychology, marking the formal recognition of AI as a significant discipline. Furthermore, AIED set the stage for collaboration between the fields of AI and education, which led to several significant developments in intelligent tutoring systems, adaptive learning technologies, and educational software [3].

The last two decades have witnessed several remarkable and fast-paced developments in the field of AIED fueled by the rapid adoption of the internet, the increase in computation power and advances in methodology [4]. Such advances led to the birth of learning analytics as a field —along with educational data mining—which all contributed to wide use of machine learning and data analytics, alongside other uses of AIED that extends to almost all fields of research and practice [5].

As AI continues to play an increasingly central role in education, concerns around transparency, and fairness have become more prominent [6]. These issues stem from the inherently complex and often opaque nature of AI systems, particularly those powered by machine learning and deep learning [7]. For instance, while AI-driven systems can offer personalized recommendations and adaptive learning experiences, their underlying decision-making processes are not always clear to educators or learners. This lack of transparency can prevent users from trusting AI systems and hinder their adoption in educational contexts [8]. Worse even, they might follow AI decisions without knowing if they are wrong or right [9].

One of the reasons for the mistrust in AI is that its performance is commonly affected by bias, as AI models are only as unbiased as the data they are trained on [6]. Historical inequities or imbalances in training datasets can perpetuate and even amplify existing disparities, leading to outcomes that disproportionately disadvantage certain groups of students [10]. For example, adaptive systems might inadvertently favor learners whose behavior aligns with the dominant patterns in the training data, leaving those with less conventional learning trajectories underserved. Such risks highlight the importance of rigorous evaluation, diverse datasets, and ethical oversight in the design and deployment of AI systems in education.

Addressing these challenges requires the integration of explainable AI (XAI) techniques, which aim to make AI outputs interpretable and accountable to users. XAI tools can provide insights into how AI systems reach their conclusions, enabling educators to critically evaluate and trust these technologies [11]. Moreover, the adoption of fairness-aware machine learning practices and inclusive design principles can help mitigate biases, ensuring that AI-driven solutions support equitable outcomes for all learners. These considerations are particularly relevant when it comes to personalized education, where the promise of tailored interventions must be balanced with a commitment to fairness and transparency.

This chapter explores the current state of artificial intelligence in education, as well as explainable AI (XAI). AI and XAI techniques and tools are illustrated in the subsequent tutorials in the book and this chapter provides an introduction for the reader to have an overarching view on their main characteristics and potential. We further introduce the concept of evaluative AI [12], which aims to address some of the limitations of XAI.

2 An Overview of AI Applications in Education

AI is an extremely broad field that permeates many disciplines, encompassing a wide range of techniques and applications that aim to replicate or enhance human intelligence. In spite of—or maybe because of—its widespread influence across fields such as healthcare, finance and, of course, education, there is no universally agreed-upon definition of AI [13], nor a standardized taxonomy to categorize its many methods and applications. In the context of education, this lack of clarity is furthered by the diverse range of technologies that fall—sometimes wrongly—under the label of AI—from rule-based systems to complex neural networks—each serving diverse purposes such as personalization, prediction, and automation. In this section, we aim to provide an overview of the key AI applications that have influenced education, presenting relevant use cases in the field. We follow the structure of the recent meta-review by Bond et al. [14], which is based on the original typology by [15], and add an additional section on AI in education research.

2.1 Adaptive Systems and Personalization

One of the main promises of AI in education is personalization. Personalized learning focuses on tailoring educational experiences to individual learners' needs, preferences, and readiness, and it can be implemented through various methods and technologies [16]. For example, adaptive systems implement personalized learning by constructing a model of each learner using artificial intelligence and machine learning as well as continuous data collection of students' activities and performance [16]. According to Zhong [17], personalized learning has been mainly implemented through learning content structuring and sequencing (organizing and ordering learning materials according to individual needs) as well as readiness support (adapting to students' performance).

Another form of personalization—closely related to adaptive systems—is the use of recommender systems. Recommender systems are algorithms designed to suggest learning resources, activities, or courses to students based on their individual profiles, past performance, and expressed preferences. According to the review by Urdaneta-Ponte et al. [18], most recommender systems in the education literature use collaborative filtering, which entails recommending resources based on similarity to other students.

In recent years, personalization has expanded beyond adaptivity and recommendation through the provision of truly personalized learning experiences, made possible thanks to the latest AI breakthroughs. Chatbots and virtual assistants currently represent one of the most prominent applications of personalization of education using AI. These tools take advantage of the advances in natural language processing—and most recently Large language Models (LLMs)—to interact with students in a conversational manner. A recent review by Labadze et al. [19] discovered that students benefit most from AI-powered chatbots in three main areas: support with homework and studying, personalized learning experiences, and skill development [20].

Other AI-powered technologies like facial recognition or mood detection are emerging as relevant tools for tailoring educational experiences [21]. These technologies can complement traditional adaptive systems through the incorporation of real-time emotional and behavioral information on students' facial expressions, tone of voice, and body language. Despite their potential, these technologies raise concerns regarding privacy, consent, and data security [22]. Furthermore, the accuracy of emotion recognition algorithms and the potential for cultural or individual biases remain critical challenges to address.

2.2 *Intelligent Tutoring Systems (ITS)*

Closely related to adaptive systems are Intelligent Tutoring Systems (ITSs). ITSs use AI techniques to provide personalized tutoring based on a model of the learner's knowledge, behavior, and learning progress [23]. ITSs aim to emulate one-on-one human tutoring through the adaptation of instructional strategies, delivering tailored feedback, and offering customized learning experiences to help students achieve specific educational goals. ITSs often include features such as error diagnosis, hint generation, and real-time adaptation to individual learner needs, typically in well-defined domains. For example, virtual patients are a form of ITS used in healthcare to practice clinical reasoning [24]. ITSs are also extensively used in programming education as a step-by-step tutor to master a programming language [25].

2.3 Profiling and Prediction

Predictive modeling has been a primary focus of AI in education. Early attempts dealt with the use of demographic and administrative data from learners to predict performance and dropouts [26]. With the birth of the educational data mining and learning analytics fields, utilizing intensive log data about learners' online activities became the norm in the field, which came accompanied to increased predictive accuracy thanks to advances in ML algorithms [27].

The goals of performing predictive modeling have been diverse. A central goal has been the early identification of students who may be at risk of failing or dropping out [28]. Based on data from previous iterations of a course, researchers attempt to predict students' performance early on in the current course to be able to conduct remedial interventions. Other less frequent goals of predictive modeling are related to career placement [29], admission [15], or student satisfaction [27].

Research suggests that predictive modeling is hardly generalizable and lacks portability between courses [30], or even between course iterations [31]. Thus, recent research has shifted the focus from predicting performance to understanding learners based on their behavioral data. Clustering algorithms have been operationalized to detect distinct patterns or profiles of engagement [32], collaboration roles [33], self-regulated learning [34], or learning strategies [35].

Lastly, recent research strands that are gathering increased attention focus on identifying and mitigating biases in AI algorithms [36], ensuring equitable treatment of all learners, and enhancing the interpretability of AI systems through XAI [11].

2.4 Assessment and Evaluation

Students often long for formative feedback that offers actionable points to improve their learning and performance. This has traditionally come at the expense of overburdening teachers with a higher workload. AI offers a lot of promise to alleviate this challenge by providing automated, personalized feedback to students [37]. Although automated assessment has been present for decades, e.g., through multiple-choice questionnaires that can be automatically graded or autograders for programming assignments, advancements in artificial intelligence, particularly in NLP, have led to the development of systems capable of analyzing student work and generating constructive feedback.

AI assessment and feedback have been especially relevant in the evaluation of artifacts that do not have a deterministic correct solution. In fact, the most widely used application of AI-powered feedback in education is automated essay scoring [38]. Another important focus has been the evaluation of programming code. Although non-AI automated assessment tools have existed for decades [39], the recent advances in AI have allowed the provision of more relevant and detailed feedback to students [40]. The main advantage of AI-powered assessment systems is that they can provide immediate feedback, allowing students to understand their mistakes and learn from them in real-time. In addition, such systems enable educators to support a larger number of students without a proportional increase in workload, making personalized education more accessible.

Beyond evaluating students' assignments and artifacts, AI has also opened new avenues for evaluating students' learning process. For instance, automated discourse coding allows educators and researchers to analyze conversations or discussion forums in order to identify patterns of collaboration, critical thinking, and engagement [41].

2.5 *AI in Education Research and Learning Analytics*

Lastly, AI has not only brought opportunities to education practice but also to education research and learning analytics. Advances in AI, such as deep learning, natural language processing (NLP), and more recently, LLMs, have significantly expanded the possibilities for analyzing unstructured data such as text, speech, and video, and facilitating new approaches to personalization and feedback in education [42, 43].

Moreover, the increasing role of AI as an educational tool has further expanded the possibilities of education and learning analytics researchers for understanding learners' behaviors in the presence of these new learning technologies [44]. For example, researchers can examine the types of prompts students pose to AI systems, providing fine-grained information about common misconceptions, knowledge gaps, or areas where additional support may be needed. These analyses can also showcase how students respond to feedback or recommendations generated by AI, and hence whether AI's guidance promotes better learning outcomes or contributes to over-reliance on AI-generated solutions [45].

3 Explainable AI

Headlines around the globe are increasingly spotlighting instances of AI bias, from skewed hiring algorithms [46] to discriminatory facial recognition systems [47], exposing the urgent need to address the unintended consequences of AI in our daily lives. These incidents highlight not only the limitations of AI but also the opaque nature of many of these systems, which makes identifying and mitigating bias a significant challenge. XAI emerges as a promising solution to shed light on the 'black box' of AI decision-making. In a nutshell, XAI can be defined as a set of methods and tools that enables humans to understand the outcomes of AI models [48]. Making the processes behind AI decisions transparent and interpretable through XAI allows all involved stakeholders—developers, users, and policymakers—to detect bias and assess fairness, and hence build trust in AI systems [49]. XAI aims to answer questions such as: *why* did the AI make this particular decision?; *how* would changes in the input data influence the decision?, and in *what* changes are needed in AI's behavior to align with ethical and societal values? This is achieved through techniques such as feature importance rankings, model-agnostic interpretability methods, and visualization tools that reveal the relationships between inputs and outputs [50].

As we have seen in the previous section, the adoption of AI tools and systems in the field of education has opened new opportunities for personalized learning, automated assessments, and early intervention for at-risk students [4]. However, as these systems increasingly influence critical decisions—such as grading, admissions, and resource allocation—they also bring the risk of embedding and perpetuating biases present in training data or algorithms. As such, XAI is of particular relevance in education, where fairness, transparency, and trust should be the basis of an equitable learning environment [11]. For instance, XAI techniques can reveal which factors are most influential in predicting a student's performance [51], enabling interventions that are data-informed yet account for individual differences. Similarly, transparency in automated grading systems ensures that assessments are objective [52], and therefore they can be trusted by students and educators. Moreover, XAI can provide students with actionable feedback that is both personalized and understandable [53], helping them detect areas for improvement.

In the remainder of this section, we provide an overview of the main concepts, techniques, and tools associated with XAI. We describe the main AI methodologies, including model-agnostic interpretability, feature importance analysis, and counterfactual reasoning, with an emphasis on how they enable transparency and accountability in AI systems.

3.1 The Trade-off Between Explainability and Performance

We start our journey into XAI by talking about explainability. We must differentiate between two broad categories of AI models: intrinsically explainable algorithms and black-box models. This distinction is critical for contextualizing the need for explainability tools and techniques discussed in later sections.

3.1.1 Intrinsically Explainable Algorithms

Intrinsically explainable (or white-box) algorithms are designed to be interpretable by their very nature [49]. These algorithms have transparent structures and straightforward mechanisms for decision-making, making it easy to understand how input features contribute to output predictions. Examples include linear regression, logistic regression, decision trees, and rule-based systems [54]. Stakeholders can

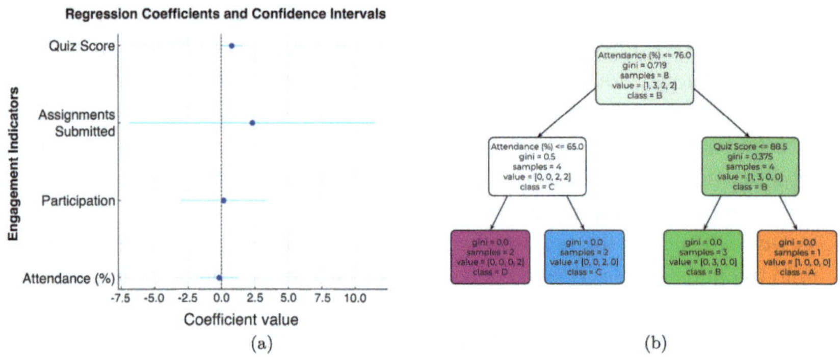

Fig. 1 Example of using intrinsically explainable algorithms to predict students' grades. (**a**) Explanation of a linear regression model. (**b**) Explanation of a decision tree model

directly observe relationships, such as the weight of a variable in a regression model (Fig. 1a) or the sequence of decisions in a tree structure (Fig. 1b).

In education, intrinsically explainable algorithms are valuable when transparency is a priority. For instance, a linear or logistic regression model predicting a student's grade based on attendance and participation provides a clear and intuitive explanation for its outcomes. Educators and other stakeholders can readily interpret the model and take actionable steps, such as encouraging students to engage in self-assessment activities.

However, intrinsically explainable models often come with limitations. Their simplicity can result in lower predictive accuracy when dealing with complex, non-linear relationships or high-dimensional data [54], which are common in educational contexts and a common problem in learning analytics research and practice. As a result, their applicability is sometimes constrained when achieving high accuracy is critical.

3.1.2 Black-Box Models

Black-box models, such as deep neural networks, gradient boosting machines, and ensemble methods [54], are characterized by their complex internal structures, which are not inherently interpretable. These models are often chosen for their ability to capture highly complex patterns and relationships in data, hence yielding higher predictive performance in many applications [49]. However, their complexity creates a significant barrier to understanding how decisions are made.

In education, black-box models might be needed when dealing with more complex unstructured data such as students' free-form text or images [26]. Although these models may achieve higher accuracy than intrinsically explainable algorithms, their opacity raises concerns. Without insights into how predictions are generated, educators and administrators may struggle to trust or act on these outputs, especially when high-stakes decisions that could be consequential for students are involved.

3.1.3 Balancing Explainability and Predictive Power

The choice between intrinsically explainable algorithms and black-box models often involves a trade-off between interpretability and performance [54]. Simpler models provide clarity and ease of use, at the expense of failing to capture the complexity of educational data. On the other hand, black-box models excel in handling such complexity but require additional methods to make their decisions interpretable [50]. XAI aims to bridge this gap by providing explainability techniques for black-box models. These tools enable stakeholders to take advantage of the high performance of black box models while maintaining transparency and accountability such as in the case of white-box models. For example, black-box models predicting student dropout can be paired with interpretability methods to reveal the main factors driving each prediction, hence allowing educators to make informed, data-driven decisions.

3.2 Global vs. Local Explainability: XAI Tools

As we have seen in the previous section, we need XAI tools to be able to explain the outcomes of black-box AI models. We can approach explainability from two complementary perspectives: global and local. Both play a crucial role in understanding and effectively applying AI systems, since both macro-level patterns and individual-level insights are essential for achieving trust and fairness. In this section, we describe each of these perspectives and the XAI tools used in each of them.

3.2.1 Global Explainability

Global explainability focuses on understanding the overall behavior of a model. It provides insights into how the model makes decisions across the entire dataset, pointing to the features that are most influential and the patterns that drive predictions. For instance, in a model designed to predict student performance (such as the ones shown in Fig. 1) global explainability might reveal that the average quiz score, attendance, and assignment submission are consistently the most important factors that predict performance. This "big picture" explanation enables educators and administrators to evaluate the model's logic and ensure it aligns with their expectations. Moreover, global explainability is instrumental in assessing whether the model behaves fairly across different demographic groups, helping to identify and address potential biases. However, its high-level nature means that it cannot explain why specific predictions are made, which limits its usefulness in scenarios requiring personalized action. It is, though, useful to take general actions that would benefit "the majority", for example changing the course design or learning materials. Chapter 6 [55] in this book offers a tutorial on global explainability using the R programming language. Below, we overview the main techniques for global explainability

3.2.1.1 Variable/Feature Importance

Feature importance measures how much each input variable contributes to a model's predictions, helping to identify the most influential factors driving its decisions [56]. In educational applications, feature importance provides actionable insights into what matters most for student outcomes. For example, a model predicting grades (see Fig. 2) might show that the frequency of forum contributions is the most important factor. This insight helps educators prioritize interventions where they are most needed, such as encouraging participation.

In models like decision trees or random forests, feature importance is calculated based on how often a variable is used in splits and how much it reduces prediction error (e.g., impurity). In linear regression, the magnitude of coefficients represents the relative weight of each variable, directly showing how much they influence the target outcome. These metrics are intrinsic to these models, making them straightforward to compute and interpret.

For black-box models, such as neural networks, gradient boosting, or ensemble methods, feature importance is not inherently available and must be derived through additional techniques. Unlike intrinsically interpretable models, black-box models process data in ways that obscure the direct contribution of individual features.

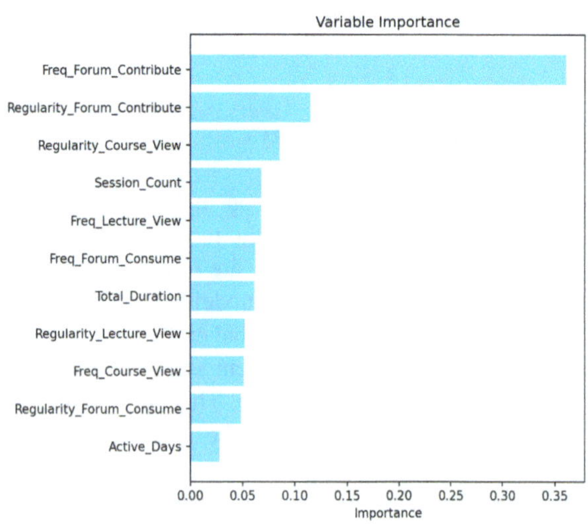

Fig. 2 Variable importance plot when predicting students' grades

To estimate feature importance, methods like permutation importance are often used [56]. This technique evaluates how model performance changes when the values of a feature are randomly shuffled, effectively breaking its relationship with the target variable. Features that cause a significant drop in performance when permuted are deemed more important. The derived feature importance remains an approximation, as black-box models can capture nonlinear interactions that are difficult to fully interpret without additional tools like SHAP or LIME. Thus, while feature importance provides valuable insights into black-box models, it often requires advanced methods to ensure reliability and transparency.

It is important to bear in mind that, while feature importance highlights *what* influences predictions, it does not fully explain *how* features influence prediction and how features interact with one another, underlining its value as a foundational but limited interpretability tool.

3.2.1.2 Partial Dependence Plots

Partial Dependence Plots (PDPs) depict how changes in a feature influence predictions while accounting for the average behavior of all other features. They isolate the effect of one feature at a time, making it easier to understand the relationship between input variables and model outcomes. For instance, Fig. 3 shows how increasing the number of reads of the forum—while the rest of the features remain the same—has barely any effect in the prediction of students' grades, while increasing the number of contributions results in a higher grade prediction. This insight allows educators to focus interventions in the variables that are most effective.

Although PDPs are useful for understanding non-linear relationships and interactions, they rely on the assumption that features are independent. In reality features like reading and writing forum contributions may be correlated, so this assumption can lead to misleading interpretations. Additionally, PDPs represent average effects, which may not fully capture individual variations in how a feature influences

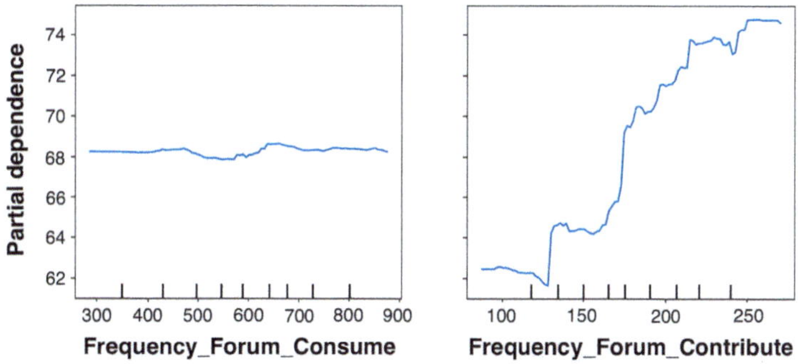

Fig. 3 PDP of frequency of forum consume and contribute

predictions. Despite these limitations, PDPs remain a valuable tool for interpreting complex models, especially in contexts like education where understanding feature contributions can inform better decisions.

3.2.1.3 SHAP Summary Plots

A SHAP summary plot combines two critical aspects of global feature importance: the magnitude of a feature's impact (ordered from top to bottom in the figure) on the predictions and the range of each feature's effects (the horizontal axis). For each feature, the plot shows the SHAP values across all instances in the dataset, where each SHAP value represents the contribution of the feature to a single prediction (i.e. individual dots) [57]. In the figure, top features such as the frequency of forum contributions have greater importance, its wide horizontal dispersion shows that the feature's contribution varies greatly across instances. Additionally, the color of each point typically reflects the feature value (e.g., high values in red and low values in blue), which helps identify whether higher or lower feature values drive predictions.

In educational contexts, SHAP summary plots can provide insights not only into which factors are most critical in predicting outcomes like grades, engagement, or dropout risk, but also about the direction towards which each feature pushes the prediction. For instance, a summary plot (Fig. 4) might reveal that frequency of forum contributions has the highest average SHAP values, indicating it is the most influential feature. The plot might also show that a higher number of active days consistently contributes to lower predictions, while regularity in viewing lectures has a more variable effect depending on other factors. This level of detail helps

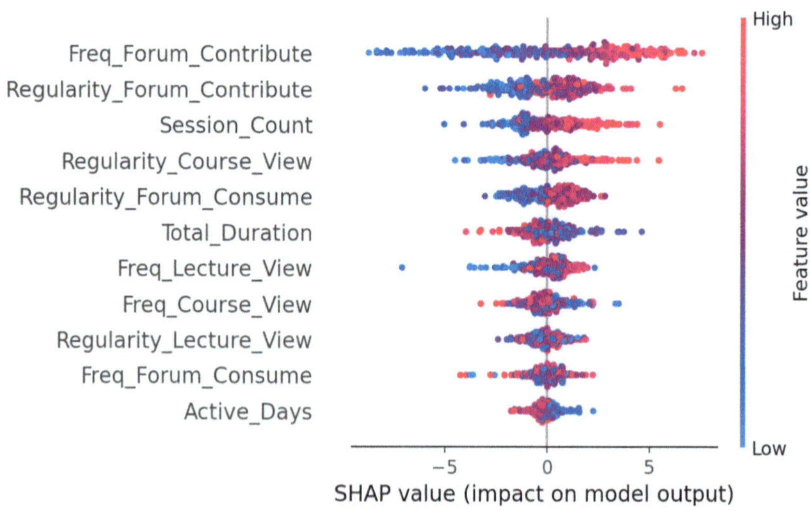

Fig. 4 SHAP summary plot

educators and administrators not only understand which features matter most but also how their impact differs across students.

It is worth mentioning that SHAP plots can also be used for explainability in other ML tasks such as unsupervised learning (e.g., clustering). For instance, [58] used clustering to detect study patterns, and leveraged XAI to discover unexpected patterns that are not apparent from the clustering alone.

Lastly, other variants of SHAP exist, such as Kernel SHAP [59], which extends SHAP by using kernel methods to efficiently approximate feature attributions, making it suitable for complex models. Permutation SHAP, another variant, evaluates the importance of features by analyzing the change in model performance when feature values are randomly permuted, providing an intuitive measure of feature impact.

3.2.2 Local Explainability

Contrary to global explainability, local explainability provides a detailed understanding of individual predictions, explaining why the model arrived at a specific outcome for a single student or instance. For example, if a student is flagged as being at risk of failing or dropping out, local explainability can identify that the prediction was driven by a drop in attendance and low quiz scores, for example. This level of granularity is of the utmost importance in education, where interventions are often targeted at individuals. Teachers, parents, counselors, and administrators can use these explanations to issue tailored interventions, such as reaching out to the student about attendance or offering extra tutoring for quizzes. Furthermore, local explainability builds trust in AI systems by making their decisions transparent and actionable, particularly in high-stakes contexts like grading or admission. Chapter 7 in this book [60] offers a tutorial on local explainability using the R programming language.

3.2.2.1 SHAP Values for Individual Predictions

SHAP values for individual predictions provide a breakdown of how each feature contributes to a specific prediction made by a machine learning model. Unlike SHAP summary plots, which summarize the overall behavior of the model across all predictions, individual SHAP values focus on explaining why the model made a particular prediction for a single instance. These explanations highlight the direction and magnitude of each feature's impact on the prediction, making them invaluable for personalized decision-making.

For an individual prediction, SHAP values calculate the contribution of each feature by comparing the model's output when the feature is included versus when it is excluded, averaged across all possible subsets of features. The SHAP values sum up to the difference between the model's baseline prediction (the average prediction for all instances) and the specific prediction for the instance. For example, in a model predicting a student's grade (see Fig. 5), if the baseline prediction is 68.35 and the

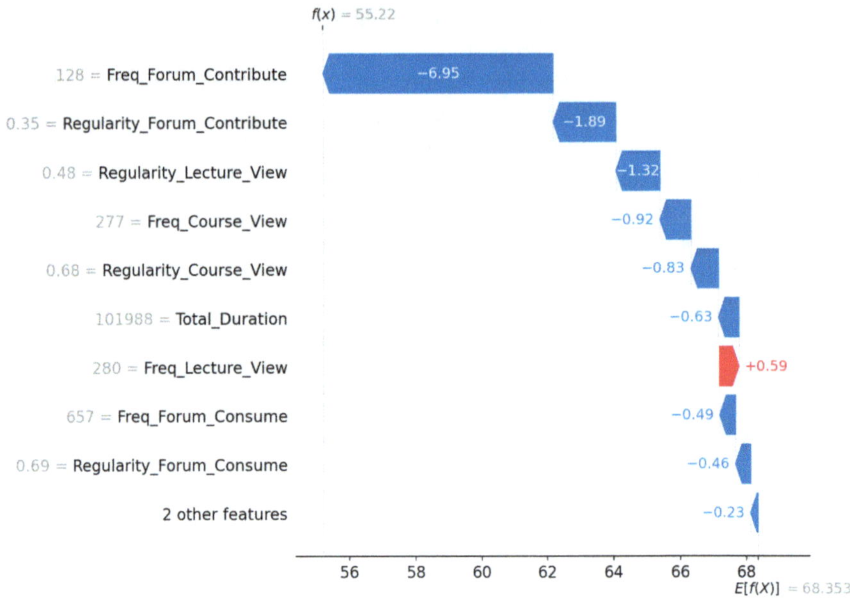

Fig. 5 SHAP instance explanations

model predicts 55.22 for the particular student under examination, the SHAP values will explain the 13-point difference by assigning contributions to features such as frequency and regularity of forum contribute, lecture view, etc. In the plot, we can see that almost all features have a negative value, thus contributing to a lower grade, except for frequency of viewing lectures.

3.2.2.2 Local Interpretable Model-Agnostic Explanations

LIME (Local Interpretable Model-Agnostic Explanations) is a method used to explain individual predictions by approximating the behavior of a complex machine learning model with a simpler, interpretable model within a localized region around the specific instance being explained [61]. Unlike SHAP, which relies on theoretical foundations from cooperative game theory, LIME focuses on practical, local interpretability by creating a surrogate model, such as linear regression, to approximate the original model's decisions in the vicinity of a single instance.

LIME works by perturbing the input data around the instance to generate a synthetic dataset. The model's predictions for these perturbed data points are then used to train a simple, interpretable model that mimics the original model locally. For instance, in a model predicting a student's grade, LIME might generate slightly varied versions of the student's features (e.g., tweaking the frequency of forum contributions and session count) and observe how the predictions change (see

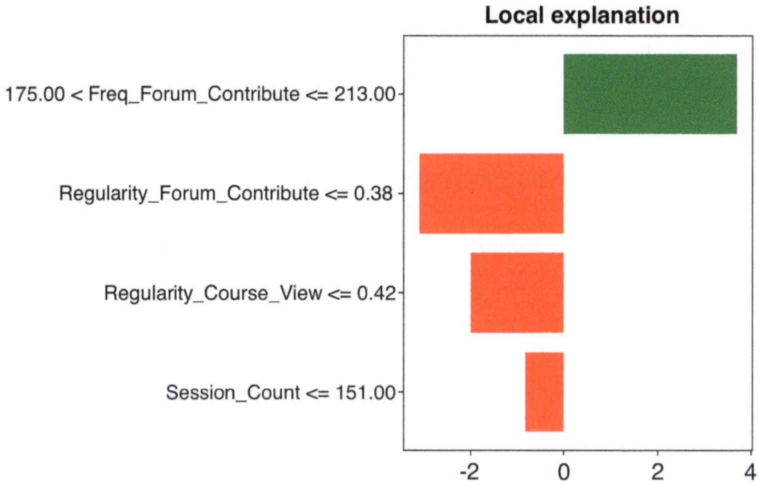

Fig. 6 LIME explanations

Fig. 6). The surrogate model identifies the main features driving the prediction, providing an interpretable explanation of the outcome.

3.2.2.3 Counterfactual Explanations

Counterfactual explanations offer a way to understand model predictions by identifying the minimal changes in input features needed to achieve a different outcome [62]. Instead of explaining why a particular prediction was made, counterfactual explanations answer the question: *What could have been different to achieve the desired result?* This makes them distinctly actionable, as they focus on what is required to alter the prediction. In other words, they offer "prescriptive analytics" in which a specific course of action is recommended [63].

A counterfactual explanation involves generating a hypothetical instance similar to the original input but modified to produce the desired prediction. For example, in an educational context, if a student is predicted to score below a passing grade, a counterfactual explanation might suggest that increasing forum participation by 20% and improving the regularity of viewing the lectures by 10% would lead to a passing grade. These explanations are of particular value when the focus is often on understanding how to improve outcomes for individual students. Counterfactual explanations are generated using optimization techniques that find the smallest or most plausible changes to input features to alter the prediction.

LLMs can play a significant role in enhancing the usability and accessibility of counterfactual explanations [63], particularly in education. Although counterfactual explanations are inherently actionable, their technical nature can make them difficult for educators or learners to interpret and apply without additional support. LLMs, with their natural language generation capabilities, can translate counterfactual outputs into easily understandable, context-specific recommendations. For instance, rather than presenting raw numerical changes in input features, an LLM can reframe the explanation as a conversational suggestion: "To improve your chances of passing, consider increasing your forum participation by 20% and attending lectures more regularly by 10%."

A related alternative to counterfactual explanations is the Contrastive Explanation Method (CEM) [64], which complements counterfactual explanations by focusing on contrastive reasoning. Rather than identifying minimal changes to alter a prediction, CEM emphasizes what features must be present (*pertinent positives*) and what features must be absent (*pertinent negatives*) for the current prediction to occur. For instance, in an educational setting, if a student is predicted to excel in a course, CEM might highlight pertinent positives such as consistent lecture engagement and forum participation, while pertinent negatives could include the absence of frequent late submissions. A question CEM aims to answer is: "Why this prediction instead of another?". Thus, CEM provides a broader understanding of model behavior, offering both prescriptive insights and a deeper contextual explanation.

3.2.2.4 Saliency Maps and Grad-CAM

Saliency maps and Grad-CAM (Gradient-weighted Class Activation Mapping) are visualization techniques primarily used to explain predictions made by deep learning models [65]. These methods highlight the parts of the input data (e.g., pixels in an image or words in a text) that are most influential in the model's decision-making process, providing localized and intuitive explanations of how the model interprets its inputs.

Saliency maps visualize how sensitive a model's prediction is to small changes in the input features. They compute the gradient of the model's output with respect to the input, capturing how each input feature contributes to the prediction. The resulting visualization highlights areas of the input data that, if altered, would most significantly change the prediction. In text data, a saliency map might underline specific words or phrases in a reflective essay that contributed most to predicting a student's understanding of a concept. Grad-CAM extends saliency maps by providing class-specific explanations. It focuses on the deeper convolutional layers of a neural network, computing gradients with respect to specific target classes. Grad-CAM generates a heatmap that overlays the input data, indicating regions that contributed most strongly to the prediction for a given class [65]. For instance, in an image classification model predicting whether a student is engaged in a video lecture, Grad-CAM might highlight that the model focused on the student's face and eyes when assessing engagement. This is extremely relevant to assess bias in AI models.

4 Why XAI May Not Be Enough

Recent research on AI in education has shown that algorithms alone cannot fully capture the full gamut of complexity of learners, learning and the way they learn. While AI can identify important variables that might affect a student's performance, it may—and often does—miss the meaningful connections to educational theories. For instance, recent work has shown that while XAI may offer insights into the factors that predict student performance, it failed to account for the heterogeneity of learners and their learning approaches. Such limitation is rather consequential given that educational interventions are often effective on average but may not be suitable for every student and such, may harm some [51, 66]. The researchers also demonstrated that when instance-level explainability was considered, it showed that AI made decisions based on incorrect predictors, leading to mis-predictions. These incorrect predictions show the inherent risk in a fully data-driven approach and the necessity for human oversight or judgment [51]. As such, a hybrid human-AI collaboration, where the expertise of educators complements—and critically evaluates—the explainability of AI might be needed. In doing so, hybrid human-AI collaboration could lead to more effective educational strategies that are responsive to the diverse needs of each learner and less prone to problematic conclusions. Such an approach would improve the reliability of AI predictions and also ensure that AI serves as an aid rather than a substitute for human expertise. It also further emphasizes the central role of critical AI literacy where AI decisions are understood and critically evaluated.

4.1 Evaluative AI: A New Perspective on XAI

Traditional XAI techniques, such as SHAP and LIME, provide explanations for the single most likely prediction or recommendation done by a system. For example, if an AI system predicts that a student is likely to drop out of a course, the XAI techniques that we have reviewed before can help understand which engagement indicators caused the AI to make that prediction. This can lead to overtrust or misinterpretation by users, especially in complex decision-making environments, since alternative hypotheses are not inspected. A recent approach by Miller [12] highlights the limitations of single outputs and introduces the concept of *evaluative AI*.

Instead of providing definitive explanations, evaluative AI presents evidence for and against multiple hypotheses, allowing users to critically assess plausible outcomes. This approach could address trust issues by ensuring that AI-driven insights, such as student performance predictions, do not dictate decisions but inform them. For example, the same AI system described earlier could present arguments supporting and refuting the influence of forum participation, encouraging educators and learners to engage in critical analysis rather than passively accepting the AI's output. This also aligns with Bearman et al. [67], who highlight the importance of developing students' evaluative judgment —the ability to appraise AI outputs and processes critically. Evaluative AI can not only enhance transparency but also support the critical engagement required for students and educators to maintain autonomy over AI-augmented decisions. Advocates for evaluative AI recognize, however, that this novel paradigm poses an additional cognitive load on the decision-makers compared with usual XAI recommendation-driven outputs, although not as much as having no explanations whatsoever [12].

Lastly, it should be noted that evaluative AI does not compete with but rather builds on XAI methods [12] and techniques such as feature importance, SHAP, and LIME. Nonetheless, new tools are needed to apply these tools in the framework of evaluative AI. Most likely, these will be more interactive tools that enable testing specific hypotheses in a similar way as local explainability is assessed. Though a system that supports inquiry-based decision-making based on evaluative AI is easy to envision for classic regression or classification models in which the outcome is heavily constrained, it is not as straightforward to picture in the context of generative AI where the possibilities are endless and therefore the need for users' evaluative judgment is greater [67].

5 Conclusion

This chapter has provided an overview of the main applications of AI in education as well as an introduction to XAI as a set of tools and methods for providing transparency and interpretability to otherwise opaque AI-driven systems, which are critical requirements for enabling trust, mitigating bias, and ensuring an ethical use of AI in educational contexts.

The recent criticism about AI with regard to bias, fairness and transparency has also raised concerns in the education field [68]. Data used in AI models often reflects existing societal inequities, such as socioeconomic, racial, or gender disparities, which can inadvertently lead to models amplifying these biases. For example, patterns in historical educational data may disadvantage underrepresented groups, further perpetuating systemic inequities in learning environments [69]. Additionally, the aggregation of data across diverse populations can obscure the unique needs of specific learner subgroups, resulting in interventions that lack contextual relevance or efficacy [70]. A further challenge lies in the opacity of data processing and decision-making pipelines, which often leave educators and learners unaware of how their data influences predictions and recommendations.

The interplay between XAI and LLMs offers a promising synergy that enhances the transparency, usability, and adaptability of AI systems in education. On the one hand, LLMs can improve the accessibility of XAI outputs by converting abstract visualizations, such as feature importance plots or decision attribution maps, into clear textual explanations. This ability ensures that the rationale behind AI-driven decisions is not only available but also comprehensible to non-technical stakeholders. In this way, LLMs serve as an interface that bridges the gap between the technical workings of XAI tools and their practical applications in educational contexts. On the other hand, XAI enhances the application of LLMs by addressing challenges related to their opacity and potential biases. It can clarify why a specific response or recommendation was generated, identify patterns of bias or unfairness in outputs, and explain the logic behind personalization or content adaptations. As these technologies continue to evolve, their integration into education, research, and other fields will further redefine workflows and learning experiences, opening new opportunities for innovation and personalization.

As AI systems increasingly influence decision-making in education, it is critical to ensure that XAI tools promote autonomy rather than replace human judgment [51]. Developing evaluative judgment, as Bearman et al. [67] argue, is essential for enabling students and educators to critically engage with AI outputs. Emerging approaches, such as evaluative AI, suggest a shift from single-point explanations to evidence-based decision support, fostering critical thinking among educators and learners. Additionally, tools like LLMs can act as interpreters of XAI outputs, bridging data literacy gaps and encouraging users to challenge AI-driven recommendations. Fostering a culture of inquiry—where explanations are evaluated, not merely accepted—is necessary for XAI to truly empower stakeholders to make informed, ethical decisions in education.

References

1. Cope B, Kalantzis M (2023) A little history of e-learning: finding new ways to learn in the PLATO computer education system, 1959–1976. Hist Educ 1–32. https://doi.org/10.1080/0046760x.2022.2141353
2. Chaudhry MA, Kazim E (2022) Artificial intelligence in education (AIEd): a high-level academic and industry note 2021. AI Ethics 2:157–165. https://doi.org/10.1007/s43681-021-00074-z
3. Guan C, Mou J, Jiang Z (2020) Artificial intelligence innovation in education: a twenty-year data-driven historical analysis. Int J Innovation Stud 4:134–147. https://doi.org/10.1016/j.ijis.2020.09.001
4. Mustafa MY, Tlili A, Lampropoulos G, Huang R, Jandrić P, Zhao J, Salha S, Xu L, Panda S, Kinshuk, López-Pernas S, Saqr M (2024) A systematic review of literature reviews on artificial intelligence in education (AIED): a roadmap to a future research agenda. Smart Learn Environ 11:1–33. https://doi.org/10.1186/s40561-024-00350-5
5. Wang S, Wang F, Zhu Z, Wang J, Tran T, Du Z (2024) Artificial intelligence in education: a systematic literature review. Expert Syst Appl 252:124167. https://doi.org/10.1016/j.eswa.2024.124167
6. Ali O, Murray PA, Momin M, Dwivedi YK, Malik T (2024) The effects of artificial intelligence applications in educational settings: challenges and strategies. Technol Forecast Soc Change 199:123076. https://doi.org/10.1016/j.techfore.2023.123076
7. Gillani N, Eynon R, Chiabaut C, Finkel K (2023) Unpacking the "black box" of AI in education. J Educ Technol Soc 26:99–111. https://www.jstor.org/stable/48707970
8. Bedué P, Fritzsche A (2022) Can we trust AI? An empirical investigation of trust requirements and guide to successful AI adoption. J Enterprise Inf Manag 35:530–549. https://doi.org/10.1108/jeim-06-2020-0233

9. Miller T (2019) Explanation in artificial intelligence: Insights from the social sciences. Artif Intell 267:1–38. https://doi.org/10.1016/j.artint.2018.07.007
10. Bogina V, Hartman A, Kuflik T, Shulner-Tal A (2022) Educating software and AI stakeholders about algorithmic fairness, accountability, transparency and ethics. Int J Artif Intell Educ 32:808–833. https://doi.org/10.1007/s40593-021-00248-0
11. Khosravi H, Shum SB, Chen G, Conati C, Tsai Y-S, Kay J, Knight S, Martinez-Maldonado R, Sadiq S, Gašević D (2022) Explainable artificial intelligence in education. Comput Educ Artif Intell 3:100074. https://doi.org/10.1016/j.caeai.2022.100074
12. Miller T (2023) Explainable AI is dead, long live explainable AI!: hypothesis-driven decision support using evaluative AI. In: 2023 ACM conference on fairness, accountability, and transparency. ACM, New York, NY, USA, pp 333–342
13. Monett D, Lewis CWP (2018) Getting clarity by defining artificial intelligence—a survey. In: Studies in applied philosophy, epistemology and rational ethics. Springer International Publishing, Cham, pp 212–214. https://doi.org/10.1007/978-3-319-96448-5_21
14. Bond M, Khosravi H, De Laat M, Bergdahl N, Negrea V, Oxley E, Pham P, Chong SW, Siemens G (2024) A meta systematic review of artificial intelligence in higher education: a call for increased ethics, collaboration, and rigour. Int J Educ Technol Higher Educ 21. https://doi.org/10.1186/s41239-023-00436-z
15. Zawacki-Richter O, Marín VI, Bond M, Gouverneur F (2019) Systematic review of research on artificial intelligence applications in higher education—where are the educators? Int J Educ Technol Higher Educ 16. https://doi.org/10.1186/s41239-019-0171-0
16. Taylor DL, Yeung M, Bashet AZ (2021) Personalized and adaptive learning. In: Innovative learning environments in STEM higher education. Springer International Publishing, Cham, pp 17–34. https://doi.org/10.1007/978-3-030-58948-6_2
17. Zhong L (2023) A systematic review of personalized learning in higher education: learning content structure, learning materials sequence, and learning readiness support. Interactive Learn Environ 31:7053–7073. https://doi.org/10.1080/10494820.2022.2061006
18. Urdaneta-Ponte MC, Mendez-Zorrilla A, Oleagordia-Ruiz I (2021) Recommendation systems for education: systematic review. Electronics 10:1611. https://doi.org/10.3390/electronics10141611
19. Labadze L, Grigolia M, Machaidze L (2023) Role of AI chatbots in education: Systematic literature review. Int J Educ Technol Higher Educ 20. https://doi.org/10.1186/s41239-023-00426-1
20. Fichten C, Pickup D, Asunsion J, Jorgensen M, Vo C, Legault A, Libman E (2022) State of the research on artificial intelligence based apps for post-secondary students with disabilities. Except Educ Int 31:62–76. https://doi.org/10.5206/eei.v31i1.14089
21. Imani M, Montazer GA (2019) A survey of emotion recognition methods with emphasis on e-learning environments. J Netw Comput Appl 147:102423. https://doi.org/10.1016/j.jnca.2019.102423
22. Andrejevic M, Selwyn N (2020) Facial recognition technology in schools: critical questions and concerns. Learn Media Technol 45:115–128. https://doi.org/10.1080/17439884.2020.1686014
23. Lin C-C, Huang AYQ, Lu OHT (2023) Artificial intelligence in intelligent tutoring systems toward sustainable education: a systematic review. Smart Learn Environ 10:1–22. https://doi.org/10.1186/s40561-023-00260-y
24. Lamti S, El Malhi M, Sekhsoukh R, Kerzazi N (2024) Intelligent tutoring system for medical students: opportunities and challenges. In: Information systems engineering and management. Springer Nature Switzerland, Cham, pp 242–251. https://doi.org/10.1007/978-3-031-66850-0_27
25. Crow T, Luxton-Reilly A, Wuensche B (2018) Intelligent tutoring systems for programming education: a systematic review. In: Proceedings of the 20th australasian computing education conference. Association for Computing Machinery, New York, NY, USA, pp 53–62. https://doi.org/10.1145/3160489.3160492

26. Shafiq DA, Marjani M, Habeeb RAA, Asirvatham D (2022) Student retention using educational data mining and predictive analytics: a systematic literature review. IEEE Access 10:72480–72503. https://doi.org/10.1109/ACCESS.2022.3188767
27. Hellas A, Ihantola P, Petersen A, Ajanovski VV, Gutica M, Hynninen T, Knutas A, Leinonen J, Messom C, Liao SN (2018) Predicting academic performance: a systematic literature review. In: Proceedings companion of the 23rd annual ACM conference on innovation and technology in computer science education. ACM, New York, NY, USA. https://doi.org/10.1145/3293881.3295783
28. Liz-Domínguez M, Caeiro-Rodríguez M, Llamas-Nistal M, Mikic-Fonte FA (2019) Systematic literature review of predictive analysis tools in higher education. Appl. Sci. (Basel, Switzerland) 9:5569. https://doi.org/10.3390/app9245569
29. Pinto AS, Abreu A, Costa E, Paiva J (2023) How machine learning (ML) is transforming higher education: a systematic literature review. J Inf Syst Eng Manag 8:21168. https://doi.org/10.55267/iadt.07.13227
30. Gašević D, Dawson S, Rogers T, Gasevic D (2016) Learning analytics should not promote one size fits all: the effects of instructional conditions in predicting academic success. Internet Higher Educ 28:68–84. https://doi.org/10.1016/j.iheduc.2015.10.002
31. Saqr M, Jovanović J, Viberg O, Gašević D (2022) Is there order in the mess? A single paper meta-analysis approach to identification of predictors of success in learning analytics. Stud Higher Educ 47:2370–2391. https://doi.org/10.1080/03075079.2022.2061450
32. López-Pernas S, Saqr M (2024) How the dynamics of engagement explain the momentum of achievement and the inertia of disengagement: a complex systems theory approach. Comput Human Behav 153:108126. https://doi.org/10.1016/j.chb.2023.108126
33. Saqr M, López-Pernas S (2022) How CSCL roles emerge, persist, transition, and evolve over time: a four-year longitudinal study. Comput Educ 104581. https://doi.org/10.1016/j.compedu.2022.104581
34. Dörrenbächer L, Perels F (2016) Self-regulated learning profiles in college students: their relationship to achievement, personality, and the effectiveness of an intervention to foster self-regulated learning. Learn Indiv Differ 51:229–241. https://doi.org/10.1016/j.lindif.2016.09.015
35. Saqr M, López-Pernas S, Jovanović J, Gašević D (2023) Intense, turbulent, or wallowing in the mire: a longitudinal study of cross-course online tactics, strategies, and trajectories. Internet Higher Educ 57:100902. https://doi.org/10.1016/j.iheduc.2022.100902
36. Baker RS, Hawn A (2022) Algorithmic bias in education. Int J Artif Intell Educ 32:1052–1092. https://doi.org/10.1007/s40593-021-00285-9
37. Zhai X, Chu X, Chai CS, Jong MSY, Istenic A, Spector M, Liu J-B, Yuan J, Li Y (2021) A review of artificial intelligence (AI) in education from 2010 to 2020. Complexity 2021:1–18. https://doi.org/10.1155/2021/8812542
38. Ramesh D, Sanampudi SK (2022) An automated essay scoring systems: a systematic literature review. Artif Intell Rev 55:2495–2527. https://doi.org/10.1007/s10462-021-10068-2
39. Ala-Mutka KM (2005) A survey of automated assessment approaches for programming assignments. Comput Sci Educ 15:83–102. https://doi.org/10.1080/08993400500150747
40. Jukiewicz M (2024) The future of grading programming assignments in education: the role of ChatGPT in automating the assessment and feedback process. Thinking Skills Creativity 52:101522. https://doi.org/10.1016/j.tsc.2024.101522
41. Siiman LA, Rannastu-Avalos M, Pöysä-Tarhonen J, Häkkinen P, Pedaste M (2023) Opportunities and challenges for AI-assisted qualitative data analysis: an example from collaborative problem-solving discourse data. In: Lecture notes in computer science. Springer Nature Switzerland, Cham, pp 87–96. https://doi.org/10.1007/978-3-031-40113-8_9
42. Cukurova M (2024) The interplay of learning, analytics and artificial intelligence in education: a vision for hybrid intelligence. Br J Educ Technol J Council Educ Technol. https://doi.org/10.1111/bjet.13514
43. Williams P (2023) AI, analytics and a new assessment model for universities. Educ Sci 13:1040. https://doi.org/10.3390/educsci13101040

44. Yan L, Martinez-Maldonado R, Gasevic D (2024) Generative artificial intelligence in learning analytics: Contextualising opportunities and challenges through the learning analytics cycle. In: Proceedings of the 14th learning analytics and knowledge conference. ACM, New York, NY, USA. https://doi.org/10.1145/3636555.3636856
45. Chen A, Xiang M, Zhou J, Jia J, Shang J, Li X, Gašević D, Fan Y (2025) Unpacking help-seeking process through multimodal learning analytics: A comparative study of ChatGPT vs human expert. Computers & education 226:105198. https://doi.org/10.1016/j.compedu.2024.105198
46. Lytton C (2024) AI hiring tools may be filtering out the best job applicants. BBC. https://www.bbc.com/worklife/article/20240214-ai-recruiting-hiring-software-bias-discrimination
47. McCallum S (2024) Payout for uber eats driver over face scan bias case. BBC News. https://www.bbc.com/news/technology-68655429
48. Ali S, Abuhmed T, El-Sappagh S, Muhammad K, Alonso-Moral JM, Confalonieri R, Guidotti R, Del Ser J, Díaz-Rodríguez N, Herrera F (2023) Explainable artificial intelligence (XAI): what we know and what is left to attain trustworthy artificial intelligence. Int J Inf Fusion 99:101805. https://doi.org/10.1016/j.inffus.2023.101805
49. Leichtmann B, Humer C, Hinterreiter A, Streit M, Mara M (2023) Effects of explainable artificial intelligence on trust and human behavior in a high-risk decision task. Comput Human Behav 139:107539. https://doi.org/10.1016/j.chb.2022.107539
50. Saeed W, Omlin C (2023) Explainable AI (XAI): a systematic meta-survey of current challenges and future opportunities. Knowl Based Syst 263:110273. https://doi.org/10.1016/j.knosys.2023.110273
51. Saqr M, López-Pernas S (2024) Why explainable AI may not be enough: predictions and mispredictions in decision making in education. Smart Learn Environ. https://doi.org/10.1186/s40561-024-00343-4
52. Kumar V, Boulanger D (2020) Explainable automated essay scoring: deep learning really has pedagogical value. Front Educ 5:572367. https://doi.org/10.3389/feduc.2020.572367
53. Afzaal M, Nouri J, Zia A, Papapetrou P, Fors U, Wu Y, Li X, Weegar R (2021) Explainable AI for data-driven feedback and intelligent action recommendations to support students self-regulation. Front Artif Intell 4:723447. https://doi.org/10.3389/frai.2021.723447
54. Yang G, Ye Q, Xia J (2022) Unbox the black-box for the medical explainable AI via multi-modal and multi-centre data fusion: a mini-review, two showcases and beyond. Int J Inf Fusion 77:29–52. https://doi.org/10.1016/j.inffus.2021.07.016
55. Saqr M, López-Pernas S (2025) Explainable artificial intelligence in education: a tutorial for identifying the variables that matter. In: Advanced learning analytics methods: AI, precision and complexity. Springer, Berlin
56. Wei P, Lu Z, Song J (2015) Variable importance analysis: A comprehensive review. Reliab Eng Syst Saf 142:399–432. https://doi.org/10.1016/j.ress.2015.05.018
57. Lundberg SM, Lee S-I (2017) A unified approach to interpreting model predictions. In: Guyon I, Luxburg UV, Bengio S, Wallach H, Fergus R, Vishwanathan S, Garnett R (eds) Advances in neural information processing systems 30. Curran Associates, Inc., Red Hook, pp 4765–4774. http://papers.nips.cc/paper/7062-a-unified-approach-to-interpreting-model-predictions.pdf
58. Tiukhova E, Vemuri P, Óskarsdóttir M, Poelmans S, Baesens B, Snoeck M (2024) Discovering unusual study patterns using anomaly detection and XAI. In: Hawaii international conference on system sciences 2024 (HICSS-57). https://aisel.aisnet.org/hicss-57/da/learning_analytics/3
59. Roshan K, Zafar A (2023) Using kernel SHAP XAI method to optimize the network anomaly detection model. https://ieeexplore.ieee.org/abstract/document/9763241/
60. Saqr M, López-Pernas S (2025) Explainable artificial intelligence: Evaluating individual and local machine learning predictions in r. In: Advanced learning analytics methods: AI, precision and complexity. Springer, Berlin
61. Ribeiro MT, Singh S, Guestrin C (2016) Why should i trust you?: Explaining the predictions of any classifier. In: Proceedings of the 22nd ACM SIGKDD international conference on knowledge discovery and data mining. ACM, New York, NY, USA. https://doi.org/10.1145/2939672.2939778

62. Guidotti R (2024) Counterfactual explanations and how to find them: literature review and benchmarking. Data Mining Knowl Discovery 38:2770–2824. https://doi.org/10.1007/s10618-022-00831-6
63. Susnjak T (2024) Beyond predictive learning analytics modelling and onto eXplainable artificial intelligence with prescriptive analytics and ChatGPT. Int J Artif Intell Educ 34:452–482. https://doi.org/10.1007/s40593-023-00336-3
64. Jacovi A, Swayamdipta S, Ravfogel S, Elazar Y, Choi Y, Goldberg Y (2021) Contrastive explanations for model interpretability. http://arxiv.org/abs/2103.01378
65. Selvaraju RR, Cogswell M, Das A, Vedantam R, Parikh D, Batra D (2017) Grad-CAM: Visual explanations from deep networks via gradient-based localization. In: Proceedings of the IEEE international conference on computer vision, pp 618–626. http://openaccess.thecvf.com/content_ICCV_2017/papers/Selvaraju_Grad-CAM_Visual_Explanations_ICCV_2017_paper.pdf
66. Saqr M, Cheng R, López-Pernas S, Beck ED (2024) Idiographic artificial intelligence to explain students' self-regulation: toward precision education. Learn Indiv Differ 114:102499. https://doi.org/10.1016/j.lindif.2024.102499
67. Bearman M, Tai J, Dawson P, Boud D, Ajjawi R (2024) Developing evaluative judgement for a time of generative artificial intelligence. Assess Eval Higher Educ 49:893–905. https://doi.org/10.1080/02602938.2024.2335321
68. Khalil M, Prinsloo P, Slade S (2023) Fairness, trust, transparency, equity, and responsibility in learning analytics. J Learn Anal 10:1–7. https://doi.org/10.18608/jla.2023.7983
69. Madaio M, Blodgett SL, Mayfield E, Dixon-Román E (2022) Beyond "fairness". In: The ethics of artificial intelligence in education, 1st edn. Routledge, New York, pp 203–239. https://doi.org/10.4324/9780429329067-11
70. Saqr M, Vogelsmeier LVDE, López-Pernas S (2024) Capturing where the learning process takes place: a person-specific and person-centered primer. Learn Indiv Differ 113:102492. https://doi.org/10.1016/j.lindif.2024.102492

Artificial Intelligence: Using Machine Learning to Predict Students' Performance

Mohammed Saqr, Kamila Misiejuk, Santtu Tikka, and Sonsoles López-Pernas

Abstract Being able to predict students' performance has been a primary driver for the adoption of learning analytics and has attracted many scientists to the field. Predictive modeling focuses on using students' data to forecast outcomes such as student grades, enabling teachers and administrators to offer just-in-time support to students at risk. This chapter uses advanced predictive methods, namely machine learning, where the goal is to predict continuous variables like grades. The chapter uses advanced and popular AI/machine learning algorithms like Random Forest, K-Nearest Neighbor, Linear Regression, Neural Networks, and Support Vector Machines. The chapter provides a practical guide to building and evaluating predictive models with R using two approaches: one is the classic approach for predictive modeling with R, and the other more modern approach using the `tidymodels` suite.

Keywords Learning analytics · Predictive modeling · Artificial intelligence · Regression

1 Introduction

Predicting student performance is one of the most common tasks in learning analytics [1]. Typically, student grades are used as an outcome variable of such modeling. Predicting a continuous variable like grades is called a regression problem, in contrast to classification that deals with nominal outcomes such as high

M. Saqr (✉) · S. López-Pernas
University of Eastern Finland, Joensuu, Finland
e-mail: mohammed.saqr@uef.fi

K. Misiejuk
FernUniversität in Hagen, Hagen, Germany

S. Tikka
University of Jyväskylä, Jyväskylä, Finland

M. Saqr, S. López-Pernas (eds.), *Advanced Learning Analytics Methods*,
https://doi.org/10.1007/978-3-031-95365-1_3

and low achievers [2]. When we do predictive modeling, we assume that historical data can be used to build a model to predict the future. In doing so, prediction goes beyond reporting what has happened but attempts to explain why an event happened, monitor what is currently happening and also attempts to foresee future developments [3]. As such, it provides a basis for actionable insights based on data or what we call data-informed decision making. This is particularly relevant for scaling up learning analytics applications. As such, prediction has become a core step within the learning analytics process that aims to improve teaching or learning [4].

There are six main steps in predictive analytics [5]. First, the problem needs to be specified (e.g., improving student performance). Second, an outcome variable is defined (e.g., student grades), and data that can potentially predict the outcome variable is collected. Several types of data can be collected: prior academic data (e.g., student admission information), demographic data (e.g., student's socio-economic status), academic data (e.g., results of formative assessment), behavioral data (e.g., student log data), and psychological data (e.g., a survey on student motivation) [6]. Third, raw data needs to be pre-processed to become suitable for modeling. This can include formatting the data or deriving features from raw data (e.g., calculating time spent in an online system by subtracting log-out time from log-in time) (see [7] for more details on data cleaning and wrangling). In order to choose the appropriate predictors for the predictive model, an exploratory analysis is conducted (see [8] for a tutorial in exploratory methods). Once the relevant indicators are selected, a predictive model can be built using one or more algorithms. The selection of an appropriate algorithm is determined by "the problem type, the nature of the outcome to be predicted, and the variables employed in the prediction" [6]. Typically, several algorithms would be compared to choose the method that models the data most accurately. The results from a predictive analysis can be used to develop preventive measures and interventions. Depending on the timing of the analysis, these interventions can be implemented on the current cohort of students, or the insights from the analytics can be used to shape future teaching and learning. Finally, it is important to build explainable prediction models—i.e., that provide an explanation for how a prediction was calculated—in order to increase the trust of the stakeholders in the model and increase awareness of what the model is actually predicting. In addition, explainable models can help validate model accuracy [9]. An example of an explainable predictive model is [10], who used data from formative assessment activities (e.g., the number of correct or incorrect answers to a question) to generate two predictions: of a grade in the midterm exam (on a scale of 0–20) and of a final grade (on a scale 0–40). The predictive model used the decision tree (DT) algorithm that supports the instructor's sense-making and helps identify student groups in need of additional support.

The goal of this chapter is to explore advanced techniques of predictive analytics to model regression problems (see [11] for an introductory tutorial in predictive analytics). It is assumed that the reader is familiar with the R programming language. If that is not the case, it is recommended to refer to previous tutorials on the basics of R [12], data cleaning [13], basic statistics [14], and visualization

[15]. The following section presents an overview of learning analytics research concerned with regression. Next, a synthetic dataset used in this tutorial is presented based on data of this study [16]. The tutorial section starts with an exploratory data analysis, including visualizing variable distribution and calculating correlations. It is followed by a description of the steps needed to create a predictive model. This tutorial explores the following predictive methods: Random Forest (RF), K-Nearest Neighbor (KNN), Linear Regression (LR), Neural Networks (NN), and Support Vector Machine (SVM). In addition, techniques to determine predictive model performance, such as Mean Absolute Error (MAE), Root-mean-square error (RMSE), and R-squared (RSQ), are presented and explained. The final part of this chapter shows how to create workflows to use multiple predictive algorithms.

2 Previous Research on Student Performance Prediction

Tomasevic et al. [17] described three categories of regression techniques: (1) similarity-based approaches that are built upon identifying similar patterns in a dataset (e.g., KNN), (2) model-based approaches are developed by estimating implicit correlation among data samples and predictors (e.g., SVM, NN, DT, LR), (3) probabilistic approaches focus on examining probability distributions in the input dataset (e.g., Bayesian LR). Regression is the most commonly applied method in predictive analysis, while predicting student performance is one of the main regression problems prevalent in learning analytics research [18]. The vast majority of learning analytics papers focus on predicting student final grades [18]. Predictions in learning analytics are conducted in two main ways: (1) formative prediction examines student features at several points in the duration of the learning activity to provide several predictions over time, and (2) summative prediction aggregates data from the whole learning activity to predict the final outcome [19–21]. In addition, different levels of learning activity can be specified, e.g., degree, year, course, or exam level [22]. Log activity data is typically combined with other data sources, such as demographic or self-reported data [18]. Current evidence suggests that engagement level expressed through participation frequencies has a positive effect on performance [18]. Prior academic data, such as grades achieved in previous courses, was found to be the most influential factor in predicting student performance. An interesting meta-analytics study conducted by [16] to examine the reliability of predictors across courses found that overall engagement with online tasks and collaborative learning activities (measured by total activity and forum indicators) showed the highest prediction ranges, indicating their reliability as metrics. Lecture reading frequency had insignificant prediction ranges, suggesting limited applicability across courses.

Predictive modeling can be challenging sometimes, especially when using low quality data for analysis or having a shallow understanding of the nature of the data and therefore can lead to untrustworthy findings. As learning processes are dynamic, using results from historical data to model current learners can be inaccurate and lead to the over-fitting or under-fitting of statistical models [23]. These issues can be mitigated by applying methods such as cross-validation or bootstrapping [4, 23, 24]. Another problem in predictive analytics is imbalanced datasets, where specific important values are not sufficiently represented in a dataset. This issue can be mitigated by over- or undersampling (for more details, see [25]). In addition, it is crucial to approach the findings from a point-in-time perspective that takes into consideration the temporality aspect of the findings' accuracy [24, 26]. Also, it is important to consider the ethical implications of implementing predictive learning analytics so as not to cause harm to students or contribute to perpetuating social injustices [27, 28]. Tomasevic et al. [17] identified three main issues with implementing predictive learning analytics. First, the issue of a cold start relates to the lack of data at the early stages of learning analytics deployment. This challenge can be mitigated by including demographic data or previous academic data. Second, the issue of scalability is connected to the increasing number of students, the complexity of algorithms used, and the computational requirements necessary to perform analytics. To address this issue, time constraints on the analysis or algorithm optimization need to be performed. Third, the issue of data sparsity refers to the lack of sufficient data about student activities, which can be caused by offline learning activities or passive interactions with available online learning systems. In order to increase available data, online learning activities can be made mandatory and embedded into a course design.

Decisions regarding which data to collect, which features to select, which algorithm to use to build a predictive model, or how to evaluate model performance depend on the available data, dataset size, and study context [16]. A substantial body of research compares the performance of different predictive algorithms. For example, [29] collected student demographic and socio-economic data, self-reported motivation level, and assessment data from two evaluations during a semester. The authors predicted the student's final grade (on a scale of 0–20) using the KNN and SVM algorithms. The performance of both algorithms was evaluated using 10-fold cross-validation and direct partition in the ratio methods, and it was found that SVM outperformed KNN. In another example, [30] collected student log data from a learning management system, such as the number of system logins the number of posts, or quiz grades, to model final exam grades using SVM and LR algorithms. The predictions were made multiple times during a semester and shown to the teacher. In this study, LR performed better than SVM, and the goodness-of-fit measure, R-squared, increased with each prediction in the semester. In addition, feature selection decreased model performance. Tomasevic et al. [17] predicted student scores on the final exam (on a scale of 0–110) using KNN, SVM, Artificial NN, DT, Bayesian Regression, and LR. The dataset includes demographic data (e.g., gender, age), engagement data (sum of clicks), and performance data (e.g., scores per assessment). Artificial NN using engagement data and performance data had the highest precision out of all models. Including demographic data did not improve model precision significantly.

Other papers focused on determining the most important features that predict the outcome variable. For example, [31] collected LMS activity data, such as login frequency or number of student posts, and student assignment and assessment grades to predict student final grades using multiple linear regression. The results showed that total login frequency in LMS, regularity of learning interval in LMS, and total assignments and assessment composites predicted the final grades. Cavazos and Garza [32] collected two datasets: (1) where historical academic grades were available and limited demographic data was used; (2) where historical academic grades were not available, while demographic and socioeconomic data were extensively used. The SVM algorithm was applied to predict students' high school final grades using both datasets. The authors reported greater errors in the case of the dataset without historical academic grades than in the dataset with historical academic grades.

Finally, the issue of transferring predictive models within the same learning context and across different learning contexts was explored in the learning analytics research [16, 33, 34]. For example, [35] examined the relationship between LMS log data and final grade (on a scale of 0–100%) after adjusting to student characteristics. In addition, separate multiple linear regression models were developed to compare different courses. The study found that the impact of LMS features varied significantly on student performance across courses. The work by [33] applied mixed effect linear regression models to analyze a sample of data from log data from 15 courses that were homogeneous regarding the institutional settings, discipline, nominal learning design, and course size. The results highlighted that the same pedagogical model and study setting did not guarantee the same predictive power among courses, as there may be differences in the practical implementation of specific pedagogical models. In addition, it was found that the overall time spent in an LMS, regular discussion forum posts, and accessing course materials regularly were significant predictors of student final course grades across all courses.

3 Performance Prediction with R

In this section, we present two approaches to implement performance prediction in R: one following a more classic procedure, and the second one following a more updated procedure based on `tidymodels`. Broadly, the workflow followed is depicted in Fig. 1, aligned with the common steps that are performed in any ML pipeline. First, we explore the data to gain an idea of the magnitude, format and content of the data, explore the relationships between the different variables, etc. This step is called exploratory data analysis. Next, we prepare the data (B) for subsequent steps. This might include removing or fixing incomplete records, converting data to different types, unifying values that are close to one another, or rescaling the data. The transformations that we need to do are commonly dictated by the type of data and the ML models that we will be using in subsequent steps. After data preparation, the next step is data splitting (C). Here, the dataset is divided into training and testing sets. The training set is used to fit the model, while the testing set is reserved for evaluating its performance. It is crucial to ensure that the testing

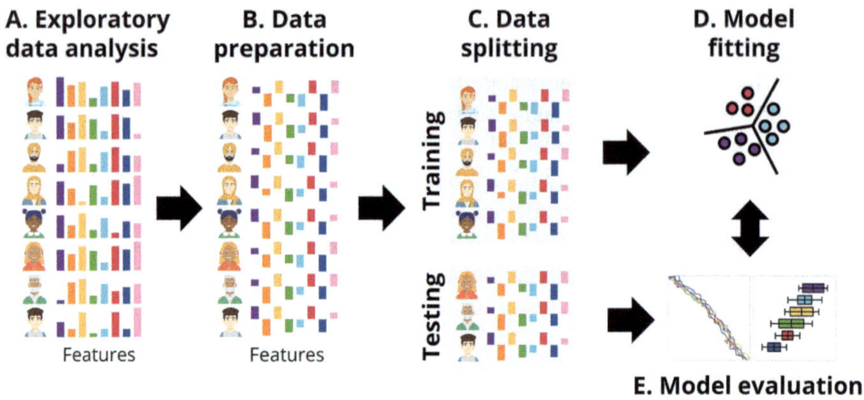

Fig. 1 ML workflow implemented in the tutorial

set remains untouched until the evaluation stage to provide an unbiased estimate of the model's generalization ability. Once the data is split, the process moves to model fitting (D). During this stage, the training data is used to train the selected machine learning models. This involves finding the optimal parameters that minimize error or maximize predictive accuracy for the given task. Depending on the algorithm, the model may identify different patterns, such as decision boundaries or relationships between features and the target variable. Finally, in the model evaluation step (E), the trained model is tested on the unseen testing data. Performance metrics that quantify the degree of error in the numeric predictions are computed to determine how well the model performs on new data. This step ensures that the model is not overfitting the training data and can generalize well to other datasets. We can repeat this process for several models with different underlying ML algorithms, or using only a subset of features of the data and compare the performance metrics among different models to select the one with the best fit.

3.1 The Dataset Used in this Chapter

The dataset we are using to illustrate the methods in this tutorial is a synthetic dataset based on the study by Jovanović et al. [33] and consists mostly of behavioral engagement indicators that can be obtained from learning management system (LMS) trace-log data. These variables capture both the engagement behavior of students (e.g., frequency and regularity of course and forum activities) and their participatory and time investment in the course (e.g., total duration, active days, session count) as well as the regularity of participation. Both time and regularity may be considered proxy indicators of cognitive engagement. We can also consider forum contributions as indicators of cognitive engagement since the context is problem-based learning, and student contributions require students to read, synthesize, critique, and formulate arguments [36].

The dataset has the following variables:

- **Frequency Variables**
  - **Freq_Course_View**: The frequency of course page views by the student.
  - **Freq_Forum_Consume**: The frequency with which the student consumes content in the forum (i.e., reads forum posts).
  - **Freq_Forum_Contribute**: The frequency with which the student contributes to the forum (i.e., posts in the forum).
  - **Freq_Lecture_View**: The frequency of lecture video views by the student.
- **Regularity Variables**
  - **Regularity_Course_View**: The consistency of the student's course page views.
  - **Regularity_Lecture_View**: The consistency of the student's lecture video views.
  - **Regularity_Forum_Consume**: The consistency of the student's forum content consumption.
  - **Regularity_Forum_Contribute**: The consistency of the student's forum contributions.
- **Time Variables**
  - **Session_Count**: The total number of sessions the student has participated in.
  - **Total_Duration**: The total duration (in seconds) of all sessions participated by the student.
  - **Active_Days**: The number of days the student was active in the course.
- **Outcome Variables**
  - **Final_Grade**: The final grade of the student in the course.

3.1.1 Exploratory Data Analysis

Exploratory data analysis (EDA) is usually performed before doing the actual machine learning tasks (Fig. 1-A). EDA includes examining the characteristics of the dataset to understand the distribution of the variables, and identify patterns and inconsistencies, if they may arise. Exploring the distribution and variance of each variable helps us identify any skewness, outliers, or anomalies in the data. Furthermore, EDA also allows us to examine the presence of missing values—or lack thereof—, handle distribution inconsistencies, and transform variables to balance their characteristics if needed. For instance, if we find that our variables have different measurement scales (one is measured in large numbers with vast variance and another in proportions and ranges between 0 and 1), we may need to transform or normalize the data to ensure that the two variables are not largely different. This process is particularly needed in some algorithms that are sensitive to discrepancies between variables. Additionally, we may have variables that are highly correlated (e.g., each is more or less a copy of the other), we may need to discard one of these similar variables as they do not offer any added information to the model and may bias the estimates. It should be mentioned that standardization will often result in less (direct) interpretability of the model, for example, when predicting final grade, there will be negative grades (below average) which might be hard to interpret, and also regression coefficients may be tricky to interpret. If standardization is carried out, the results may be presented on the original scales for better interpretation.

The first step in EDA is to load the R packages and use them to explore the data. We will need several R packages. Firstly, we will need `tidyverse`: a set of R packages designed to make it easy to read, edit, manipulate and visualize data through a large collection of functions for filtering, selecting and modifying the data [37]. As a collection of packages, `tidyverse` includes packages such as `dplyr` and `tidyr` for data manipulation and `ggplot2` for data visualization, among others. The `skimr` package provides an easy, yet comprehensive, summary of a dataset with more detail than most R built-in summary functions [38]. The `skimr` package offers central tendency statistics like mean and median as well as the number and proportion of missing values for each variable, making it easy to get an overview of the data and detect any anomalies or patterns quickly. Besides, `skimr` calculates variance measures, maximum, minimum and displays a miniature histogram for each variable. The `correlation` package is an easy-to-use package for estimating and visualizing several types of correlations. In particular, it will be used to create a heatmap to understand relationships between variables [39]. We will also use the package `rio` to import the data [40]. Lastly, we will use the `performance` package to evaluate our predictive models [41].

The next code starts by loading the R packages and reading the data.

```
# Load required libraries
library(tidyverse)
library(correlation)
library(skimr)
library(rio)
library(performance)

student_data <- import("https://github.com/lamethods/data2/raw/main/lms/lms.csv")
```

The previous R code imports student data from a CSV file. Next, the following EDA tasks will be performed: (1) provide a detailed summary of the data using `skimr`, (2) create histograms for all variables to examine their distribution, (3) calculate and visualize correlations between variables, and (4) extract and display correlations with the final grade of the course (`Final_Grade`).

In our EDA, first, we get a quick and comprehensive overview of the dataset using the `skim` function from the `skimr` package. The function provides detailed information such as the number of rows, number of columns, type of variables, and most importantly, the number of missing values (`n_missing`). `skimr` also provides basic statistics for each variable like mean, standard deviation (sd), minimum (p0) and maximum (p100), first quartile (p25), median (p50), third quartile (p75), and a small histogram of the data (Table 1).

```
# 1. A detailed summary using skimr
skim(student_data)
```

Name	student_data
Number of rows	285
Number of columns	12
Column type frequency	
Numeric	12
Group variables	None

The results of the `skim(student_data)` command show that the data has 285 observations and no missing data in any variable (Table 1). However, the variables have large differences in variance. For instance, the frequency variables are different from the regularity variables which are all below 1 and even more different if we compared them to the duration which has a mean of 83150.98 (SD = 23818.16). This information tells us that data may need to be normalized if the algorithm that we choose is sensitive to differences in variance.

The second step is to inspect the distribution of all variables (Fig. 2) using the following code, which starts by using `pivot_longer(everything())` to reshape our data, putting all variables into a single column. This makes it easier to create a faceted plot (multiple plots). We use `ggplot2` to create histograms for each variable. The `facet_wrap(~ name, scales = "free")` line creates a separate plot for each variable, and `scales = "free"` allows each plot to have its own horizontal and vertical scales, which is often necessary when dealing with variables of different types or ranges and scales of measurement. For a tutorial about data visualization with R, refer to [15].

```
# 2. Histograms of all variables
student_data |>
  pivot_longer(everything()) |>
  ggplot(aes(x = value)) +
  geom_histogram(bins = 25, fill = "skyblue", color = "black") +
  facet_wrap(~ name, scales = "free") +
  theme_minimal()
```

The third step is to look at the correlations between variables (Fig. 3). This allows us to see if there are any variables that are highly correlated to each other. If there are variables that are perfectly or almost perfectly correlated, they might be redundant or linear combinations of one another and therefore, do not add much to the model.

Table 1 Summary of the dataset created using `skimr`

skim_type	skim_variable	n_missing	complete_rate	numeric.mean	numeric.sd	numeric.p0	numeric.p25	numeric.p50	numeric.p75	numeric.p100	numeric.hist
Numeric	Freq_Course_View	0	1	223.79	59.33	51.00	187.00	222.00	264.00	356.00	
Numeric	Freq_Forum_Consume	0	1	582.55	175.55	84.00	465.00	589.00	703.00	991.00	
Numeric	Freq_Forum_Contribute	0	1	178.07	55.57	17.00	144.00	175.00	213.00	347.00	
Numeric	Freq_Lecture_View	0	1	201.87	56.68	32.00	162.00	202.00	247.00	346.00	
Numeric	Regularity_Course_View	0	1	0.51	0.14	0.15	0.42	0.52	0.61	0.90	
Numeric	Regularity_Lecture_View	0	1	0.48	0.15	0.07	0.38	0.49	0.57	0.80	
Numeric	Regularity_Forum_Consume	0	1	0.49	0.14	0.13	0.41	0.50	0.58	0.95	
Numeric	Regularity_Forum_Contribute	0	1	0.48	0.16	0.03	0.38	0.47	0.59	0.90	
Numeric	Session_Count	0	1	189.54	52.83	53.00	151.00	191.00	229.00	306.00	
Numeric	Total_Duration	0	1	83150.98	23818.16	25055.00	68632.00	83667.00	98373.00	147827.00	
Numeric	Active_Days	0	1	16.06	4.15	5.00	14.00	16.00	19.00	31.00	
Numeric	Final_Grade	0	1	68.62	8.86	44.73	62.42	68.66	75.26	90.89	

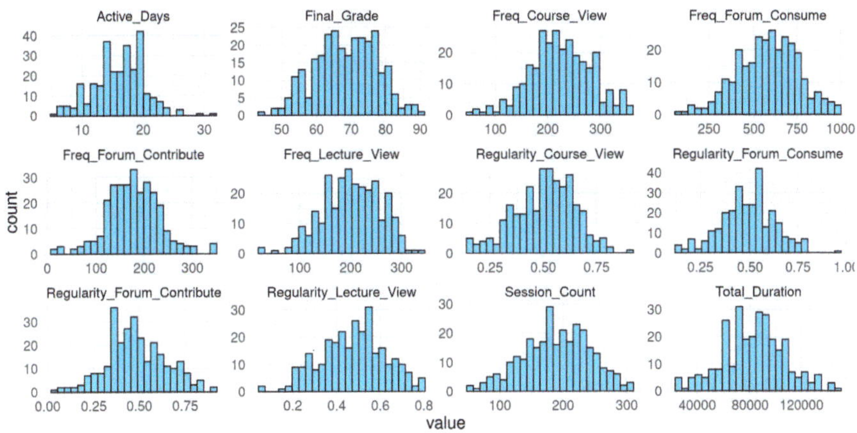

Fig. 2 Histograms of all variables

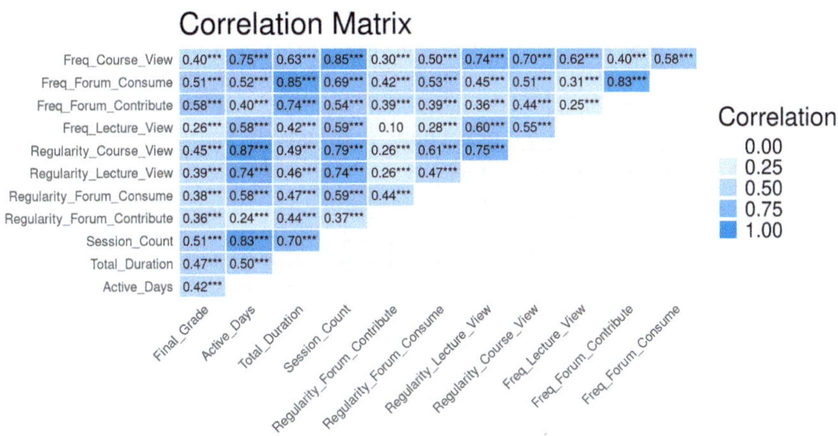

Fig. 3 Correlations between `Final_Grade` and other variables

The heatmap shows that most variables have correlations below the 0.8 level, and only a few cross this level. We may also examine the correlation with the final grade (our outcome variable that we want to predict) to explore variable importance (Table 2).

```
# 3. Relationship between Variables and `Final_Grade`
# Calculate correlations with `Final_Grade`
correlations <- correlation(student_data, method = "pearson")
correlations |> summary() |> plot() +
  theme_minimal() +
  theme(axis.text.x = element_text(angle = 45, vjust = 1, hjust = 1))
```

```
# Extract and display correlations with `Final_Grade`
final_grade_correlations <- correlations |>
  as.data.frame() |>
  filter(Parameter2 == "Final_Grade") |>
  arrange(desc(r))

# Print the correlations with `Final_Grade`
final_grade_correlations
```

3.1.2 Data Preparation

The next stage in our workflow is data preparation (Fig. 1-B), where we clean and transform our data to fix any problems and inconsistencies and to get it ready for subsequent steps. We can see two potential issues with our data: data with large differences in variance (e.g., duration and regularity) as well as potential high collinearity in some variables. We may need to standardize the data (subtract the mean and divide by the standard deviation) to make all variables have a comparable scale (Table 3). The code in the next chunk does exactly that: it simply scales all numeric variables. You may need to re-run the EDA code again to verify:

```
# Standardize numeric columns in the student_data data frame
student_data_standardized <- student_data |>
  mutate(across(
    where(is.numeric),  # Select all numeric columns
    ~scale(.) |> # Standardize each column (M=0, SD=1)
      as.vector() # Convert to vector
  ))

# Use skimr package to get a summary of the standardized data
skim(student_data_standardized)
```

We may also need to investigate the multicollinearity issue further. While multicollinearity is not a commonly discussed problem in machine learning, it is particularly important in explainable AI. If present, multicollinearity can lead to unstable estimates of model coefficients and inflated standard errors. One way to detect multicollinearity is by calculating the variance inflation factor (VIF). VIF measures the amount of inflation in the variance of a coefficient due to its correlation with other predictors. A high VIF value indicates that a variable is highly correlated with one or more other variables, suggesting multicollinearity. While there is no consensus about the VIF cutoff, in practice, a VIF value greater than 5 or 10 is often used as a cutoff to indicate multicollinearity. VIF is commonly computed in regression models, so we fit a regression model in the following code and then estimate the VIF from it. The code fits a linear regression model to predict the

Table 2 Correlations between variables and Final_Grade

Parameter1	Parameter2	r	CI	CI_low	CI_high	t	df_error	p	Method	n_Obs
Freq_Forum_Contribute	Final_Grade	0.58	0.95	0.50	0.65	12.0	283	0	Pearson correlation	285
Freq_Forum_Consume	Final_Grade	0.51	0.95	0.42	0.59	10.0	283	0	Pearson correlation	285
Session_Count	Final_Grade	0.51	0.95	0.42	0.59	10.0	283	0	Pearson correlation	285
Total_Duration	Final_Grade	0.47	0.95	0.38	0.56	9.1	283	0	Pearson correlation	285
Regularity_Course_View	Final_Grade	0.45	0.95	0.35	0.54	8.4	283	0	Pearson correlation	285
Active_Days	Final_Grade	0.42	0.95	0.32	0.51	7.8	283	0	Pearson correlation	285
Freq_Course_View	Final_Grade	0.40	0.95	0.30	0.50	7.4	283	0	Pearson correlation	285
Regularity_Lecture_View	Final_Grade	0.39	0.95	0.29	0.48	7.1	283	0	Pearson correlation	285
Regularity_Forum_Consume	Final_Grade	0.38	0.95	0.28	0.48	7.0	283	0	Pearson correlation	285
Regularity_Forum_Contribute	Final_Grade	0.36	0.95	0.25	0.45	6.4	283	0	Pearson correlation	285
Freq_Lecture_View	Final_Grade	0.26	0.95	0.15	0.37	4.5	283	0	Pearson correlation	285

Table 3 Summary of the standardized dataset

skim_type	skim_variable	n_missing	complete_rate	numeric.mean	numeric.sd	numeric.p0	numeric.p25	numeric.p50	numeric.p75	numeric.p100	numeric.hist
Numeric	Freq_Course_View	0	1	0	1	−2.9	−0.62	−0.03	0.68	2.2	
Numeric	Freq_Forum_Consume	0	1	0	1	−2.8	−0.67	0.04	0.69	2.3	
Numeric	Freq_Forum_Contribute	0	1	0	1	−2.9	−0.61	−0.06	0.63	3.0	
Numeric	Freq_Lecture_View	0	1	0	1	−3.0	−0.70	0.00	0.80	2.5	
Numeric	Regularity_Course_View	0	1	0	1	−2.5	−0.62	0.09	0.73	2.8	
Numeric	Regularity_Lecture_View	0	1	0	1	−2.8	−0.66	0.09	0.63	2.2	
Numeric	Regularity_Forum_Consume	0	1	0	1	−2.6	−0.60	0.05	0.62	3.3	
Numeric	Regularity_Forum_Contribute	0	1	0	1	−2.8	−0.64	−0.08	0.66	2.6	
Numeric	Session_Count	0	1	0	1	−2.6	−0.73	0.03	0.75	2.2	
Numeric	Total_Duration	0	1	0	1	−2.4	−0.61	0.02	0.64	2.7	
Numeric	Active_Days	0	1	0	1	−2.7	−0.50	−0.02	0.71	3.6	
Numeric	Final_Grade	0	1	0	1	−2.7	−0.70	0.00	0.75	2.5	

outcome variable (`Final_Grade`) using student LMS indicators as predictors. After fitting the model, it pipes the result into the `check_collinearity()` function from the `performance` package. The results of the VIF show that we have only four variables with moderate VIF values, so we can proceed with the analysis with no exclusion of any variable, given that we do not have a considerable collinearity problem.

```
# Fit a linear regression model and check for multicollinearity
lm(Final_Grade ~ Freq_Course_View + Freq_Lecture_View + Freq_Forum_Consume +
                 Freq_Forum_Contribute + Regularity_Course_View +
                 Regularity_Lecture_View + Regularity_Forum_Consume +
                 Regularity_Forum_Contribute + Session_Count +
                 Total_Duration + Active_Days,
   data = student_data_standardized) |>  # Use standardized data for the model
check_collinearity()    # Check for multicollinearity among predictors
```

```
# Check for Multicollinearity

Low Correlation

                       Term  VIF    VIF 95% CI Increased SE Tolerance Tolerance 95% CI
           Freq_Course_View 4.57 [3.77, 5.59]         2.14      0.22     [0.18, 0.26]
          Freq_Lecture_View 1.96 [1.68, 2.34]         1.40      0.51     [0.43, 0.59]
      Freq_Forum_Contribute 3.74 [3.11, 4.56]         1.93      0.27     [0.22, 0.32]
    Regularity_Lecture_View 3.09 [2.59, 3.75]         1.76      0.32     [0.27, 0.39]
   Regularity_Forum_Consume 2.05 [1.76, 2.46]         1.43      0.49     [0.41, 0.57]
Regularity_Forum_Contribute 1.43 [1.27, 1.69]         1.20      0.70     [0.59, 0.79]
             Total_Duration 4.75 [3.92, 5.81]         2.18      0.21     [0.17, 0.26]

Moderate Correlation

                  Term  VIF    VIF 95% CI Increased SE Tolerance Tolerance 95% CI
    Freq_Forum_Consume 6.57 [5.38, 8.08]         2.56      0.15     [0.12, 0.19]
Regularity_Course_View 5.17 [4.26, 6.34]         2.27      0.19     [0.16, 0.23]
         Session_Count 7.31 [5.97, 9.01]         2.70      0.14     [0.11, 0.17]
           Active_Days 5.79 [4.75, 7.10]         2.41      0.17     [0.14, 0.21]
```

3.2 *Tutorial 1: A Classic Approach to Predictive Modeling*

Having explored and prepared the data, we now proceed to the next step of predicting students' performance using machine learning. We will predict the final grades using the variables that represent the students' engagement. The steps include loading the necessary R packages, splitting the data, fitting the model, and evaluating the model's performance. First, we load the required R packages: the `randomForest` package, which is used to build the random forest model, the `rsample` package, which will be used to split the data into training and testing sets, and the `yardstick` package to facilitate model evaluation by calculating performance metrics.

Second, we split the dataset into two parts: a **training** set, which is used to train the machine learning model (perform the prediction), and a **testing** set (a holdout set for evaluation). The main rationale for data splitting is to use it to assess the performance of the estimated machine learning model on new, unseen data (the testing dataset). When keeping a testing dataset aside, we can evaluate how well the trained model generalizes to new data instances and estimate its accuracy. After all, we are creating the model with the hope that the same results will be obtainable in the future with new data. Furthermore, if we perform the prediction on all of the data, the model may—and most probably will—overfit. Overfitting occurs when a model learns the training data too well, including noise or irrelevant features, resulting in poor performance on new data. By having a separate testing set held out, we are increasing the likelihood that the model is learning the relevant patterns within the data not the noise. In practice, if a good model performs well on both the training and testing datasets, it is a good indication that the model is not overfitting and is likely to perform well in other situations with new data. Lastly, having a separate testing data set is rather valuable when our goal is to compare multiple machine learning algorithms to determine which model performs better on new data.

The code below uses the function `initial_split` function from the `rsample` package to perform an 80/20 split, where 80% of the data is used to train the model (perform the prediction), and the remaining 20% is kept aside as a testing set. While there is no consensus on the split ratio with many others using 70/30 and closer values, 80/20 is a reasonable starting point as it provides sufficient data for training while reserving a reasonable amount for evaluation.

Third, we build the random forest model using the `randomForest` function from the `randomForest` package. The random forest algorithm is an ensemble learning method that creates multiple decision trees and combines their predictions to improve accuracy and control overfitting. This combination of multiple models helps reduce overfitting by averaging out the errors, resulting in an—arguably—more robust and generalizable model. More importantly, the random forest algorithm is an explainable model and provides feature importance scores, which can be used to identify the most relevant features for the problem at hand. In our case, it can help explain which variables help predict students' performance. In this model, we predict the final grade (`Final_Grade`) based on the engagement variables such as the frequency and regularity of course views, lecture views, forum consumption and contributions, session count, total duration of activities, and the number of active days. We specify the number of trees (`ntree`) to be 1000, a parameter that controls the complexity and performance of the model.

After fitting the model, we print the model summary using the functions `print` and `importance` (which shows the explanatory variables). The summary provides the model call, insights into the model's performance, including error rates and the proportion of variance explained. Variable importance shows how much each predictor has contributed to the prediction accuracy of students' performance. In doing so, it allows us to understand which features are more associated with higher grades. More accurately, which features the random forest algorithm associated with higher grades. Learning the important variables allows us to gain insights of what can help in improving teaching and learning based on the results of our model.

Fourth, we use the trained model with new unseen data instances that we held out (the testing dataset) to perform predictions. In other words, the already trained model will estimate the final grades in the testing data. In this step, we use the `predict` function to apply the model to the test dataset to generate predictions of final grades. These predictions will be compared against the actual grades to evaluate how close or far the algorithm was able to predict the grades and by that we evaluate the model's performance. The closer the predictions are to the actual grade values, the better the performance of the algorithm. If the predicted grades differ considerably from the actual grades, it is an indication that the model is not good enough. It might be that we do not have enough data to make reliable predictions or, on the contrary, that our model is overfitted to the training data.

In the fifth step, we evaluate the model performance, given that there are multiple ways to assess performance, we will use different indicators, each of them provides different insights. Among the most common metrics when evaluating a regression machine learning model are the root mean squared error (RMSE), R-squared (R^2), and mean absolute error (MAE). RMSE is a measure of the differences between predicted and actual values. It is calculated as the square root of the average of the squared differences between predicted and actual values. RMSE is always a positive value (because the errors are squared), where lower RMSE values indicate a better fit of the model to the data. Further, RMSE is sensitive to outliers because RMSE squares the errors, inflating the weight of larger errors.

R-squared (R^2 or rsq) measures the proportion of variance in the dependent variable (the final grade in our case) that is explained by the independent variables. R^2 values range from 0 to 1, where values of 1 indicate that the model explains all the variability and R^2 of 0 indicates that the model explains none (bad model). However, although R^2 is popular, it can be misleading, pointing to erroneous conclusions. A high R^2 does not necessarily mean that the model is good; it only means that the model explains a large portion of the variance in the dependent variable. A major issue is that R^2 only measures the goodness of fit between the model and the data, but it does not account for the complexity or interpretability of the model. Adding more independent variables to a model, regardless of their relevance, will always increase or at least not decrease the R^2 value, leading to a more complex model that may be hard to interpret. As such, high R^2 values may give a false impression of a model's predictive power. Furthermore, a high R^2 does not guarantee that the model will perform well on new, unseen data. Other metrics such as RMSE or MAE, are more useful in this regard. MAE measures the absolute average magnitude of the errors in a set of predictions (i.e., without considering the direction of the difference). It is calculated as the average of the absolute differences between predicted and actual values. Unlike RMSE, MAE is not sensitive to outliers (since no squaring is performed).

The sixth step in our analysis is creating better and enhanced model performance evaluation. In that, we visualize the predicted versus actual grades to see how the model fared. We also plot the variable importance in a better way using the ggplot2 package as well as the residuals. Here are the steps of the tutorial in detail:

3.2.1 Loading the Necessary Libraries

The code below starts by loading the necessary libraries. The randomForest [42] library is used to build and work with random forest models, while rsample [43] aids in splitting the dataset into training and testing subsets. The yardstick [44] package is used for evaluating the performance of the model, and ggplot2 is employed for creating visualizations, although the latter is not explicitly mentioned in the code given that it is part of tidyverse [37] that we loaded before. Next, we set the seed function to a number set.seed(256) to ensure that our results would repeat in a similar way when re-running the code, e.g., data splitting would be consistent across different runs of the code.

```
# Step 1: Load necessary libraries
library(randomForest)  # For building the Random Forest model
library(rsample)       # For data splitting
library(yardstick)     # For model evaluation

# Set seed for reproducibility
set.seed(256)
```

3.2.2 Splitting the Dataset

Then, in the second step, we split the dataset (student_data_standardized) into training and testing sets (Fig. 1-C). This is done using the initial_split function from the rsample package, which divides the data into an 80% training set and a 20% testing set. The training function extracts the training subset, while the testing function retrieves the testing subset and they are both assigned to two data frames train_data and test_data respectively.

```
# Step 2: Split the data into training and testing sets
# Using initial_split from rsample package for an 80/20 split
data_split <- initial_split(student_data_standardized, prop = 0.8)
train_data <- training(data_split)
test_data <- testing(data_split)
```

3.2.3 Creating and Fitting the Model

In the third step, we build and fit the random forest model (Fig. 1-D). The `randomForest` function is used to create this model, where `Final_Grade` is predicted based on the engagement predictor variables such as `Freq_Course_View`, `Freq_Lecture_View`, and `Session_Count`. The model is configured to use 1000 trees, which is specified by the `ntree` parameter. In Random Forest models, specifying a large number of trees enhances performance by reducing variance and improving stability through the aggregation of predictions from 1000 diverse trees. Each tree provides a unique perspective due to its training on random data subsets, and averaging their predictions minimizes overfitting. Yet, whereas more trees may improve model performance, balancing the number of trees is important, as performance gains diminish—or disappear—beyond a certain point. Also, setting a very high number of trees incurs computational costs, increases the risk of overfitting, and training time.

After fitting the model, the code prints a summary of the random forest model using `print(rf_model)`. This summary provides insights into the model's performance and structure. Additionally, `importance(rf_model)` is called to extract and display the importance of each variable in the model, helping to identify which predictors contribute most to the model's predictions.

```
# Step 3: Create and fit a Random Forest model
# Building the Random Forest model with 1000 trees
rf_model <- randomForest(Final_Grade ~
                           Freq_Course_View + Freq_Lecture_View + Freq_Forum_Consume +
                           Freq_Forum_Contribute + Regularity_Course_View +
                           Regularity_Lecture_View + Regularity_Forum_Consume +
                           Regularity_Forum_Contribute + Session_Count +
                           Total_Duration + Active_Days,
                         data = train_data, ntree = 1000)

# Print model summary and variable importance
print(rf_model)
```

```
\begin{verbatim}
Call:
 randomForest(formula = Final_Grade ~ Freq_Course_View + Freq_Lecture_View +
 Freq_Forum_Consume + Freq_Forum_Contribute + Regularity_Course_View +
 Regularity_Lecture_View + Regularity_Forum_Consume + Regularity_Forum_Contribute +
 Session_Count + Total_Duration + Active_Days, data = train_data,
 ntree = 1000)
              Type of random forest: regression
                    Number of trees: 1000
No. of variables tried at each split: 3

          Mean of squared residuals: 0.69
                    % Var explained: 34
```

```
# Printing the model summary and variable importance
importance(rf_model)
```

```
                              IncNodePurity
Freq_Course_View                       17.0
Freq_Lecture_View                      14.5
Freq_Forum_Consume                     24.0
Freq_Forum_Contribute                  47.7
Regularity_Course_View                 18.7
Regularity_Lecture_View                13.1
Regularity_Forum_Consume               13.7
Regularity_Forum_Contribute            23.7
Session_Count                          21.6
Total_Duration                         22.0
Active_Days                             8.9
```

3.2.4 Evaluating the Model's Performance

The next stage in our workflow is to evaluate how well our model predicts performance (Fig. 1-E). In Step 4, we make predictions on the test data using our trained Random Forest model by using the `predict` function, which generates a set of predicted values based on the features in `test_data` and the algorithm estimated before.

```
# Step 4: Make predictions on the test data
# Making predictions based on the test data using the trained model
predictions <- predict(rf_model, newdata = test_data)
```

In Step 5, we add these predictions to the test dataset, creating a new data frame called `evaluation_data` that includes both the actual and predicted grades. We then use the `metrics` function from the yardstick `package` to evaluate our model's performance by comparing the predicted grades with the actual ones, and we print the performance results to obtain performance metrics RMSE, MAE, and R-squared.

```
# Step 5: Evaluate the model's performance
# Adding predictions to the test data for evaluation
evaluation_data <- bind_cols(test_data, Predicted_Grade = predictions)

# Evaluating model performance
performance_metrics <- evaluation_data |>
  metrics(truth = Final_Grade, estimate = Predicted_Grade)

# Print the model performance metrics
print(performance_metrics)
```

```
# A tibble: 3 x 3
  .metric .estimator .estimate
  <chr>   <chr>          <dbl>
```

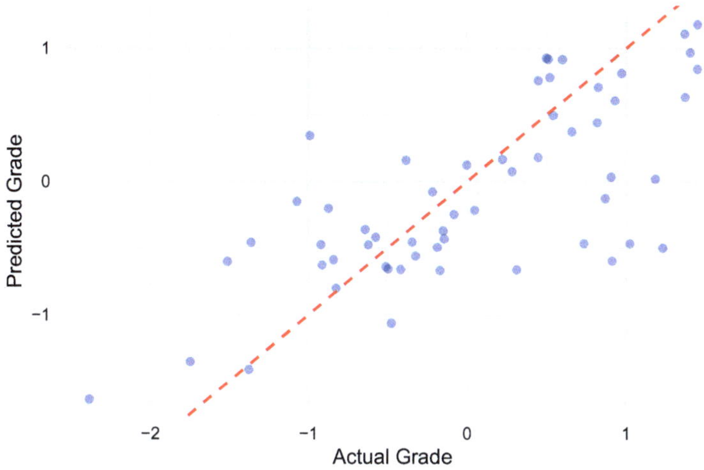

Fig. 4 Scatter plot comparing predicted vs. actual grades

```
1 rmse    standard       0.631
2 rsq     standard       0.525
3 mae     standard       0.480
```

In step 6, we produce two visualizations. The first visualization plots the actual versus predicted grades (Fig. 4). Examining the distance between the residual and the fitted line gives an idea about how the model fared. The other plot is an improved plot of the important variables (Fig. 5).

```
# Step 6: Visualize predicted vs actual grades
ggplot(evaluation_data, aes(x = Final_Grade, y = Predicted_Grade)) +
  geom_point(color = "blue", alpha = 0.5) +
  geom_abline(intercept = 0, slope = 1, color = "red", linetype = "dashed") +
  labs(x = "Actual Grade", y = "Predicted Grade") +
  theme_minimal()
```

```
# Enhanced variable importance plot
# Extracting variable importance from the Random Forest model
variable_importance <- importance(rf_model)
var_imp_df <- data.frame(Variable = rownames(variable_importance),
                         Importance = variable_importance[, 1])

# Plotting variable importance
ggplot(var_imp_df, aes(x = reorder(Variable, Importance), y = Importance)) +
  geom_bar(stat = "identity", fill = "skyblue") +
  coord_flip() +
  labs(x = "Variable", y = "Importance") +
  theme_minimal()
```

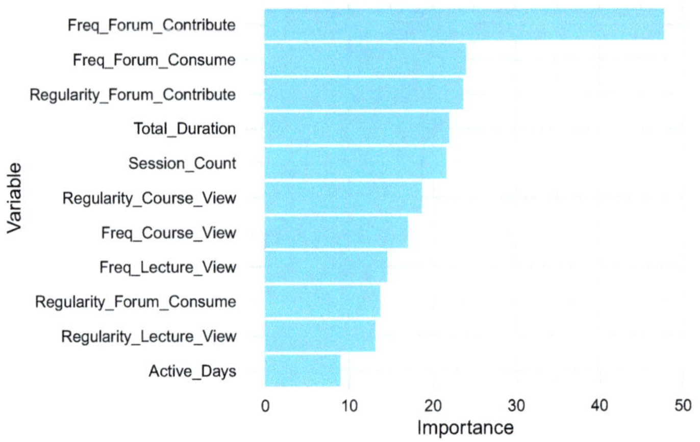

Fig. 5 Variable importance

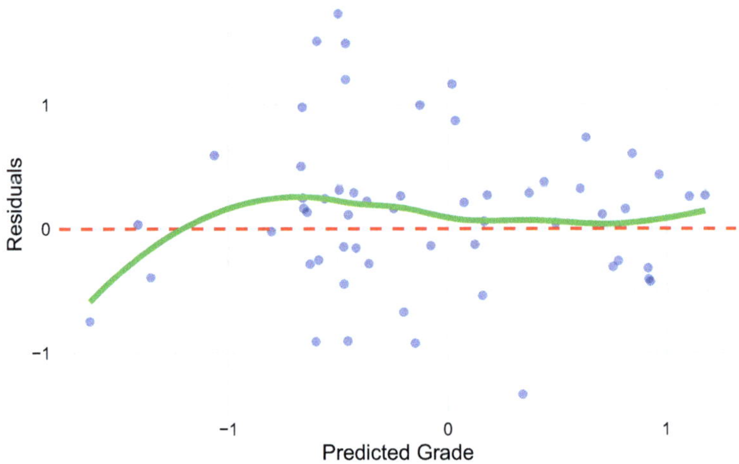

Fig. 6 Residual plot for random forest model with trend line

This code snippet demonstrates how to create and interpret residual plots for a random forest model. The process begins by calculating residuals, which are the differences between the actual `Final_Grade` and the `Predicted_Grade`. These residuals are then visualized in two separate plots using `ggplot2`. The first plot is a basic scatter plot of residuals against predicted grades, with each point representing a prediction (Fig. 6). A horizontal red dashed line at zero helps identify any systematic over- or under-prediction. The second plot builds upon the first by adding a green smoothed trend line, which can reveal non-linear patterns in the residuals. Ideally, the residuals should be randomly scattered around the zero line with no discernible pattern, indicating that the model's errors are randomly distributed and there are no underlying biases or issues with the model.

```
# Calculate residuals
evaluation_data$residuals <-
  evaluation_data$Final_Grade - evaluation_data$Predicted_Grade

# Add a smoothed line to show trends
ggplot(evaluation_data, aes(x = Predicted_Grade, y = residuals)) +
  geom_point(color = "blue", alpha = 0.5) +
  geom_hline(yintercept = 0, color = "red", linetype = "dashed") +
  geom_smooth(method = "loess", color = "green", se = FALSE) +
  labs(x = "Predicted Grade", y = "Residuals") +
  theme_minimal()
```

3.2.5 Other Algorithms

In the same way, we can execute other algorithms. Here we use a simpler algorithm (linear regression) given that it assumes that the relation between predictors and predicted grades is linear. Also, linear regression model does not need tuning (e.g., number of trees etc.). The way variable importance is calculated also differs: Random forest provides an inherent measure of feature importance, whereas for linear regression, we use the absolute values of the coefficients for importance. The evaluation metrics and visualization steps remain largely the same, although the results and interpretations would likely differ due to the fundamental differences in the models. Please also note the differences and similarities in variable importance between the two algorithms (Fig. 7).

```
# Step 1: Create and fit a Linear Regression model
lm_model <- lm(Final_Grade ~
               Freq_Course_View + Freq_Lecture_View + Freq_Forum_Consume +
               Freq_Forum_Contribute + Regularity_Course_View +
               Regularity_Lecture_View + Regularity_Forum_Consume +
               Regularity_Forum_Contribute + Session_Count +
               Total_Duration + Active_Days,
               data =train_data)

# Print model summary
print(summary(lm_model))
```

```
Call:
lm(formula = Final_Grade ~ Freq_Course_View + Freq_Lecture_View +
    Freq_Forum_Consume + Freq_Forum_Contribute + Regularity_Course_View +
    Regularity_Lecture_View + Regularity_Forum_Consume + Regularity_Forum_Contribute +
    Session_Count + Total_Duration + Active_Days, data = train_data)

Residuals:
    Min      1Q  Median      3Q     Max
-2.4421 -0.5135  0.0325  0.5422  1.9094
```

```
Coefficients:
                             Estimate Std. Error t value   Pr(>|t|)    
(Intercept)                   -0.0112     0.0546   -0.21      0.837    
Freq_Course_View               0.0525     0.1159    0.45      0.651    
Freq_Lecture_View             -0.0077     0.0761   -0.10      0.920    
Freq_Forum_Consume            -0.1954     0.1444   -1.35      0.177    
Freq_Forum_Contribute          0.5361     0.1039    5.16 0.00000056 ***
Regularity_Course_View         0.1178     0.1228    0.96      0.339    
Regularity_Lecture_View       -0.0486     0.0964   -0.50      0.615    
Regularity_Forum_Consume       0.0745     0.0795    0.94      0.350    
Regularity_Forum_Contribute    0.1087     0.0634    1.72      0.088 .  
Session_Count                  0.2997     0.1403    2.14      0.034 *  
Total_Duration                -0.0705     0.1228   -0.57      0.566    
Active_Days                   -0.1176     0.1281   -0.92      0.360    
---
Signif. codes:  0 '***' 0.001 '**' 0.01 '*' 0.05 '.' 0.1 ' ' 1

Residual standard error: 0.82 on 216 degrees of freedom
Multiple R-squared:  0.391, Adjusted R-squared:  0.36 
F-statistic: 12.6 on 11 and 216 DF,  p-value: <0.0000000000000002
```

```r
# Step 2: Make predictions on the data
predictions_lm <- predict(lm_model, newdata = test_data)

# Step 3: Evaluate the model's performance
# Adding predictions to the data for evaluation
evaluation_data_lm <- bind_cols(test_data,
                                Predicted_Grade_lm = predictions_lm)

# Evaluating model performance
performance_metrics_lm <- evaluation_data_lm |>
  metrics(truth = Final_Grade, estimate = Predicted_Grade_lm)

# Print the model performance metrics
print(performance_metrics_lm)
```

```
# A tibble: 3 x 3
  .metric .estimator .estimate
  <chr>   <chr>          <dbl>
1 rmse    standard       0.625
2 rsq     standard       0.524
3 mae     standard       0.478
```

```r
# Step 4: Visualize predicted vs actual grades
ggplot(evaluation_data_lm, aes(x = Final_Grade, y = Predicted_Grade_lm)) +
  geom_point(color = "blue", alpha = 0.5) +
  geom_abline(intercept = 0, slope = 1, color = "red", linetype = "dashed") +
  labs(x = "Actual Grade", y = "Predicted Grade") +
  theme_minimal()

# Step 5: Variable importance plot
# Extracting variable importance from the Linear Regression model
variable_importance_lm <- abs(coef(lm_model)[-1])  # Exclude intercept
var_imp_df_lm <- data.frame(Variable = names(variable_importance_lm),
                            Importance = variable_importance_lm)
```

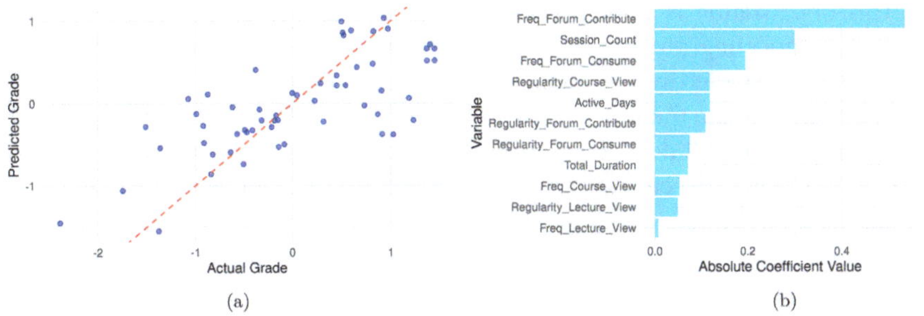

Fig. 7 Linear regression plots. (**a**) Scatter plot comparing actual grades to predicted grades. (**b**) Variable importance

```
# Plotting variable importance
ggplot(var_imp_df_lm, aes(x = reorder(Variable, Importance), y = Importance)) +
  geom_bar(stat = "identity", fill = "skyblue") +
  coord_flip() +
  labs(x = "Variable", y = "Absolute Coefficient Value") +
  theme_minimal()
```

3.3 Tutorial 2: A Modern Approach to Predictive Modelling Using `tidymodels`

Having used the traditional way, now we introduce a more modern approach of doing machine learning in R using `tidymodels`. The `tidymodels` framework provides a consistent interface for modeling in R by integrating several packages that work together seamlessly throughout the modeling process. The `tidymodels` package eliminates the need to learn and remember the various syntax and functions required by different modeling packages. Instead of having to switch between different paradigms and approaches for each package, you can use the same set of tools and functions. This uniformity speeds up model development and enhances code readability and maintainability.

Additionally, `tidymodels` offers integrated tools for tasks like pre-processing, resampling, and performance evaluation, further simplifying and accelerating the entire modeling workflow. For instance, the `parsnip` package in `tidymodels` allows us to define models in a consistent manner, regardless of the underlying algorithm (e.g., `randomForest`, `ranger`, `xgboost`). It provides functions like `rand_forest()` to specify the model, `set_engine()` to choose the computational backend, and `set_mode()` to define whether the model is used for regression or classification.

This organization is achieved through the use of workflows, which combines the entire modeling process in a single, manageable object, ensuring that each component is handled systematically. The tidy approach includes several steps: (1) splitting the data into training and testing sets (as done before), (2) defining a model specification (we will do a Random Forest model) using the standard approach of `parsnip` package, (3) creating a workflow combining the model specifications, (4) fitting the model to the training data, (5) making predictions on the test data, and (6) evaluating the model's performance and visualizing the results. This tutorial will guide you through these steps, using `parsnip` to create a random forest model, evaluate its performance, and visualize the results. Below, we describe each step in detail:

1. **Data Splitting**: This step is done in the same way we did before.
2. **Selecting the predictor and target variables**: The formula specifies the relationship between the target variable and the predictor variables. In regression tasks, this formula includes the dependent variable (e.g., `Final_Grade`) and the independent variables—in our case, engagement indicators—such as `Freq_Course_View` and `Total_Duration`. Defining the formula clearly outlines which variables are used to predict the outcome, ensuring that the model is correctly specified. We can also reuse the formula in other models without having to specify the arguments again.
3. **Model Specification with `parsnip`**: Using `parsnip`, we define our random forest model with specific parameters. The `rand_forest()` function allows us to set the models parameters. The most important parameters are number of trees and number of variables to try in each split. In our model, we set number of trees (`trees`) to 1000 and the number of variables to sample at each split (`mtry`) to 3. Setting the engine to `"randomForest"` and the mode to `"regression"`, we configure the model to perform regression tasks using the random forest algorithm.
4. **Creating the Workflow**: A workflow is an object that combines together the model specification and the formula. This organized approach helps manage different components of the modeling process efficiently. Using the `workflow()` function, we add the random forest model specification and formula to create a complete modeling pipeline.
5. **Fitting the Model**: The model is fitted to the training data using the `fit()` function.
6. **Making Predictions and Evaluating Performance**: After fitting the model, we use it to make predictions on the test data as we did before. Similarly, for evaluation, these predictions are compared to the actual values. We calculate performance metrics such as R-squared (`rsq`), MAE (`mae`), and RMSE (`rmse`) using the `metrics()` function from the `yardstick` package. We will visually assess the model's performance using plots as we have done before.

3.3.1 Load the Necessary Libraries

First, we load the necessary libraries. The `tidymodels` package [45] is a meta-package (a collection of packages) that loads several other packages useful for modeling, including `parsnip`, `rsample`, `recipes`, and more. Note that we will not run the data splitting code since it was already split before.

```
# Load necessary libraries
library(tidymodels) #Loading tidymodels loads all the necessary packages for estimation.
# Set seed for reproducibility
set.seed(256)

# Step 1: Split the data into training and testing sets, this step will not be
# run as it is already done before. In case you are running this code only,
# you may need to uncomment it.
#
# data_split_tidy <- initial_split(student_data_standardized, prop = 0.8)
# train_data <- training(data_split_tidy)
# test_data <- testing(data_split_tidy)
```

3.3.2 Select the Predictor and Target Variables

In `tidymodels`, the models are specified using a formula that defines the target variable, and the predictors following a specific syntax: `target_variable ~ predictor_1 + predictor_2 + ... + predictor_n`, where `target_variable` is the variable we want to predict. In our case, it is `Final_Grade`. In turn, `predictor_1`, `predictor_2`, ..., `predictor_n` are the predictor variables that will be used to make the prediction. In our case, these are the engagement indicators. See below how the formula is defined for our model in the same way.

```
# Step 2: Define the formula
# We define the formula to specify the relationship between the target variable
# 'Final_Grade' and the predictor variables
formula_tidy <- Final_Grade ~ Freq_Course_View + Freq_Lecture_View +
                             Freq_Forum_Consume + Freq_Forum_Contribute +
                             Regularity_Course_View + Regularity_Lecture_View +
                             Regularity_Forum_Consume + Regularity_Forum_Contribute +
                             Session_Count + Total_Duration + Active_Days
```

3.3.3 Define the Random Forest Model Specification

In this step, we specify the random forest model we plan to estimate by defining the type of model, the engine (the computational back-end), and the hyperparameters (the model configurations). This process enables the creation of straightforward and modular machine learning pipelines (we can change any of these components easily). The package `parsnip` encompassed within `tidymodels` ecosystem provides a unified interface for defining and fitting various statistical and machine learning models. To specify a random forest model in `parsnip`, we use the `rand_forest()` function. This function allows us to define the type of model and set hyperparameters without immediately fitting the model. This separation of concerns is beneficial for the reusability of the code. The `mode` argument is set to `"regression"` because we are predicting a continuous outcome (`Final_Grade`). The `trees = 1000` argument specifies the number of trees in the forest. The `mtry` argument represents the number of predictors in each split. Setting this to `tune()` indicates that we want to tune this hyperparameter, meaning we will search for the optimal value during the model training process. The `set_engine()` function specifies the computational backend to use for fitting the model. In this case, we use `"ranger"`, which is a fast implementation of random forests suitable for large datasets.

```
# Step 3: Define a random forest model specification with `ntree` and `mtry` parameters
# We specify the random Forest model with 1000 trees and 3 variables randomly
# sampled at each split
rf_specification_tidy <- rand_forest(trees = 1000, mtry = tune()) |>
  set_mode("regression") |>
  set_engine("ranger")
```

In the `tidymodels` framework, we integrate this formula into a workflow to streamline the modeling process. The `workflow()` function initializes a new workflow. The `add_model` function allows us to specify the model which is the random forest in our case. And `add_formula` function allows us to add the formula to the workflow, specifying the relationship between the target variable and predictor variables. The workflow helps ensure that all necessary components for training the model are combined in a single object. This makes it easier to apply or modify the model, and helps maintain a tidy, reproducible code.

```
# Step 4: Create the workflow. We create a workflow that combines the model
# specification with the formula
rf_workflow_tidy <- workflow() |>
  add_model(rf_specification_tidy) |>
  add_formula(formula_tidy)
```

The next steps are similar to our previous approach with small changes like using the pipe '|>' to enhance model readability. First, using the workflow, we fit the random forest model (corresponding to Fig. 1-D) using the previously defined workflow and training data (step 5). Using the fitted model, we make predictions on the test data and combines these predictions with the original test data to evaluate the model performance (step 6, corresponding to Fig. 1-E). Next, we calculate performance metrics (R-squared, MAE, and RMSE) by comparing the predicted values to the actual values and then print the results. The last part of the code visualizes the predicted versus actual grades as before (Fig. 8).

```
# Step 5: Fit the random forest model
# We fit the random forest model to the training data using the workflow
rf_fitting_tidy <- rf_workflow_tidy |>
  fit(data = train_data)
```

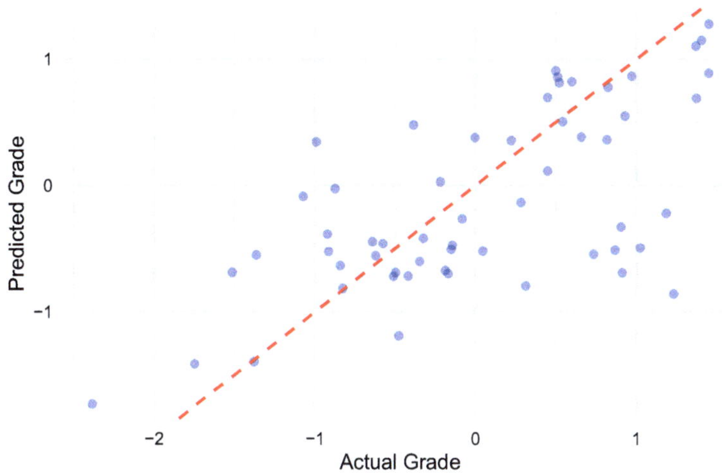

Fig. 8 Predicted vs. Actual Grades

```
# Step 6: Make predictions on the test data and evaluate the model's performance
# We make predictions on the test data, calculate performance metrics, and
# visualize the results
predictions_tidy <- predict(rf_fitting_tidy, new_data = test_data) |>
  bind_cols(test_data)

# Calculate performance metrics: R-squared, MAE, and RMSE
performance_metrics_tidy <- predictions_tidy |>
  metrics(truth = Final_Grade, estimate = .pred)

# Print the model performance metrics
print(performance_metrics_tidy)
```

```
# A tibble: 3 x 3
  .metric .estimator .estimate
  <chr>   <chr>          <dbl>
1 rmse    standard       0.705
2 rsq     standard       0.431
3 mae     standard       0.531
```

```
# Scatter plot comparing actual grades to predicted grades
ggplot(predictions_tidy, aes(x = Final_Grade, y = .pred)) +
  geom_point(color = "blue", alpha = 0.5) +
  geom_abline(intercept = 0, slope = 1, color = "red", linetype = "dashed") +
  labs(x = "Actual Grade", y = "Predicted Grade") +
  theme_minimal()
```

3.3.4 Multiple Algorithms

The uniform framework of tidymodels enables easier estimation and comparison of multiple models and in fact several configurations of the same algorithm if needed. The next code uses tidymodels to build and evaluate five machine learning algorithms namely: linear regression, support vector machine (SVM), random forest, K-nearest neighbors (KNN), and a neural network. Using tidymodels saves a lot of time in working with diverse packages and interfaces for each algorithm. Given that we already used similar code before, and we already loaded and split the data, we briefly describe the code for the estimation.

3.3.4.1 Model Specification

First, we define model specifications using the parsnip package. The linear_reg function is used to specify a linear regression model with the lm engine. For SVM , we use the svm_rbf function and set the engine to kernlab, specifying that it will be used for regression. The random forest model is specified using rand_forest, setting the number of trees to 1000 and allowing the number of variables randomly sampled at each split (mtry) to be tuned. The nearest_neighbor function specifies a KNN model with 5 neighbors, and the engine is set to kknn. Finally, for the neural network, the mlp function specifies a model with 10 hidden units and 100 epochs, using the nnet engine, set for regression.

```
# Linear Regression specification
lr_specification <- linear_reg() |> set_engine("lm")

# SVM specification
svm_specification <- svm_rbf() |> set_engine("kernlab") |>
  set_mode("regression")

# Random Forest specification with ntree and mtry parameters
rf_specification <- rand_forest(trees = 1000, mtry = tune()) |>
  set_engine("randomForest") |>
  set_mode("regression")

# KNN specification
knn_specification <- nearest_neighbor(neighbors = 5) |>
  set_engine("kknn") |>
  set_mode("regression")

# Neural Network specification
nn_specification <- mlp(hidden_units = 10, epochs = 100) |>
  set_engine("nnet") |>
  set_mode("regression")
```

3.3.4.2 Create Workflows

Next, we create workflows for each model specification using the `workflow` function. This involves adding the model specification and the formula to the workflow. The formula defines the relationship between the predictors and the target variable `Final_Grade`.

```
# Linear Regression workflow
lr_workflow <- workflow() |>
  add_model(lr_specification) |>
  add_formula(formula_tidy)

# SVM workflow
svm_workflow <- workflow() |>
  add_model(svm_specification) |>
  add_formula(formula_tidy)

# Random Forest workflow
rf_workflow <- workflow() |>
  add_model(rf_specification) |>
  add_formula(formula_tidy)

# KNN workflow
knn_workflow <- workflow() |>
  add_model(knn_specification) |>
  add_formula(formula_tidy)

# Neural Network workflow
nn_workflow <- workflow() |>
  add_model(nn_specification) |>
  add_formula(formula_tidy)
```

3.3.4.3 Fit the Models

We then fit each model to the training data using the `fit` function. This step trains the models using the specified workflows and the training dataset.

```
# Fit the Linear Regression model
lr_fitting <- lr_workflow |>
  fit(data = train_data)

# Fit the SVM model
svm_fitting <- svm_workflow |>
```

```
  fit(data = train_data)

# Fit the Random Forest model
rf_fitting <- rf_workflow |>
  fit(data = train_data)

# Fit the KNN model
knn_fitting <- knn_workflow |>
  fit(data = train_data)

# Fit the Neural Network model
nn_fitting <- nn_workflow |>
  fit(data = train_data)
```

3.3.4.4 Calculate Fit Indices and Residuals

Lastly, we create `calculate_metrics` function to assess the performance of each fitted model. It takes the model fit object and test data as inputs, makes predictions on the test data, and binds these predictions to the test data (Table 4). Performance metrics such as R-squared, mean absolute error (MAE), and root mean squared error (RMSE) are calculated using the `metrics` function from the `yardstick` package.

```
# Function to calculate metrics and residuals
calculate_metrics <- function(model_fitting, test_data, truth_col = "Final_Grade") {
  # Make predictions
  predictions <- predict(model_fitting, new_data = test_data) |>
    bind_cols(test_data)

  # Calculate residuals
  residuals <- predictions |>
    mutate(residuals = !!sym(truth_col) - .pred)

  # Calculate performance metrics
  performance_metrics <- residuals |>
    metrics(truth = !!sym(truth_col), estimate = .pred)

  list(performance_metrics = performance_metrics, residuals = residuals)
}

# Calculate metrics and residuals for each model
lr_results <- calculate_metrics(lr_fitting, test_data)
svm_results <- calculate_metrics(svm_fitting, test_data)
rf_results <- calculate_metrics(rf_fitting, test_data)
knn_results <- calculate_metrics(knn_fitting, test_data)
nn_results <- calculate_metrics(nn_fitting, test_data)

# Combine predictions and residuals
lr_residuals <- lr_results$residuals |> mutate(model = "Linear Regression")
svm_residuals <- svm_results$residuals |> mutate(model = "SVM")
rf_residuals <- rf_results$residuals |> mutate(model = "Random Forest")
knn_residuals <- knn_results$residuals |> mutate(model = "KNN")
```

Table 4 Performance metrics by model

.metric	.estimator	.estimate	Model
mae	Standard	0.48	Linear regression
mae	Standard	0.51	SVM
mae	Standard	0.53	Random forest
mae	Standard	0.60	KNN
mae	Standard	1.11	Neural network
rmse	Standard	0.62	Linear regression
rmse	Standard	0.67	SVM
rmse	Standard	0.70	Random forest
rmse	Standard	0.75	KNN
rmse	Standard	1.35	Neural network
rsq	Standard	0.52	Linear regression
rsq	Standard	0.45	SVM
rsq	Standard	0.43	Random forest
rsq	Standard	0.35	KNN
rsq	Standard	0.05	Neural network

```
nn_residuals <- nn_results$residuals |> mutate(model = "Neural Network")

combined_residuals <- bind_rows(lr_residuals, svm_residuals,
                               rf_residuals, knn_residuals, nn_residuals)

# Extract and combine performance metrics
performance_metrics <- bind_rows(
  lr_results$performance_metrics |> mutate(model = "Linear Regression"),
  svm_results$performance_metrics |> mutate(model = "SVM"),
  rf_results$performance_metrics |> mutate(model = "Random Forest"),
  knn_results$performance_metrics |> mutate(model = "KNN"),
  nn_results$performance_metrics |> mutate(model = "Neural Network")
) |> arrange(.metric)

# Print performance metrics
print(performance_metrics)
```

We can plot the performance metrics to compare them across the models (Fig. 9).

```
# Plot performance metrics
performance_metrics |>
  ggplot(aes(x = model, y = .estimate, fill = model)) +
  geom_bar(stat = "identity", position = "dodge") +
  facet_wrap(~ .metric, scales = "free") +
  labs(x = "Model",
       y = "Metric Value") +
  theme_minimal() +
  theme(axis.text.x = element_text(angle = 45, hjust = 1))
```

We can also create a scatter plot comparing actual grades to predicted grades, providing a visual assessment of the model's performance (Fig. 10).

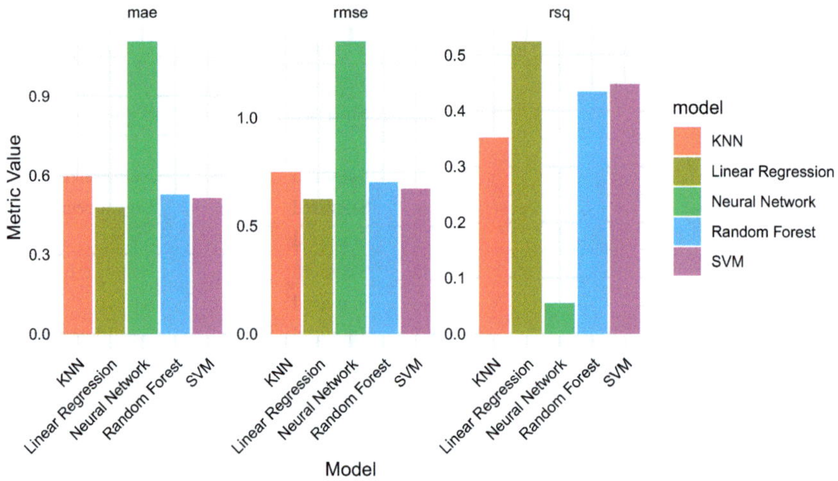

Fig. 9 Comparing performance metrics across models

```
combined_residuals |>
  ggplot(aes(x = Final_Grade, y = .pred, color = model)) +
  geom_point() +
  # Add a linear model fit line
  geom_smooth(method = "lm", se = FALSE, linetype = "dashed", color = "black") +
  facet_wrap(~ model) +
  labs(x = "Actual Final Grade",
       y = "Predicted Final Grade") +
  theme_minimal() +
  theme(legend.position = "bottom")
```

You can evaluate the results of each model by comparing it against the other models and choose the best performing model with lowest error rates.

4 Discussion and Conclusions

This chapter has provided a comprehensive exploration of predictive modeling in the context of learning analytics, where we focus on the practical application of machine learning techniques to forecast student performance. We have systematically mapped the main steps of the predictive modeling process, from data preparation and exploratory analysis to the building, evaluation, and interpretation of various models. The chapter explored different traditions from traditional machine learning, to the modern tidy approach and comparison across several models (linear regression, random forests, support vector machines, K-nearest neighbors, and neural networks). Furthermore, we investigated the use of various evaluation metrics—RMSE, MAE, and R-squared—to estimate model accuracy,

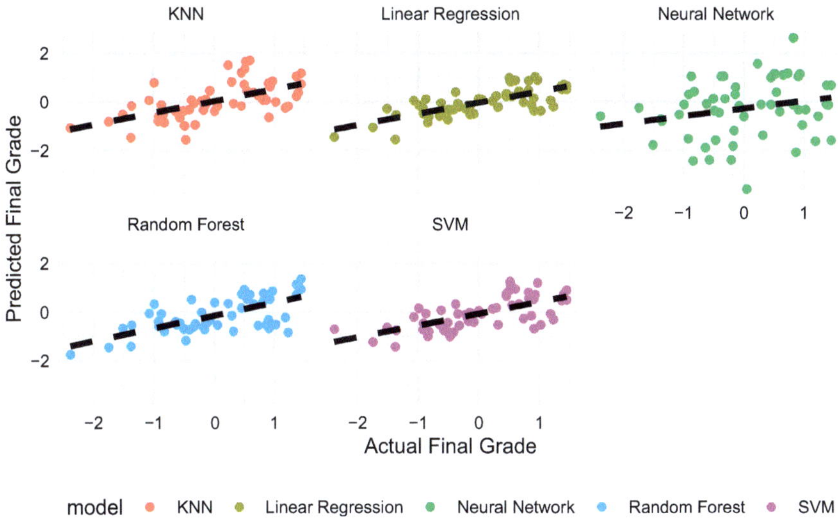

Fig. 10 Plot actual vs. predicted with a line matching the slope of actual vs. predicted

which can help us understand the trade-offs associated with each algorithm. We also shown how to visualize model predictions against actual outcomes and examine residual plots to assess the model fit and highlight areas that needed further consideration.

Several important points need to be emphasized here. First, the importance of data and variable selection according to a predefined learning theory that can offer a platform for variable inclusion and model results interpretation for more discussion of this issue, please refer to the vast literature on theory and predictive analytics. Second, the significance of careful data preprocessing cannot be overstated; even the most advanced algorithms cannot compensate for poor-quality data. Third, the absence of a single "best" algorithm became apparent, highlighting how algorithm choice is always dataset-specific, goal-oriented, and depends on the desired interpretability of the model. Fourth, we recognized that predictive models are tools to enhance decision-making, but do not replace human judgment and expertise, particularly in educational contexts, which require ethical considerations.

Finally, this chapter has demonstrated that the practical application of data science techniques can help to inform and improve educational practices. However, it also highlighted the importance of understanding the limitations of predictive models and machine learning in general [46]. Worth also noting that while explanation and transparency of the machine learning models are useful, they still can do deliver results that are more data driven than plausible or aligned with theory [21]. Other Issues like overfitting, data quality, and ethical considerations need to be carefully addressed to ensure responsible and meaningful use of predictive analytics in education. Finally, the chapter emphasized that predictive modeling is not just about achieving high accuracy, but also about gaining actionable insights into factors influencing student success and informing interventions and improving learning outcomes. We also need to emphasize that these results are aggregates of the whole dataset and therefore, it can be misleading and lacks full representation of each and everyone [47, 48].

References

1. López-Pernas S, Oliveira E, Song Y, Saqr M (2025) AI, explainable AI and evaluative AI: an introduction to informed data-driven decision-making in education. In: Saqr M, López-Pernas S (eds) Advanced learning analytics methods: AI, precision and complexity. Springer Nature Switzerland, Cham
2. Saqr M, Misiejuk K, Tikka S, López-Pernas S (2025) Artificial intelligence: using machine learning to classify students and predict low achievers. In: Saqr M, López-Pernas S (eds) Advanced learning analytics methods: AI, precision and complexity. Springer Nature Switzerland, Cham
3. Eckerson WW (2007) Predictive analytics. Extending the Value of Your Data Warehousing Investment TDWI Best Practices Report 1:1–36
4. Pardo A (2014) Designing learning analytics experiences. In: Learning analytics: from research to practice. Springer, Berlin, pp 15–38
5. Brooks C, Thompson C (2017) Predictive modelling in teaching and learning. Handbook of learning analytics. SOLAR, Owings Mills, pp 61–68
6. Sghir N, Adadi A, Lahmer M (2023) Recent advances in predictive learning analytics: a decade systematic review (2012–2022). Educ Inf Technol 28:8299–8333
7. Kopra J, Tikka S, Heinäniemi M, López-Pernas S, Saqr M (2024) An r approach to data cleaning and wrangling for education research. In: Learning analytics methods and tutorials: a practical guide using r. Springer Nature Switzerland Cham, pp 95–119
8. Tikka S, Kopra J, Heinäniemi M, López-Pernas S, Saqr M (2024) Introductory statistics with r for educational researchers. In: Learning analytics methods and tutorials: a practical guide using r. Springer Nature Switzerland Cham, pp 121–150
9. Dabhade P, Agarwal R, Alameen K, Fathima A, Sridharan R, Gopakumar G (2021) Educational data mining for predicting students' academic performance using machine learning algorithms. Mater Today Proc 47:5260–5267
10. Pardo A, Mirriahi N, Martinez-Maldonado R, Jovanovic J, Dawson S, Gašević D (2016) Generating actionable predictive models of academic performance. In: Proceedings of the sixth international conference on learning analytics & knowledge, pp 474–478
11. Jovanovic J, López-Pernas S, Saqr M (2024) Predictive modelling in learning analytics: a machine learning approach in r. In: Learning analytics methods and tutorials: a practical guide using r. Springer Nature Switzerland Cham, pp 197–229
12. Tikka S, Kopra J, Heinäniemi M, López-Pernas S, Saqr M (2024) Getting started with R for education research. In: Saqr M, López-Pernas S (eds) Learning analytics methods and tutorials: a practical guide using r. Springer, Berlin
13. Kopra J, Tikka S, Heinäniemi M, López-Pernas S, Saqr M (2024) An R approach to data cleaning and wrangling for education research. In: Saqr M, López-Pernas S (eds) Learning analytics methods and tutorials: a practical guide using r. Springer, Berlin
14. Tikka S, Kopra J, Heinäniemi M, López-Pernas S, Saqr M (2024) Introductory statistics with R for educational researchers. In: Saqr M, López-Pernas S (eds) Learning analytics methods and tutorials: a practical guide using r. Springer, Berlin

15. López-Pernas S, Misiejuk K, Tikka S, Kopra J, Heinäniemi M, Saqr M (2024) Visualizing and reporting educational data with R. In: Learning analytics methods and tutorials. Springer Nature Switzerland, Cham, pp 151–194. https://doi.org/10.1007/978-3-031-54464-4/_6
16. Saqr M, Jovanovic J, Viberg O, Gašević D (2022) Is there order in the mess? A single paper meta-analysis approach to identification of predictors of success in learning analytics. Stud Higher Educ 47:2370–2391. https://doi.org/10.1080/03075079.2022.2061450
17. Tomasevic N, Gvozdenovic N, Vranes S (2020) An overview and comparison of supervised data mining techniques for student exam performance prediction. Comput Educ 143:103676
18. Du X, Yang J, Shelton BE, Hung J-L, Zhang M (2021) A systematic meta-review and analysis of learning analytics research. Behav Inf Technol 40:49–62
19. Namoun A, Alshanqiti A (2020) Predicting student performance using data mining and learning analytics techniques: a systematic literature review. Appl Sci 11:237
20. Saqr M, Fors U, Tedre M (2017) How learning analytics can early predict under-achieving students in a blended medical education course. Med Teacher 39:757–767
21. Saqr M, López-Pernas S (2024) Why explainable AI may not be enough: predictions and mispredictions in decision making in education. Smart Learn Environ. https://doi.org/10.1186/s40561-024-00343-4
22. Alyahyan E, Düştegör D (2020) Predicting academic success in higher education: literature review and best practices. Int J Educ Technol Higher Educ 17:3
23. Mathrani A, Susnjak T, Ramaswami G, Barczak A (2021) Perspectives on the challenges of generalizability, transparency and ethics in predictive learning analytics. Comput Educ Open 2:100060
24. Winne PH (2020) Construct and consequential validity for learning analytics based on trace data. Comput Human Behav 112:106457
25. Avelino JG, Cavalcanti GD, Cruz RM (2024) Resampling strategies for imbalanced regression: a survey and empirical analysis. Artif Intell Rev 57:82
26. Saqr M, Nouri J (2020) High resolution temporal network analysis to understand and improve collaborative learning. Proceedings of the tenth international conference on learning analytics & knowledge, pp 314–319. https://doi.org/10.1145/3375462.3375501
27. Rets I, Herodotou C, Gillespie A (2023) Six practical recommendations enabling ethical use of predictive learning analytics in distance education. J Learn Anal 10:149–167
28. Saqr M (2017) Big data and the emerging ethical challenges. Int J Health Sci 11:1
29. Al-Shehri H, Al-Qarni A, Al-Saati L, Batoaq A, Badukhen H, Alrashed S, Alhiyafi J, Olatunji SO (2017) Student performance prediction using support vector machine and k-nearest neighbor. In: 2017 IEEE 30th canadian conference on electrical and computer engineering (CCECE). IEEE, pp 1–4
30. Nguyen VA (2024) Applying learning analytics to predict the student's learning outcome based on online learning activities. In: Proceedings of the 2024 10th international conference on frontiers of educational technologies, pp 140–146
31. Jo I-H, Yu T, Lee H, Kim Y (2015) Relations between student online learning behavior and academic achievement in higher education: a learning analytics approach. In: Emerging issues in smart learning. Springer, Berlin, pp 275–287
32. Cavazos R, Garza SE (2018) Learning models for student performance prediction. In: Advances in computational intelligence: 16th mexican international conference on artificial intelligence, MICAI 2017, enseneda, mexico, october 23–28, 2017, proceedings, part II 16. Springer, pp 171–182
33. Jovanović J, Saqr M, Joksimović S, Gašević D (2021) Students matter the most in learning analytics: the effects of internal and instructional conditions in predicting academic success. Comput Educ 172:104251. https://doi.org/10.1016/j.compedu.2021.104251
34. Conijn R, Snijders C, Kleingeld A, Matzat U (2017) Predicting student performance from LMS data: a comparison of 17 blended courses using moodle LMS. IEEE Trans Learn Technol 10:17–29. https://doi.org/10.1109/tlt.2016.2616312
35. Gašević D, Dawson S, Rogers T, Gasevic D (2016) Learning analytics should not promote one size fits all: the effects of instructional conditions in predicting academic success. Internet Higher Educ 28:68–84

36. Saqr M, López-Pernas S, Helske S, Hrastinski S (2023) The longitudinal association between engagement and achievement varies by time, students' profiles, and achievement state: a full program study. Comput Educ 199:104787. https://doi.org/10.1016/j.compedu.2023.104787
37. Wickham H, Averick M, Bryan J, Chang W, McGowan LD, François R, Grolemund G, Hayes A, Henry L, Hester J, Kuhn M, Pedersen TL, Miller E, Bache SM, Müller K, Ooms J, Robinson D, Seidel DP, Spinu V, Takahashi K, Vaughan D, Wilke C, Woo K, Yutani H (2019) Welcome to the tidyverse. J Open Source Softw 4:1686. https://doi.org/10.21105/joss.01686
38. Waring E, Quinn M, McNamara A, Arino de la Rubia E, Zhu H, Ellis S (2022) Skimr: Compact and flexible summaries of data. https://CRAN.R-project.org/package=skimr
39. Makowski D, Ben-Shachar M, Patil I, Lüdecke D (2020) Methods and algorithms for correlation analysis in R. J Open Source Softw 5:2306. https://doi.org/10.21105/joss.02306
40. Chan C, Leeper TJ, Becker J, Schoch D (2023) Rio: A swiss-army knife for data file i/o. https://cran.r-project.org/package=rio
41. Lüdecke D, Ben-Shachar M, Patil I, Waggoner P, Makowski D (2021) Performance: an R package for assessment, comparison and testing of statistical models. J Open Source Softw 6:3139. https://doi.org/10.21105/joss.03139
42. Liaw A, Wiener M (2002) Classification and regression by randomForest. R News 2:18–22. https://CRAN.R-project.org/doc/Rnews/
43. Frick H, Chow F, Kuhn M, Mahoney M, Silge J, Wickham H (2024) Rsample: General resampling infrastructure. https://CRAN.R-project.org/package=rsample
44. Kuhn M, Vaughan D, Hvitfeldt E (2024) Yardstick: Tidy characterizations of model performance. https://CRAN.R-project.org/package=yardstick
45. Kuhn M, Wickham H (2020) Tidymodels: A collection of packages for modeling and machine learning using tidyverse principles.. https://www.tidymodels.org
46. Mustafa MY, Tlili A, Lampropoulos G, Huang R, Jandrić P, Zhao J, Salha S, Xu L, Panda S, Kinshuk, López-Pernas S, Saqr M (2024) A systematic review of literature reviews on artificial intelligence in education (AIED): a roadmap to a future research agenda. Smart Learn Environ 11. https://doi.org/10.1186/s40561-024-00350-5
47. Saqr M, Cheng R, López-Pernas S, Beck ED (2024) Idiographic artificial intelligence to explain students' self-regulation: toward precision education. Learn Indiv Differ 114:102499. https://doi.org/10.1016/j.lindif.2024.102499
48. Saqr M, Tlili A, López-Pernas S (2025) Automating individualized machine learning and AI prediction using AutoML: the case of idiographic predictions. In: Saqr M, López-Pernas S (eds) Advanced learning analytics methods: AI, precision and complexity. Springer Nature Switzerland, Cham

Artificial Intelligence: Using Machine Learning to Classify Students and Predict Low Achievers

Mohammed Saqr, Kamila Misiejuk, Santtu Tikka, and Sonsoles López-Pernas

Abstract This chapter addresses the classification of at-risk students in educational settings using machine learning approaches in R. Transitioning from regression-based predictions of students' grades covered by the previous chapter, the focus here shifts to identifying broader categories of academic performance, such as low achievers or potential dropouts. Early identification of such students enables timely interventions, one of the main goals of learning analytics. The process is first illustrated through a Random Forest classifier, using engagement indicators to classify students into high and low achievers. The chapter demonstrates the complete modeling workflow, including data preparation, model training, and evaluation using performance metrics. Additionally, the `tidymodels` framework is explored as a more modern alternative that enables easy comparison with other AI / machine learning algorithms like Naive Bayes or Support Vector Machine.

Keywords Learning analytics · Predictive modeling · Classification · Machine learning · Artificial intelligence

1 Introduction

In the previous chapter [1], we implemented machine learning to predict students' grades as a continuous outcome, i.e., predicted the numerical final grades, which is often referred to as a regression problem. However, more often than not, we are not interested in the specific grade but rather in a broad range or category such as

M. Saqr (✉) · S. López-Pernas
University of Eastern Finland, Joensuu, Finland
e-mail: mohammed.saqr@uef.fi

K. Misiejuk
FernUniversität in Hagen, Hagen, Germany

S. Tikka
University of Jyväskylä, Jyväskylä, Finland

M. Saqr, S. López-Pernas (eds.), *Advanced Learning Analytics Methods*,
https://doi.org/10.1007/978-3-031-95365-1_4

drop-outs (students who abandon the course) or low achievers (students who will score below a certain grade or fail the course). In that, we want to identify whether a student is at risk of dropping out or failing a course and, therefore, may need support. Hence, in this chapter we address this achievement classification problem. The most common achievement classification problem is to classify or in fact predict low achievers [2]. This is commonly done with early course data given that there will be an opportunity for remedial and support before the end of the course. Similar to the previous chapter, we will start by demonstrating the code for a random forest model [1]. Random forest is chosen because it is robust and offers insights (explanation) into the important variables that drove the algorithm to take a decision [3]. The analysis process involves a sequence of steps similar to the previous chapter with two notable differences. First, we have to prepare the data in a slightly different manner and the evaluation and the metrics follow a different approach given that it is a classification problem. Also, similar to the previous chapter, we will begin with the traditional approach and then we will proceed with the `tidymodels` approach.

2 Previous Research on Identifying At-risk Students

There are several aims in research on the identification of at-risk students: (1) to develop a model to identify the at-risk students, (2) to identify attributes or indicators for the identification of at-risk students, (3) to develop and evaluate intervention methods to prevent student dropout or failure [4–6].

Several studies compared the performance of several algorithms to predict at-risk students. For example, [7] reported that Naïve Bayes (NB) outperformed k-Nearest Neighbors (KNN), Decision Tree (DT), Random Forest (RF), and Support Vector Machines (SVM). Queiroga et al. [8] developed a predictive model using genetic algorithms that performed better than DT, RF, Multilayer Perceptron, Logistic Regression (LG), and the AdaBoost algorithm (ADA). The Deep Long Short-Term Memory Model by [9] outperformed LG and artificial neural networks. RF was the best-performing algorithm in [10] compared to SVM, KNN, ET, ADA, Gradient boosting classifier, and Deep Feed Forward Neural Network, in [11] compared to SVM, KNN, DT, ADA, Extreme Gradient, Bayesian Classifier, and Linear Discriminant Analysis, and in [12] compared to DT, SVM and NB. Azcona et al. [13] reported that KNN had a higher performance than LG, SVM with different kernels, DT, and RF. Finally, instead of comparing different methods, [14] combined five base classifiers (NB, SVM, KNN, Repeated Incremental Pruning to Produce Error Reduction classifier and C4.5 classifier) into the Majority Voting ensemble classifier. As the diversity of potential classification algorithms and the differing performance results indicate, the portability of predictive models from one context to another can be challenging [12, 15]. In addition, model interpretability and transparency are important in the practical implementations of predictive models. However, there is a trade-off in predictive accuracy and the ability to generate generalizable models of low transparency models to take into consideration [15].

The selection of data and development of appropriate indicators used in a predictive model is a subject of extensive research. Most studies use student log data to determine student performance [e.g., [8, 16, 17]]. Several studies complement student activity data with additional data sources. For example, [18] included not only student log data but also library book loans or card swipes into buildings. Bayazit et al. [7] used the pre-test scores and formative assessment scores in their predictive model. Wolff et al. [19] found that demographic data improved prediction. Queiroga et al. [20] included a bias analysis in order to include three protected demographic attributes into their predictive model. Choi et al. [21] collected clicker data, as the learning management system was not popular to use by students. It is important to remember that every predictive model needs to be complemented by human judgment and contextual information. The research reported in [22], where the authors focused on examining false positive and false negative errors of a predictive model, showed that there are many events that can influence student behavior and performance that are not captured by the predictive models, such as family or work responsibilities and health issues.

The typical definitions of at-risk students include students who are likely to drop out of a class or students who may fail a particular exam. A vast body of research reported indicators of struggling students identified by their predictive models. For example, [14] found that the strongest predictors of performance are connected to the successful completion of a set of course activities, such as watching videos or completing self-assessment quizzes. Similarly, [6] reported that student engagement and the consistency of using online resources were the strongest performance predictors. Adnan et al. [10] found that clickstream and assessment variables had the most significant impact on student performance. Jayaprakash et al. [12] reported the highest predictive power of formative assessment scores and cumulative GPA. Gkontzis et al. [23] highlighted the importance of the first written assignment and quiz for student attrition. A similar finding was reported in [11], who reported the average grade obtained in the first period of the first semester as the strongest dropout predictor. Other conceptualizations of at-risk students include tracking when a student suddenly changes their typical behavior in an online learning system [19] or defining a specific threshold to consider a student at risk, e.g., scores below 5% over the passing mark [6].

The temporal dimension is crucial in developing predictive models. It includes several aspects, such as how often predictions should be calculated, how long the data should be collected to ensure accurate predictions, when predictions should take place in order to be actionable, or how long a student should be inactive to trigger an intervention. For example, [14] used data from the first cohort of students to develop a classifier for another cohort of students. Queiroga et al. [8] generated a prediction model every two weeks, while [18] generated an alert of no-engagement after a period of two weeks of inactivity. Others calculate predictions every week e.g., [24, 25] or every four weeks [12] or sometimes even on daily basis [26]. The intervals can also vary depending on the purpose of the predictive model. Azcona et al. [13] developed predictions of passing or failing the next formative exam once a week, and, in addition, the classifiers from the first six weeks of classes were trained to predict the outcome of the mid-term exam, and the classifiers from weeks 7 to 12 were used to predict the end-of-semester's exam outcome. Aljohani et al. [9] developed a classifier that has shown around 90% accuracy within the first 10 weeks of student interaction in an online learning environment, [23] reported a satisfactory prediction probability after week 5, while [10] showed almost 80% precision and accuracy score at 20% of the course length.

The predictions can be displayed directly to a student in the form of a dashboard or an alert. Another option is to inform a teacher about struggling students in a teacher-facing dashboard or through notifications [27]. Learning analytics dashboards that include predictive analytics are not widely implemented [28]. In addition, these dashboards often present only the prediction in the form of a visualization (e.g., signal lights) or percentage score without any explanation [29]. However, there is evidence that awareness of being at risk is not enough to improve student outcomes, and predictions need to be connected to an intervention [12]. The interventions can have different forms. For example, based on the analytics shown in a teacher-facing dashboard, a teacher can contact at-risk students personally to provide additional support [30]. Automated interventions embedded within the predictive analytics can include an adaptive feedback system that sends students custom messages based on their performance [13] or sending student-specific recommendations for learning materials and additional mentoring [12]. The evaluation of the interventions can be measured in reduced drop-out or increased performance. A comprehensive implementation of predictive analytics includes several iterations of improving the predictive model, adjusting an effective and transparent representation of the predictive results, and evaluating several intervention approaches [31].

3 Classifying Students with R

In this section we present two approaches to implement classification with R: one following a more classic procedure (tutorial 1), and the second one following a more updated procedure (tutorial 2) based on `tidymodels`. Broadly, the workflow followed is depicted in Fig. 1, aligned with the common steps that are performed in any ML pipeline. First, we explore the data to gain an idea of the magnitude, format and content of the data, explore the relationships between the different variables, etc. This step is called exploratory data analysis. Next, we prepare the data (B) for subsequent steps. This might include removing or fixing incomplete records, converting data to different types, unifying values that are close to one another, or rescaling the data. The transformations that we need to do are commonly dictated by the type of data and the ML models that we will be using in subsequent steps. After data preparation, the next step is data splitting (C). Here, the dataset is divided into training and testing sets. The training set is used to fit the model, while the testing set is reserved for evaluating its performance. It is crucial to ensure that the testing set remains untouched until the evaluation stage to provide an unbiased estimate of

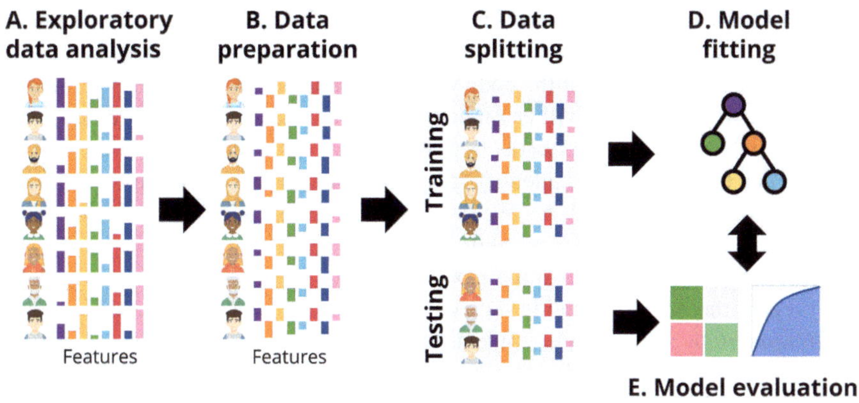

Fig. 1 ML workflow implemented in the tutorial

the model's generalization ability. Once the data is split, the process moves to model fitting (D). During this stage, the training data is used to train the selected machine learning models. This involves finding the optimal parameters that minimize error or maximize predictive accuracy for the given task. Depending on the algorithm, the model may identify different patterns, such as decision boundaries or relationships between features and the target variable. Finally, in the model evaluation step (E), the trained model is tested on the unseen testing data. Performance metrics that quantify the accuracy and precision of the model are computed to determine how well the model performs on new data. This step ensures that the model is not overfitting the training data and can generalize well to other datasets. We can repeat this process for several models with different underlying ML algorithms (tutorial 3), or using only a subset of features of the data and compare the performance metrics among different models to select the one with the best fit.

3.1 Tutorial 1: A Traditional Classification Approach with Random Forest

The first example follows the workflow depicted in Fig. 1—We will estimate a Random Forest model to classify students into ***high achievers and low achievers*** based on their final grades and the engagement indicators as predictors. It is assumed that the reader is familiar with the R programming language. If that is not the case, it is recommended to refer to previous tutorials on the basics of R [32], data cleaning [33], basic statistics [34], and visualization [35]. In this chapter we will use a dataset that was used in the previous chapter [1] and based on this paper [17].

In **Step 1**, we start by loading the necessary R packages. We use `randomForest` [3] to estimate a random forest model, `tidyverse` [36] for data manipulation and visualization, `rsample` [37] for splitting the data, `caret` [38] for model evaluation, and pROC to visualize this evaluation. We also use the package `rio` [39] to import the data. To ensure reproducibility, we set the seed to a random number *(1410)*. Setting the seed ensures that the code will produce identical results every time it is run. For instance, the data will be randomly split in the same way each time.

The next step would be conducting exploratory data analysis (EDA) to better understand the dataset and identify possible relationships between variables or problems in the data (Fig. 1-A). We implemented this step in detail with the same dataset in Chapter 3 [1] so we will not repeat it here for the sake of brevity. Therefore, we can move on to the next step, which is data preparation (Fig. 1-B). We prepare the data by creating a new binary classification target variable named ***Achievement*** by categorizing students into *High_Achievers* and *Low_Achievers* based on their `Final_Grade`. Those who are above the median (top 50%) will be classified as high achievers and the others will be classified as lower achievers. This categorical variable is coded as a factor as it is required by several functions. Of course, there are several ways to label students as at risk; for instance, based on a threshold score of 60% or even higher.

In **Step 2**, we split the dataset into training and testing sets using an 80/20 split where we will use 80% of the data to train the model and 20% for evaluation (corresponding to Fig. 1-C). We use the argument `strata = "Achievement"` to make sure that both the training and testing datasets have balanced classes of low and high achievers.

Having prepared the data we will then, define and fit the random forest model (Fig. 1-D) to predict `Achievement` based on engagement indicators e.g., lecture views, forum activity, session counts, total duration of activities, and active days.

Lastly, to evaluate the model (Fig. 1-E), we make predictions on the test dataset using the fitted model. For binary classification models, the evaluation commonly includes a confusion matrix as well as other metrics e.g., accuracy, precision, and recall as well as a visual evaluation through the receiver operating characteristic (ROC) curve, which we will explain in context with the estimation code.

The first chunk of code loads the necessary libraries and the data that we will use for building the model and sets a random seed as described earlier.

```
# Load necessary libraries
library(randomForest)  # For Random Forest model
library(tidyverse)     # For data manipulation and visualization
library(rsample)    # For data splitting and modeling workflow
library(caret)      # For model evaluation
library(pROC) # For visualizing receiver operating characteristic (ROC curves)
library(rio)  # For importing data files

# Set seed for reproducibility
set.seed(1410)
```

3.1.1 Preparing the Data

As a first step in our analysis, we prepare the data (Fig. 1-B) by creating a binary target variable, `Achievement`, which categorizes students based on their `Final_Grade`. Specifically, students with grades above the median are labeled as "High Achievers", and those below are labeled as "Low Achievers". This transformation is performed using the `mutate` function, ensuring the target variable is a factor with appropriate levels.

```
# Step 1: Load and prepare the data
raw_data <- import("https://github.com/lamethods/data2/raw/main/lms/lms.csv")

# Create a binary classification target variable
median_grade <- median(raw_data$Final_Grade)
student_data <- raw_data |>
  mutate(Achievement = factor(ifelse(Final_Grade > median_grade,
                                     "High_Achievers", "Low_Achievers"),
                              levels = c("Low_Achievers", "High_Achievers")))
```

3.1.2 Splitting the Data into Training and Testing Sets

In the second step, we split the dataset into training and testing sets (Fig. 1-C) using an 80/20 split. To maintain the class distribution, we use stratified sampling by passing the option `strata` to the `initial_split` function which takes care of distributing the achievement classes (high and low achievers) in the training and testing sets in a balanced way. In doing so, we ensure that both the training and testing sets are representative of the overall class distribution and we create a more balanced model.

```
# Step 2: Split the data into training and testing sets
data_split <- initial_split(student_data, prop = 0.8, strata = "Achievement")
train_data <- training(data_split)
test_data <- testing(data_split)
```

3.1.3 Creating and Training the Model

In **Step 3**, we create and fit a random forest model using the `randomForest` function (Fig. 1-D). The model predicts the `Achievement` variable based on engagement indicators (e.g., course views, lecture views, forum consumption and contribution), regularity indicators (Regularity Course View, Regularity Lecture_View, Regularity Forum Consume, Regularity Forum Contribute), and time indicators (session counts, total duration of activities, active days) [16]. The model is specified using a formula syntax: `target_variable ~ predictor_1 + predictor_2 + ... + predictor_n`. To ensure model robustness, we set `ntree` to 1000 (please refer to the previous chapter for more details on the number of trees [1]). Given that we are also interested in explaining the results, we set the `importance` argument to `TRUE` to estimate the importance of each variable.

```
# Step 3: Create and fit a Random Forest model
rf_model <- randomForest(
  Achievement ~ Freq_Course_View + Freq_Lecture_View + Freq_Forum_Consume +
    Freq_Forum_Contribute + Regularity_Course_View +
    Regularity_Lecture_View + Regularity_Forum_Consume +
    Regularity_Forum_Contribute + Session_Count +
    Total_Duration + Active_Days,
  data = train_data,
  ntree = 1000,
  importance = TRUE)
```

3.1.4 Evaluating the Model

In **Step 4**, we apply the model that we just trained to a new dataset (the test set) to evaluate its performance on new data (unseen by the model) that we held out. This allows us to evaluate how the model perform, or in other words, how likely the results we obtained to replicate in the future. A good model is expected to perform well on both learning and testing data. For the evaluation purpose, we use the `predict` function to generate predictions from our trained random forest model, `rf_model` and the `test_data`. The `predict` function generates two types of predictions: the probabilities and the classes.

1. Probability Predictions: estimate the probability of a student belonging to each achievement category (Low Achievers or High Achievers), we save these probabilities in a `predictions_prob` variable. To get these probabilities we specify the argument `type = "prob"` argument.
2. Class Predictions: predict the most likely achievement category for each student i.e., whether the student is predicted as high or low achiever. We save these class predictions in the `predictions_class` variable.

In the last step, and for convenience, we add these prediction as extra columns within the original test dataset to make it easier for further analysis.

```
# Step 4: Make predictions on the test set for the probabilities and the classes
predictions_prob <- predict(rf_model, test_data, type = "prob")
predictions_class <- predict(rf_model, test_data)

# Add predictions and probabilities to the test dataset
test_data <- test_data |>
  mutate(Predicted_Class = predictions_class,
         Probability_Low_Achievers = predictions_prob[, "Low_Achievers"],
         Probability_High_Achievers = predictions_prob[, "High_Achievers"])
```

In **Step 5**, we evaluate the model's performance (Fig. 1-E). For this, we use the `confusionMatrix` function from the `caret` package, which provides a detailed evaluation, including the confusion matrix and several other metrics. The confusion matrix is a table that summarizes the model's predictions compared to the actual outcomes. Specifically, it shows the counts of correctly predicted positive cases (true positives), correctly predicted negative cases (true negatives), incorrectly predicted positive cases (false positives), and incorrectly predicted negative cases (false negatives).

For example, in our case:

- **True Positives (TP)**: 20 Low-achieving students were correctly classified as "Low Achievers".
- **True Negatives (TN)**: 23 High-achieving students were correctly classified as "High Achievers".
- **False Positives (FP)**: 9 Low-achieving students were incorrectly classified as "High Achievers".
- **False Negatives (FN)**: 6 High-achieving students were incorrectly classified as "Low Achievers".

3.1.4.1 Evaluation Metrics

The `confusionMatrix` package produces several classification metrics; we will explain them here with equations in light of our results.

- **Accuracy:** Measures the proportion of correct predictions (both true positives and true negatives) out of the total number of cases. In our case, `accuracy` was 0.7414 indicating that approximately 74.14% of the classifications were correct. The 95% CI (Confidence Interval) indicates that the interval (0.6096, 0.8474) likely contains the true accuracy. The accuracy is calculated according to the following equation:

$$\text{Accuracy} = \frac{TP + TN}{TP + TN + FP + FN} = \frac{20 + 23}{20 + 23 + 9 + 6} = 0.7414$$

- **Kappa:** Cohen's Kappa is a measure of agreement between the observed accuracy and the accuracy expected by chance. In our case, the kappa value was 0.483 indicating a moderate agreement beyond chance. As a practical rule, Kappa of 0–0.20 is considered slight, 0.21–0.40 as fair, 0.41–0.60 as moderate, 0.61–0.80 as substantial, and 0.81–1 as almost perfect.

$$\kappa = \frac{\text{Accuracy} - \text{Expected Accuracy}}{1 - \text{Expected Accuracy}} = \frac{0.7414 - 0.5}{1 - 0.5} = 0.4828$$

- **Mcnemar's Test P-Value:** This test assesses if FP and FN differ significantly, the high p-value (0.789) in our case suggests no significant difference between the error rates for either class.

3.1.4.2 Sensitivity and Specificity

- **Sensitivity** (True Positive Rate) measures the model's ability to correctly identify "Low Achievers" (the positive class in this case). A value of 0.6897 indicates that the model were able to correctly identify 68.97% of actual "Low Achievers."

$$\text{Sensitivity} = \frac{TP}{TP + FN} = \frac{20}{20 + 6} = 0.769$$

- **Specificity** (True Negative Rate) measures the model's ability to correctly identify "High Achievers" (the negative class). A value of 0.7931 suggests that the model correctly identified 79.31% of actual "High Achievers."

$$\text{Specificity} = \frac{TN}{TN + FP} = \frac{23}{23 + 9} = 0.72$$

Please note that the model was able to identify more of high achievers than low achievers, this is probably due to the fact that high achievers generate more data that can be used to identify them.

3.1.4.3 Predictive Values

- **Positive Predictive Value (PPV or Precision):** The proportion of positive results that are true positives. The PPV of 0.7692 indicates that 76.92% of the predicted "Low Achievers" are actual "Low Achievers".

$$\text{PPV} = \frac{TP}{TP + FP} = \frac{20}{20 + 9} = 0.689$$

- **Negative Predictive Value (NPV):** The proportion of negative results that are true negatives. The NPV of 0.7188 indicates that 71.88% of the predicted "High Achievers" are actual "High Achievers".

$$\text{NPV} = \frac{TN}{TN + FN} = \frac{23}{23 + 6} = 0.793$$

3.1.4.4 Prevalence and Detection

These metrics provide insights into the distribution of the positive class (Low Achievers) in the actual data and in the model's predictions.

Prevalence The proportion of the total number of cases that are positives. The prevalence of 0.5 indicates that 50% of the students are "Low Achievers". The detection rate of 0.3448 indicates that 34.48% of the total students were correctly identified as "Low Achievers". The detection prevalence of 0.4655 indicates that 44.83% of the students were predicted to be "Low Achievers".

The code below computes the confusion_matrix and prints it.

```
# Step 5: Evaluate the model
# Create confusion matrix
confusion_matrix <- confusionMatrix(test_data$Predicted_Class, test_data$Achievement)

# Print confusion matrix and other metrics
print(confusion_matrix)
```

```
Confusion Matrix and Statistics

                Reference
Prediction        Low_Achievers High_Achievers
  Low_Achievers              21              6
  High_Achievers              8             23

               Accuracy : 0.7586
                 95% CI : (0.6283, 0.8613)
    No Information Rate : 0.5
    P-Value [Acc > NIR] : 5.025e-05

                  Kappa : 0.5172

 Mcnemar's Test P-Value : 0.7893

            Sensitivity : 0.7241
            Specificity : 0.7931
         Pos Pred Value : 0.7778
         Neg Pred Value : 0.7419
             Prevalence : 0.5000
         Detection Rate : 0.3621
   Detection Prevalence : 0.4655
      Balanced Accuracy : 0.7586

       'Positive' Class : Low_Achievers
```

3.1.4.5 Visual Evaluation

In **Step 6** we evaluate the model performance visually. We use a `fourfold` plot to visualize the confusion matrix, in which the circles in each quadrant are proportional

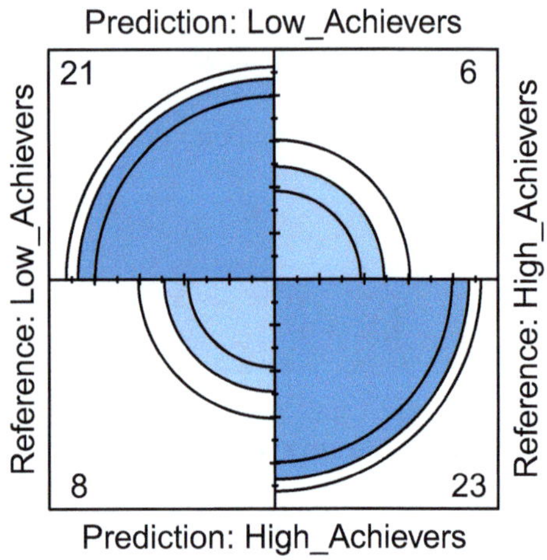

Fig. 2 Confusion matrix

to the counts in the matrix (Fig. 2). The additional circumferences in each quadrant represent the confidence intervals of each count.

```
library(caret)
# Step 6: Visualize results
fourfoldplot(confusion_matrix$table)
roc_obj <- roc(test_data$Achievement, predictions_prob[, "Low_Achievers"])
```

Next we plot the ROC curve which is a graphical representation of model's performance across all possible threshold values (Fig. 3). The ROC curve plots the True Positive Rate (Sensitivity) against the False Positive Rate (1 − Specificity). In other words, the ROC curve helps visualize the trade-off between sensitivity and specificity. A good model would have higher values of sensitivity and specificity resulting in a curve above the diagonal line while a bad model would produce a diagonal line. Another related value is the area under the curve (AUC). The AUC quantifies the overall ability of the model to discriminate between positive and negative classes, value ranges from 0 to 1 where AUC = 0.5 indicates the model performs no better than random guessing, values below 0.5 indicate that the model performs worse than random guessing and AUC values more than 0.5 indicates the model performs better than guessing (assuming that there are the same number of high achievers than low achievers). In our case, the value was 0.76 which is a moderate performance.

```
# Plot ROC curve
plot(roc_obj, asp = NA)
auc_value <- auc(roc_obj)
print(paste("AUC:", auc_value))
```

```
[1] "AUC: 0.762187871581451"
```

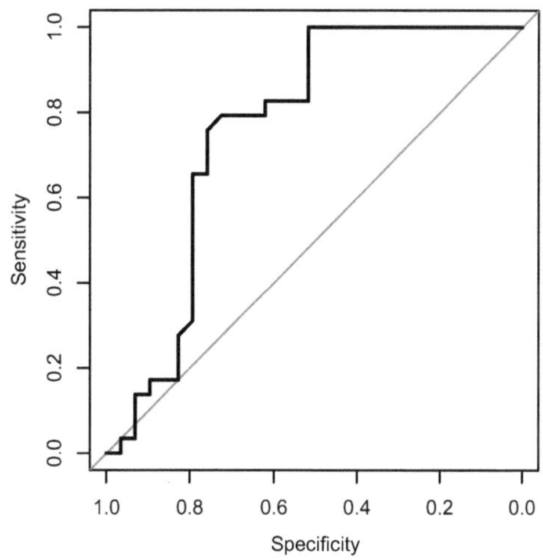

Fig. 3 ROC curve for random forest model

3.1.4.6 Explainability

Given the importance of interpretability of machine learning models, it is important to understand which of the features of our dataset contribute to the final prediction [40, 41]. For this purpose, we can extract the variable importance and visualize them (Fig. 4). The code below extracts variable `importance` values from the fitted random forest model (`rf_model`). The `importance` values contain two metrics: `MeanDecreaseAccurac` and `MeanDecreaseGini`. The `MeanDecreaseAccuracy` measures how much the accuracy of the model decreases when a particular variable is excluded. Higher values indicate that the variable is more important for maintaining the accuracy of the model. For instance, `Freq_Forum_Contribute` has the highest `MeanDecreaseAccuracy` value of 39.6879041, suggesting it is the most critical variable for the model's accuracy. In contrast, `Freq_Lecture_View` has a negative value (– 0.2569615), indicating that excluding this variable slightly improves the model's accuracy. Please note that a more detailed explanation of model explainability is offered in next chapters with more comprehensive methods [42, 43].

The `MeanDecreaseGini` measures the total decrease in node impurity that a variable contributes across all the trees in the forest. Higher values signify greater importance. For example, `Freq_Forum_Contribute` again shows high importance with a `MeanDecreaseGini` value of 21.348017, emphasizing its importance for predicting students' achievement. On the other hand, `Freq_Lecture_View` has a lower `MeanDecreaseGini` value of 5.959581, indicating it is less influential in reducing impurity.

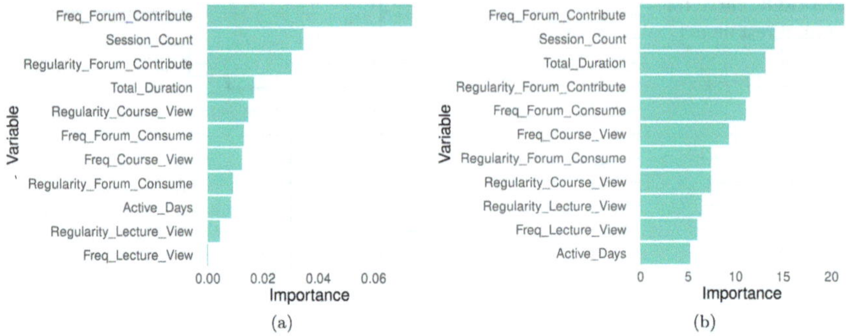

Fig. 4 Variable importance for the random forest model. (**a**) `MeanDecreaseAccuracy`. (**b**) `MeanDecreaseGini`

```
# Extract variable importance from the fitted model
importance_values <- rf_model$importance

# Convert to a data frame for plotting
importance_df <- as.data.frame(importance_values)
importance_df$Variable <- rownames(importance_df)
```

```
# Plot variable importance for MeanDecreaseAccuracy
ggplot(importance_df, aes(x = reorder(Variable, MeanDecreaseAccuracy),
                          y = MeanDecreaseAccuracy)) +
  geom_bar(stat = "identity", fill = "turquoise") +
  coord_flip() +
  labs(x = "Variable",
       y = "Importance") +
  theme_minimal()

# Plot variable importance for MeanDecreaseGini
ggplot(importance_df, aes(x = reorder(Variable, MeanDecreaseGini),
                          y = MeanDecreaseGini)) +
  geom_bar(stat = "identity", fill = "turquoise") +
  coord_flip() +
  labs(x = "Variable",
       y = "Importance") +
  theme_minimal()
```

3.2 Tutorial 2: An Alternative Implementation of Random Forests with `tidymodels`

In this section, we present an alternative to the process described in the previous example using `tidymodels` [44] instead of the classic approach. The `tidymodels` framework offers a unified streamlined interface for machine learning in R that makes it easier to specify models, update them, and compare one against another. Instead of working with different packages for different models, each with its own syntax, `tidymodels` employs a consistent syntax across all stages of estimation. All the more so, when we need to estimate several models as it is the case in our tutorial, `tidymodels` makes it easier to estimate and maintain the code. The main packages within `tidymodels` include `rsample` [37] for data splitting, `parsnip` for a specifying models [45], `recipes` for data pre-processing and feature engineering [46], `workflows` for combining pre-processing, modeling, and `yardstick` [47] for model evaluation. A more detailed discussion of `tidymodels` can be found the in the previous chapter.

Performing the classification with `tidymodels` follows a sequence of steps similar to what we had before with some differences. In general, the first three steps (loading packages, preparing the data and splitting) are the same as before but later steps uses the standard approach of defining models, workflows, training and evaluation. Below is a detailed description of the code and estimation steps.

3.2.1 Preparing the Data

The first steps are similar to the traditional approach: we load the necessary packages `tidyverse` and `tidymodels`; then we prepare the data (Fig. 1-B) by creating a binary target variable, which we call ***Achievement*** based on `Final_Grade` to classify students to low and high achievers.

```
library(tidymodels)

# Set seed for reproducibility
set.seed(1410)

# Load and prepare the data
# Assuming student_data is already loaded into the environment
student_data <- raw_data |>
  mutate(Achievement = factor(ifelse(Final_Grade > median(Final_Grade),
                                    "High_Achievers", "Low_Achievers"),
                              levels = c("Low_Achievers", "High_Achievers")))
```

3.2.2 Splitting the Data into Training and Testing Sets

Next, we split the data into training and testing sets (Fig. 1-C). The training set is then used to make the actual predictions and the testing test is used for model evaluation. We opt again here for stratified sampling `strata = Achievement` to preserve class distribution (balanced proportion of high and low achievers in the training and testing datasets).

```
# Step 2: Split the data into training and testing sets
data_split <- initial_split(student_data, prop = 0.8, strata = Achievement)
train_data <- training(data_split)
test_data <- testing(data_split)
```

3.2.3 Creating a Recipe

Unlike the previous chapter [1], where we used a formula, we extend it here by using a *recipe*. A recipe offers more options like including other data preparation and pre-processing steps. In our case, we define a recipe with a formula and also a function for normalizing the predictors so that all variables are on a similar scale. For that, we use `step_normalize(all_predictors())` function which will create new variables with standard deviation of one and a mean of zero.

```
# Create a recipe
rf_recipe <- recipe(Achievement ~ Freq_Course_View + Freq_Lecture_View +
                    Freq_Forum_Consume + Freq_Forum_Contribute +
                    Regularity_Course_View + Regularity_Lecture_View +
                    Regularity_Forum_Consume + Regularity_Forum_Contribute +
                    Session_Count + Total_Duration + Active_Days,
                    data = train_data) |>
  step_normalize(all_predictors())
```

3.2.4 Creating the Model

Next, we define the random forest model specifications. In this specification, we define the `rf_spec` object which will store the configuration for our model. The specification sets the `trees` parameter to 1000 which means the random forest will consist of 1000 decision trees. A larger number of trees generally leads to more stable results. The `mtry` parameter is set to 5, which means that during each split in the decision trees, the algorithm will randomly select 5 predictors (from the engagement indicators). This randomness helps prevent overfitting and ensures diversity among the trees. Then, we specify the engine for our random forest using the `set_engine("ranger")`** function. The `ranger` engine is a fast and efficient engine that can handle large datasets. Most importantly, we specify `set_mode("classification")`, which means that the random forest model will predict a categorical outcome (high or low achievers), compare this to regression in the previous chapter.

```
# Create a random forest model specification
rf_spec <- rand_forest(trees = 1000, mtry = 5) |>
  set_engine("ranger", importance = "impurity") |>
  set_mode("classification")
```

3.2.5 Creating a Workflow

The next step is to create a workflow which is a container that holds the recipe and the model specifications that we just defined. Workflows provide a way to chain these components together and execute them in a specific order when fitting the algorithm. To create a workflow for a random forest model, we begin by initializing the workflow object using the function `workflow()` and adding the recipe `rf_recipe` that that we just created containing the formula and the pre-processing steps. Next, we specify the model that we just defined `rf_spec`. In the following step, we fit the model (Fig. 1-D) to the training data by passing it to the `fit` function and the workflow and we store the fitted model to the object `rf_fit_tidy`. `rf_fit_tidy` contains data about how the model predicts the data.

To evaluate the model's performance (Fig. 1-E), we generate predictions on the test dataset `test_data` using the trained model `rf_fit_tidy`. We enhance our test dataset with these predictions by computing three new columns and adding them to the test_data: `Predicted_Class_tidy`, which contains the predicted class for each observation; `Probability_Low_Achievers_tidy`, which gives the probability of an observation being classified as a "Low Achiever"; and `Probability_High_Achievers_tidy`, which provides the probability of an observation being classified as a "High Achiever". The resulting augmented dataset combines original features and the predictions and their probabilities which will make it easier to plot and evaluate the data later.

```
# Create a workflow
rf_workflow <- workflow() |>
  add_recipe(rf_recipe) |>
  add_model(rf_spec)

# Fit the model
rf_fit_tidy <- rf_workflow |>
  fit(data = train_data)

# Make predictions on the test set
predictions_prob_tidy <- predict(rf_fit_tidy, test_data, type = "prob")
predictions_class_tidy <- predict(rf_fit_tidy, test_data)

# Add predictions and probabilities to the test dataset
test_data <- test_data |>
  mutate(
    Predicted_Class_tidy = predictions_class_tidy$.pred_class,
    Probability_Low_Achievers_tidy = predictions_prob_tidy$.pred_Low_Achievers,
    Probability_High_Achievers_tidy = predictions_prob_tidy$.pred_High_Achievers)
```

3.2.6 Evaluating the Model

In the next step, we evaluate the performance of the model. The process involves creating a confusion matrix and computing several evaluation metrics as before. The first part of the code generates a confusion matrix. This matrix provides a summary of how well the model is performing in distinguishing between the classes.

In the code below, we create the confusion matrix using the `conf_mat()` function. Computing the evaluation metrics in `yardstick` [47] has two steps (1) defining the metrics and then computing them. So, we define the custom set of evaluation metrics using the `metric_set` function. This set includes the common classification metrics: accuracy, sensitivity, specificity, F1 score, balanced accuracy, positive predictive value, and negative predictive value.

Finally, we compute these custom metrics and add them the test data. The function calculates all the specified metrics, comparing the true `Achievement` values against the `Predicted_Class_tidy` predictions. The `event_level = "first"` parameter specifies which class should be considered the positive class for binary classification metrics, i.e., low achievers. The resulting detailed metrics are stored in 'detailed_metrics_tidy' and then printed.

The results of the evaluation shows that the accuracy of the model is 0.741 which means that 74.1% of "High Achievers" and "Low Achievers" were correctly classified. The sensitivity of 0.793 means that in 79.3% of the times, the model correctly identified students who are actually "Low Achievers" and labeled them as such. Specificity shows how the model identified the other category of students: "High Achievers" and identifying them which was approximately 69.0%. The F-measure, around 0.754, combines sensitivity and precision into a single metric, balancing the model's ability to identify "High Achievers" with how accurate those predictions are. This score indicates a good balance between detecting "High Achievers" and avoiding false positives. In case you are interested, the F-measure is computed as the harmonic mean of precision and recall following this formula.

$$\text{F1-Score} = 2 \cdot \frac{\text{Precision} \cdot \text{Recall}}{\text{Precision} + \text{Recall}} = 2 \cdot \frac{0.7692 \cdot 0.6897}{0.7692 + 0.6897} = 0.7273$$

Positive Predictive Value (PPV) of 0.719 indicates how often the model's prediction of "Low Achievers" is correct which was accurate in 71.9% of the time. Negative Predictive Value (NPV) was 0.769, showing the accuracy of predictions for "High Achievers was correct 76.9% of the time. **Balanced Accuracy** was 0.741 and reflects the model's overall performance across both classes by equally weighting the ability to correctly identify"Low Achievers" and "High Achievers."

```
# Confusion Matrix
conf_mat_tidy <- conf_mat(test_data, truth = Achievement,
                          estimate = Predicted_Class_tidy)
print(conf_mat_tidy)
```

```
                Truth
Prediction       Low_Achievers High_Achievers
  Low_Achievers             20              6
  High_Achievers             9             23
```

```
# Custom metric set
custom_metrics_tidy <- metric_set(accuracy, sens, yardstick::spec,
                                  f_meas, bal_accuracy, ppv, npv)

# Compute all metrics
detailed_metrics_tidy <- test_data |>
  custom_metrics_tidy(
    truth = Achievement,
    estimate = Predicted_Class_tidy,
    event_level = "first")

print(detailed_metrics_tidy)
```

```
# A tibble: 7 x 3
  .metric      .estimator .estimate
  <chr>        <chr>          <dbl>
1 accuracy     binary         0.741
2 sens         binary         0.690
3 spec         binary         0.793
4 f_meas       binary         0.727
5 bal_accuracy binary         0.741
6 ppv          binary         0.769
7 npv          binary         0.719
```

In the last step, we estimate and plot the AUC and ROC curve. The visualization and estimation follow exactly the same way we did before (Fig. 5). The only difference is that we use `roc_curve()` function from the `yardstick` package [47]. The function takes two important arguments, the truth argument which represents the class we predicted and the probability of the class we are focusing on which is lower achievers in our case.

```
# Compute ROC and AUC using yardstick
roc_data_tidy <- test_data |>
  roc_curve(truth = Achievement, Probability_Low_Achievers_tidy)

auc_value_tidy <- test_data |>
  roc_auc(Achievement, Probability_Low_Achievers_tidy) |>
  pull(.estimate)

# Plot ROC Curve using ggplot2
ggplot(roc_data_tidy, aes(x = 1 - specificity, y = sensitivity)) +
  geom_line(color = "blue") +
  geom_abline(linetype = "dashed", color = "red") +
  labs(x = "False Positive Rate", y = "True Positive Rate") +
  annotate("text", x = 0.75, y = 0.25,
           label = paste("AUC =", round(auc_value_tidy, 3)), size = 5) +
  theme_minimal()
```

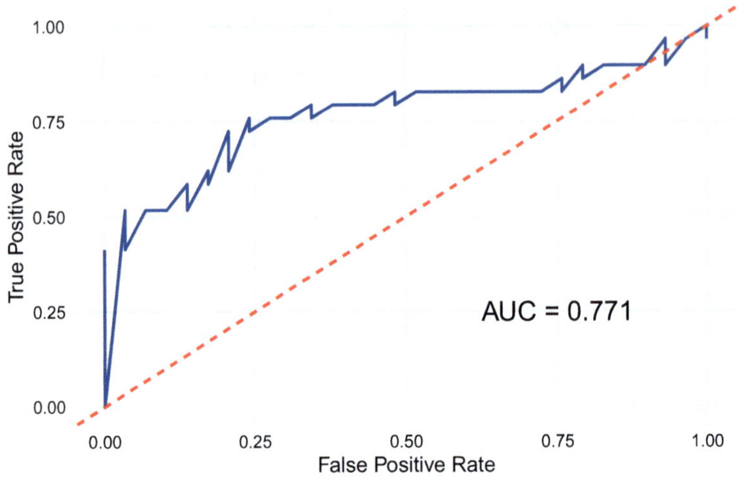

Fig. 5 ROC curve for random forest model

3.2.7 Explainability

To extract the variable importance we have to pull the data from the model. In the code below we pull the variable importance data, create a data frame and then use `ggplot` to visualize it. As you can see, variable importance can be plotted and their interpretation could help guide us understand what actions students take that may be associated with better performance (Fig. 6).

```
# Extract the ranger model from the workflow
importance_values <- pull_workflow_fit(rf_fit_tidy)$fit$variable.importance

# Convert to a data frame for plotting
importance_df <- as.data.frame(importance_values)
importance_df$Variable <- rownames(importance_df)
colnames(importance_df) <- c("Importance", "Variable")

# Plot variable importance
ggplot(importance_df, aes(x = reorder(Variable, Importance), y = Importance)) +
  geom_bar(stat = "identity", fill = "turquoise") +
  coord_flip() +
  labs(x = "Variable", y = "Importance") +
  theme_minimal() +
  theme(
    axis.text.y = element_text(color = "black"),
    axis.title = element_text(color = "black"),
    plot.title = element_text(size = 14, face = "bold", color = "black"))
```

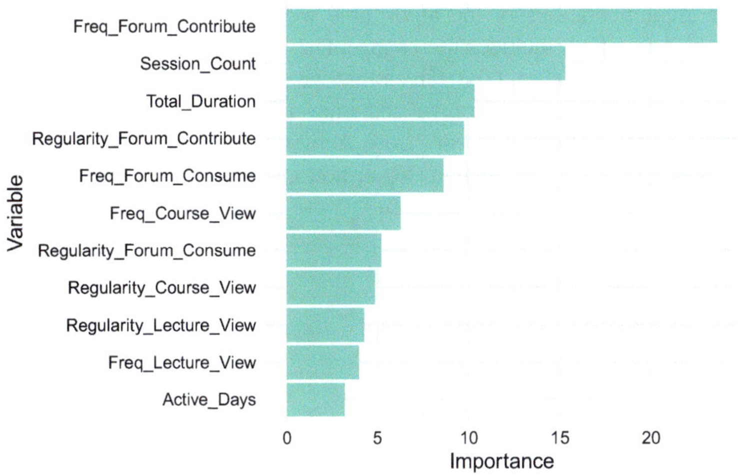

Fig. 6 Variable importance (`tidymodels`)

3.3 *Tutorial 3: Evaluating Multiple Models with* `Tidymodels`

In the next example, we will take advantage of `tidymodels` framework to fit, evaluate and compare several models. In most practical cases, we will need to estimate several models, evaluate them, compare them and choose the one that performs better.

In the next example, we will estimate several machine learning models using the `tidymodels` ecosystem to streamline the process. The aim—as we did before—is to classify students into high or low achievers based on their engagement indicators. The process will follow a systematic sequence: loading the packages and the data, preparing the data, splitting it into training and testing sets, creating a recipe, defining model specifications, creating workflows, fitting the models, making predictions, evaluating the models, and finally visualizing the results. Given that we are estimating 13 models, we will create functions to save time rather than repeating the same process 13 times in each step, these functions will help automate the repetitive tasks. The following code will explain these steps as follows:

3.3.1 Preparing the Data

The first step is similar in each code we discussed before, in which we load the required packages. Also, you may be prompted to install missing packages if some dependencies are not loaded. The required packages here are the packages that provide the *engines* for the algorithms we are estimating e.g., (`ranger`, `xgboost`, `kernlab`, `glmnet`, `nnet`, `discrim`, `naivebayes`, `randomForest`, `baguette`), and modeling and evaluation tools (`yardstick`, `tidymodels`). The engines are the software implementations of the algorithms we are estimating. Then we import the student data which includes the engagement indicators and the final grade and prepare it for analysis (Fig. 1-B). Given that we are doing a classification task, we need a binary outcome, as before, we add a new column to the data that divides the students into "High_Achievers" or "Low_Achievers" based on whether their final grade is above or below the median (top or bottom 50%).

```
# Step 1: Load and prepare the data
# Assuming student_data is already loaded into the environment
student_data <- raw_data |>
  mutate(Achievement = factor(ifelse(Final_Grade > median(Final_Grade),
                                     "High_Achievers", "Low_Achievers"),
                              levels = c("Low_Achievers", "High_Achievers")))
```

3.3.2 Splitting the Data into Training and Testing Sets

Then, we split the data into training and testing sets (Fig. 1-C). We use `initial_split` from `rsample` with the option `strata` to ensure an 80/20 split while stratifying by the `Achievement` column to maintain the proportion of high and low achievers in both sets. These steps are already familiar by now.

```
# Step 2: Split the data into training and testing sets
data_split <- initial_split(student_data, prop = 0.8, strata = Achievement)
train_data <- training(data_split)
test_data <- testing(data_split)
```

3.3.3 Creating a Recipe

In this step, we create a recipe that specifies the predictors (engagement indicators) and the target variable (achievement). The recipe also includes a step to normalize all predictor variables, so that they are measured on the same scale. This normalization step is particularly important for algorithms that are sensitive to the scale of the input features to ensure that features contribute equally to the model. Given that we will use identical recipe for all models, we won't have to write multiple recipes, this is in fact, one of the powerful features of `tidymodels`.

```
model_recipe <- recipe(Achievement ~ Freq_Course_View + Freq_Lecture_View +
                       Freq_Forum_Consume + Freq_Forum_Contribute +
                       Regularity_Course_View + Regularity_Lecture_View +
                       Regularity_Forum_Consume + Regularity_Forum_Contribute +
                       Session_Count + Total_Duration + Active_Days,
                       data = train_data) |>
  step_normalize(all_predictors())
```

3.3.4 Create All Model Specifications

In this step, we specify the machine learning models through the `parsnip` package interface [45]. This involves specifying each algorithm and setting the engine and the parameters. We selected some popular algorithms for demonstration purposes. However, for a full list of supported algorithms, please refer to the `parsnip` [45] documentation. Below is the list of algorithms we will demonstrate, along with their respective engines and packages:

1. **Random Forest**: Random Forest is an ensemble learning method that builds multiple decision trees and merges their predictions by averaging for more accurate and stable results. It is implemented using two engines: `ranger` from the `ranger` package [48], known for its speed and performance, and `randomForest` from the `randomForest` package, which is the original R implementation [3].
2. **XGBoost (eXtreme Gradient Boosting)**: XGBoost is an efficient and scalable implementation of the gradient boosting framework, optimizing performance by iteratively improving model accuracy. It is implemented using the `xgboost` engine from the `xgboost` package [49].
3. **Support Vector Machines (SVM)**: SVM is a powerful classification algorithm that finds the best boundary (hyperplane) to separate different classes in the data. It is implemented through the `kernlab` engine from the `kernlab` package [50].
4. **Logistic Regression**: Logistic Regression is a well-known classification algorithm that models the probability of a binary outcome. It uses the `glm` engine from base R.
5. **K-Nearest Neighbors (KNN)**: KNN is a simple, non-parametric algorithm that classifies a data point based on the majority class of its nearest neighbors. It is implemented using the `kknn` engine from the `kknn` package [51].
6. **Neural Networks**: Neural Networks are computational models inspired by the human brain, capable of modeling complex patterns in data through layers of interconnected nodes. They are implemented using the `nnet` engine from the `nnet` package [52].
7. **Decision Trees**: Decision Trees split the data into branches to help make decisions based on the features of the data. They are implemented using the `rpart` engine from the `rpart` package [53].
8. **Naive Bayes**: Naive Bayes is a probabilistic algorithm based on Bayes' theorem, assuming independence between features. It is implemented through the `naivebayes` engine from the `naivebayes` package [54].
9. **Linear Discriminant Analysis (LDA)**: LDA is used for classification, finding the linear combinations of features that best separate different classes. It uses the `MASS` engine from the `MASS` package [55].
10. **Bagged Trees**: Bagged Trees is an ensemble method that improves the stability and accuracy of machine learning algorithms by combining multiple models. It uses the `rpart` engine from the `rpart` package [53].

11. **Multivariate Adaptive Regression Splines (MARS)**: MARS is a non-linear regression technique that models relationships by fitting piecewise linear regressions. It is implemented using the `earth` engine from the `earth` package [56].
12. **Bayesian Additive Regression Trees (BART)**: BART is a Bayesian approach to machine learning that creates a sum-of-trees model for both regression and classification tasks. It is implemented using the `dbarts` engine from the `dbarts` package.

These models cover a wide range of popular approaches, all implemented under the `parsnip` framework for consistent and streamlined usage.

Below is the code that loads the required packags and creates a function specifying each model and its engine, ensuring the full name of the algorithm is used for better display in later steps.

```
library(tidymodels) # laod the tidymodels framework and its packages
# Load the required packages for each algorithm
# Random Forest
library(ranger) # Engine: ranger, known for its speed and performance
library(randomForest) # Engine: randomForest, the original R implementation
# XGBoost (eXtreme Gradient Boosting)
library(xgboost) # Engine: xgboost, gradient boosting framework
# Support Vector Machines (SVM)
library(kernlab) # Engine: kernlab, a powerful classification algorithm
# Logistic Regression
# Base R includes glm, no additional package needed
# K-Nearest Neighbors (KNN)
library(kknn)  # Engine: kknn, a simple, non-parametric classification algorithm
# Neural Networks
library(nnet) # Engine: nnet, inspired by the human brain for complex patterns
# Decision Trees
library(rpart) # Engine: rpart, recursive partitioning for classification trees
# Naive Bayes
library(discrim) # Engine: naivebayes, based on Bayes' theorem
# Linear Discriminant Analysis (LDA)
library(MASS)  # Engine: MASS, classification via linear combinations of features
# Bagged Trees
# Uses the same package as Decision Trees: rpart
# Multivariate Adaptive Regression Splines (MARS)
library(earth) # Engine: earth, non-linear regression with piecewise linear fits
# Bayesian Additive Regression Trees (BART)
library(dbarts) # Engine: dbarts, Bayesian approach creating a sum-of-trees model
library(baguette)  # Engine: rpart, an ensemble of decision trees
library(parsnip) # Interface for `tidymodels`

set_classification <- function(x, engine) {
  x |> set_engine(engine) |> set_mode("classification")
}

# Function to specify each model and its engine
create_model_specs <- function() {
  list(
    "Random Forest (ranger)" = rand_forest() |> set_classification("ranger"),
    "XGBoost" = boost_tree() |> set_classification("xgboost"),
    "SVM (RBF)" = svm_rbf() |> set_classification("kernlab"),
    "Logistic Regression" = logistic_reg() |> set_classification("glm"),
    "K-Nearest Neighbors" = nearest_neighbor() |> set_classification("kknn"),
    "Neural Network" = mlp() |> set_classification("nnet"),
    "Decision Tree" = decision_tree() |> set_classification("rpart"),
    "Naive Bayes" = naive_Bayes() |> set_classification("naivebayes"),
    "Linear Discriminant Analysis" = discrim_linear() |> set_classification("MASS"),
```

```
    "Bagged Tree" = bag_tree() |> set_classification("rpart"),
    "Random Forest (randomForest)" = rand_forest() |> set_classification("randomForest"),
    "MARS" = mars() |> set_classification("earth"),
    "BART" = parsnip::bart() |> set_classification("dbarts")
  )
}

# Create the model specifications
model_specs <- create_model_specs()
```

3.3.5 Creating the Workflows and Fitting the Data

Rather than creating a list of 13 workflows, we define a function that automates the process and we use `lapply` to apply this function to all list items. The function helps easy specification of the models with their respective engines and parameters and the results will be a list which we will also use in further analysis in the same way.

The code below has two key functions: `create_workflows` and `fit_models`. The `create_workflows` function builds a workflow for each model specification in the provided `model_specs` list, using the specified model recipe and `lapply` to iterate over the list items. This results in a list of workflows. Then, the `fit_models` function is used to fit each of these workflows to the training data (Fig. 1-D), again using `lapply` to apply the function to every workflow in the list.

```
create_workflows <-  function(spec) {
  workflow() |> add_recipe(model_recipe) |> add_model(spec)
}

model_workflows <- lapply(model_specs, create_workflows)

### Fitting the models

fit_model <- function(workflow) {
  fit(workflow, data = train_data)
}

model_fits <- lapply(model_workflows, fit_model)
```

3.3.6 Evaluating the Models

Once the models are trained, we can make predictions on the test data. Similar to the previous steps, we create a `make_predictions` function, and then apply this function to each element of the `model_fits` list (created in the previous step). The function uses `predict` to get both class predictions and probabilities, and bind them with the test data for late steps in the evaluation evaluation.

We then, evaluate the models (Fig. 1-E) with a custom function `evaluate_model` using several metrics as before. Our evaluation will includes computing a confusion matrix, the evaluation metrics (accuracy, sensitivity, specificity, F1 score, balanced accuracy, positive predictive value (PPV), and negative predictive value (NPV)) as well as ROC curve for each algorithm. Given that the function creates several objects for all the model, the function is a but more sophisticated, and uses list lists.

```
make_predictions <- function(fit) {
  predict(fit, test_data) |>
    bind_cols(predict(fit, test_data, type = "prob")) |>
    bind_cols(test_data)
}

model_preds <- lapply(model_fits, make_predictions)

evaluate_model <- function(pred, model_name) {
  # Confusion Matrix
  conf_mat <- conf_mat(pred, truth = Achievement, estimate = .pred_class)

  # Other metrics (Class metrics only)
  metrics <- metric_set(accuracy, sens, yardstick::specificity,
                        f_meas, bal_accuracy, ppv, npv)
  model_performance <- metrics(pred, truth = Achievement,
                               estimate = .pred_class)

  # ROC Curve
  roc_curve <- roc_curve(pred, truth = Achievement, .pred_Low_Achievers) |>
    mutate(model = model_name)

  list(conf_mat = conf_mat, performance = model_performance, roc = roc_curve)
}

# Evaluate each model and store results
evaluate_all_models <- function(preds, model_names) {
  mapply(evaluate_model, preds, model_names, SIMPLIFY = FALSE)
}

evaluation_results <- evaluate_all_models(model_preds, names(model_preds))
```

3.3.6.1 Combining and Plotting Results

We then combine the performance metrics and ROC curve data for all models into single data frames. These data frames will be used to generate visualizations and for comparison if needed.

```
combine_performance_metrics <- function(evaluation_results) {
  performance_df <- do.call(rbind, lapply(evaluation_results, function(res) {
    res$performance
  }))
  performance_df$model <- rep(names(evaluation_results),
                              times = sapply(evaluation_results, function(res) {
    nrow(res$performance)
  }))
```

```
  performance_df
}

performance_df <- combine_performance_metrics(evaluation_results)

combine_roc_curves <- function(evaluation_results) {
  do.call(rbind, lapply(evaluation_results, function(res) {
    res$roc
  }))
}

roc_df <- combine_roc_curves(evaluation_results)

extract_confusion_matrices <- function(evaluation_results) {
  lapply(evaluation_results, function(res) {
    res$conf_mat
  })
}

conf_mat_list <- extract_confusion_matrices(evaluation_results)
```

We define a color palette for plotting and then create plots for performance metrics (Fig. 7) and ROC curves (Fig. 8).

```
palette <- c("darkgreen", "green", "cyan", "blue", "purple", "magenta", "pink",
             "red", "orange", "yellow", "darkgoldenrod4", "grey", "black" )

performance_df |>
  dplyr::select(model, .metric, .estimate) |>
  pivot_longer(cols = .estimate, names_to = "metric", values_to = "value") |>
  ggplot(aes(y = .metric, x = value, fill = model)) +
  geom_bar(stat = "identity", position = "dodge") +
  scale_fill_manual(values = palette, name = "Metric") +
  theme_minimal() +
  labs(x = "Value", y = "Model")
```

```
ggplot(roc_df, aes(x = 1 - specificity, y = sensitivity, color = model)) +
  geom_line(size = 0.5) +
  geom_abline(linetype = "dashed") +
  scale_color_manual(values = palette, name = "Metric") +
  theme_minimal() +
  labs(x = "1 - Specificity", y = "Sensitivity") +
  theme(legend.position = "right")
```

Finally, we plot the confusion matrices for each model using a heatmap (Fig. 9).

```
plot_confusion_matrix <- function(conf_mat, model_name) {
  autoplot(conf_mat, type = "heatmap") +
    scale_fill_gradient(low = "white", high = "blue") +
    theme_minimal() +
    labs(title = paste(model_name), fill = "Count")
}

library(gridExtra)

# List to store ggplot objects
plot_list <- list()
```

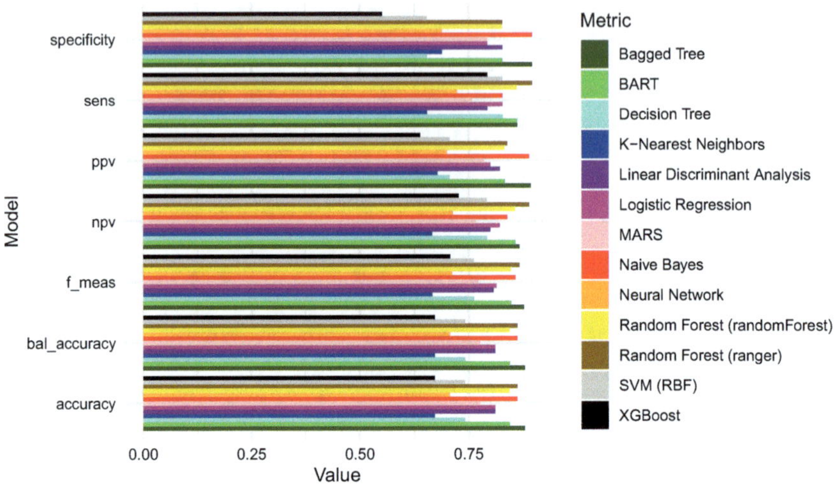

Fig. 7 Model performance metrics

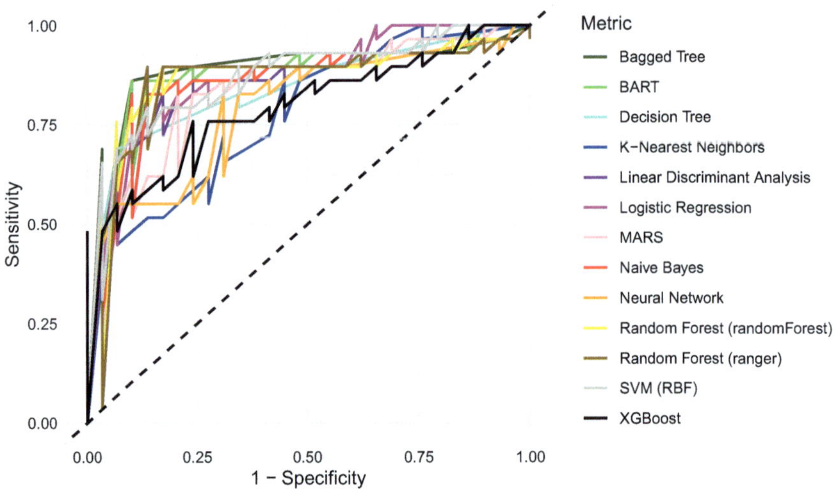

Fig. 8 ROC curves for different models

```
# Generate and store each confusion matrix plot
for (model_name in names(conf_mat_list)) {
  conf_mat_plot <- plot_confusion_matrix(conf_mat_list[[model_name]], model_name)
  plot_list[[model_name]] <- conf_mat_plot
}

# Combine all plots into one grid
# Adjust ncol based on how many columns you want
grid.arrange(grobs = plot_list, ncol = 3, nrow = 5)
```

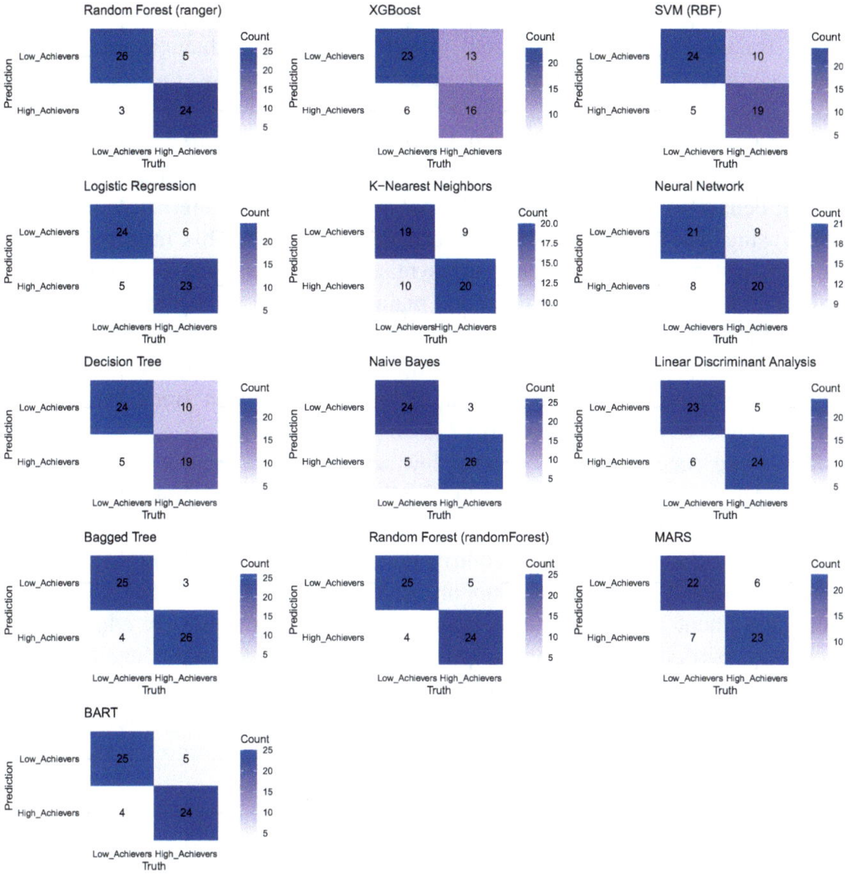

Fig. 9 Confusion matrices of each model

You can compare the algorithms and evaluate which algorithm has better performance and choose the best model based on this analysis.

4 Discussion

This chapter explained the application of machine learning, specifically classification techniques, to identify at-risk students in educational settings. We transitioned from the regression-based prediction of student grades discussed in the previous chapter [1], to a focus on identifying risky categories of academic performance namely low achievers.

The chapter demonstrated a detailed tutorial on using the Random Forest algorithm for classification. It covered the entire workflow, including data preparation—

creating a binary target variable and splitting data into training and testing sets—model training, applying the model to new data, and evaluation using metrics like accuracy, precision, recall and sensitivity. Additionally, the importance of model interpretability was highlighted by showcasing how to extract and visualize variable importance in a Random Forest model. This helps in understanding which factors contribute most significantly to the classification, offering insights into student behaviors that correlate with success or failure. An alternative implementation using the `tidymodels` framework was presented. This modern approach provides a streamlined and consistent syntax for various machine learning tasks, making it easier to build, compare, and maintain models. Using `tidymodels`, we demonstrated how to automate the process of fitting, predicting, and evaluating thirteen different classification algorithms, including Random Forest, XGBoost, SVM, Logistic Regression, KNN, Neural Networks, Decision Trees, Naive Bayes, LDA, Bagged Trees, MARS, and BART.

If we can accurately early predict low achievers, we can identify those who are struggling early in the course. This allows for proactive interventions, preventing students from falling too far behind and potentially dropping out [17, 57]. Understanding the factors that contribute to a student being classified as "at-risk" can inform personalized support strategies [42, 43, 58]. For example, if low engagement with forum discussions is a significant predictor, educators can encourage participation or encourage students to interact with their peers. For teachers, they can use the classification models to identify struggling students before they fail a course, allowing them to reach out, offer additional support, and address any underlying issues that may be hindering progress. The insights gained from variable importance analysis can help teachers tailor their feedback and resources to address specific student needs. Furthermore, these models may help teachers allocate their time and resources more efficiently by focusing on students who need the most support—particularly valuable in large classes where providing individualized attention can be challenging. also, educators can use the data to initiate conversations about academic progress, identify areas where students are struggling, and collaboratively develop strategies for improvement.

However, it is important to acknowledge potential drawbacks associated with predictive modeling in education. One significant concern is the risk of profiling students and inaccurately labeling them as low achievers when they are not. Misclassifications can have detrimental effects on students' self-esteem, motivation, and overall educational experience. False positives—students incorrectly identified as at-risk—may receive unnecessary interventions, which can lead to stigmatization or decreased expectations from teachers and peers.

Therefore, it is important to use predictive models as supportive tools rather than judgments. Integrating human insight with predictive analytics ensures a more nuanced understanding of each student's needs. We can't stress enough how ethical considerations must be at the forefront when implementing predictive analytics where transparency about how predictions are made, ensuring data privacy, and actively working to mitigate biases in the models are essential steps.

References

1. Saqr M, Misiejuk K, Tikka S, López-Pernas S (2025) Artificial intelligence: using machine learning to predict students' performance. In: Saqr M, López-Pernas S (eds) Advanced learning analytics methods: AI, precision and complexity. Springer Nature Switzerland, Cham
2. Jovanovic J, López-Pernas S, Saqr M (2024) Predictive modelling in learning analytics: a machine learning approach in r. In: Learning analytics methods and tutorials: a practical guide using r. Springer Nature Switzerland Cham, pp 197–229
3. Liaw A, Wiener M (2002) Classification and regression by randomForest. R News 2:18–22. https://CRAN.R-project.org/doc/Rnews/
4. Na KS, Tasir Z (2017) Identifying at-risk students in online learning by analysing learning behaviour: a systematic review. In: 2017 IEEE conference on big data and analytics (ICBDA). IEEE, pp 118–123
5. Oliveira CF de, Sobral SR, Ferreira MJ, Moreira F (2021) How does learning analytics contribute to prevent students' dropout in higher education: a systematic literature review. Big Data Cognitive Comput 5:64
6. Saqr M, Fors U, Tedre M (2017) How learning analytics can early predict under-achieving students in a blended medical education course. Med Teacher 39:757–767
7. Bayazit A, Apaydin N, Gonullu I (2022) Predicting at-risk students in an online flipped anatomy course using learning analytics. Educ Sci 12:581
8. Queiroga EM, Lopes JL, Kappel K, Aguiar M, Araújo RM, Munoz R, Villarroel R, Cechinel C (2020) A learning analytics approach to identify students at risk of dropout: a case study with a technical distance education course. Appl Sci 10:3998
9. Aljohani NR, Fayoumi A, Hassan S-U (2019) Predicting at-risk students using clickstream data in the virtual learning environment. Sustainability 11:7238
10. Adnan M, Habib A, Ashraf J, Mussadiq S, Raza AA, Abid M, Bashir M, Khan SU (2021) Predicting at-risk students at different percentages of course length for early intervention using machine learning models. IEEE Access 9:7519–7539
11. Gonzalez-Nucamendi A, Noguez J, Neri L, Robledo-Rella V, García-Castelán RMG (2023, October). Predictive analytics study to determine undergraduate students at risk of dropout. In: Frontiers in Education, vol 8. Frontiers Media SA, p 1244686
12. Jayaprakash SM, Moody EW, Lauría EJ, Regan JR, Baron JD (2014) Early alert of academically at-risk students: an open source analytics initiative. J Learn Anal 1:6–47
13. Azcona D, Hsiao I-H, Smeaton AF (2019) Detecting students-at-risk in computer programming classes with learning analytics from students' digital footprints. User Modeling User-Adapted Interaction 29:759–788
14. Anagnostopoulos T, Kytagias C, Xanthopoulos T, Georgakopoulos I, Salmon I, Psaromiligkos Y (2020) Intelligent predictive analytics for identifying students at risk of failure in moodle courses. In: Intelligent tutoring systems: 16th international conference, ITS 2020, athens, greece, june 8–12, 2020, proceedings 16. Springer, pp 152–162
15. Mathrani A, Susnjak T, Ramaswami G, Barczak A (2021) Perspectives on the challenges of generalizability, transparency and ethics in predictive learning analytics. Comput Educ Open 2:100060
16. Saqr M, Jovanovic J, Viberg O, Gašević D (2022) Is there order in the mess? A single paper meta-analysis approach to identification of predictors of success in learning analytics. Stud Higher Educ 47:2370–2391. https://doi.org/10.1080/03075079.2022.2061450
17. Jovanović J, Saqr M, Joksimović S, Gašević D (2021) Students matter the most in learning analytics: the effects of internal and instructional conditions in predicting academic success. Comput Educ 172:104251. https://doi.org/10.1016/j.compedu.2021.104251
18. Foster E, Siddle R (2020) The effectiveness of learning analytics for identifying at-risk students in higher education. Assess Eval Higher Educ 45:842–854
19. Wolff A, Zdrahal Z, Nikolov A, Pantucek M (2013) Improving retention: predicting at-risk students by analysing clicking behaviour in a virtual learning environment. In: Proceedings of the third international conference on learning analytics and knowledge, pp 145–149

20. Queiroga EM, Batista Machado MF, Paragarino VR, Primo TT, Cechinel C (2022) Early prediction of at-risk students in secondary education: a countrywide k-12 learning analytics initiative in uruguay. Information 13:401
21. Choi SP, Lam SS, Li KC, Wong BT (2018) Learning analytics at low cost: at-risk student prediction with clicker data and systematic proactive interventions. J Educ Technol Soc 21:273–290
22. Hlosta M, Herodotou C, Papathoma T, Gillespie A, Bergamin P (2022) Predictive learning analytics in online education: a deeper understanding through explaining algorithmic errors. Comput Educ Artif Intell 3:100108
23. Gkontzis AF, Kotsiantis S, Panagiotakopoulos CT, Verykios VS (2022) A predictive analytics framework as a countermeasure for attrition of students. Interactive Learn Environ 30:1028–1043
24. Cechinel C, Dos Santos MDF, Barrozo C, Schardosim JE, Vila E de, Ramos V, Primo T, Munoz R, Queiroga EM (2021) A learning analytics dashboard for moodle: Implementing machine learning techniques to early detect students at risk of failure. In: 2021 XVI latin american conference on learning technologies (LACLO). IEEE, pp 130–136
25. Saqr M, Nouri J (2020) High resolution temporal network analysis to understand and improve collaborative learning. Proceedings of the tenth international conference on learning analytics & knowledge, pp 314–319. https://doi.org/10.1145/3375462.3375501
26. Saqr M, Cheng R, López-Pernas S, Beck ED (2024) Idiographic artificial intelligence to explain students' self-regulation: toward precision education. Learn Indiv Differ 114:102499. https://doi.org/10.1016/j.lindif.2024.102499
27. Kaliisa R, Misiejuk K, López-Pernas S, Khalil M, Saqr M (2024) Have learning analytics dashboards lived up to the hype? A systematic review of impact on students' achievement, motivation, participation and attitude. Proceedings of the 14th learning analytics and knowledge conference, pp 295–304. https://doi.org/10.1145/3636555.3636884
28. Paulsen L, Lindsay E (2024) Learning analytics dashboards are increasingly becoming about learning and not just analytics-a systematic review. Educ Inf Technol 1–30
29. Ramaswami G, Susnjak T, Mathrani A, Umer R (2023) Use of predictive analytics within learning analytics dashboards: a review of case studies. Technol Knowl Learn 28:959–980
30. Saqr M, Fors U, Tedre M, Nouri J (2018) How social network analysis can be used to monitor online collaborative learning and guide an informed intervention. PLOS ONE 13:e0194777. https://doi.org/10.1371/journal.pone.0194777
31. Rienties B, Cross S, Zdrahal Z (2017) Implementing a learning analytics intervention and evaluation framework: what works?. In: Kei Daniel B (ed) Big data and learning analytics in higher education. Springer, Cham. https://doi.org/10.1007/978-3-319-06520-5_10
32. Tikka S, Kopra J, Heinäniemi M, López-Pernas S, Saqr M (2024) Getting started with R for education research. In: Saqr M, López-Pernas S (eds) Learning analytics methods and tutorials: a practical guide using r. Springer, Berlin
33. Kopra J, Tikka S, Heinäniemi M, López-Pernas S, Saqr M (2024) An R approach to data cleaning and wrangling for education research. In: Saqr M, López-Pernas S (eds) Learning analytics methods and tutorials: a practical guide using r. Springer, Berlin
34. Tikka S, Kopra J, Heinäniemi M, López-Pernas S, Saqr M (2024) Introductory statistics with R for educational researchers. In: Saqr M, López-Pernas S (eds) Learning analytics methods and tutorials: a practical guide using r. Springer, Berlin
35. López-Pernas S, Misiejuk K, Tikka S, Kopra J, Heinäniemi M, Saqr M (2024) Visualizing and reporting educational data with R. In: Learning analytics methods and tutorials. Springer Nature Switzerland, Cham, pp 151–194. https://doi.org/10.1007/978-3-031-54464-4/_6
36. Wickham H, Averick M, Bryan J, Chang W, McGowan LD, François R, Grolemund G, Hayes A, Henry L, Hester J, Kuhn M, Pedersen TL, Miller E, Bache SM, Müller K, Ooms J, Robinson D, Seidel DP, Spinu V, Takahashi K, Vaughan D, Wilke C, Woo K, Yutani H (2019) Welcome to the tidyverse. J Open Source Softw 4:1686. https://doi.org/10.21105/joss.01686

37. Frick H, Chow F, Kuhn M, Mahoney M, Silge J, Wickham H (2024) Rsample: General resampling infrastructure. https://CRAN.R-project.org/package=rsample
38. Kuhn, Max (2008) Building predictive models in r using the caret package. J Stat Softw 28:1–26. https://doi.org/10.18637/jss.v028.i05
39. Chan C, Leeper TJ, Becker J, Schoch D (2023) Rio: A swiss-army knife for data file i/o. https://cran.r-project.org/package=rio
40. Saqr M, López-Pernas S (2024) Why explainable AI may not be enough: predictions and mispredictions in decision making in education. Smart Learn Environ. https://doi.org/10.1186/s40561-024-00343-4
41. López-Pernas S, Oliveira E, Song Y, Saqr M (2025) AI, explainable AI and evaluative AI: an introduction to informed data-driven decision-making in education. In: Saqr M, López-Pernas S (eds) Advanced learning analytics methods: AI, precision and complexity. Springer Nature Switzerland, Cham
42. Saqr M, López-Pernas S (2025) Explainable artificial intelligence in education: a tutorial for identifying the variables that matter. In: Saqr M, López-Pernas S (eds) Advanced learning analytics methods: AI, precision and complexity. Springer Nature Switzerland, Cham
43. Saqr M, López-Pernas S (2025) Individualized explainable artificial intelligence: A tutorial for identifying local and individual predictions. In: Saqr M, López-Pernas S (eds) Advanced learning analytics methods: AI, precision and complexity. Springer Nature Switzerland, Cham
44. Kuhn M, Wickham H (2020) Tidymodels: A collection of packages for modeling and machine learning using tidyverse principles. https://www.tidymodels.org
45. Kuhn M, Vaughan D (2024) Parsnip: A common API to modeling and analysis functions. https://CRAN.R-project.org/package=parsnip
46. Kuhn M, Wickham H, Hvitfeldt E (2024) Recipes: Preprocessing and feature engineering steps for modeling. https://CRAN.R-project.org/package=recipes
47. Kuhn M, Vaughan D, Hvitfeldt E (2024) Yardstick: Tidy characterizations of model performance. https://CRAN.R-project.org/package=yardstick
48. Wright MN, Ziegler A (2017) ranger: A fast implementation of random forests for high dimensional data in C++ and R. J Stat Softw 77:1–17. https://doi.org/10.18637/jss.v077.i01
49. Chen T, He T, Benesty M, Khotilovich V, Tang Y, Cho H, Chen K, Mitchell R, Cano I, Zhou T, Li M, Xie J, Lin M, Geng Y, Li Y, Yuan J (2024) Xgboost: Extreme gradient boosting. https://CRAN.R-project.org/package=xgboost
50. Karatzoglou A, Smola A, Hornik K, Zeileis A (2004) Kernlab—an S4 package for kernel methods in R. J Stat Softw 11:1–20. https://doi.org/10.18637/jss.v011.i09
51. Schliep K, Hechenbichler K (2016) Kknn: Weighted k-nearest neighbors. https://CRAN.R-project.org/package=kknn
52. Venables WN, Ripley BD (2002) Modern applied statistics with s, 4th edn. Springer, New York https://www.stats.ox.ac.uk/pub/MASS4/
53. Therneau T, Atkinson B (2022) Rpart: Recursive partitioning and regression trees. https://CRAN.R-project.org/package=rpart
54. Majka M (2024) Naivebayes: High performance implementation of the naive bayes algorithm in r. https://CRAN.R-project.org/package=naivebayes
55. Venables WN, Ripley BD (2002) Modern applied statistics with s, 4th edn. Springer, New York. https://www.stats.ox.ac.uk/pub/MASS4/
56. Friedman JH (1991) Multivariate adaptive regression splines. Ann Statist 19(1):1–67. https://doi.org/10.1214/aos/1176347963
57. Saqr M, López-Pernas S (2021) The longitudinal trajectories of online engagement over a full program. Comput Educ 175:104325. https://doi.org/10.1016/j.compedu.2021.104325
58. Khosravi H, Shum SB, Chen G, Conati C, Tsai Y-S, Kay J, Knight S, Martinez-Maldonado R, Sadiq S, Gašević D (2022) Explainable artificial intelligence in education. Comput Educ Artif Intell 3:100074

Comparative Analysis of Regularization Methods for Predicting Student Certification

Tian Li, Feifei Han, Jiesi Guo, and Jinran Wu

Abstract This chapter uses predicting student certification in online courses hosted on edX to demonstrate how to apply three powerful regularization methods, namely Least Absolute Shrinkage and Selection Operator (LASSO), Smoothly Clipped Absolute Deviation (SCAD), and Minimax Concave Penalty (MCP) in constructing a predictive model. These advanced statistical methods are especially valuable in scenarios where we have a multitude of potential predictors and need to select appropriate variables to establish viable predictive models.

Keywords Regularization methods · Comparison · Prediction · Online courses in edX

1 Introduction

In the past decade, the educational landscape has undergone a seismic shift with the advent and proliferation of online learning platforms [1]. Among these, Massive Open Online Courses (MOOCs) have emerged as a transformative force, democratizing access to high-quality education on a global scale [2]. First emerged in 2008, Massive Open Online Courses (MOOCs) provide anyone with an internet connection and low-cost access to educational materials and activities delivered by experts. Like other forms of adult education, the establishment of MOOCs aims to

T. Li
Faculty of Engineering, Architecture and Information Technology, The University of Queensland, Brisbane, QLD, Australia

F. Han (✉)
Institute for Learning Sciences & Teacher Education, Australian Catholic University, Sunnybank, QLD, Australia
e-mail: Feifei.Han@acu.edu.au

J. Guo · J. Wu
Institute for Positive Psychology and Education, Australian Catholic University, Sydney, NSW, Australia

M. Saqr, S. López-Pernas (eds.), *Advanced Learning Analytics Methods*,
https://doi.org/10.1007/978-3-031-95365-1_5

help learners realize their potential and apply newly acquired knowledge and skills to improve their lives [3]. Students enroll in MOOCs for various reasons, including personal interest, career advancement, academic support, or social connections [4]. Unlike traditional online or in-person adult education courses, MOOCs allow students to access courses anywhere and time. Despite their potential, MOOCs have faced many challenges, such as accreditation and certification (Sanchez-Gordon), instructional design [5], types of learners [6], as well as dropout rates [7].

At the forefront of this revolution stands edX Online Courses & Programs, a pioneering platform founded in May 2012 under the administration of the Massachusetts Institute of Technology (MIT). As a leader in the MOOC space, the courses on the edX are, offered by world-renowned institutions, schools, non-profit organizations, and corporations across a diverse array of subjects, from subjects in science, technology, engineering, and math (STEM) to social sciences and humanities. edX has opened the doors of elite education to millions of learners worldwide through its commitment to accessibility and rigorous academic standards, making it a beacon of innovation in online education.

In the traditional classrooms where the teaching staff interact with students face-to-face [8], this allows them to have direct perceptions of students' levels of engagement, hence, targeted strategies can be implemented in advance to increase student engagement and to prevent students from dropping out.

In the online context, with the advancement of learning analytics to capture large volumes of digital trace data, which are combined with various demographic data, researchers and educators can predict students' tendency to drop out earlier in the course by applying advanced data mining and statistical techniques [9]. This also aids in understanding and improving the quality of student learning experiences [10].

One of the important indicators of student completion of online courses hosted on edX is to obtain an official certification of course. Hence, understanding factors impacting students obtaining course certification may provide course developers valuable information on strategies to improve the design of the course, online pedagogies, and methods of delivery to foster student online engagement and improve retention rate.

It shall be noted that many references like Jovanovic et al. [11] and Gray and Perkins [12] studied machine learning methods, such as random forests, to predict student course performance. However, our method differs as we focus on high-dimensional predictors and apply regularization techniques to highlight key predictors associated with learning outcome predictions. Specifically, we aim to create an elegant predictive model by using regularization to eliminate redundant predictors. Moreover, this chapter uses predicting student certification in online courses hosted on edX to demonstrate how to apply three powerful regularization methods, namely Least Absolute Shrinkage and Selection Operator (LASSO) [13], Smoothly Clipped Absolute Deviation (SCAD) [14], and Minimax Concave Penalty (MCP) [15] in constructing a predictive model. These advanced statistical methods are especially valuable in scenarios where we have a multitude of potential predictors and need to select appropriate variables to establish viable predictive models. Therefore, we aim to achieve the following objectives:

1. Illustrate the strengths of LASSO, SCAD, and MCP in constructing predictive models using predicting obtaining course certification in edX courses as an example. The three advanced regularization methods each have unique strengths in the process of establishing predictive models. Through applying them to a specific context—success in obtaining course certification in edX courses, it is easy to understand the relative effectiveness of these methods, providing researchers, educators, and course designers hands-on knowledge so that they can select appropriate methods by considering the unique characteristics of each method and make similar predictions of their online courses;
2. Uncover the key factors contributing to student certification in edX courses: We hope to identify the most key predictors of student success by analyzing a wide range of variables. This knowledge can inform both learners and educators, helping to create more effective learning strategies and support systems;
3. Identify the most important predictors of student success: Through variable selection and model interpretation, we aim to pinpoint the factors that strongly influence certification outcomes. This information can be invaluable for course designers, instructors, and platform developers in optimizing the online learning experience;
4. Provide actionable insights to inform course design and student support strategies: By understanding the key drivers of student success, we can offer evidence-based recommendations for improving course structure, content delivery, and support mechanisms on edX and similar platforms.

2 Methodologies

In this section, we will describe three regularization methods (i.e., LASSO, SCAD, and MCP) and the cross-validation method for hyperparameter tuning. In addition, we also will introduce some popular classification evaluation metrics.

2.1 Regularization Methods for High-Dimensional Data

In the context of high-dimensional data, where the number of predictors (p) is large relative to the number of observations (n), traditional statistical methods often fail due to overfitting, multicollinearity, and computational issues. Regularization methods have emerged as powerful tools to address these challenges by introducing a penalty term to the loss function, effectively shrinking some coefficient estimates towards zero and, in some cases, setting them exactly to zero [16].

These methods perform simultaneous variable selection and coefficient estimation, making them particularly useful in scenarios where identifying the most important predictors is as crucial as making accurate predictions. The general form of a penalized regression problem (with regression coefficients β) can be written as [17, 18]:

$$\min_{\beta}\{L(\beta) + \lambda P(\beta)\}$$

where $L(\beta)$ is the loss function (e.g., negative log-likelihood for logistic regression), $P(\beta)$ is the penalty function, and λ is the tuning parameter controlling the strength of regularization. Moreover, we will delve deeper into each of the three regularization methods (LASSO, SCAD, and MCP) we implemented in the following contents.

Remark In this study, we focus on a classification task using regularization methods, specifically through three types of regularized logistic regression models. We need to adjust the first component for continuous response variables, replacing the logistic loss function with alternatives like least squares or robust losses such as Huber [19] or exponential loss [20]. Multicollinearity is not a concern in our regularization approach. Regularization methods inherently address multicollinearity by applying penalties that shrink the coefficients of correlated predictors, effectively reducing redundancy and enhancing model stability.

2.1.1 LASSO

LASSO, introduced by Tibshirani [13], adds an L1 penalty to the loss function:

$$P(\beta) = ||\beta||_1 = \Sigma|\beta_j|.$$

The L1 penalty has the property of shrinking some coefficients exactly to zero, effectively performing variable selection. This is due to the geometry of the L1 ball, which intersects with the contours of the loss function at the axes, leading to sparse solutions [16]. Key properties of LASSO are listed: (a) Performs both variable selection and regularization; (b) Produces sparse models, which are often easier to interpret; (c) Can handle high-dimensional data ($p > n$); and (d) Tends to select one variable from a group of correlated predictors. However, LASSO has some limitations: (a) In the $p > n$ case, it selects most n variables before saturation; (b) It can be biased for large coefficients; and (c) It does not have the oracle property (consistent variable selection).

2.1.2 SCAD

SCAD, proposed by Fan and Li [14], is a non-convex penalty that aims to overcome some of the limitations of LASSO. The SCAD penalty is defined as:

$$P(\beta) = \begin{cases} \lambda|\beta| & \text{if } |\beta| \leq \lambda \\ -\frac{\beta^2 - 2a\lambda|\beta| + \lambda^2}{2(a-1)} & \text{if } \lambda < |\beta| \leq a\lambda \\ \frac{(a+1)\lambda^2}{2} & \text{if } |\beta| > a\lambda \end{cases},$$

where $a > 2$ is a tuning parameter, often set to 3.7 as suggested by Fan and Li [14]. Moreover, key properties of SCAD are given as (a) Overcomes the bias issue of LASSO for large coefficients; (b) Possesses the oracle property (asymptotically performs as well as if the true model were known in advance); c) Produces sparse solutions; (d) Continuous in β, which leads to more stable solutions. However, SCAD also has some challenges: (a) Non-convex optimization problem, which can be computationally more demanding; and (b) Requires tuning of two parameters (λ and a).

2.1.3 MCP

MCP introduced by Zhang [15], is another non-convex penalty designed to address the limitations of LASSO. The MCP is defined as:

$$P(\beta) = \begin{cases} \lambda|\beta| - \frac{\beta^2}{2\gamma} & \text{if } |\beta| \leq \gamma\lambda \\ \frac{\gamma\lambda^2}{2} & \text{if } |\beta| > \gamma\lambda \end{cases},$$

where $\gamma > 1$ is a tuning parameter controlling the concavity of the penalty. It shall be noted that some key properties of MCP are listed as (a) Like SCAD, it overcomes the bias issue of LASSO for large coefficients; (b) Possesses the oracle property; (c) Produces sparse solutions; (d) Provides a smooth transition between the region of penalization and the constant region. However, there are still challenges of MCP: (a) Non-convex optimization problem; and (b) Requires tuning of two parameters (λ and γ).

It shall be noted that both SCAD and MCP are part of a broader class of concave penalty functions that aim to combine the variable selection properties of L1 penalties with the unbiasedness of L2 penalties for large coefficients [21]. In practice, the choice between these methods often depends on the specific characteristics of the data and the goals of the analysis. LASSO is often preferred for its simplicity and convexity, while SCAD and MCP can provide better theoretical properties and potentially more accurate coefficient estimates, especially for large effects.

2.2 K-Fold Cross-Validation for Hyperparameter Tuning

K-Fold Cross-Validation (CV) is a widely used method for hyperparameter tuning in statistical learning, providing a robust estimate of model performance across different data subsets [22]. The process begins by setting aside an independent test set and dividing the remaining data into K equally sized folds. Hyperparameter tuning is then performed by systematically exploring a predefined hyperparameter space. For each hyperparameter combination, the model is trained on $K-1$ folds and validated on the remaining fold, rotating through all K folds. The average performance across these K iterations is used to assess each hyperparameter set. This approach helps select hyperparameters that generalize well to unseen data, reducing overfitting risk.

The K-Fold CV process for hyperparameter tuning typically involves several steps: defining a hyperparameter grid or search space, performing the cross-validation loop for each hyperparameter combination, selecting the best-performing hyperparameters, and finally training and evaluating the model on the entire dataset and held-out test set, respectively. The hyperparameter search can be conducted using various methods, including grid search, random search [23], or more advanced techniques like Bayesian optimization [24]. While this method offers advantages such as robust performance estimation and efficient use of limited data, it also presents challenges. These include computational cost, especially for large datasets or complex models, and the potential instability of results depending on the specific folds used.

2.3 Classification Evaluation Metrics

In classification problems, especially binary classification, several metrics are commonly used to evaluate model performance. These metrics provide different perspectives on the model's effectiveness and are derived from the confusion matrix. Understanding these metrics is crucial for properly assessing and comparing classification models. Before diving into the metrics, it is important to understand the confusion matrix in Table 1, which is the foundation for these metrics.

Based on the confusion matrix, five popular classification measurement metrics (Accuracy, Precision, Recall, F1 score, and AUC) are derived as below [25]:

Table 1 The illustration of the confusion matrix

	Predicted positive	Predicted negative
Actual positive	True positive (TP)	False negative (FN)
Actual negative	False positive (FP)	True negative (TN)

1. **Accuracy** is the proportion of correct predictions (both true positives and true negatives) among the total number of cases examined as:

$$\text{Accuracy} = \frac{\text{TP} + \text{TN}}{\text{TP} + \text{TN} + \text{FP} + \text{FN}}.$$

2. **Precision** is the proportion of true positive predictions among all positive predictions and is formulated as:

$$\text{Precision} = \frac{\text{TP}}{\text{TP} + \text{FP}}.$$

3. **Recall** is the proportion of actual positive instances that were correctly identified and is formulated as:

$$\text{Recall} = \frac{\text{TP}}{\text{TP} + \text{FN}}.$$

4. **F1 score** is the harmonic mean of precision and recall, providing a single score that balances both metrics and the mathematical formula is:

$$\text{F1} = \frac{2 \times (\text{Precision} \times \text{Recall})}{\text{Precision} + \text{Recall}}.$$

5. **AUC** (Area Under the receiver operating characteristic curve) is a performance metric that represents the probability that a randomly chosen positive instance is ranked higher than a randomly chosen negative instance. It is calculated as the area under the Receiver Operating Characteristic curve. A higher AUC value indicates better model performance in distinguishing between the classes.

It shall be noted that for imbalanced datasets, in cases where classes are imbalanced, accuracy can be misleading. For example, in a dataset with 95% negative cases and 5% positive cases, a model that always predicts negative would have 95% accuracy but would be useless for identifying positive cases. Therefore, an F1 score (or AUC) that combines Precision and Recall is suggested to measure the classification performance for imbalanced datasets.

3 Data Description and Exploratory Analysis

This section will present the dataset chosen for our illustration and conduct a preliminary exploration using basic visualization techniques.

3.1 Dataset Overview

Our dataset comes from an online learning platform and contains information about students enrolled in various courses. The key variables in our analysis include:

- `certified`: Whether the student received certification (our target variable);
- `course_id`: Identifier for different courses;
- `explored`: Whether the student explored the course content;
- `gender`: Student's gender;
- `nevents`: Number of events (interactions) by the student;
- `ndays_act`: Number of days the student was active;
- `nplay_video`: Number of video plays;
- `nchapters`: Number of chapters accessed;
- `nforum_posts`: Number of forum posts made by the student;
- `final_cc_cname_DI`: Student's country name.

It is important to note that the response variable in our study is "certified", with the other variables serving as predictors, as applied in Yuan et al. [26].

3.2 Data Preparation and Exploratory Data Analysis

We begin our analysis by loading the necessary libraries and preparing our data:

```
library(ggplot2)
library(patchwork)
library(dplyr)
library(caret)
library(ncvreg)
library(pROC)

repo <- "https://github.com/lamethods/data2/raw/refs/heads/main/"
dataset <- "mooc/HXPC13_DI_v3_11-13-2019.RDS"

mydata <- readRDS(url(paste0(repo, dataset)))
data <- mydata[,c("certified","course_id","explored","gender",
                  "nevents","ndays_act","nplay_video","nchapters",
                  "nforum_posts","final_cc_cname_DI")]
data[data == ""] <- NA
data <- na.omit(data)
```

The following code snippet loads our dataset, selects the relevant columns, replaces empty strings with NA values, and removes any rows with missing data. Here, we shall notice our target of this chapter is to illustrate how to employ high-dimensionality reduction approaches for identifying key predictors (factors) relevant to the response (whether certified or not), thus we remove all the samples with missing data.

```
table(data$certified, data$course_id)
```

```
   HarvardX/PH207x/2012_Fall HarvardX/PH278x/2013_Spring
 0                     17671                       11006
 1                      1724                         616
```

```
table(data$certified, data$gender)
```

```
       f     m
 0 13077 15600
 1  1078  1262
```

```
table(data$certified, data$explored)
```

```
       0     1
 0 26027  2650
 1    60  2280
```

```
# Ensure 'course_id' is a factor
data$course_id <- as.factor(data$course_id)
# Rename specific course names
levels(data$course_id)[levels(data$course_id) ==
                        "HarvardX/PH207x/2012_Fall"] <- "c_id1"
levels(data$course_id)[levels(data$course_id) ==
                        "HarvardX/PH278x/2013_Spring"] <- "c_id2"
```

Before diving into our predictive modeling, it is crucial to understand the distribution and relationships within our data. We use visualization techniques with two R packages "ggplot2" [27] and "patchwork" [28] to gain these insights.

```
# Certification rates by course
plot1 <- ggplot(data, aes(x = course_id, fill = factor(certified))) +
  geom_bar(position = "fill") +
  labs(y = "proportion", title = "course_id", fill = "certified") +
  theme(axis.text.x = element_text(angle = 45, hjust = 1),
        plot.title = element_text(hjust = 0.5))
# Certification rates by gender
plot2 <- ggplot(data, aes(x = gender, fill = factor(certified))) +
  geom_bar(position = "fill") +
  labs(y = "proportion", title = "gender", fill = "Certified") +
  theme(plot.title = element_text(hjust = 0.5))
# Certification rates by exploration status
plot3 <- ggplot(data, aes(x = explored, fill = factor(certified))) +
  geom_bar(position = "fill") +
  labs(y = "proportion", title = "explored", fill = "Certified") +
  theme(plot.title = element_text(hjust = 0.5))
```

```
# Combine the plots
combined_plot <- plot1 | plot2 | plot3
combined_plot + plot_layout(guides = 'collect')
```

This code creates three bar plots reported in Fig. 1 showing the certification rates by course (categorical variable), gender (binary variable), and exploration status (binary variable). These visualizations help us understand how certification rates vary across different subgroups in our dataset.

Moreover, we can obtain the detailed numerical results of the certification rates by course, gender, and exploration status with the following codes:

```
# course_id
table(data$certified,data$course_id)
```

```
   c_id1 c_id2
 0 17671 11006
 1  1724   616
```

```
# gender
table(data$certified,data$gender)
```

```
      f     m
 0 13077 15600
 1  1078  1262
```

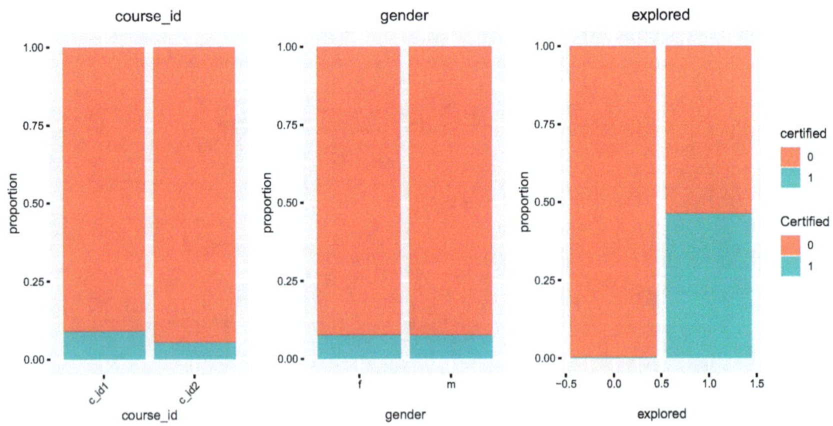

Fig. 1 The visualization of the certification rates by course, gender, and exploration status

```
# explored
table(data$certified,data$explored)
```

```
      0     1
0 26027  2650
1    60  2280
```

In addition, different from the former visualization for binary and categorical variables, here, we also create boxplots to visualize the relationship between certification and various numeric variables (Fig. 2):

```
data_long <- data.frame(
  certified = rep(data$certified, 5),
  metric = factor(rep(c("nevents", "ndays_act", "nchapters", "nplay_video",
                        "nforum_posts"), each = nrow(data))),
  value = c(data$nevents, data$ndays_act, data$nchapters, data$nplay_video, data$nforum_posts)
)
# Create a data frame manually for faceting
faceted_plot <- ggplot(data_long, aes(x = factor(certified), y = value, fill = factor(certified))) +
  geom_boxplot() +
  facet_wrap(~ metric, scales = "free_y", ncol = 2) +
  labs(title = "Metrics by Certification Status", x = "Certified", y = "Value", fill = "Certified") +
  theme(plot.title = element_text(hjust = 0.5),
        strip.text = element_text(size = 12),
        legend.position = "bottom",
        aspect.ratio = 0.45) # Adjust the aspect ratio to make plots taller or wider

print(faceted_plot)
```

These boxplots in Fig. 2 reveal the differences in various engagement metrics between certified and non-certified students, highlighting that certified users generally show higher levels of engagement across multiple dimensions. In each boxplot, variables such as `nevents`, `ndays_act`, and `nchapters` display distinct patterns between certified and non-certified individuals, with certified students showing higher medians and wider distributions in these metrics. This pattern suggests that active and sustained participation—through frequent events, consistent activity days, and more completed chapters—is associated with a greater likelihood of certification. However, the patterns for `nplay_video` and `nforum_posts` are less clear, as both certified and non-certified students show similar distributions with low levels of forum participation overall. This lack of differentiation implies that while direct interaction with course material is crucial, forum activity and video play frequency may not significantly impact certification outcomes. Therefore, we will further analyze the visualization results with advanced regression approaches to better understand these relationships and quantify the extent to which each metric influences the likelihood of certification.

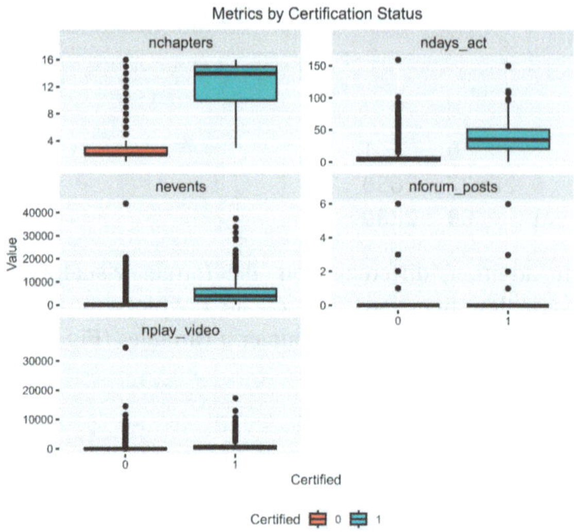

Fig. 2 The visualization of the certification outcomes by `nevents`, `ndays_act`, `nchapters`, `nplay_video`, and `nforum_posts`

4 Data Modelling and Numerical Results

4.1 Data Preprocessing

In regularized regression for high-dimensional predictors, the optimization objective combines two components: the negative log-likelihood (for logistic regression) and the penalty function, with a hyperparameter lambda balancing them. In practice, predictors can have different scales, which may complicate the hyperparameter tuning process by impacting the search for an optimal lambda within a specified range. Therefore, it is necessary to scale the numeric variables to ensure they are on a consistent scale, improving the stability and efficiency of tuning. Additionally, converting categorical variables to factors allows the model to correctly interpret them as categorical, which is essential for meaningful analysis and accurate results. Therefore, before applying our regularization methods, we need to preprocess our data. We scale the numeric variables and convert categorical variables to factors:

```
data2 <- as.data.frame(scale(data[,c("nevents", "ndays_act", "nplay_video",
                                     "nchapters", "nforum_posts")]))
data2$certified <- as.factor(data$certified)
data2$course_id <- as.factor(data$course_id)
data2$explored <- as.factor(data$explored)
data2$gender <- as.factor(data$gender)
data2$final_cc_cname_DI <- as.factor(data$final_cc_cname_DI)
```

4.2 Model Preparation

We split our data into training (80%) and test (20%) sets:

```
set.seed(123)
train_indices <- createDataPartition(data2$certified, p = 0.8, list = FALSE)
train_data <- data2[train_indices, ]
test_data <- data2[-train_indices, ]
```

Considering the potential impacts of interactions between predictors, we construct many interaction terms in our statistical modeling. With such some interaction terms, we can clearly illustrate the effectiveness of the regularization approaches for handling high-dimensional predictors in regression modeling. We prepare our design matrices with interaction terms and response vectors:

```
y_train <- train_data$certified
X_train <- model.matrix(certified ~ course_id + gender + final_cc_cname_DI +
                          nevents + ndays_act +nplay_video + explored + nchapters +
                          nforum_posts + final_cc_cname_DI +
                          (course_id + nevents + ndays_act + nplay_video +
                             explored + nchapters + nforum_posts) * gender +
                          (nevents + ndays_act + nplay_video + explored +
                             nchapters + nforum_posts) * course_id,
                        data = train_data)[,-1]

y_test <- test_data$certified
X_test <- model.matrix(certified ~ course_id + gender + final_cc_cname_DI +
                         nevents + ndays_act + nplay_video + explored + nchapters +
                         nforum_posts + final_cc_cname_DI +
                         (course_id + nevents + ndays_act + nplay_video +
                            explored + nchapters + nforum_posts) * gender +
                         (nevents + ndays_act + nplay_video + explored +
                            nchapters + nforum_posts) * course_id,
                       data = test_data)[,-1]
```

4.3 Cross-validation and Model Fitting

We implement three regularization methods: LASSO, SCAD, and MCP. These methods are particularly useful for high-dimensional data where we have a large number of predictors relative to the number of observations. They perform variable selection by shrinking some coefficients to exactly zero, effectively removing those predictors from the model. We implement these methods using the "ncvreg" package [29] in R.

We use k-fold cross-validation to select the optimal regularization parameter (lambda) for each method with the folder number 5. Our implementation uses the Area Under the ROC Curve (AUC) as the performance metric:

```
# Function to perform cross-validation using AUC and fit the model
cv_and_fit_auc <- function(X, y, penalty, nfolds = 5) {
  set.seed(123)  # for reproducibility
  # Fit the model using the entire path
  fit <- ncvreg(X, y, family = "binomial", penalty = penalty, standardize = FALSE)
  # Perform k-fold cross-validation
  folds <- sample(rep(1:nfolds, length.out = length(y)))
  aucs <- matrix(0, nrow = nfolds, ncol = length(fit$lambda))

  for (i in 1:nfolds) {
    test_indices <- which(folds == i)
    X_train_cv <- X[-test_indices, ]
    y_train_cv <- y[-test_indices]
    X_test_cv <- X[test_indices, ]
    y_test_cv <- y[test_indices]
    # Fit model on training data
    cv_fit <- ncvreg(X_train_cv, y_train_cv, family = "binomial",
                     penalty = penalty, lambda = fit$lambda,standardize = FALSE)
    # Calculate AUC for each lambda
    for (j in 1:length(cv_fit$lambda)) {
      beta <- coef(cv_fit)[, j]
      # Ensure beta is a numeric vector
      beta <- as.numeric(beta)
      # Initialize vector to store probabilities
      probs <- numeric(nrow(X_test_cv))
      # Loop over each row in X_test_cv
      for (k in 1:nrow(X_test_cv)) {
        # Extract the predictor vector for the current row
        x_row <- X_test_cv[k, , drop = FALSE]
        # Calculate the linear predictor (including intercept)
        linear_predictor <- cbind(1, x_row) %*% beta
        # Calculate the predicted probability
        probs[k] <- 1 / (1 + exp(-linear_predictor))
      }
      # Calculate AUC for the current lambda
      roc_curve <- roc(y_test_cv, probs, quiet = TRUE)
      aucs[i, j] <- auc(roc_curve)
    }
  }
  # Calculate mean AUCs across folds
  mean_aucs <- colMeans(aucs)
  # Determine the optimal lambda
  optimal_lambda_index <- which.max(mean_aucs)
  optimal_lambda <- fit$lambda[optimal_lambda_index]
  # Extract coefficients at the optimal lambda
  coef <- coef(fit)[, optimal_lambda_index]
  return(list(fit = fit, coef = coef, optimal_lambda = optimal_lambda,
              mean_aucs = mean_aucs))
}
```

The cv_and_fit_auc() function performs cross-validation and model fitting to select an optimal regularization parameter (lambda) using AUC as the performance metric. We first fit a model across a range of lambda values using the ncvreg function, which applies a penalized regression technique. The data is split into nfolds subsets for cross-validation, wherein each fold, the function trains a model on the training data and evaluates it on the test data. The AUC for each value of lambda is calculated by looping through the test set and computing predicted probabilities, which are then used to generate the ROC curve. After calculating the mean AUC across all folds for each lambda, the optimal lambda is selected based on the highest average AUC. The function returns the fitted model, the coefficients at the optimal lambda, the selected lambda, and the mean AUCs across folds, providing an effective method for model tuning and evaluation.

We then fit our models with cv_and_fit_auc() function:

```
lasso_results <- cv_and_fit_auc(X_train, y_train, "lasso")
scad_results <- cv_and_fit_auc(X_train, y_train, "SCAD")
mcp_results <- cv_and_fit_auc(X_train, y_train, "MCP")
```

It shall be noted that the computational efficiency of the model fitting depends on the sample size and the dimensionality of the predictors. To speed up the fitting process, some parallel computing techniques are suggested.

4.4 Results

We evaluate our models on the test set using various performance metrics. The function `calculate_test_performance()` evaluates a binary classification model's performance by taking model coefficients (`coef`), test predictors `X_test`, and true labels (`y_test`) as inputs. First, we ensure the data formats and dimensions are correct. Then, we calculate linear predictor values and use the logistic function to compute predicted probabilities for each test instance. These probabilities are converted to binary predictions using a threshold of 0.5. Next, the function computes various performance metrics, including accuracy, precision, recall, and F1 score using a confusion matrix, as well as the AUC using the ROC curve to measure the model's ability to distinguish between classes. Finally, all calculated metrics are returned in a list for easy interpretation, providing a comprehensive evaluation of the model's predictive performance.

```
# Function to calculate test performance metrics
calculate_test_performance <- function(coef, X_test, y_test) {
  # Ensure X_test is a matrix
  X_test <- as.matrix(X_test)
  # Ensure coef is a numeric vector
  coef <- as.numeric(coef)
  # Check that coef length matches the number of columns in X_test + 1
  # (for the intercept)
  if (length(coef) != (ncol(X_test) + 1)) {
    stop("Length of coef must be equal to the number of predictors
         plus one for the intercept.")
  }
  # Initialize vectors to store linear predictors and probabilities
  linear_predictors <- numeric(nrow(X_test))
  probs <- numeric(nrow(X_test))
  # Loop over each row in X_test
  for (i in 1:nrow(X_test)) {
    # Extract the predictor vector
```

```
    x_row <- X_test[i, , drop = FALSE]
    # Calculate the linear predictor (including intercept)
    linear_predictors[i] <- cbind(1, x_row) %*% coef
    # Calculate the predicted probability
    probs[i] <- 1 / (1 + exp(-linear_predictors[i]))
  }
  # Convert probabilities to binary predictions
  predictions <- ifelse(probs > 0.5, 1, 0)
  # Calculate performance metrics
  confusion <- confusionMatrix(factor(predictions), factor(y_test))
  accuracy <- confusion$overall['Accuracy']
  precision <- confusion$byClass['Pos Pred Value']
  recall <- confusion$byClass['Sensitivity']
  f1_score <- 2 * (precision * recall) / (precision + recall)
  # Calculate AUC
  roc_curve <- roc(y_test, probs, quiet = TRUE)
  auc <- auc(roc_curve)
  return(list(accuracy = accuracy, precision = precision, recall = recall,
              f1_score = f1_score, auc = auc))
}
```

Now, we employ the function `calculate_test_performance()` with parameter estimates from three different regularization methods.

```
# Calculate test performance for each method
lasso_performance <- calculate_test_performance(lasso_results$coef, X_test, y_test)
scad_performance <- calculate_test_performance(scad_results$coef, X_test, y_test)
mcp_performance <- calculate_test_performance(mcp_results$coef, X_test, y_test)
```

Let's compare the performance of our three models on the test set:

```
Test Performance Metrics:
-------------------------------------------------------

LASSO:
Accuracy:  0.9234694
Precision: 0.8973384
Recall:    0.7744361
F1 Score:  0.8314607
AUC:       0.9508162

SCAD:
Accuracy:  0.9210204
Precision: 0.8888889
Recall:    0.7744361
F1 Score:  0.8278146
AUC:       0.9502024

MCP:
Accuracy:  0.9210204
Precision: 0.8888889
Recall:    0.7744361
```

```
F1 Score:  0.8278146
AUC:       0.9502024
```

All three models perform well, with accuracy above 92% and AUC above 0.95. LASSO slightly outperforms SCAD and MCP in terms of accuracy and AUC, while all three models have the same recall. The high AUC values indicate that all models are good at distinguishing between certified and non-certified students. Furthermore, one of the key advantages of these regularization methods is their ability to perform variable selection. Let's compare which variables were selected by each method:

```
# Compare selected variables
compare_selection <- function(lasso, scad, mcp) {
  all_vars <- unique(c(names(lasso), names(scad), names(mcp)))
  selection <- data.frame(Variable = all_vars,
          LASSO = ifelse(all_vars %in% names(lasso)[lasso != 0], "Selected",
          ↪ ""),
          SCAD = ifelse(all_vars %in% names(scad)[scad != 0], "Selected", ""),
          MCP = ifelse(all_vars %in% names(mcp)[mcp != 0], "Selected", ""))
  return(selection)
}

variable_selection <- compare_selection(lasso_results$coef,
                                        scad_results$coef,
                                        mcp_results$coef)
# Print the first 5 rows of the variable_selection data frame
print(head(variable_selection, 5))
```

Here, we output the first 5 rows of the variable selection by using these three regularization methods where empty means unselected:

```
Variable                        LASSO     SCAD      MCP
---------------------------------------------------------
(Intercept)                     Selected  Selected  Selected
course_idc_id2                  Selected  Selected
genderm
final_cc_cname_DIBangladesh
final_cc_cname_DIBrazil         Selected  Selected
```

This comparison reveals which predictors are considered important by each method, and the whole results are reported in Table 2.
Moreover, we can obtain the number of selected variables:

```
num_vars <- c(sum(lasso_results$coef != 0) - 1,
              sum(scad_results$coef != 0) - 1,
              sum(mcp_results$coef != 0) - 1)
cat("\nNumber of selected variables:\n")
cat("LASSO:", num_vars[1], "\n")
cat("SCAD:", num_vars[2], "\n")
cat("MCP:", num_vars[3], "\n")
```

Table 2 Results of variable selection with three different methods

	Variable	LASSO	SCAD	MCP
1	(Intercept)	Selected	Selected	Selected
2	course_idc_id2	Selected	Selected	
3	genderm			
4	final_cc_cname_DIBangladesh			
5	final_cc_cname_DIBrazil	Selected	Selected	
6	final_cc_cname_DICanada	Selected		
7	final_cc_cname_DIChina	Selected		
8	final_cc_cname_DIColombia	Selected		
9	final_cc_cname_DIEgypt	Selected		
10	final_cc_cname_DIFrance	Selected		
11	final_cc_cname_DIGermany	Selected		
12	final_cc_cname_DIGreece	Selected		
13	final_cc_cname_DIIndia	Selected	Selected	Selected
14	final_cc_cname_DIIndonesia			
15	final_cc_cname_DIJapan			
16	final_cc_cname_DIMexico			
17	final_cc_cname_DIMorocco			
18	final_cc_cname_DINigeria	Selected	Selected	Selected
19	final_cc_cname_DIOther Africa			
20	final_cc_cname_DIOther East Asia			
21	final_cc_cname_DIOther Europe	Selected		
22	final_cc_cname_DIOther Middle East/Central Asia			
23	final_cc_cname_DIOther North & Central Amer., Caribbean	Selected		
24	final_cc_cname_DIOther Oceania	Selected		
25	final_cc_cname_DIOther South America			
26	final_cc_cname_DIOther South Asia			
27	final_cc_cname_DIPakistan	Selected	Selected	
28	final_cc_cname_DIPhilippines	Selected		
29	final_cc_cname_DIPoland			
30	final_cc_cname_DIPortugal	Selected	Selected	
31	final_cc_cname_DIRussian Federation	Selected		
32	final_cc_cname_DISpain	Selected	Selected	Selected
33	final_cc_cname_DIUkraine	Selected		
34	final_cc_cname_DIUnited Kingdom			
35	final_cc_cname_DIUnited States	Selected		
36	final_cc_cname_DIUnknown/Other	Selected		
37	nevents	Selected	Selected	Selected
38	ndays_act	Selected	Selected	Selected
39	nplay_video	Selected	Selected	Selected
40	explored1	Selected	Selected	

(continued)

Table 2 (continued)

41	nchapters	Selected	Selected	Selected
42	nforum_posts	Selected		
43	course_idc_id2:genderm			
44	genderm:nevents			
45	genderm:ndays_act	Selected		
46	genderm:nplay_video			
47	genderm:explored1			
48	genderm:nchapters	Selected		
49	genderm:nforum_posts			
50	course_idc_id2:nevents	Selected	Selected	Selected
51	course_idc_id2:ndays_act	Selected		
52	course_idc_id2:nplay_video	Selected	Selected	Selected
53	course_idc_id2:explored1	Selected		Selected
54	course_idc_id2:nchapters	Selected		
55	course_idc_id2:nforum_posts	Selected		

This comparison shows how aggressive each method is in terms of variable selection. Typically, LASSO tends to select more predictors than SCAD and MCP, which are often more aggressive in their variable selection.

```
Number of selected variables:
--------------------------------------------------------
LASSO: 36
SCAD:  14
MCP:   10
```

Based on the provided results, the MCP method demonstrates superior performance in variable selection while maintaining competitive predictive accuracy. MCP selected only 10 variables, significantly fewer than LASSO (36) and SCAD (14), indicating its effectiveness in achieving a more parsimonious model. Despite using fewer predictors, MCP matched SCAD's performance across all metrics (Accuracy: 0.9210, Precision: 0.8889, Recall: 0.7744, F1 Score: 0.8278, AUC: 0.9502) and was only marginally behind LASSO in accuracy and AUC. This suggests that MCP effectively identified the most influential predictors, resulting in a simpler model with comparable predictive power. The ability to achieve such performance with fewer variables highlights MCP's strength in balancing model complexity and predictive accuracy, making it the preferable choice for this dataset.

Remark It shall be noted that although the original dataset contained only 10 variables, we constructed an additional 45 interaction terms for our modeling, which increased the dimensionality. As shown in Table 2, all three regularization methods effectively shrank redundant predictors to zero, with the MCP method being particularly efficient, retaining only 10 predictors. This demonstrates the effectiveness of regularization for variable selection in educational data modeling. It is also important to note that regularization methods tend to be even more effective when the number of predictors significantly exceeds the sample size, especially in extremely high-dimensional settings.

4.5 Implications for Online Learning

The Minimax Concave Penalty (MCP) method identified 10 key predictors for online learning outcomes, offering a concise yet comprehensive model. These predictors reflected the critical role of student engagement, including metrics such as total events, active days, video plays, and chapters accessed. The model also highlighted specific country effects of India, Nigeria, and Spain, suggesting that geographical or cultural backgrounds had an impact on learning outcomes. Notably, the interaction terms between course ID and engagement metrics were selected, indicating that the impact of student engagement varies across different courses. This parsimonious selection of variables provides valuable insights into the most significant factors affecting online learning success.

These findings have important implications for online education stakeholders. The results underscore the necessity of promoting active participation and regular interaction with course content, particularly video lectures. The model suggests that tailored approaches may be beneficial for students from specific countries, addressing unique challenges or leveraging strengths associated with these geographical contexts. The variation in engagement effects across courses indicates that customized strategies for course design and student support could be more effective than a uniform approach. Interestingly, the absence of demographic variables like gender in the final model implies that engagement metrics are more predictive of success than these demographic variables in this context. These insights can guide educators, course designers, and platform developers to embed desirable elements (e.g., gamification or animation) into video lectures to improve students' online engagement [30, 31], which may, in turn, help students from diverse geographical regions to complete the course successfully.

5 Conclusion

This chapter has made significant progress in understanding the dynamics of student success in online education, particularly within the edX platform. Through our quantitative analysis employing LASSO, SCAD, and MCP regularization methods, we have successfully identified key factors contributing to obtaining certification in edX courses. The comparative evaluation of these advanced regularization techniques has highlighted their respective strengths in this context, which researchers can easily adapt to construct predictive models for similar contexts in online education.

Our findings have identified the most crucial predictors of student success, emphasizing key aspects of student engagement, course design, and geographical factors. These insights provide practical guidance for course design and student support. By applying these evidence-based results, course designers, instructors, and platform developers can enhance the online learning experience and potentially improve student outcomes and certification rates. This study lays a strong foundation for optimizing digital learning platforms and supporting diverse learners worldwide as online education evolves.

Furthermore, in the field of education, regularization methods can help researchers identify the most influential factors impacting various cognitive (e.g., academic achievement, development of generic skills) and affective learning outcomes (e.g., learning engagement, course satisfaction, motivation in learning), even when dealing with numerous potential predictors. In particular, regularization methods are useful in understanding learners' experiences in online/blended learning via selecting key factors from the large volume of observational data, which records detailed and nuanced learner behaviors in a time-stamped manner.

Data and Code Availability The data and code used in this study are publicly available. They can be accessed at the following GitHub repository: https://github.com/wujrtudou/learning_analytics.git. The repository includes all relevant datasets and the codes used for data preprocessing, model training, and analysis.

References

1. Papadakis S (2023) MOOCs 2012–2022: an overview. Adv Mobile Learn Edu Res 3:682–693
2. Deng R, Benckendorff P, Gannaway D (2019) Progress and new directions for teaching and learning in MOOCs. Comput Educ 129:48–60
3. Knowles MS (1970) The modern practice of adult education: andragogy versus pedagogy. Cambridge Adult Education, Englewood Cliffs
4. Brooker A, Corrin L, De Barba P, Lodge J, Kennedy G (2018) A tale of two MOOCs: how student motivation and participation predict learning outcomes in different MOOCs. Aust J Educ Technol 34.
5. Spector JM, Johnson TE, Young PA (2014) An editorial on research and development in and with educational technology. Educ Technol Res Develop 62:1–12
6. Bingol I, Kursun E, Kayaduman H (2020) Factors for success and course completion in massive open online courses through the lens of participant types. Open Praxis 12:223–239
7. Xiong Y, Li H, Kornhaber ML, Suen HK, Pursel B, Goins DD (2015) Examining the relations among student motivation, engagement, and retention in a MOOC: a structural equation modeling approach. Global Educ Rev 2:23–33
8. Han F, Ellis RA (2023) Self-reported and digital-trace measures of computer science students' self-regulated learning in blended course designs. Educ Inf Technol 28:13253–13268
9. Han F, Ellis RA, Pardo A (2022) The descriptive features and quantitative aspects of students' observed online learning: how are they related to self-reported perceptions and learning outcomes? IEEE Trans Learn Technol 15:32–41
10. Siemens G, Gasevic D (2012) Guest editorial-learning and knowledge analytics. J Educ Technol Soc 15:1–2

11. Jovanovic J, López-Pernas S, Saqr M (2024) Predictive modelling in learning analytics: a machine learning approach in R. In: Learning analytics methods and tutorials: a practical guide using r. Springer Nature Switzerland, Cham, pp 197–229
12. Gray CC, Perkins D (2019) Utilizing early engagement and machine learning to predict student outcomes. Comput Educ 131:22–32
13. Tibshirani R (1996) Regression shrinkage and selection via the LASSO. J R Stat Soc Series B Stat Methodol 58:267–288
14. Fan J, Li R (2001) Variable selection via nonconcave penalized likelihood and its oracle properties. J Am Stat Assoc 96:1348–1360
15. Zhang C-H (2010) Nearly unbiased variable selection under minimax concave penalty. Ann Stat 894–942
16. Hastie T, Tibshirani R, Wainwright M (2015) Statistical learning with sparsity. Monographs Stat Appl Probab 143:8
17. Fu L, Wang Y-G, Wu J (2024) Recent advances in longitudinal data analysis. Model Anal Longitudinal Data 50:173
18. Wang Y-G, Wu J, Hu Z-H, McLachlan GJ (2023) A new algorithm for support vector regression with automatic selection of hyperparameters. Pattern Recogn 133:108989
19. Huber PJ (1973) Robust regression: asymptotics, conjectures and Monte Carlo. Ann Stat 799–821
20. Jiang Y, Wang Y-G, Fu L, Wang X (2019) Robust estimation using modified Huber's functions with new tails. Technometrics 61:111–122
21. Zou H, Li R (2008) One-step sparse estimates in nonconcave penalized likelihood models. Ann Stat 36:1509
22. Hastie T, Tibshirani R, Friedman JH, Friedman JH (2009) The elements of statistical learning: data mining, inference, and prediction. Springer, Berlin
23. Bergstra J, Bengio Y (2012) Random search for hyper-parameter optimization. J Mach Learn Res 13
24. Snoek J, Larochelle H, Adams RP (2012) Practical Bayesian optimization of machine learning algorithms. Adv Neural Inf Process Syst 25
25. Hand DJ (2012) Assessing the performance of classification methods. Int Stat Rev 80:400–414
26. Yuan J, Qiu X, Wu J, Guo J, Li W, Wang Y-G (2024) Integrating behavior analysis with machine learning to predict online learning performance: a scientometric review and empirical study. Preprint. arXiv:240611847
27. Wickham H (2011) ggplot2. Wiley Interdisciplinary Rev Comput Stat 3:180–185
28. Pedersen TL (2019) Package "patchwork." R package http://CRAN R-project org/package= patchwork Cran
29. Breheny P, Breheny MP (2024) Package "ncvreg"
30. Chans GM, Portuguez Castro M (2021) Gamification as a strategy to increase motivation and engagement in higher education chemistry students. Computers 10:132
31. Looyestyn J, Kernot J, Boshoff K, Ryan J, Edney S, Maher C (2017) Does gamification increase engagement with online programs? A systematic review. PloS One 12:e0173403

Explainable Artificial Intelligence in Education: A Tutorial for Identifying the Variables that Matter

Mohammed Saqr and Sonsoles López-Pernas

Abstract Despite the potential of integrating machine learning (ML) and artificial intelligence capabilities into educational settings, there are several challenges that hamper their widespread use and adoption. Among these challenges is that these technologies often function as "opaque-box" models. This lack of transparency can undermine trust, fairness, and accountability. To address this, explainability methods are essential for understanding how models, e.g., predict at-risk students, grade essays, or identify plagiarism. This chapter demonstrates the use of several techniques to explain ML models in educational contexts through a tutorial covering both regression (predicting student grades) and classification (identifying high versus low achievers). We describe how variable-importance measures, partial dependence plots, and accumulated local effects may help educators interpret the outcomes of predictive models, increasing transparency and trust.

Keywords Learning analytics · Explainable artificial intelligence · Machine learning · DALEX

1 Introduction

Machine learning (ML) models and artificial intelligence (AI) applications are increasingly improving their performance and efficiency, often at the expense of becoming more complex [1]. A common drawback of complex models is that they often operate as "opaque boxes," making it difficult to understand the grounds behind their decisions or predictions [1, 2]. This lack of transparency can be problematic, especially in critical applications where trust, fairness, and accountability are indispensable. In the context of education, we must understand how models predict student performance, recommend learning resources, or profile

M. Saqr (✉) · S. López-Pernas
University of Eastern Finland, Joensuu, Finland
e-mail: mohammed.saqr@uef.fi

M. Saqr, S. López-Pernas (eds.), *Advanced Learning Analytics Methods*,
https://doi.org/10.1007/978-3-031-95365-1_6

students according to their behavior [2–4]. Otherwise, without model transparency, there is a risk of reinforcing biases, misinterpreting student actions and abilities, or making unfair decisions about placements and interventions [1]. Explainability ensures that researchers, educators and administrators can trust AI outcomes and make informed decisions [2, 5].

Variable-importance measures hold significant value in ML explainability. They can transform models from opaque to white-box ones where the inner workings are understandable, i.e., we can understand how variables explain the model results [6]. In our case that entails knowing which variables are more associated with students' performance; in other words, what were the learning activities that students did so that they may score higher grades. Further, if we identify the variables that have little or no influence on model predictions, we can simplify our models by excluding irrelevant features. Moreover, model explainability allows us to assess the validity of a model against a theory or hypothesis or discover the factors that may be involved in affecting our outcome beyond our theoretical model. In doing so, explainability helps us to get actionable insights from the data, build informed decisions and expand our theoretical knowledge [6–8].

In this chapter, we provide a tutorial on ML explainability in the context of education using two examples. First, we describe a case study on regression using a well-known example: predicting students' grades. Secondly, we describe the same procedure for classification. It is assumed that the reader is familiar with the R programming language. If that is not the case, it is recommended to refer to previous tutorials on the basics of R [9], data cleaning [10], basic statistics [11], visualization [12], and prediction [14].

2 Explainable AI in Education

The use of explainable AI in educational research is not widespread but it is becoming increasingly prevalent. Studies often employ inherently interpretable algorithms, such as decision trees, to provide global explanations [6]. An example is the work by [15], who used logistic regression to predict factors that would allow to early predict dropout. Since this is not possible for less interpretable algorithms — which often yield higher accuracy—other studies rely on explainable AI to interpret the results of predictive models. For instance, Pellagatti et al. [16] used random forests to predict student dropout. The authors used variable importance to identify which features contributed the most to the predictions, and partial dependence plots to assess how variations in these features influenced the results. Saqr & López-Pernas [2] predicted students' grades using several algorithms and examined the explainability of each model, the most important variables (variable importance) and how they contributed to the predicted grade (partial dependence plot). The authors also used Shapley Additive explanations to visualize the distribution of contribution for each feature to the predictions across students. In a similar study, Tiukhova et al. [17] used Shapley Additive explanations to explore the stability of the most important features that predict student success in a Naïve Bayes predictive model. Please, refer to Chap. 2 in this book for an in-depth explanation of xAI [1].

3 A Tutorial on Global xAI Explainability Using DALEX

In this section, we provide a tutorial on ML explainability in the context of education using two examples. First, we describe a case study on regression using a well-known example: predicting students' grades. Secondly, we describe the same procedure for classification. The general workflow followed in each of the examples is the usual ML pipeline, depicted in Fig. 1, enhanced with the xAI techniques.

3.1 DALEX for Student Grade Prediction

For our first example, we use opaque box model and we show how it can be explained (i.e., white-boxed) using the `DALEX` (Descriptive mAchine Learning EXplanations) approach [18]. We demonstrate the process of building and evaluating a Support Vector Machine (SVM) regression model using the `kernlab` engine and analyzing its performance. The estimation follows what we have learned in the previous chapters e.g., [13, 14] to create a predictive model. Briefly, we start by loading the necessary libraries and importing the student data as we mentioned before, the data is based on the study [19] and have been briefly explained in [13] and represent students' engagement indicators e.g., reading lectures, time spent online etc. This corresponds to the exploratory data analysis (EDA) described in Fig. 1-A.

In the following code, we performed the necessary steps for data preparation (Fig. 1-B). We begin by standardizing the numeric columns to ensure all features are on the same scale. Then, we split the data into training and testing sets for model evaluation (Fig. 1-C). Then, we define the formula that specifies the relationship

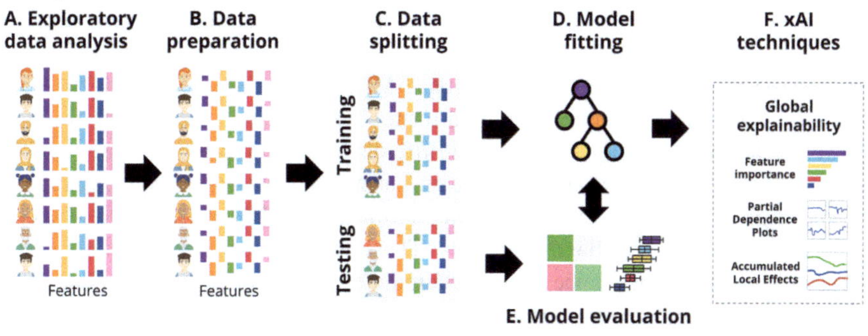

Fig. 1 xAI workflow implemented in the tutorial

between the target variable (`Final_Grade`) and the predictor variables (engagement indicators). Finally, we fit the SVM model using the radial kernel on the training data (Fig. 1-D).

```
set.seed(50)
# Load necessary libraries
library(tidyverse)
library(rsample)
library(e1071)
library(DALEX)
library(rio)

# Import the data
student_data <- import("https://github.com/lamethods/data2/raw/main/lms/lms.csv")

# Standardize the numeric columns
student_data_standardized <- student_data |>
  mutate(across(where(is.numeric), ~scale(.) |> as.vector()))

# Split the data into training and testing sets
data_split <- initial_split(student_data_standardized, prop = 0.8)
train_data <- training(data_split)
test_data <- testing(data_split)

# Define the formula to specify the relationship between the target variable
# and the predictor variables
formula <- Final_Grade ~ Freq_Course_View + Freq_Lecture_View +
  Freq_Forum_Consume + Freq_Forum_Contribute + Regularity_Course_View +
  Regularity_Lecture_View + Regularity_Forum_Consume +
  Regularity_Forum_Contribute + Session_Count + Total_Duration + Active_Days

# Fit the SVM model
svm_fit <- svm(formula,
               data = train_data,
               kernel = "radial")
```

This time, we will not proceed to evaluate the model (Fig. 1-E) using the traditional methods discussed in the previous chapters [13, 14], because `DALEX` has its own evaluation and explanation approach that makes this process easy and straightforward. Instead, we will create an explainer. An `explainer` is the first step in evaluating and explaining the model. Given that models create different output formats, an `explainer` standardizes model outputs making them uniform across different algorithms. In other words, the `explainer` object combines all the necessary components for evaluating a ML model (i.e., the model, the data, and the target variable) into a single uniform object. Whether comparing multiple models or analyzing a single model over different datasets, the `explainer` provides a standardized way to approach these tasks. Further, when we create the `explainer` object, it gives us access to the full potential of the `DALEX` capabilities and access to an extensive suite of diagnostic tools. These tools include variable importance measures, partial dependence plots, accumulated local effects plots, break down plots, SHAP values, *ceteris paribus* profiles, and more.

Creating an `explainer` is a straightforward process. We use the `explain` function with a few arguments. The first argument is the model, where we pass the svm_fit object representing the trained SVM regression model. The data argument specifies the holdout dataset which will be used for evaluating the model (test_data) excluding the target variable Final_Grade. The y argument tells DALEX what is the target variable that we are predicting and we specify that by passing the target variable (test_data$Final_Grade). The next arguments are optional, the label argument assigns a readable label to the explainer object which will be helpful in identifying the model in plots and summaries, especially when working with multiple models. The verbose argument specifies the level of details of the function's output. When verbose = FALSE, the function suppresses output messages.

```
set.seed(50)

library(DALEX)
# Create an explainer with DALEX
explainer_svm <- explain(
  svm_fit,
  data = test_data |> dplyr::select(-Final_Grade),
  y = test_data$Final_Grade,
  label = "SVM",
  verbose = FALSE
)
print(explainer_svm)
```

```
Model label:  SVM
Model class:  svm.formula,svm
Data head  :
  Freq_Course_View Freq_Forum_Consume Freq_Forum_Contribute Freq_Lecture_View
1       -1.4795970         -1.6037982             -1.476917        -0.1035753
2        0.3237727          0.9709293              2.284040         0.9374399
  Regularity_Course_View Regularity_Lecture_View Regularity_Forum_Consume
1            -1.04549094               0.0204884               -0.7421165
2             0.08799424              -0.2511021                0.8384102
  Regularity_Forum_Contribute Session_Count Total_Duration Active_Days
1                  -1.3787562    -1.4299170     -1.8650048  -0.7372311
2                   0.7258669     0.7089812      0.8587155   0.2254762
```

Alternatively (in the alternative code chunk below), given that DALEX works also with tidymodels. The same result can be also achieved using this package, where we can specify the model, create a workflow and then, fit the model as we have seen in the previous chapters [13] and then create an explainer.

```
# Alternative code chunk
set.seed(50)

#Alternative model estimation for SVM

library(tidymodels)
# Define SVM model specification with a radial basis function kernel
```

```
svm_specification <- svm_rbf() |>
  set_mode("regression") |>
  set_engine("kernlab")

# Create the workflow. Combine the model specification with the formula
svm_workflow <- workflow() |>
  add_model(svm_specification) |>
  add_formula(formula)

# Fit the SVM model
svm_fit_tidy <- svm_workflow |>
  parsnip::fit(data = train_data)

# Create an explainer with DALEX
explainer_svm_tidy <- explain(
  svm_fit_tidy,
  data = test_data |> dplyr::select(-Final_Grade),
  y = test_data$Final_Grade,
  label = "SVM",
  verbose = FALSE)

# print(explainer_svm_tidy) # Uncomment to see the output
```

Having created the `explainer`, we can use it with `DALEX` for a range of uses and, most importantly, for model explaining.

3.1.1 Model Evaluation

`DALEX` offers an easy interface for calculating model performance. To evaluate a model, we simply call the function `model_performance()` which takes one argument, the `explainer` object (`explainer_svm`) which we created earlier. The `model_performance()` function calculates the performance metrics for the regression task: Mean Squared Error (MSE), Root Mean Squared Error (RMSE), R-squared (R^2), and Mean Absolute Error (MAE). The result of this function call is assigned to a new object called `model_performance_svm` which we print to display the calculated performance metrics.

As the results show, the model's performance is moderate, with an R^2 of about 0.41 meaning it captures some of the patterns in the data, but there is still a significant amount of unexplained variance. The RMSE of about 0.71 gives us an idea of the typical prediction errors (remember that the grades were scaled). The distribution of residuals suggests that the model's errors are fairly balanced between over-predictions and under-predictions, which is good. However, there are some larger errors at the extremes (as seen in the 0 and 100% percentiles of the residuals).

```
set.seed(50)

# Calculate and print model performance metrics for the SVM model
model_performance_svm <- model_performance(explainer_svm)
print(model_performance_svm)
```

```
Measures for:  regression
mse        : 0.5556946
rmse       : 0.7454493
r2         : 0.4063128
mad        : 0.4442718

Residuals:
         0%         10%         20%         30%         40%         50%
-1.64093297 -0.58085583 -0.41026001 -0.07698939  0.07855870  0.28133263
        60%         70%         80%         90%        100%
 0.34303022  0.51622444  0.76114088  1.06787159  1.81745294
```

To visualize these metrics, we use DALEX to create several types of performance plots (Fig. 2).

The **Lift Curve** demonstrates how well the model ranks predictions compared to random guessing; a higher curve indicates better performance. The curve starts with a brief dip near zero positive rate, indicating a small initial under-performance then rises sharply to a peak lift of about 4.5 at around 10% positive rate. This suggests the model is particularly effective at identifying the top 10% of most likely positive instances. From this peak, the curve gradually declines as the positive rate increases, maintaining a lift above 1 (random selection) for most of its range (i.e., continues to outperform random selection until very high positive rates.). The curve's shape, starting high and slowly descending, is characteristic of a well-performing predictive model. This lift curve suggests that the SVM model has good discriminating power and could be particularly useful in scenarios where identifying the most likely positive cases is important.

The **Residuals Boxplot** visualizes the spread and outliers of residuals, ideally showing minimal outliers and a symmetric distribution around zero. The red dot represents the RMSE, a measure of the model's average prediction error. Its position around 0.7–0.8, suggesting moderate prediction accuracy. Overall, this boxplot indicates that while the SVM model's residuals are somewhat symmetrically distributed around zero, there is a slight tendency for the model to underpredict (positive residuals) more often than overpredict.

The **Residuals Histogram** provides a frequency distribution of residuals, ideally normal around zero. The residuals histogram reveals that most residuals are not very centered around zero. The histogram shows an asymmetrical distribution with more residuals on the positive side, suggesting a potential bias in the model where it tends to underpredict more frequently than overpredicting. The concentration is noticeable most at around 1, suggests a tendency for the model to underpredict grades.

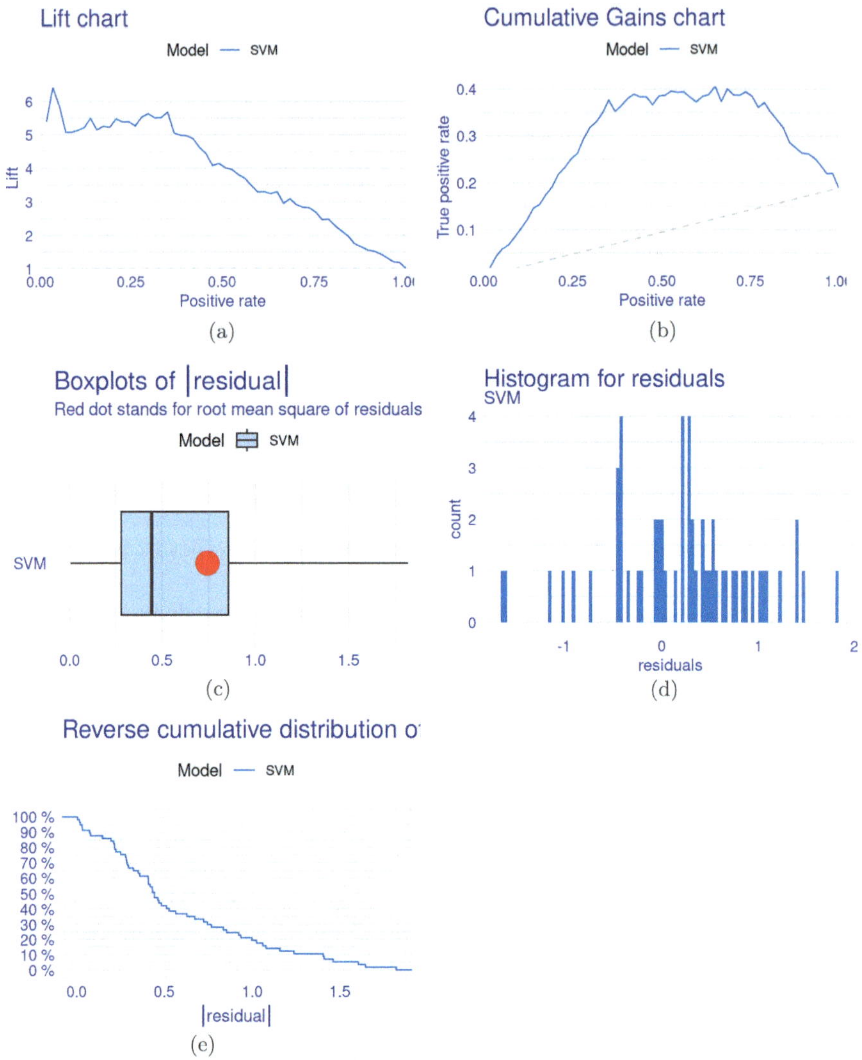

Fig. 2 Performance plots. (**a**) Lift curve—SVM model. (**b**) Cumulative gain chart—SVM model. (**c**) Residuals boxplot—SVM model. (**d**) Residuals histogram—SVM model. (**e**) Empirical cumulative distribution function—SVM model

The **Cumulative Gain Chart** shows the cumulative gain achieved by the model across percentiles, with curves above the diagonal suggesting better predictive power. Finally, the **Empirical Cumulative Distribution Function (ECDF)** plot illustrates the cumulative probability of residuals, helping to understand their distribution across values.

While visualizing model performance using plots can be beneficial, especially for comparing multiple models, interpreting single-model plots might be challenging. However, performance plots may not provide actionable insights to help tune the model or adjust data.

```
set.seed(50)

# Generate various model performance plots
plot(model_performance_svm, geom = "lift")
plot(model_performance_svm, geom = "gain")
plot(model_performance_svm, geom = "boxplot")
plot(model_performance_svm, geom = "histogram")
plot(model_performance_svm, geom = "ecdf")
```

3.1.2 Model Explainability

The `DALEX` way of explaining models is built on Leo Breiman's variable-importance measure for random forests [20]. The method uses permutation to assess the significance of explanatory variables. The algorithm changes a variable's values several times and assesses the resulting impact on model performance. The greater the degradation in performance, the more important the variable is to the prediction task. These methods—while straightforward—offer an effective and robust method for model-agnostic explanations. In doing so, it can be applied to explain any model regardless of the model explainability.

To compute and visualize the importance of our SVM model, we use the `model_parts` function. `model_parts` is the main function for generating global (i.e., model level) model explanations (contrast those to local explanation in [21]). The function accepts several arguments, but only one is required: the explainer object we previously created (`explainer_svm`). In its simplest form, you can call `model_parts(explainer_svm)`, with all other options set to their default values. For instance, `loss_function`, is set by default to `loss_root_mean_square` indicating that the drop in RMSE will be used to measure the model's performance.

We store the results of the `model_parts` in the `vi_svm` object which contains the calculated variable importance and then pass it to the `plot` function. Other arguments may need to be specified, for instance, the `B` argument (defaults to 10) controls the number of permutations used in the calculation, higher numbers (e.g., 500) are always recommended to increase the robustness of the results but may be computationally demanding. Also, the `type` argument can be specified to change the type of feature importance. By default (`"raw"`), provides raw drop loss values or `"ratio"` to return the ratio of drop loss to the full loss in model, while `"difference"` returns drop loss in the full model. As the results show (Fig. 3), the most important variables are *contributing to forums*, *regularity in course view*, and in *contributing to forums* as well as *session count and duration*. These variables are consistent across all methods and they represent cognitive engagement (contributing to problem solving through forums) and time investment in the course, all are consistent with the hypothesis that cognitive engagement is more likely to result in higher achievement.

```r
set.seed(50)

# Default model
vi_svm <- model_parts(explainer_svm)
plot(vi_svm)

# Same as before with all arguments specified
vi_svm <- model_parts(explainer_svm, loss_function = loss_root_mean_square,
                      B = 100)
plot(vi_svm)

# other variations: difference and ratio
vi_svm_difference <- model_parts(explainer_svm, loss_function = loss_root_mean_square,
                                 B = 100, type = "difference")
plot(vi_svm_difference)

vi_svm_ratio <- model_parts(explainer_svm, loss_function = loss_root_mean_square,
                            B = 100, type = "ratio")
plot(vi_svm_ratio)
```

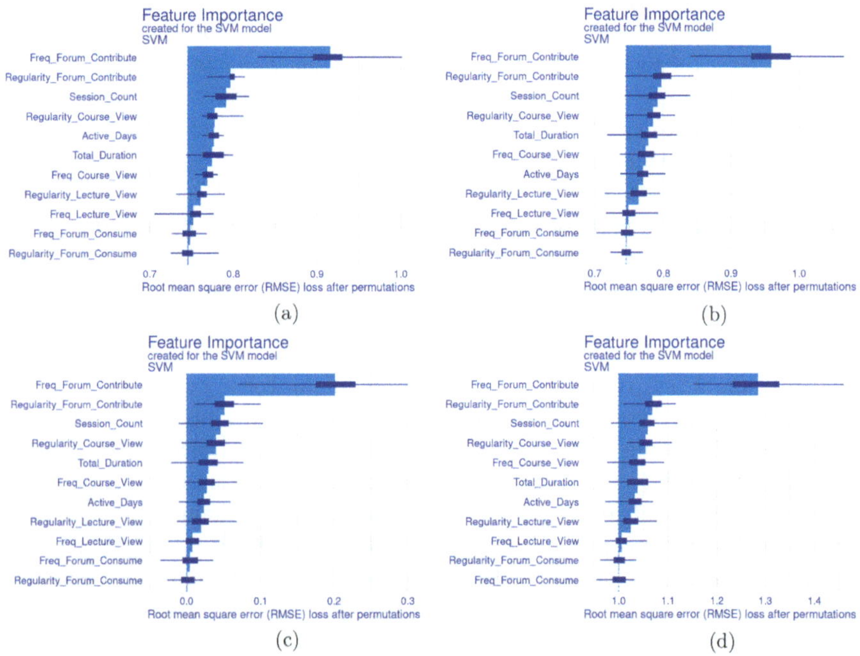

Fig. 3 Global explanations. (**a**) Default model. (**b**) All arguments specified. (**c**) Difference. (**d**) Ratio

3.1.2.1 Creating Custom Loss Functions for DALEX

Whereas the default is loss function in DALEX is based on RMSE, we can also use different metrics to assess the model loss. In the next code chunk, we show how to build two custom functions. These custom functions are used to compute variable importance for the SVM model. In particular, we build two custom loss functions: one for MAE, which calculates the mean of absolute differences between observed and predicted values, and another for MAD, which computes the median of these absolute differences. The results are visualized by creating plots for the MAE-based and MAD-based variable importance (Fig. 4). The choice of loss function can significantly impact the interpretation of variable importance, so researchers need to select a function that aligns with the data and analysis goals.

```
set.seed(50)

# Load necessary libraries
library(DALEX)
library(ggplot2)

# Define the custom MAE loss function
loss_mae <- function(observed, predicted) {
  # Compute the Mean Absolute Error
  mean(abs(observed - predicted))
}

# Define the custom MAD loss function
loss_mad <- function(observed, predicted) {
  # Compute the Median Absolute Deviation
  median(abs(observed - predicted))
  #mean(abs(predicted - median(predicted))) another version
}

# Compute variable importance using the custom MAE loss function
vi_svm_mae <- model_parts(explainer_svm, loss_function = loss_mae)

# Compute variable importance using the custom MAD loss function
vi_svm_mad <- model_parts(explainer_svm, loss_function = loss_mad)

# Plot the results for MAE
plot(vi_svm_mae)

# Plot the results for MAD
plot(vi_svm_mad)
```

3.1.3 Model Selection: Creating and Evaluating Several ML Models

We have previously explored how to estimate, evaluate, and explain a single ML model. However, explainability techniques allow us to make an informed decision

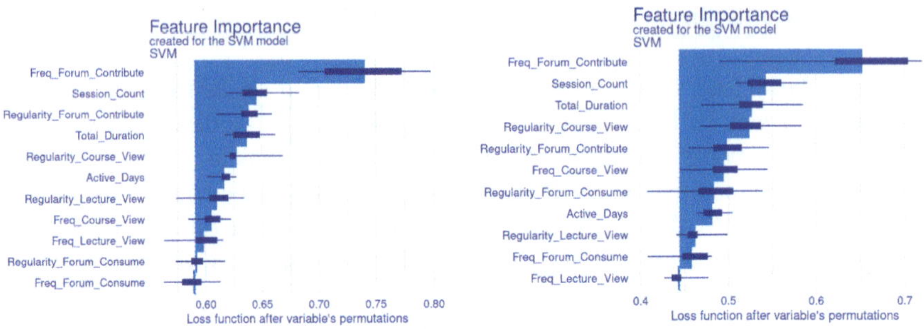

(a) (b)

Fig. 4 Variable Importance with Custom loss functions. (**a**) MAE. (**b**) MAD

about the most suitable model beyond simple performance metrics. To showcase this potential, we will estimate several ML models, evaluate their performance, and then select and explain the best-performing model. Since we have already covered the detailed process of estimating multiple ML models in the previous chapters [13, 14], we will not revisit that process in depth. The following code estimates seven models:

- **Linear Regression:** Fits a linear model to our data using the `lm` function available in base R.
- **Decision Tree:** Implements a decision tree model suitable for regression tasks with the `rpart` package [22].
- **Random Forest:** Builds a forest model using the `randomForest` package [23].
- **Support Vector Machine:** Estimates a support vector machine model using the `svm` function of the `e1071` package [24].
- **K-Nearest Neighbors:** Estimates a k-nearest neighbors model via the `kNN` function [25].
- **Gradient Boosting:** Introduces gradient boosting with the `gbm` package [26].
- **XGBoost:** Trains an efficient and scalable version of gradient boosting with the `xgboost` package [27].

For the XGBoost model, a special approach is required. First, we convert our data into matrix format using `model.matrix`, excluding the first column to avoid the dummy variable. We then create `train_matrix` and `test_matrix` for our training and test sets and extract the target variable into `train_label`. Finally, the `xgboost` function is used to train the model.

```
set.seed(50)

# Load necessary packages
library(tidyverse)
library(caret)
library(rpart)
library(randomForest)
library(e1071)
library(kknn)
library(gbm)
```

```
library(xgboost)
library(DALEX)

# Linear Regression
linear_model <- lm(formula, data = train_data)

# Decision Tree
decision_tree_model <- rpart(formula, data = train_data, method = "anova")

# Random Forest
random_forest_model <- randomForest(formula, data = train_data, ntree = 100)

# Support Vector Machine
svm_model <- svm(formula, data = train_data)

# k-Nearest Neighbors
knn_model <- train(formula, data = train_data, method = "kknn", tuneLength = 5)

# Gradient Boosting Machine
gbm_model <- gbm(formula, data = train_data, distribution = "gaussian",
                 n.trees = 100, interaction.depth = 3, shrinkage = 0.1, cv.folds = 5)

# XGBoost
train_matrix <- model.matrix(formula, data = train_data)[,c(-1)]
test_matrix <- model.matrix(formula, data = test_data)[,c(-1)]

train_label <- train_data$Final_Grade
xgboost_model <- xgboost(data = train_matrix, label = train_label, nrounds = 100,
                         objective = "reg:squarederror", verbose = 0)
```

3.1.4 Multiple Models

3.1.4.1 Creating Multiple Explainers

In our previous step, we constructed several ML models. To understand how each model makes predictions, we need to create an explainer for each model. We do so using the `explain` function to generate these explainers in the same way we did before. We begin by specifying the model, the test data (excluding the target variable `Final_Grade`), and a descriptive label to identify the model when multiple models are visualized or compared.

The below code creates an `explainer_linear` for the linear regression model using the test data without the `Final_Grade` column, and set a label indicating `"Linear Regression"`. We then replicate this process for the decision tree, random forest, SVM, k-nearest neighbors, gradient boosting, and XGBoost models, creating respective explainers for each. For XGBoost, we use the pre-created test data matrix, the XGBoost model, test data matrix. Finally, we use the print function to provide summary information about the explainer and allow us to verify its correct configuration.

```r
set.seed(50)

# Create `DALEX`explainers
explainer_linear <- explain(
  model = linear_model, y = test_data$Final_Grade,
  data = dplyr::select(test_data, -Final_Grade),
  label = "Linear Regression", verbose = FALSE
)

explainer_decision_tree <- explain(
  model = decision_tree_model, y = test_data$Final_Grade,
  data = dplyr::select(test_data, -Final_Grade),
  label = "Decision Tree", verbose = FALSE
)

explainer_random_forest <- explain(
  model = random_forest_model, y = test_data$Final_Grade,
  data = dplyr::select(test_data, -Final_Grade),
  label = "Random Forest", verbose = FALSE
)

explainer_svm <- explain(
  model = svm_model, y = test_data$Final_Grade,
  data = dplyr::select(test_data, -Final_Grade),
  label = "SVM", verbose = FALSE
)

explainer_knn <- explain(
  model = knn_model$finalModel, y = test_data$Final_Grade,
  data = dplyr::select(test_data, -Final_Grade),
  label = "k-NN", verbose = FALSE
)

explainer_gbm <- explain(
  model = gbm_model, y = test_data$Final_Grade,
  data = dplyr::select(test_data, -Final_Grade),
  label = "GBM", verbose = FALSE
)

explainer_xgboost <- explain(
  model = xgboost_model, y = test_data$Final_Grade,
  data = test_matrix, label = "XGBoost", verbose = FALSE
)
```

3.1.4.2 Evaluation of Multiple Models

In the next step, we evaluate the performance of the multiple ML models (Fig. 1-E). For each model of our models (linear regression, decision tree, random forest, support vector machine, k-nearest neighbors, gradient boosting machine, and XGBoost), the code calculates performance metrics using the `model_performance` function. The process is simply a replication of what we have seen for the single model. The `model_performance` function then generates the performance metrics for each model. The performance metrics are then stored in separate variables for each model, allowing for easy comparison and analysis of how well each model performs on the given dataset. We extract and combine these metrics in a table for comparison (Table 1).

```
set.seed(50)

# Calculate performance metrics for each model
performance_linear <- model_performance(explainer_linear)
performance_decision_tree <- model_performance(explainer_decision_tree)
performance_random_forest <- model_performance(explainer_random_forest)
performance_svm <- model_performance(explainer_svm)
performance_knn <- model_performance(explainer_knn)
performance_gbm <- model_performance(explainer_gbm)
performance_xgboost <- model_performance(explainer_xgboost)

# Combine performance metrics into a single table
performance_metrics <- rbind(
  `Linear Regression` = as.array(performance_linear$measure),
  `Decision Tree` = performance_decision_tree$measures,
  `Random Forest` = performance_random_forest$measures,
  `SVM` = performance_svm$measures$r2,
  `KNN` = performance_knn$measures,
  `GBM` = performance_gbm$measures,
  `XGBoost` = performance_xgboost$measures)

# Print the combined dataframe
print(performance_metrics)
```

Table 1 Performance metrics of each model

Model	MSE	RMSE	r2	MAD
Linear regression	0.56	0.75	0.41	0.48
Decision tree	0.63	0.79	0.33	0.48
Random forest	0.51	0.71	0.46	0.42
SVM	0.41	0.41	0.41	0.41
KNN	0.57	0.75	0.39	0.54
GBM	0.49	0.70	0.47	0.42
XGBoost	0.50	0.71	0.46	0.39

The performance metrics for these models reveal interesting insights into their predictive capabilities. XGBoost emerges as the top performer, closely followed by Gradient Boosting Machine (GBM) and Random Forest. This pattern suggests that ensemble methods are good for this dataset, likely due to their ability to capture complex relationships and reduce overfitting. Interestingly, Linear Regression model performs rather well compared to the more complex algorithms like SVM and K-Nearest Neighbors (KNN). This suggests that while there are certainly non-linear aspects to the data, there are also significant linear relationships. The KNN model's performance is somewhat moderate, with a high Mean Absolute Deviation (MAD). This could indicate that the model is sensitive to outliers although we performed feature scaling. It is clear that no single model dominates across all metrics. Moreover, the respectable performance of Linear Regression reminds us that simpler models shouldn't be discounted, especially when interpretability is a concern. The plots of the model's performance confirm these findings (Fig. 5).

```
set.seed(50)

# Create combined plots
# Lift plot
plot(performance_linear, performance_decision_tree, performance_random_forest,
     performance_svm, performance_knn, performance_gbm, performance_xgboost,
    geom = "lift")

# Gain plot
plot(performance_linear, performance_decision_tree, performance_random_forest,
     performance_svm, performance_knn, performance_gbm, performance_xgboost,
     geom = "gain")

# Residuals plot
plot(performance_linear, performance_decision_tree, performance_random_forest,
     performance_svm, performance_knn, performance_gbm, performance_xgboost,
     geom = "boxplot"
)

# ecdf curve
plot(performance_linear, performance_decision_tree, performance_random_forest,
     performance_svm, performance_knn, performance_gbm, performance_xgboost,
     geom = "ecdf"
)

# histogram
plot(performance_linear, performance_decision_tree, performance_random_forest,
     performance_svm, performance_knn, performance_gbm, performance_xgboost,
     geom = "histogram"
)
```

The lift plot shows a lot of overlap, same with the cumulative gain chart. Of course, we see some models perform well across some ranges but nothing consistent. The box plot of residuals show that XGBoost, GBM, and Random Forest have the best performance, with the smallest and most compact boxes, indicating lower and more consistent residuals. Their median lines (the vertical line inside each box) are closest to zero, suggesting more accurate predictions overall. The red dots represent

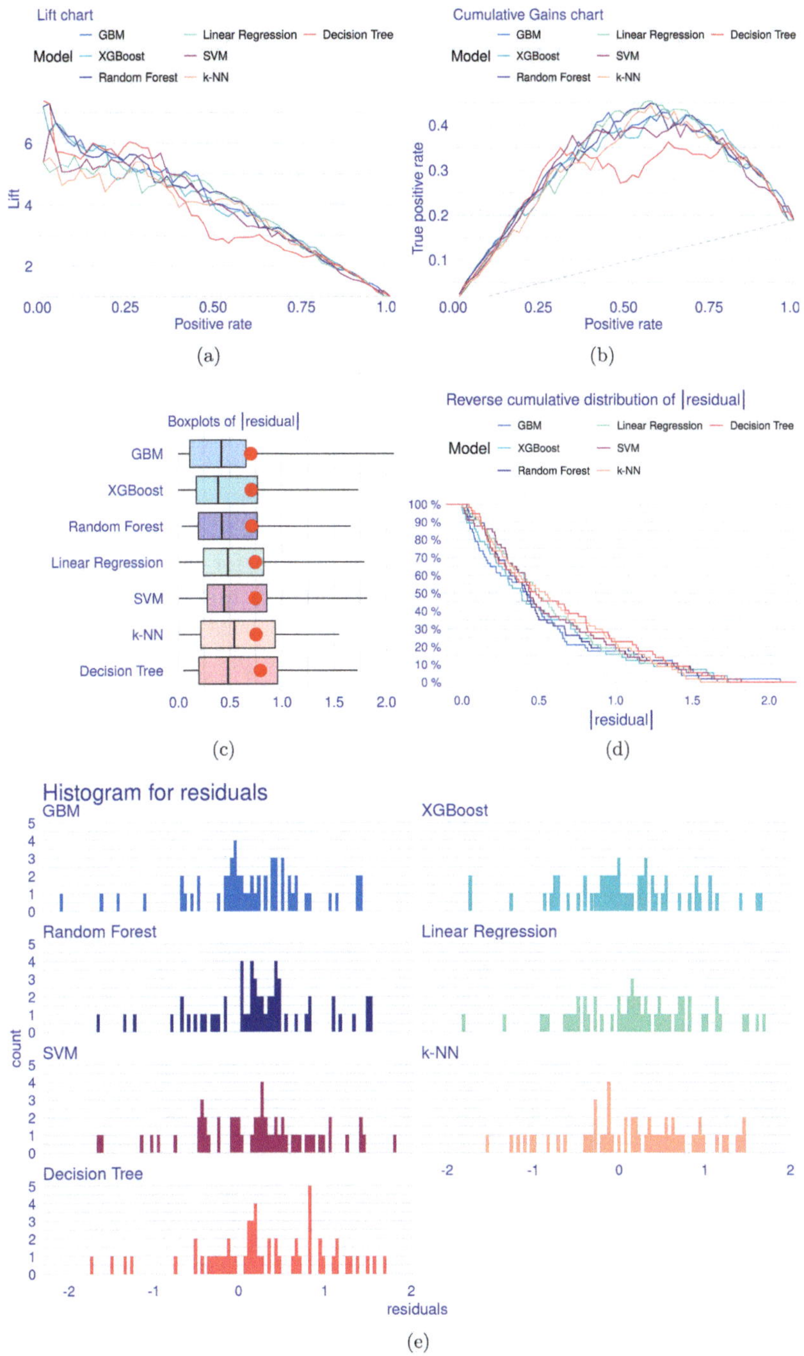

Fig. 5 All models. (**a**) Lift curves. (**b**) Cumulative gain charts. (**c**) Residuals. (**d**) Reverse cumulative distribution. (**e**) Histogram plot

the root mean square of residuals for each model, providing a single metric for comparison. These dots align well with the overall box plot distributions, confirming the relative performance of the models. Overall, this visualization reinforces the strengths of ensemble methods like XGBoost and GBM for this particular dataset. The reverse cumulative distribution plot offers a visual comparison of residual performance across the ML models. The graph's leftward shift indicates better model performance, with XGBoost, GBM, and Random Forest as top performers, aligning with earlier metric analyses. Linear Regression and SVM show similar, middling performance, while k-NN and Decision Tree lag behind. The y-axis represents the percentage of predictions with residuals exceeding the x-axis value which gives an idea about the error distribution.

To conclude, it seems that XGBoost is the best performing model, and so we will demonstrate how to explain it using the same methods we learnt before.

3.1.4.3 Explaining the XGBoost Model

The next step replicates what we have already done for the SVM model. The code computes and visualizes the important variables using the default and the custom loss functions. The variable importance is calculated three times: for RMSE (the default function), MAD, and MAE (the custom loss functions). We use the `model_parts` function with the `explainer_xgboost` to generate variable importance scores. The parameter `B=500` indicates that 500 permutations are performed to assess importance. The results are stored in `vi_xgboost`, `vi_xgboost_mad`, and `vi_xgboost_mae`, corresponding to the different loss functions. Finally, these results are plotted with titles indicating the loss function used (Fig. 6).

```
set.seed(50)

# Compute and plot variable importance using the custom RMSE loss function
vi_xgboost <- model_parts(explainer_xgboost, loss_function = loss_root_mean_square,
                          B=500, type = "ratio")
plot(vi_xgboost)

# Compute and plot variable importance using the custom MAD loss function
vi_xgboost_mad <- model_parts(explainer_xgboost, loss_function = loss_mad,  B=500)
plot(vi_xgboost_mad)

# Compute and plot variable importance using the custom MAE loss function
vi_xgboost_mae <- model_parts(explainer_xgboost, loss_function = loss_mae,B=500)
plot(vi_xgboost_mae)
```

3.1.5 Partial-dependence Profiles

Two types of plots can help explain impact of different engagement variables on students' grades (Fig. 1-F): Partial Dependence Plots (PDPs) and Accumulated Local Effects (ALE) plots. Both plots aim to illustrate how the predicted outcome

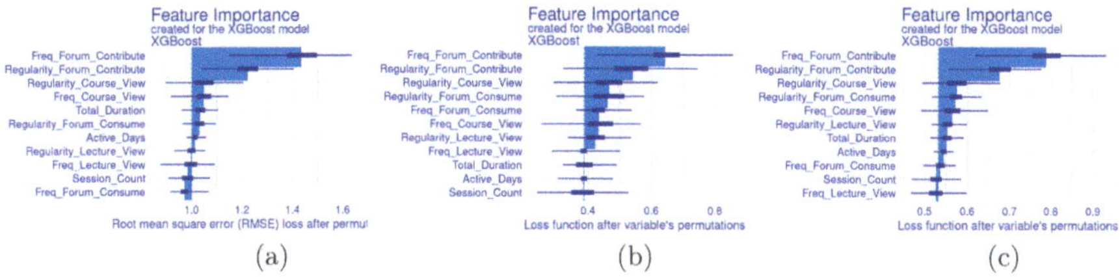

Fig. 6 XGBoost Variable importance using custom loss function. (**a**) RMSE. (**b**) MAD. (**c**) MAE

(grades) changes in response to variations in specific predictor variables (such as engagement indicators), but they do so in slightly different ways, each with its own strengths.

3.1.5.1 Partial Dependence Plots (PDPs)

PDPs are designed to show the average effect of a single feature on the predicted outcome. This is useful for understanding the overall trend or relationship between a specific feature (e.g., frequency of forum contributions) and the outcome (grades). For instance, in the generated PDPs, you might observe that increasing the frequency of forum contributions has a positive impact on the predicted grades, indicating that students who participate more in forums tend to achieve higher grades. Conversely, a flat line in a PDP, such as the one you might see for the "active days" variable, suggests that changes in this feature do not significantly affect the outcome. In general, the regularity of activity often seems more impactful than raw frequency or duration. For instance, regularity of course and lecture viewing show positive associations with grade predictions. The PDPs can be generated by the code using `model_profile` and `plot(pdp_xgboost)` with the test data excluding the outcome variable.

3.1.5.2 Accumulated Local Effects (ALE) Plots

ALE plots offer a more refined alternative to PDPs, particularly when dealing with correlated features (Fig. 7). Unlike PDPs, which show the average effect of a feature, ALE plots focus on the local changes in predictions as the feature varies. This approach accounts for the potential interactions between features and provides a more accurate interpretation when features are correlated. ALE can be created in the same way, but with the option `type = "accumulated"`. This makes ALE plots particularly useful in complex educational datasets where many engagement variables might be interrelated.

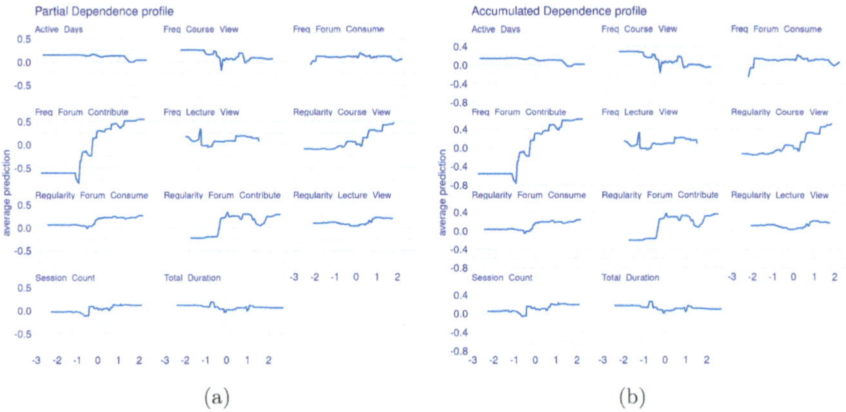

Fig. 7 PDPs vs. ALE plots. (**a**) Partial dependence plots. (**b**) Accumulated local effects (ALE) plots

```
# Partial Dependence Plots
pdp_xgboost <- model_profile(explainer_xgboost,
                  variables = colnames(test_data)[colnames(test_data) != "Final_Grade"])
plot(pdp_xgboost) + labs(subtitle = "")

# Accumulated Local Effects (ALE) Plots
ale_xgboost <- model_profile(explainer_xgboost, type = "accumulated",
                  variables = colnames(test_data)[colnames(test_data) != "Final_Grade"])
plot(ale_xgboost) + labs(subtitle = "")
```

The ALE plots might show that, even after accounting for other correlated behaviors, frequent contributions to forums still have a significant positive effect on grades, reinforcing the importance of this type of engagement. PDPs and ALEs can also be plotted while clustering the data (Fig. 8). For instance the k=3 argument in the code specifies that the feature values should be divided into three clusters for calculating Accumulated Local Effects (ALE). This means that the ALE plots will show how the predicted outcome changes as the feature varies within each of these three clusters, providing a view of the heterogeneity of predictor outcome effect.

```
# Accumulated Local Effects (ALE) Plots
ale_xgboost <- model_profile(explainer_xgboost, type = "accumulated",
                  variables = colnames(test_data)[colnames(test_data) != "Final_Grade"],
                  k = 3)
plot(ale_xgboost) + labs(subtitle = "")
```

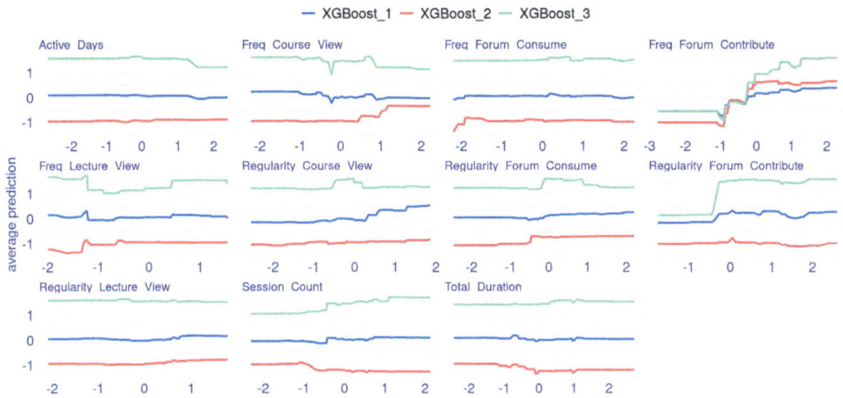

Fig. 8 ALE plot with 3 clusters

3.2 *Explaining a Classification Model*

Explaining a classification model goes almost the same way as regression with the necessary modifications, e.g., specifying the correct target. The next code estimates a Gradient Boost Model (GBM) in the same way we explained before in the classification chapter. The code begins by loading the necessary libraries, setting a random seed is set to ensure consistent results. Data is imported from a CSV file, and a new binary classification variable, `Achievement`, is created based on whether a student's `Final_Grade` is above or below the median (top 50% of achievement level). The `Achievement` variable is then converted to numeric format to be suitable for model training (Fig. 1-B). The dataset is split into training and testing sets (Fig. 1-C), with 80% allocated for training while ensuring class distribution is preserved. Finally, a GBM model is fitted using the engagement predictors to estimate the binary outcome (Fig. 1-D). The model is configured with a Bernoulli distribution for binary classification, 1000 trees, a tree depth of 3, a learning rate of 0.01, and a bagging fraction of 0.7. These values are chosen for demonstration.

```
# Load necessary libraries
library(gbm)
library(rio)
library(dplyr)
library(caret)
library(DALEX)
library(rsample)

# Set seed for reproducibility
set.seed(50)

# Load data
student_data <- import("../student_data.csv")

# Create binary classification target variable
student_data <- student_data |>
  mutate(Achievement = ifelse(Final_Grade > median(Final_Grade),
```

```
                               "High_Achievers", "Low_Achievers")) |>
  mutate(Achievement = factor(Achievement,
                               levels = c("High_Achievers", "Low_Achievers")))

# Convert target to numeric for model fitting
student_data$Achievement_numeric <- as.numeric(student_data$Achievement) - 1

# Split data into training and testing sets
data_split <- initial_split(student_data, prop = 0.8, strata = Achievement_numeric)
train_data <- training(data_split)
test_data <- testing(data_split)

# Fit GBM model
gbm_model <- gbm(
  Achievement_numeric ~  Freq_Course_View + Freq_Lecture_View + Freq_Forum_Consume +
    Freq_Forum_Contribute + Regularity_Course_View + Regularity_Lecture_View +
    Regularity_Forum_Consume + Regularity_Forum_Contribute +
    Session_Count + Total_Duration + Active_Days,
  data = train_data,
  distribution = "bernoulli",  # For binary classification
  n.trees = 1000,
  interaction.depth = 3,
  shrinkage = 0.01,
  bag.fraction = 0.7
)
```

Similar to what we did before, we create an explainer for the fitted Gradient Boosting Model (GBM). The `explain` function takes the estimated `gbm_model` and the test data (excluding the target variable and the final grade). The explainer is labeled as "GBM Model" for identification. Subsequently, the model's performance is assessed using the `model_performance` function, which computes the classification performance metrics (Fig. 1-E) based on the explainer object. The results are then printed out to provide insights into how well the GBM model performs on the test data.

```
# Create `DALEX` explainer
explainer_gbm <- explain(
  model = gbm_model,
  data = test_data[, 1:11],
  y = test_data$Achievement_numeric,
  label = "GBM Model"
)
```

```
Preparation of a new explainer is initiated
  -> model label       :  GBM Model
  -> data              :  58  rows  11  cols
  -> target variable   :  58  values
  -> predict function  :  yhat.gbm  will be used (  default  )
  -> predicted values  :  No value for predict function target column.
                          (  default  )
  -> model_info        :  package gbm , ver. 2.2.2 , task classification
                          (  default  )
  -> predicted values  :  numerical, min =  0.01791613 , mean =  0.5135735 ,
```

```
                              max =  0.9899749
  -> residual function :  difference between y and yhat (  default  )
  -> residuals         :  numerical, min =  -0.9362661 , mean =  -0.01357354 ,
                              max =  0.9794141
  A new explainer has been created!
```

```
# Model performance
model_performance_gbm <- model_performance(explainer_gbm)
print(model_performance_gbm)
```

```
Measures for:  classification
recall     : 0.7931034
precision  : 0.7666667
f1         : 0.779661
accuracy   : 0.7758621
auc        : 0.8133175

Residuals:
         0%          10%          20%          30%          40%          50%
-0.936266132 -0.537579796 -0.400499445 -0.149711621 -0.054921553 -0.003945525
        60%          70%          80%          90%         100%
 0.051416919  0.101556426  0.251119244  0.505825858  0.979414109
```

- The *recall* of 79.31% means that the model correctly identifies 79.31% of the low achievers, showing strong sensitivity in detecting students who might struggle.
- The *precision* of 76.67% shows that, out of all the students predicted to be low achievers, 76.67% were correctly classified, highlighting the model's reliability.
- The *F1-score* of 77.97%, which balances both recall and precision, suggests the model performs well in identifying low achievers overall.
- The *accuracy* of 77.59% indicates that the model correctly classifies 77.59% of all students, whether they are high or low achievers.
- The *AUC* of 81.33% further confirms the model's ability to effectively distinguish between low and high achievers, with a score over 80% generally indicating strong performance.

We can also visualize the results by using the `plot` function with the `model_performance_gbm` object and specifying the desired geometry (`geom`). The resulting plots (Fig. 9) show that the ROC curve is well above the diagonal (the dashed line), indicating good model performance with high true positive rates even at low false positive rates, suggesting strong discriminating power. The residuals boxplot is fairly symmetrical around the zero line, which is favorable. The lift curve starts high (around 2) and gradually decreases, remaining above 1 throughout, meaning the model consistently outperforms random selection in identifying positive cases. The cumulative gain chart rises steeply at first and then flattens, significantly outperforming the diagonal baseline, indicating that the model effectively identifies a large proportion of positive cases early in its ranked predictions. The precision-recall curve shows high precision at low recall values, followed by a gradual decline as recall increases, suggesting a trade-off between

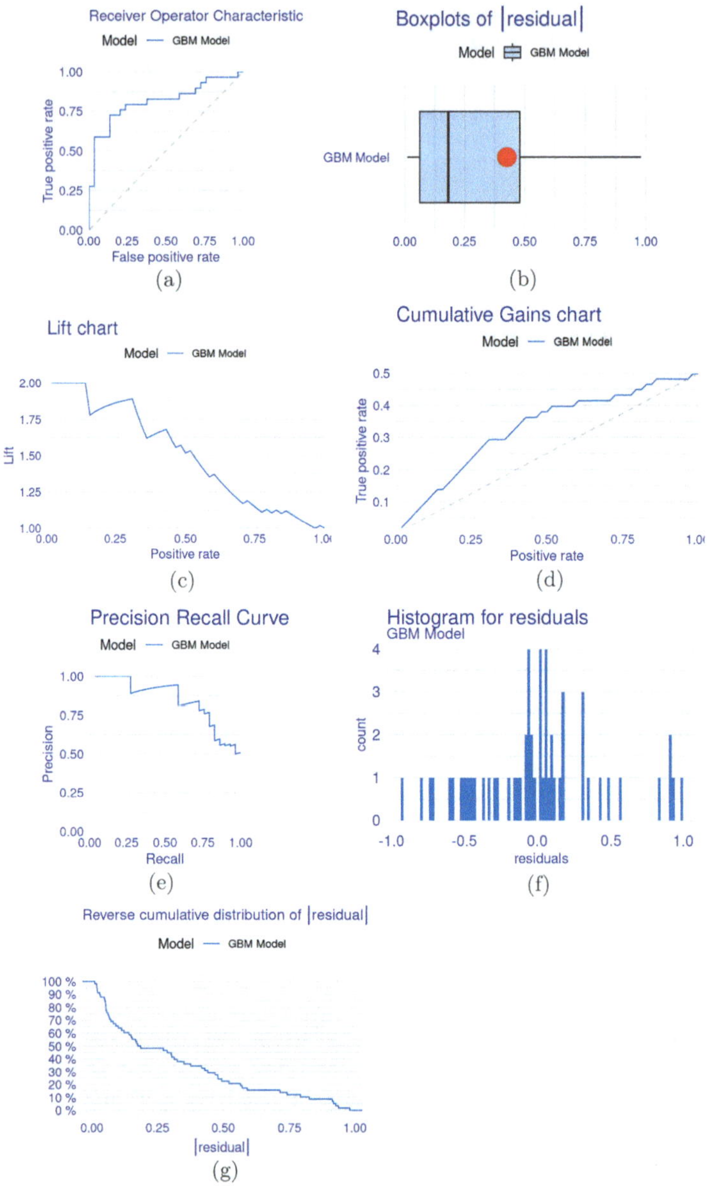

Fig. 9 GBM plots. (**a**) ROC curve—GBM model. (**b**) Residuals Boxplot—GBM model. (**c**) Lift curve—GBM model. (**d**) Cumulative gain chart—GBM model. (**e**) Precision-recall curve—GBM model. (**f**) Residuals histogram—GBM model. (**g**) Empirical cumulative distribution function—GBM model

precision and recall. The residuals histogram distribution is roughly symmetrical and centered near zero, which is desirable, with some outliers. Lastly, the empirical cumulative distribution curve indicates that a large proportion of residuals are concentrated around zero, with relatively few extreme values.

```
# Create individual plots
plot(model_performance_gbm, geom = "roc")
plot(model_performance_gbm, geom = "boxplot")
plot(model_performance_gbm, geom = "lift")
plot(model_performance_gbm, geom = "gain")
plot(model_performance_gbm, geom = "prc")
plot(model_performance_gbm, geom = "histogram")
plot(model_performance_gbm, geom = "ecdf")
```

Explaining the model (Fig. 1-F) is performed in the same way we did before, we use the `model_parts` function with the explainer object `explainer_gbm`. The only parameter we pass is the number of permutations `B = 500` to make the results more robust. The default method computes variable importance using the AUC loss function which measures how much the AUC would decrease if that variable were removed from the model. We see that the forum contribute variables (frequency and regularity) as well as session count are the variables with most importance (Fig. 10).

```
# Variable importance with AUC loss
vi_gbm <- model_parts(explainer_gbm, B = 500)
plot(vi_gbm)
```

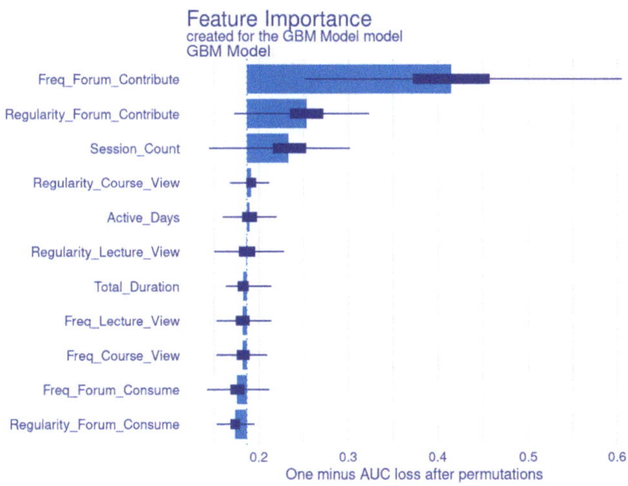

Fig. 10 Variable importance with AUC loss

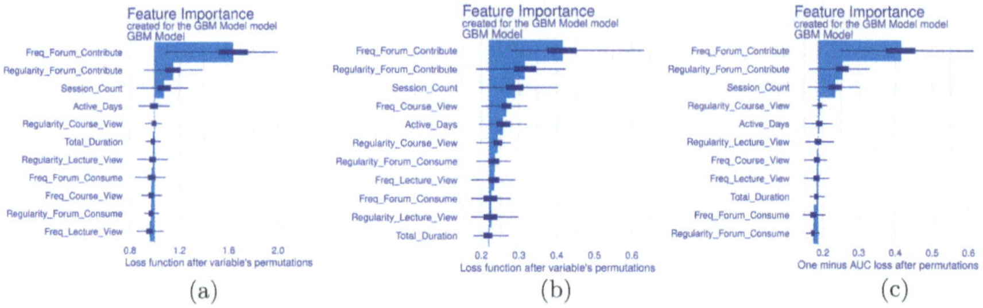

Fig. 11 GBM variable importance with custom loss functions. (**a**) Custom logloss loss. (**b**) F1 score loss. (**c**) Default loss (Logloss)

While standard metrics like AUC are useful, sometimes you may want to evaluate the model using custom loss functions that better align with your specific goals (Fig. 11). Here, we define two custom loss functions: `loss_logloss`, which computes the logarithmic loss (or logloss), and `loss_f1_score`, which is based on the F1 score.

- **Logarithmic Loss (Logloss)**: This function measures how well the predicted probabilities match the actual outcomes. Lower values indicate better performance.
- **F1 Score**: The F1 score balances precision and recall, especially important when dealing with imbalanced datasets. Here, we invert the F1 score so that lower values represent better performance, aligning with the convention used by `model_parts`.

```
# Custom loss functions (same as in the previous code)}
loss_logloss <- function(observed, predicted) {
  -mean(observed * log(predicted) + (1 - observed) * log(1 - predicted))
}

loss_f1_score <- function(observed, predicted) {
  predicted_class <- ifelse(predicted > 0.5, 1, 0)
  TP <- sum(observed == 1 & predicted_class == 1)
  FP <- sum(observed == 0 & predicted_class == 1)
  FN <- sum(observed == 1 & predicted_class == 0)
  F1 <- 2 * TP / (2 * TP + FP + FN)
  return(1 - F1)  # return 1 - F1 to keep lower values better
}

# Variable importance with custom loss functions
vi_gbm_logloss <- model_parts(explainer_gbm, loss_function = loss_logloss,
                              B = 500, type = "ratio")
plot(vi_gbm_logloss)

vi_gbm_f1 <- model_parts(explainer_gbm, loss_function = loss_f1_score, B = 500)
plot(vi_gbm_f1)

vi_gbm_default <- model_parts(explainer_gbm, B = 500)
plot(vi_gbm_default)
```

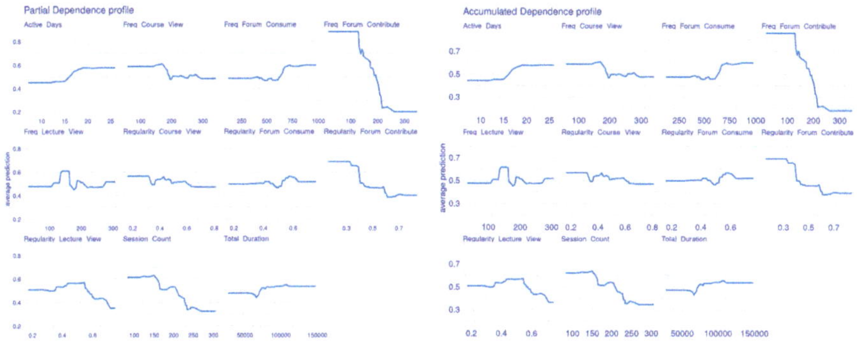

(a) (b)

Fig. 12 GBM Partial dependence and ALE plots. (**a**) PDP. (**b**) ALE plot

We can also generate the PDP and ALE plots for the GBM model (Fig. 12)

```
# Partial Dependence Plots (PDP)
pdp_gbm <- model_profile(
  explainer_gbm,
  variables = colnames(test_data[, 1:11])
)
plot(pdp_gbm) + labs(subtitle = "")

# Accumulated Local Effects (ALE) Plots
ale_gbm <- model_profile(
  explainer_gbm,
  type = "accumulated",
  variables = colnames(test_data[, 1:11])
)
plot(ale_gbm) + labs(subtitle = "")
```

4 Discussion and Conclusions

In this chapter, we went through a tutorial of xAI. In particular, we demonstrated various techniques for making the inner workings of machine learning models clear and transparent. In doing so, we can understand and reveal the variables that significantly influence student outcomes, whether predicting grades or identifying at-risk learners. We demonstrated several variable importance measures, partial dependence plots, and accumulated local effects. This transparency is very important for ensuring fairness, accountability, and the effective implementation of AI in educational settings, empowering stakeholders to make informed decisions and build a deeper understanding of the underlying learning processes. Most importantly, with transparency comes trust and reliance. If the stakeholders understand the inner workings of an algorithm, they will be fast to implement it and adopt such methods in their institutions [28].

However, while xAI provides a powerful lens for understanding model behavior, it is essential to acknowledge its inherent limitations and the important role of human interpretation [2, 5, 29, 30]. As our results suggest, even detailed explanations of individual algorithmic decisions might fall short when it comes to achieving genuine personalization or providing truly individualized support. Furthermore, instances where algorithms rely on seemingly "wrong predictors" to generate explanations highlight the potential pitfalls of a purely data-driven approach [2]. This underscores that while AI can offer valuable insights and identify patterns, it cannot and must not operate in isolation.

As such we can say that the effectiveness of AI in education hinges on a collaborative approach where human expertise and contextual understanding complement the analytical power of algorithms [2]. The ability to interpret explanations, identify potential biases or limitations, and integrate domain knowledge remains paramount, for a more detailed discussion please refer to the second chapter of this book [1].

References

1. López-Pernas S, Oliveira E, Song Y, Saqr M (2025) AI, explainable AI and evaluative AI: an introduction to informed data-driven decision-making in education. In: Saqr M, López-Pernas S (eds) Advanced learning analytics methods: AI, precision and complexity. Springer Nature Switzerland, Cham
2. Saqr M, López-Pernas S (2024) Why explainable AI may not be enough: predictions and mispredictions in decision making in education. Smart Learn Environ. https://doi.org/10.1186/s40561-024-00343-4
3. Sghir N, Adadi A, Lahmer M (2023) Recent advances in predictive learning analytics: a decade systematic review (2012–2022). Educ Inf Technol 28:8299–8333
4. Na KS, Tasir Z (2017) A systematic review of learning analytics intervention contributing to student success in online learning. In: 2017 international conference on learning and teaching in computing and engineering (LaTICE). IEEE, pp 62–68
5. Saqr M, Cheng R, López-Pernas S, Beck ED (2024) Idiographic artificial intelligence to explain students' self-regulation: toward precision education. Learn Indiv Differ 114:102499. https://doi.org/10.1016/j.lindif.2024.102499
6. Khosravi H, Shum SB, Chen G, Conati C, Tsai Y-S, Kay J, Knight S, Martinez-Maldonado R, Sadiq S, Gašević D (2022) Explainable artificial intelligence in education. Comput Educ Artif Intell 3:100074
7. Alyahyan E, Düştegör D (2020) Predicting academic success in higher education: literature review and best practices. Int J Educ Technol Higher Educ 17:3
8. Biecek P, Burzykowski T (2021) Explanatory model analysis: Explore, explain, and examine predictive models, 1st edn. Chapman & Hall/CRC, Philadelphia. https://www.routledge.com/Explanatory-Model-Analysis-Explore-Explain-and-Examine-Predictive-Models/Biecek-Burzykowski/p/book/9780367693923
9. Tikka S, Kopra J, Heinäniemi M, López-Pernas S, Saqr M (2024) Getting started with R for education research. In: Saqr M, López-Pernas S (eds) Learning analytics methods and tutorials: a practical guide using r. Springer, Berlin

10. Kopra J, Tikka S, Heinäniemi M, López-Pernas S, Saqr M (2024) An R approach to data cleaning and wrangling for education research. In: Saqr M, López-Pernas S (eds) Learning analytics methods and tutorials: a practical guide using r. Springer, Berlin
11. Tikka S, Kopra J, Heinäniemi M, López-Pernas S, Saqr M (2024) Introductory statistics with R for educational researchers. In: Saqr M, López-Pernas S (eds) Learning analytics methods and tutorials: a practical guide using r. Springer, Berlin
12. López-Pernas S, Misiejuk K, Tikka S, Kopra J, Heinäniemi M, Saqr M (2024) Visualizing and reporting educational data with R. In: Learning analytics methods and tutorials. Springer Nature Switzerland, Cham, pp 151–194. https://doi.org/10.1007/978-3-031-54464-4/_6
13. Saqr M, Misiejuk K, Tikka S, López-Pernas S (2025) Artificial intelligence: Using machine learning to predict students' performance. In: Saqr M, López-Pernas S (eds) Advanced learning analytics methods: AI, precision and complexity. Springer Nature Switzerland, Cham
14. Saqr M, Misiejuk K, Tikka S, López-Pernas S (2025) Artificial intelligence: using machine learning to classify students and predict low achievers. In: Saqr M, López-Pernas S (eds) Advanced learning analytics methods: AI, precision and complexity. Springer Nature Switzerland, Cham
15. López-Pernas S, Kleimola R, Väisänen S, Hirsto L (2022) Early detection of dropout factors in vocational education: a large-scale case study from finland. In: Proceedings of the 1st finnish learning analytics and artificial intelligence in education conference (FLAIEC22)
16. Pellagatti M, Masci C, Ieva F, Paganoni AM (2021) Generalized mixed-effects random forest: a flexible approach to predict university student dropout. Stat Anal Data Mining 14:241–257
17. Tiukhova E, Vemuri P, Flores NL, Islind AS, Óskarsdóttir M, Poelmans S, Baesens B, Snoeck M (2024) Explainable learning analytics: assessing the stability of student success prediction models by means of explainable AI. Decision Support Syst 182:114229
18. Biecek P (2018) DALEX: Explainers for complex predictive models in r. J Mach Learn Res 19:1–5. https://jmlr.org/papers/v19/18-416.html
19. Saqr M, Jovanovic J, Viberg O, Gašević D (2022) Is there order in the mess? A single paper meta-analysis approach to identification of predictors of success in learning analytics. Stud Higher Educ 47:2370–2391. https://doi.org/10.1080/03075079.2022.2061450
20. Breiman L (2001) Random forests. Mach Learn 45:5–32. https://doi.org/10.1023/A:1010933404324
21. Saqr M, López-Pernas S (2025) Individualized explainable artificial intelligence: a tutorial for identifying local and individual predictions. In: Saqr M, López-Pernas S (eds) Advanced learning analytics methods: AI, precision and complexity. Springer Nature Switzerland, Cham
22. Therneau T, Atkinson B (2022) Rpart: Recursive partitioning and regression trees. https://CRAN.R-project.org/package=rpart
23. Liaw A, Wiener M (2002) Classification and regression by randomForest. R News 2:18–22. https://CRAN.R-project.org/doc/Rnews/
24. Meyer D, Dimitriadou E, Hornik K, Weingessel A, Leisch F (2023) e1071: Misc functions of the department of statistics, probability theory group (formerly: E1071), TU wien. https://CRAN.R-project.org/package=e1071
25. Schliep K, Hechenbichler K (2016) Kknn: Weighted k-nearest neighbors. https://CRAN.R-project.org/package=kknn
26. Ridgeway G, Developers G (2024) Gbm: Generalized boosted regression models. https://CRAN.R-project.org/package=gbm
27. Chen T, He T, Benesty M, Khotilovich V, Tang Y, Cho H, Chen K, Mitchell R, Cano I, Zhou T, Li M, Xie J, Lin M, Geng Y, Li Y, Yuan J (2024) Xgboost: Extreme gradient boosting. https://CRAN.R-project.org/package=xgboost
28. Mustafa MY, Tlili A, Lampropoulos G, Huang R, Jandrić P, Zhao J, Salha S, Xu L, Panda S, Kinshuk, López-Pernas S, Saqr M (2024) A systematic review of literature reviews on artificial intelligence in education (AIED): a roadmap to a future research agenda. Smart Learn Environ 11. https://doi.org/10.1186/s40561-024-00350-5

29. Ribeiro MT, Singh S, Guestrin C (2016) Why should I trust you?: Explaining the predictions of any classifier. In: Proceedings of the 22nd ACM SIGKDD international conference on knowledge discovery and data mining. ACM, New York, NY, USA. https://doi.org/10.1145/2939672.2939778
30. Jovanovic J, López-Pernas S, Saqr M (2024) Predictive modelling in learning analytics: a machine learning approach in r. In: Learning analytics methods and tutorials: a practical guide using r. Springer Nature Switzerland Cham, pp 197–229

Individualized Explainable Artificial Intelligence: A Tutorial For Identifying Local and Individual Predictions

Mohammed Saqr and Sonsoles López-Pernas

Abstract In the context of explainable artificial intelligence, global explanations provide aggregate insights about the performance and factors influencing a machine learning model, as we have seen in the previous chapter. However, local explanations are needed to understand the factors influencing a specific decision that affect an individual. For instance, local explanations could help teachers understand why a certain student was flagged as at-risk of dropping out a course, fostering transparency and trust. This chapter highlights the need for local explanations in educational contexts and explores three key techniques: Break Down plots, SHAP (SHapley Additive exPlanations), and LIME (Local Interpretable Model-agnostic Explanations)—implemented. Practical examples demonstrate how these methods address prediction interpretability, identify critical features, and support targeted interventions

Keywords Learning analytics · Explainable artificial intelligence · Machine learning · SHAP · LIME · DALEX

1 Introduction to Local Explanations

One of the key objectives of learning analytics is to predict—and understand—students' performance to identify which indicators of students' behavior can be used to predict student success [1]. Understanding these indicators not only helps in making accurate predictions but, more importantly, allows educators and administrators to explain the decision-making process of predictive algorithms. For instance, if a model predicts a student's final grade, we need to know which factors (such as frequency of course views, participation in forums, or regularity of lecture viewing) influenced that prediction. If a model predicts a low grade for a student

M. Saqr (✉) · S. López-Pernas
University of Eastern Finland, Joensuu, Finland
e-mail: mohammed.saqr@uef.fi

M. Saqr, S. López-Pernas (eds.), *Advanced Learning Analytics Methods*,
https://doi.org/10.1007/978-3-031-95365-1_7

despite high engagement, we need to understand if the timing or irregularity of their engagement might be the reason. This transparency is a prerequisite for trust and for making informed interventions to support students [2–4] .

Transparency and trust are even more important in scenarios where decisions need to be justifiable on a case-by-case basis or where model predictions have consequences such as issuing an intervention or changing a curriculum [2, 4]. To have these case-by-case explanations, we would need local explanations (single instance explanations). Unlike global explanations, which provide an aggregate overview of the model across the entire dataset (see Chapter [3]), local explanations explain why the model made a specific prediction for a particular student.

In the previous chapter [3], we have seen how we can use explainable artificial intelligence (xAI) to obtain global explanations from predictive models using the `DALEX` R package. Though global explanations give an idea about what works for most students (i.e., feature importance), they do not accurately capture the individual differences or the variability among students [5]. In the present chapter, we go one step further and leverage the same technology to obtain local explanations. To be able to follow the tutorial presented in this chapter, it is assumed that the reader is familiar with the R programming language. If that is not the case, it is recommended to refer to previous tutorials on the basics of R [6], data cleaning [7], basic statistics [8], visualization [9], and prediction [10–12].

2 Main Methods for Local Explanations

Several methods exist for local explanations, each offering a different perspective on how the model arrives at a particular prediction. We will focus here on the options offered by the `DALEX` package due to its simplicity and our previous experience with it. More importantly, DALEX combines several methods and offers them within its unified framework. Further, `DALEX` is model-agnostic, meaning it can be applied to any type of predictive model. We will discuss three methods here [13], namely, **Break Down**, **SHAP (SHapley Additive exPlanations)**, and **LIME (Local Interpretable Model-agnostic Explanations)**. We will go over the three methods briefly here, but further detailed explanation is given later in context. Also, refer to [2] for an overview of xAI.

Break Down Method decomposes a prediction into contributions from individual features, starting from a baseline (often the average prediction). Break Down is simple to interpret, making it easy for educators who need to quickly understand why a particular grade was predicted. However, its simplicity comes with some downsides. Most importantly, it assumes feature contributions are independent, which may not always be the case, especially when student behaviors are interrelated in complex ways.

SHAP values offer a more robust approach that is insensitive to variable order and accounts for variable interactions. Based on game theory, SHAP values fairly distribute the prediction's deviation from the baseline across all features [14]. This is particularly useful when indicators interact in ways that significantly impact the prediction. For example, the combination of frequent forum contributions and regular lecture viewing might have a different impact than these behaviors considered independently. However, SHAP is computationally intensive, which can be a drawback for large datasets or complex models.

LIME is another popular method that approximates the examined models with a simpler, interpretable model to decipher their complex decisions [15]. LIME does so by perturbing the features slightly and observing the changes in predictions. However, LIME's explanations can be sensitive to the choice of parameters, and the simple model it uses might not fully capture the complexity of the original model.

3 Local Explanations in the Existing Research

Though local explainability is a powerful tool to make ML predictions understandable, this technique has not been yet capitalized on in educational research and practice. Very few instances have been reported in which local explainability has been operationalized. For instance, Jang et al. [16] analyzed data from seven Korean courses and employed SHapley Additive exPlanation (SHAP) to visualize two instances where students were identified as at-risk. Their findings highlighted low engagement with homework as the most critical factor in both cases. Similarly, Nagy and Molontay [17] leveraged SHAP to explain instances of student dropout through detailed case studies. Other researchers have used local explanations to focus on specific subsets of students. For example, Lin et al. [18] examined students who improved versus those who did not, while Adnan et al. [19] concentrated on students who failed. Saqr and López-Pernas [4] used local explainability to diagnose why students that have been erroneously predicted as low achievers or high achievers were mispredicted. These studies demonstrate the utility of local explanations but remain limited to isolated cases of explainability. Notably, the full potential of single-instance explanations has yet to be realized. Current research focuses primarily on identifying factors associated with specific outcomes but stops short of addressing mispredictions or offering actionable insights for improvement. Developing methods that use algorithms to suggest interventions to help students achieve their desired outcomes represents a critical gap in the field.

4 Local Explanations

In this section, we provide a tutorial on local explainability in the context of education using two examples. First, we describe a case study on regression using a well-known example: predicting students' grades. Secondly, we describe the same

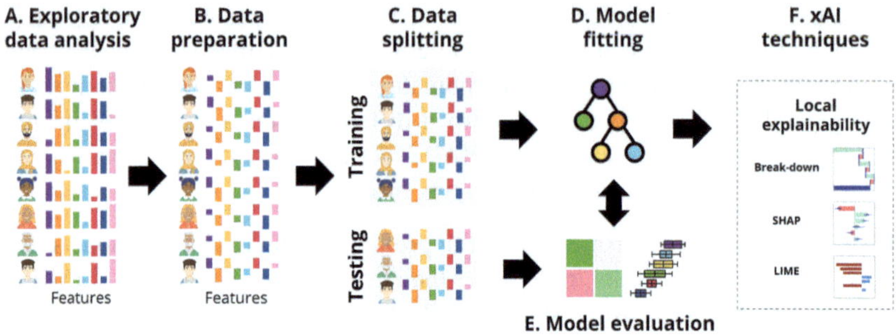

Fig. 1 xAI workflow implemented in the tutorial

procedure for classification. The general workflow followed in each of the examples is the usual ML pipeline, depicted in Fig. 1, that we have followed in previous chapters [11, 12, 20] enhanced with the local xAI techniques.
First, we load the necessary libraries and we import the data. The data contains engagement indicators based on the data in this paper [21] and discussed briefly in [11].

4.1 Building a Machine Learning Model for Explanation

The initial steps involve data importing, preparation and creating a machine learning model. In this example we will build a Support Vector Machine (SVM) model and explain a single instance using different methods. First, to ensure that the analysis is reproducible, we start by setting a seed using `set.seed(265)`, which guarantees that the random processes, such as data splitting, produce the same results every time the code is run.

```
# Set a seed for reproducibility of results
set.seed(265)
```

Next, the dataset is pre-processed by centering the predictor variables (Fig. 1–B). Centering involves subtracting the mean from each value so that the variables have a mean of zero. In this case, variables (e.g., `Freq_Course_View`, `Freq_Lecture_View`, and `Active_Days`) so the mean=0 corresponding to the average of all students, negative values indicate below the mean (lower activity) and positive values (higher activities). Thus, centering will make it easier for us to interpret the values by comparing to the reference average.

```
# Center the selected predictor variables in the dataset. This involves
# subtracting the mean from each value, making the mean of each variable zero
student_data_centered <- student_data |>
  mutate(across(Freq_Course_View:Active_Days, ~scale(., scale = FALSE) |> as.vector()))
```

After preparing the data, we split it into training and testing (Fig. 1-C) sets using `initial_split`. The dataset is divided into two parts: 80% of the data (`train_data`) will be used to train the model, and the remaining 20% (`test_data`) will be reserved for testing and validation.

```
# Split the centered dataset into training (80%) and testing (20%) sets
data_split <- initial_split(student_data_centered, prop = 0.8)
train_data <- training(data_split)  # Extract the training set
test_data <- testing(data_split)    # Extract the testing set
```

We then define the model formula where `Final_Grade` is the target variable to be predicted, and the predictors e.g., `Freq_Course_View`, `Freq_Lecture_View`, `Session_Count`, and `Active_Days`. Finally, we fit the SVM model (Fig. 1-D) using the radial kernel, which is a popular choice for capturing non-linear relationships in the data. We use the `svm()` function from the `e1071` package to train the model using the specified formula.

```
# Define the formula specifying the relationship between the target variable
#  (Final_Grade) and the predictors (Freq_Course_View, Freq_Lecture_View, etc.)
formula <- Final_Grade ~ Freq_Course_View + Freq_Lecture_View + Freq_Forum_Consume +
  Freq_Forum_Contribute + Regularity_Course_View +
  Regularity_Lecture_View + Regularity_Forum_Consume +
  Regularity_Forum_Contribute + Session_Count +
  Total_Duration + Active_Days

# Fit a Support Vector Machine (SVM) model using the radial kernel
# The model is trained on the training dataset with the specified formula
svm_fit_radial <- svm(formula, data = train_data, kernel = "radial")
```

After fitting the SVM model (`svm_fit_radial`) on the training data, the next step would be to evaluate its performance (Fig. 1-E) and use global explainability techniques to understand the model's general functioning. Since we covered that in Chapter 5 [20], we will move on to using local explainability techniques (Fig. 1-F) to interpret and understand the model's predictions. The `DALEX` package provides tools for model explanation, and to use `DALEX` [22], we need to create an explainer. As we have learned in the previous chapter [3], the explainer object encapsulates the model, the data it was trained on, and the corresponding target variable. Once the explainer is created, we can use it to generate local explanations with the Break Down method, SHAP values, or LIME.

The explainer is created using the `explain()` function, which takes several arguments: the `model` argument is set to `svm_fit_radial`, which is the SVM model trained using the radial kernel; the `data` argument is the test dataset, excluding the target variable `Final_Grade` (achieved by using `test_data[,-12]`); the `y` argument specifies the true values of the target variable `Final_Grade`, allowing `DALEX` to compare the model's predictions with the actual outcomes, and the `label` argument provides a name for the model in the explainer, useful for comparing multiple models, in this case labeled as "SVM Radial". With the explainer set up, we can now use various `DALEX` functions to assess the behavior of the SVM model, such as understanding which features are most important or how the model behaves on individual predictions.

```
# Create an explainer for the SVM model using the DALEX package
explainer_svm <- explain(
  model = svm_fit_radial,    # The trained SVM model with a radial kernel
  data = test_data[,-12],    # Test data excluding the target variable (Final_Grade)
  y = test_data$Final_Grade, # True values of the target variable for comparison
  label = "SVM Radial"       # Label for the model in the explainer
)
```

```
Preparation of a new explainer is initiated
  -> model label       :  SVM Radial
  -> data              :  57  rows  11  cols
  -> target variable   :  57  values
  -> predict function  :  yhat.svm  will be used (  default  )
  -> predicted values  :  No value for predict function target column. (  default  )
  -> model_info        :  package e1071 , ver. 1.7.16 , task regression (  default  )
  -> predicted values  :  numerical, min =  55.05479 , mean =  68.49092 , max =  80.54996
  -> residual function :  difference between y and yhat (  default  )
  -> residuals         :  numerical, min =  -21.71365 , mean =  -1.440045 , max =  13.08022
  A new explainer has been created!
```

```
# Check the explainer
print(explainer_svm)
```

```
Model label:  SVM Radial
Model class:  svm.formula,svm
Data head  :
  Freq_Course_View Freq_Forum_Consume Freq_Forum_Contribute Freq_Lecture_View
1         8.210526           5.449123              8.926316          28.12982
2        35.210526          99.449123            -15.073684          63.12982
  Regularity_Course_View Regularity_Lecture_View Regularity_Forum_Consume
1           -0.007578947              0.08301754              -0.09329825
2            0.112421053              0.21301754               0.04670175
  Regularity_Forum_Contribute Session_Count Total_Duration Active_Days
1                 -0.02273684      16.45614       8545.021   0.9368421
2                 -0.08273684      60.45614       9076.021   5.9368421
```

4.2 Breakdown Profiles

Breakdown profiles provide a clear and simple way to examine how each feature contributes to a model's prediction. The process begins with a baseline value (often the mean prediction of the average outcome, which serves as a reference point). The contribution of each feature is then calculated by assessing how much the prediction changes when that feature is added to the baseline, assuming all other features remain constant. For example, the plot starts with the average grade and then shows how much a variable—such as the frequency of course views or the number of active days—affects the prediction to be higher or lower. Since the process is sequential, the order in which features affect the results. If the model first adds the contribution of one feature, then another, it could lead to overestimation or underestimation of the impact of the feature that follows. Therefore, it is important to interpret the breakdown results with this in mind, especially in models where features might interact.

Whereas breakdown profiles are simple and easy to understand, they have some disadvantages. Breakdown methods typically assume that features contribute independently to the prediction, which might not hold true in complex models where interactions between features are common. It is not hard to assume that session count and active days are related or have some interactions. This simplification can lead to erroneous interpretations as the combined effect of interacting features may not be accurately represented by summing individual contributions. Another disadvantage is the issue of order dependence which we already mentioned. Different orders of predictors can produce different breakdowns, making it challenging to ascertain the true impact of each feature. This can complicate the interpretation, especially in models with highly correlated features, where the order in which features are considered may significantly alter the results presented by the plot.

To visualize the results of the breakdown process, we will use waterfall plots. In waterfall plots, each bar represents the contribution of a feature showing how it increases or decreases the predicted value as we move from the baseline (usually the average prediction) to the final prediction. The bars can be green (positive) or red (negative) contribution and is proportional to the magnitude of contribution of the feature, see Fig. 2 for an example.

4.2.1 Creating and Visualizing Breakdown Results Through Waterfall Plots

To demonstrate how the Breakdown method and waterfall plots explain the contribution of different features to a model's prediction for a specific student, we will select a single instance from the test dataset. Here, we choose the fifth row from `test_data`, which we obtain it by subsetting the 5th row, and storing it in the `Instance5` object.

```
Instance5 <- test_data[5, ] # Select any other observation from `test_data` if needed
```

To decompose the prediction for `Instance5`, we use the `predict_parts()` function, which provides a detailed, instance-specific explanation of how each feature affects the predicted grade. To do so, we call `predict_parts()` with the following arguments: `explainer = explainer_svm` which contains the SVM model and necessary data; `new_observation = Instance5`, specifying that we want to explain the predictions for `Instance5`; and `type = "break_down"`, indicating that we want to use the Break Down method for the explanation. The `predict_parts()`function returns a detailed breakdown of the model's prediction.

```
# Set a seed for reproducibility
set.seed(265)

breakdown_explanation_5 <- predict_parts(
  explainer = explainer_svm,
  new_observation = Instance5,
  type = "break_down"
)
```

To print the breakdown results, we use `print(breakdown_explanation_5)`, where positive values indicate that a feature increases the prediction, and negative values indicate a downward effect (Table 1).

```
print(breakdown_explanation_5) # Print the breakdown explanation to the console
```

We can also create a visual representation of these contributions using `plot(breakdown_explanation_5)`, making it easier to interpret how each feature influences the prediction.

```
# Plot the breakdown explanation to visualize the contribution of each feature
plot(breakdown_explanation_5)
```

Figure 2 below illustrates the breakdown plot for `Instance5` predicted grade. The bars represent the contribution of each predictor to the final grade prediction. Red bars indicate factors that decreased the prediction, while green bars show factors that increased it. The purple bar on the far right reflects the final predicted grade,

Table 1 Breakdown explanation to assess the contribution of each feature to the prediction of instance 5

	Contribution
SVM Radial: intercept	68.491
SVM Radial: Freq_Forum_Contribute = −46.07	−2.767
SVM Radial: Regularity_Forum_Contribute = −0.1327	−2.324
SVM Radial: Freq_Course_View = 132.2	0.435
SVM Radial: Session_Count = 107.5	1.101
SVM Radial: Total_Duration = −13480	0.015
SVM Radial: Regularity_Lecture_View = 0.07302	−0.054
SVM Radial: Active_Days = −2.063	−0.201
SVM Radial: Freq_Forum_Consume = −84.55	0.442
SVM Radial: Regularity_Course_View = −0.007579	−0.238
SVM Radial: Regularity_Forum_Consume = −0.0133	−0.026
SVM Radial: Freq_Lecture_View = 34.13	−0.008
SVM Radial: prediction	64.868

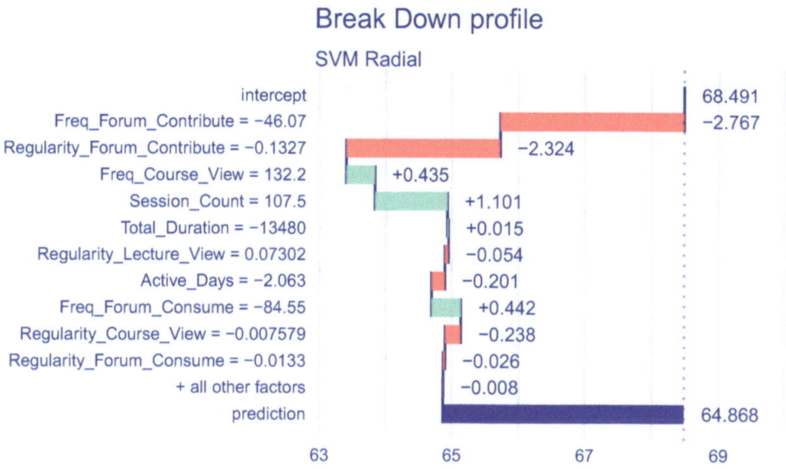

Fig. 2 Breakdown explanation to visualize the contribution of each feature to the prediction of instance 5

which is lower than the baseline due to the cumulative negative impact of several factors, particularly related to forum participation and activity levels.

Let us go through the results of breakdown explanation of (`Instance5`). The baseline prediction (the intercept) is 68.491, which represents the average grade prediction across all students in the dataset. For `Instance5`, the final predicted grade is 64.868, a notable decrease from the baseline prediction. The breakdown plot helps us understand why and how the predicted grades deviated below this average.

We see that several negative indicators impacted the student's predicted grade. The largest decrease came from low frequency of forum contributions (−46.07), which resulted in a −2.767 decrease in the predicted grades. Furthermore, the student's irregularity of forum contributions (−0.1327) contributed to a further reduction of −2.324 grades. Another negative predictor was the student's low number of active days (−2.063), which negatively affected the grade prediction by −0.201 points. Interestingly, despite the low frequency of forum consumption (−84.55), it contributed positively to the predicted grade by 0.442 points. This finding suggests that values below the average can still be a positive indicator for predicting better grades. In other words, the average may be higher than what the algorithm considers "enough". There were also several positive engagement behaviors that increased the student's predicted grade. The frequency of course views (132.2) positively influenced the prediction by 0.435 grades, and the session count (107.5) contributed even more with an increase of 1.101 grades. The impact of other engagement indicators was minimal or close to 0. For instance, the total duration of engagement (−13,480) had a negligible effect on the prediction, adding only 0.015 grades. Additionally, indicators such as regularity of lecture viewing (0.07302), frequency of lecture viewing (34.13), regularity in course viewing

(−0.007579), and regularity in forum consumption (−0.0133) showed minor effects with contributions.

The following equation shows how each feature adds to or subtracts from the initial baseline of **68.491**, resulting in the final predicted grade of **64.868**.

$$\begin{aligned}
&-2.767\ (\text{Freq_Forum_Contribute})\\
&-2.324\ (\text{Regularity_Forum_Contribute})\\
&+0.435\ (\text{Freq_Course_View})\\
&+1.101\ (\text{Session_Count})\\
&+0.015\ (\text{Total_Duration})\\
&-0.054\ (\text{Regularity_Lecture_View})\\
&-0.201\ (\text{Active_Days})\\
&+0.442\ (\text{Freq_Forum_Consume})\\
&-0.238\ (\text{Regularity_Course_View})\\
&-0.026\ (\text{Regularity_Forum_Consume})\\
&-0.008\ (\text{Freq_Lecture_View})\\
&= \mathbf{64.868}
\end{aligned}$$

4.2.2 Order of Variables

One of the problems of breakdown plots is that they are influenced by the order different variables where different orders may result in different results. To demonstrate these changes, let us change the order of the variables that we have. For that, we will shuffle the regularity features to come first, then followed by time and later come the frequency features. We do so, by specifying a new object and we store the order to it `Custom_order`.

```
set.seed(265)

# Define a custom order for the features in the breakdown analysis
Custom_order <- c("Regularity_Course_View", "Regularity_Lecture_View",
                  "Regularity_Forum_Consume", "Regularity_Forum_Contribute",
                  "Session_Count", "Total_Duration", "Active_Days",
                  "Freq_Course_View", "Freq_Lecture_View",
                  "Freq_Forum_Consume", "Freq_Forum_Contribute")

# Select the fifth instance from the test data for local explanation
Instance5 <- test_data[5, ]

# Generate a breakdown explanation using the custom order of features
breakdown_explanation_5_ordered <- predict_parts(
  explainer = explainer_svm,              # The model explainer
```

```
  new_observation = Instance5,          # The instance to explain
  order = Custom_order,                 # Apply the custom feature order
  type = "break_down"                   # Use the breakdown method
)

# Display and plot the breakdown explanation
print(breakdown_explanation_5_ordered)
```

```
plot(breakdown_explanation_5_ordered)
```

We can see there are differences between our original case, and the shuffled example (Table 2, Fig. 3). The most significant difference comes from Freq_Forum_Contribute where it reduces the prediction by 2.767 in the original example while in the shuffled order example, it resulted in 1.364 grade drop. Similarly, Regularity_Forum_Contribute showed a more negative impact in the original example (−2.324) than in the shuffled ordered example (−2.040). Whereas Total_Duration has almost no effect in the first example (0.015), it turned negative in the shuffled example reducing the grade by −0.728. Session_Count contributes positively in both cases but to different extents, adding 1.101 in the first example and 0.758 in the second. Freq_Course_View also shifts from a small positive contribution in the first example (0.435) to a slight negative impact in the second (−0.355). Finally, Active_Days remains consistently negative but with similar small effects in both cases. Overall, while we saw differences as well as consistency, the fact that the importance and contribution of variables differ with changing order of variables makes some of our conjectures unreliable. To further show the magnitude of the problem we have plotted different variations of the variable order in Fig. 4. Therefore, more consistent methods in the next sections.

Table 2 Breakdown explanation of instance 5 in a custom order

	Contribution
SVM Radial: intercept	68.491
SVM Radial: Regularity_Course_View = −0.007579	−0.122
SVM Radial: Regularity_Lecture_View = 0.07302	0.209
SVM Radial: Regularity_Forum_Consume = −0.0133	−0.036
SVM Radial: Regularity_Forum_Contribute =−0.1327	−2.040
SVM Radial: Session_Count = 107.5	0.758
SVM Radial: Total_Duration = −13480	−0.728
SVM Radial: Active_Days = −2.063	−0.184
SVM Radial: Freq_Course_View = 132.2	−0.355
SVM Radial: Freq_Lecture_View = 34.13	0.117
SVM Radial: Freq_Forum_Consume = −84.55	0.122
SVM Radial: Freq_Forum_Contribute = −46.07	−1.364
SVM Radial: prediction	64.868

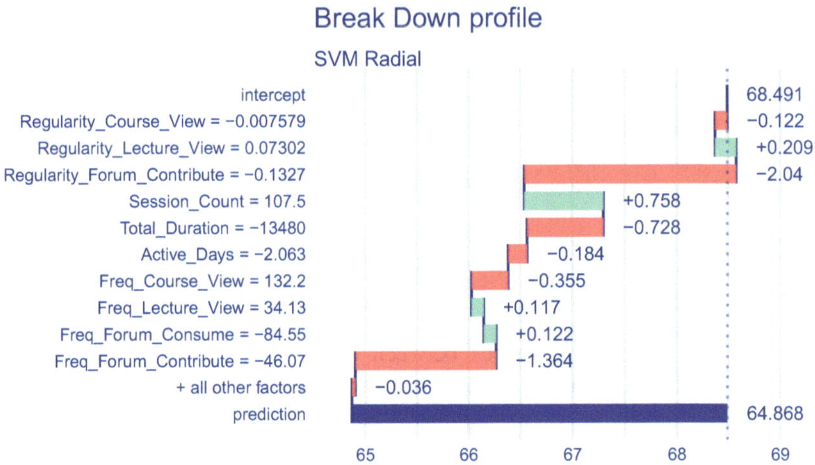

Fig. 3 Breakdown explanation to visualize the contribution of each feature to the prediction of instance 5 providing the features in a custom order

```
set.seed(265)

library(patchwork)  # For combining plots

# Generate 12 different plots with different feature orders
plots <- list()
for (i in 1:length(Custom_order)) {
  random_order <- sample(Custom_order)
  # Compute breakdown explanation with the current order
  breakdown_explanation_ordered <- predict_parts(
    explainer = explainer_svm,
    new_observation = Instance5,
    order = random_order,  # Apply the random feature order
    type = "break_down"
  )

  # Plot the breakdown explanation
  plots[[i]] <- plot(breakdown_explanation_ordered) +
    ggtitle(paste("Breakdown Explanation - Order", i))
}

# Combine all plots into a single image
combined_plot <- wrap_plots(plots, ncol = 3)  # Adjust ncol as needed
print(combined_plot)
```

Fig. 4 Breakdown explanation to visualize the contribution of each feature for different feature orders

4.3 SHAP (Shapley Additive Explanations) Values: A Detailed Explanation

SHAP (Shapley Additive Explanations) is a popular and more robust method for explaining model predictions with roots coming from game theory principles. By design, SHAP allocates "credit" or "blame" to each feature in a model's prediction in a balanced manner. By doing so, SHAP ensures that the allocation of importance to each feature reflects their true impact on the model's output. The contribution of each feature is considered by treating the prediction process like a game, where the features are like players cooperating or working together to make a prediction. Further, in order to ensure that the SHAP results are robust, the SHAP values are calculated by considering all possible combinations (subsets) of features and measuring how the inclusion of the feature changes the prediction in each combination. For example: In our model, the SHAP value for lecture viewing frequency would measure how much this feature changes the prediction compared to when it is not considered, across all possible feature combinations. If lecture viewing frequency has a SHAP value of +2, this means that the lecture viewing frequency contributes 2 points above the base prediction towards the final predicted grade.

4.3.1 Advantages of SHAP Values

Unlike methods like the breakdown profile, SHAP does not depend on the order in which features are evaluated. SHAP achieves so by considering all possible permutations of feature order and their interactions. This makes SHAP more robust, and replicable, i.e., one is more likely to obtain the same results when repeating the same procedure and with even a different order. Furthermore, SHAP values ensure that every feature is given "credit" (or "blame") for its contribution in a fair way. This fairness comes from the way SHAP credit is distributed evenly across all features. Another advantage of SHAP is that their capability to capture complex variable interactions. SHAP values can capture interaction effects between features. If two features are jointly influencing a prediction in a significant way, SHAP can model this interaction effectively by considering all feature combinations. The last but not least, is that SHAP values can offer both global and local interpretations. In local Interpretation (such as our single instance here), SHAP values explains exactly why the model in this particular instance made a certain prediction, showing the contribution of each feature to that prediction. Aggregating SHAP values across multiple instances (Global Interpretation) gives the overall importance of features for the a dataset or a group of data-points.

4.3.2 Using SHAP for Single Instance Explanation

Estimating SHAP values and plotting them is similar to the same method we used before for estimating breakdown values with one difference that we need to tell `predict_parts()` function to use SHAP for calculating the variable contributions. As the code shows below, we start by selecting `Instance5` for explanation. We use the same function `predict_parts()` and we set the (`type = "shap"`). In the same way, we can use the `print(shap_explanation)` to print the SHAP values (Table 3) to the console, and use the `plot(shap_explanation)` to visualize the SHAP values (Fig. 5).

```
# Compute SHAP explanation for a single instance
# 'type = "shap"' specifies that we want to use SHAP (SHapley Additive
# exPlanations) to explain the prediction.
shap_explanation <- predict_parts(
  explainer = explainer_svm,      # The explainer object for the SVM model
  new_observation = Instance5,    # The specific instance that we want to explain
  type = "shap"                   # Use the SHAP method for generating the explanation
)

# Print the SHAP explanation
print(shap_explanation)
```

```
# Plot the SHAP explanation
plot(shap_explanation)
```

Table 3 SHAP explanation for instance 5

	Min	q1	Median	Mean	q3	Max
Active_Days = −2.063	−0.50	−0.22	−0.20	−0.20	−0.22	0.17
Freq_Course_View = 132.2	−1.48	−1.21	−0.54	−0.47	−1.21	0.64
Freq_Forum_Consume = −84.55	−0.29	0.10	0.15	0.16	0.10	0.44
Freq_Forum_Contribute = −46.07	−3.11	−2.66	−1.99	−2.06	−2.66	−1.20
Freq_Lecture_View = 34.13	−0.26	−0.07	−0.01	−0.01	−0.07	0.28
Regularity_Course_View = −0.007579	−0.59	−0.35	−0.25	−0.29	−0.35	0.00
Regularity_Forum_Consume = −0.0133	−0.29	−0.13	−0.10	−0.08	−0.13	0.33
Regularity_Forum_Contribute =−0.1327	−2.20	−1.63	−1.34	−1.37	−1.63	−0.57
Regularity_Lecture_View = 0.07302	−0.26	0.04	0.14	0.14	0.04	0.43
Session_Count = 107.5	−0.10	0.71	0.78	0.78	0.71	1.26
Total_Duration = −13,480	−1.12	−0.36	−0.18	−0.22	−0.36	0.14

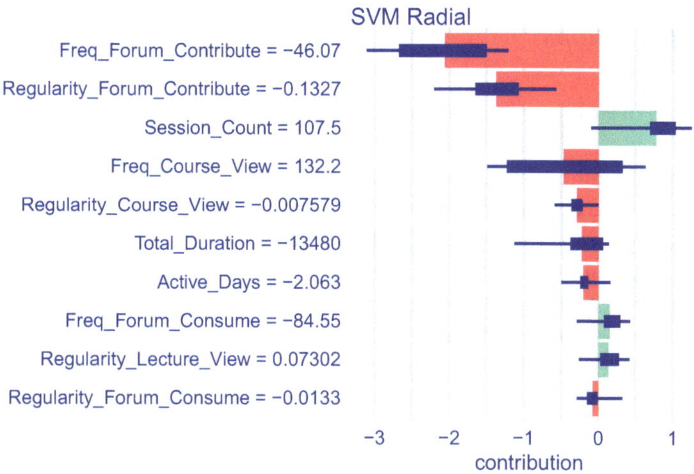

Fig. 5 SHAP explanation for instance 5

Let us decompose the SHAP values in our final example to examine the individual contributions of each variable. As the figure shows, the largest negative contribution comes from Freq_Forum_Contribute, which resulted in a drop of −2.77 points from the baseline of 68.49 shown as a red bar. Following this, Regularity_Forum_Contribute reduced the predicted grade by −2.32 points, also shown as a red bar on the negative side of the plot (left of the 0). Both indicators reflect that low frequency and irregular engagement in collaborative learning activities may decrease the student's predicted achievement. This is of course because, the model has *learnt* that students who contribute to the forums always score better grades. In contrast, Session_Count increases the predicted grade by 1.10 points, Freq_Course_View by 0.44 points. Regularity_Course_View reduces the prediction by −0.12 points. Additionally, low Total_Duration contributes a small negative value of −0.73 points, while low values of Active_Days decreases −0.20 points. Regarding lecture engagement, Regularity_Lecture_View has a low positive contribution of 0.21 points, while Freq_Lecture_View slightly increases the prediction by 0.12 points. These findings suggest that regularity in engaging with lectures is a stronger predictor of performance than the sheer number of views. Lastly, forum consumption shows an interesting dynamic: although Freq_Forum_Consume is low, it adds 0.44 points to the prediction, whereas Regularity_Forum_Consume slightly decreases the prediction by −0.04 points. To put this in an equation, we start by the average model prediction of 68.49 and add or substract each variable contribution. Prediction $= 68.49 - 2.77 - 2.32 + 1.10 + 0.44 - 0.12 - 0.73 - 0.20 + 0.21 + 0.12 + 0.44 - 0.04 = 64.87$.

4.4 LIME (Local Interpretable Model-Agnostic Explanations) Explanations

LIME works in a different way than the previous methods. In that, LIME creates an approximation of the model's behavior by creating a "simpler model" [15]. LIME does so by generating perturbations i.e., slight modifications to the variables of the instance being explained and then analyzes how these modifications affect the model's predictions. In doing so, LIME constructs a simple, interpretable model (such as linear regression) that mimics the behavior of the more complex model. This approach allows LIME to estimate the contribution of each feature. The contributions are then ranked and visualized. Unlike SHAP, which provides globally consistent feature contributions (i.e., the contributions sum up to the prediction), LIME focuses more on local behavior and might provide different interpretations.

The code is different here, since we use a different function, the `predict_surrogate` from the DALEXtra package to generate a LIME explanation and the *surrogate* model. But before using `predict_surrogate`, we need to ensure the necessary `DALEXtra` methods are loaded as shown below. Then we pass `explainer_svm` object, which contains the SVM model and the instance (`Instance5`), the `n_features` argument specifying the number of features contributing to the prediction will be considered, the `n_permutations` sets the number of perturbations around the instance to 1000 to accurately approximate the model's local behavior. The `type = "lime"` argument specifies that LIME is the method used for generating the explanation. Finally, the plot function visualizes the results.

```
# Ensure the necessary DALEXtra methods are loaded
model_type.dalex_explainer <- model_type.dalex_explainer
predict_model.dalex_explainer <-  predict_model.dalex_explainer

# Generate LIME explanation using DALEXtra's predict_surrogate function
Lime_Explainer <- predict_surrogate(
  explainer = explainer_svm,     # The explainer object for the SVM model
  new_observation = Instance5,   # The specific instance (Instance5) to be explained
```

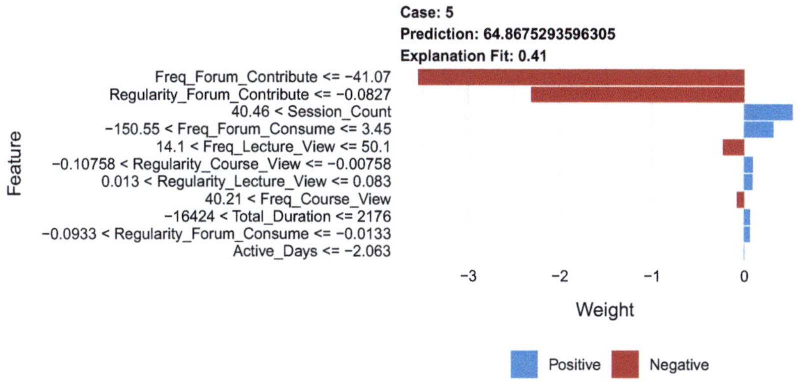

Fig. 6 LIME explanation for instance 5

```
  n_features = 12,              # Number of features to include in the explanation
  n_permutations = 1000,        # Number of permutations for the surrogate model
  type = "lime"                 # Specify that we want a LIME explanation
)

# Plot the LIME explanation
plot(Lime_Explainer)
```

The plot shown in Fig. 6 provides LIME explanations of how the model estimated the final grade (64.87) of instance 5. First, the explanation fit is 0.41 indicating how well the surrogate model approximates the original model's decision. We can see that the student low `Freq_Forum_Contribute` have the highest negative influence on the predicted grade reducing it by nearly 3 points. So did `Regularity_Forum_Contribute` which reduced slightly less than 3 points. The positive contributions are indicated by blue bars. We can see that the `Session_Count` is the strongest positive variable, adding about 1.5 points to the prediction. Other features like `Freq_Forum_Consume` and `Active_Days` had smaller but still notable positive contributions.

4.4.1 Multiple Instances

Similar to single instances, we can estimate the SHAP values for multiple instances. In the next example the code proceeds to making predictions on the test data. In these examples, we zoom down to students who were mis-predicted and differ from their actual grades by more than 10 points. Once these students are identified, we separate them into two groups: those whose grades were over-predicted (Fig. 7) and those whose grades were under-predicted (Fig. 8). We then calculate the Aggregated SHAP values for both groups of students. These values can help us identify which features most significantly contributed to the over-prediction or under-prediction of

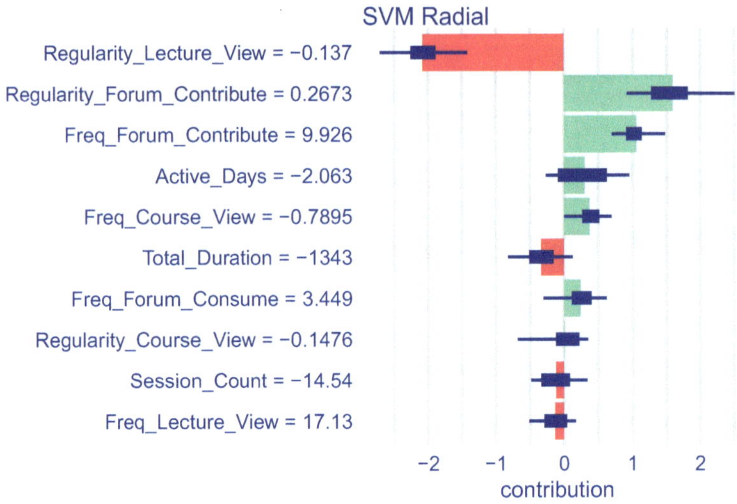

Fig. 7 SHAP for increased grades

grades. You can examine both charts and you will see, for instance, that students who were over predicted were highly active.

```
# Predict the Final_Grade on the test dataset
predictions <- predict(svm_fit_radial, newdata = test_data)

# Calculate residuals (the difference between actual grades and predicted grades)
residuals <- test_data$Final_Grade - predictions

# Define the threshold for significant prediction errors (10 points)
threshold <- 10

# Identify students whose predicted grades were increased by more than the threshold
increased_grades <- test_data |>
  mutate(Residual = residuals) |>
  filter(Residual <= threshold)

# Identify students whose predicted grades were decreased by more than the threshold
decreased_grades <- test_data |>
  mutate(Residual = residuals) |>
  filter(Residual >= threshold)
```

```
# Aggregate SHAP values for students with increased grades
shap_increased <- predict_parts(
  explainer = explainer_svm,
  new_observation = increased_grades |> dplyr::select(-Final_Grade, -Residual),
  type = "shap",
  B = 50  # Number of Monte Carlo simulations for SHAP values
)

# Plot the aggregated SHAP values for increased grades
plot(shap_increased)
```

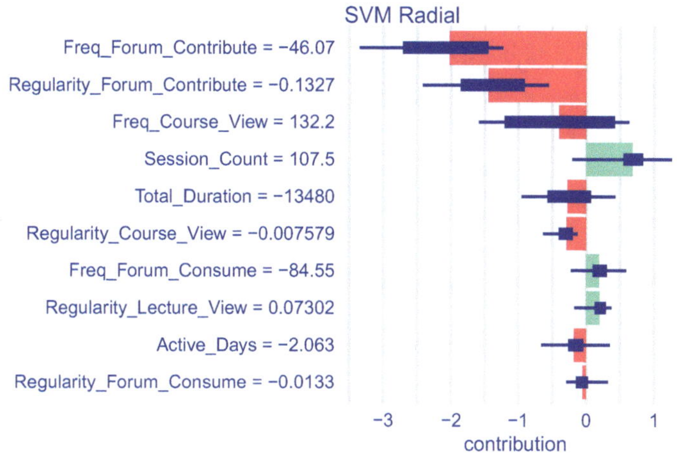

Fig. 8 SHAP for decreased grades

```
# Aggregate SHAP values for students with decreased grades
shap_decreased <- predict_parts(
  explainer = explainer_svm,
  new_observation = decreased_grades |> dplyr::select(-Final_Grade, -Residual),
  type = "shap",
  B = 50  # Number of Monte Carlo simulations for SHAP values
)

# Plot the aggregated SHAP values for decreased grades
plot(shap_decreased)
```

4.5 Explanation of a Classifier Model

The process of explaining a single instance in classifier models is almost the same. The below code creates a classification model for explaining student achievement (high or low achievement) and then creates an explanation of instance 10. In the code below, we start by centering the data as we did before (step 1). Then, we create a binary classification target, `Achievement` from `Final_Grade` where "Low_Achievers" are considered if they score below the median grade (step 2). Then, we split the data into training and testing sets, ensuring stratification by `Achievement` to maintain consistent proportions of high and low achievers in both subsets (step 3). Then, we estimate an SVM model using the `e1071` package, with `probability = TRUE` to produce probability estimates for each prediction to help us in evaluating the model (step 4). Lastly, we create the explainer using `explain()` function, which prepares the model and test data for generating local explanations (step 5).

```
# Step 1: Center the numeric variables in the dataset
student_data_centered <- student_data |>
  mutate(across(Freq_Course_View:Active_Days, ~scale(., scale = FALSE) |> as.vector()))

# Step 2: Create a binary classification target variable
median_grade <- median(student_data_centered$Final_Grade)
student_data_centered <- student_data_centered |>
  mutate(Achievement = ifelse(Final_Grade < median_grade,
                              "Low_Achievers", "High_Achievers")) |>
  mutate(Achievement = factor(Achievement,
                              levels = c("High_Achievers","Low_Achievers")))

# Step 3: Split the data into training and testing, stratifying by the target variable
data_split <- initial_split(student_data_centered, prop = 0.8, strata = Achievement)
train_data <- training(data_split)
test_data <- testing(data_split)

# Step 4: Train an SVM model using e1071
svm_model_Classify <- svm(
  Achievement ~ Freq_Course_View + Freq_Lecture_View + Freq_Forum_Consume +
    Freq_Forum_Contribute + Regularity_Course_View +
    Regularity_Lecture_View + Regularity_Forum_Consume +
    Regularity_Forum_Contribute + Session_Count +
    Total_Duration + Active_Days,
  data = train_data,
  probability = TRUE)

# Step 5: Create explainer
explainer_svm_Classify <- explain(
  svm_model_Classify,
  data = test_data |> dplyr::select(-Final_Grade, -Achievement),
  y = test_data$Achievement,
  label = "SVM Classification",
  verbose = FALSE
)
```

Using the same code as earlier, we create the local explanation for a single student (Fig. 9).

```
# Select a single instance for local explanation
Instance10 <- test_data[10, ]  # You can select any observation from test_data

# Local interpretation using Break Down method
breakdown_Classify <- predict_parts(
  explainer = explainer_svm_Classify,
  new_observation = Instance10 |> dplyr::select(-Final_Grade, -Achievement),
  type = "break_down")

# Display and plot the breakdown explanation
plot(breakdown_Classify)

# Local interpretation using SHAP method
Shap_Classify <- predict_parts(
  explainer = explainer_svm_Classify,
  new_observation = Instance10 |> dplyr::select(-Final_Grade, -Achievement),
  type = "shap")

# Display and plot the SHAP explanation
plot(Shap_Classify)
```

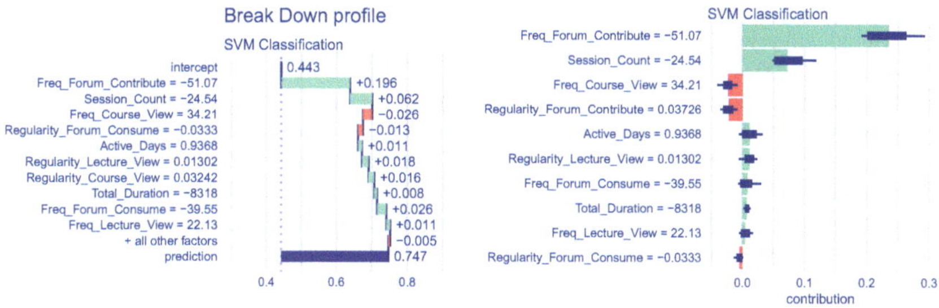

(a) (b)

Fig. 9 Local explanations for the classification results of student 10. (**a**) Breakdown. (**b**) SHAP

5 Conclusions

In conclusion, this chapter has demonstrated the role of instance-level explanations in moving beyond aggregate insights to understand the unique factors influencing individual predictions within machine learning models. Unlike global explanations that offer a broad overview, these local techniques, exemplified by Break Down plots, SHAP values, and LIME enable educators and stakeholders to dissect the specific reasons behind a particular outcome for a given student which helps enhance transparency and trust.

However, while instance-level predictions allow us to understand parts or instances of an aggregated model they should not be thought of as idiographic models. For idiographic modeling of individual students, we need to implement models that uses the student as a unit of analysis [23]. Put another way, the whole model data comes from the same student which will be demonstrated in a separate chapter [24]. For more reading about idiographic modeling and what they mean, readers are advised to refer to the idiographic chapter [25].

References

1. Johnson L, Smith R, Willis H, Levine A, Haywood K (2011) The 2011 horizon report. In: New media consortium. New Media Consortium, Austin. http://www.nmc.org. http://files.eric.ed.gov/fulltext/ED515956.pdf
2. López-Pernas S, Oliveira E, Song Y, Saqr M (2025) AI, explainable AI and evaluative AI: An introduction to informed data-driven decision-making in education. In: Saqr M, López-Pernas S (eds) Advanced learning analytics methods: AI, precision and complexity. Springer Nature Switzerland, Cham
3. Saqr M, López-Pernas S (2025) Explainable artificial intelligence in education: a tutorial for identifying the variables that matter. In: Saqr M, López-Pernas S (eds) Advanced learning analytics methods: AI, precision and complexity. Springer Nature Switzerland, Cham

4. Saqr M, López-Pernas S (2024) Why explainable AI may not be enough: predictions and mispredictions in decision making in education. Smart Learn Environ. https://doi.org/10.1186/s40561-024-00343-4
5. Bobrowicz K, López-Pernas S, Teuber Z, Saqr M, Greiff S (2024) Prospects in the field of learning and individual differences: examining the past to forecast the future using bibliometrics. Learn Individ Differ 109:102399. https://doi.org/10.1016/j.lindif.2023.102399
6. Tikka S, Kopra J, Heinäniemi M, López-Pernas S, Saqr M (2024) Getting started with R for education research. In: Saqr M, López-Pernas S (eds) Learning analytics methods and tutorials: a practical guide using r. Springer, Berlin
7. Kopra J, Tikka S, Heinäniemi M, López-Pernas S, Saqr M (2024) An R approach to data cleaning and wrangling for education research. In: Saqr M, López-Pernas S (eds) Learning analytics methods and tutorials: a practical guide using r. Springer, Berlin
8. Tikka S, Kopra J, Heinäniemi M, López-Pernas S, Saqr M (2024) Introductory statistics with R for educational researchers. In: Saqr M, López-Pernas S (eds) Learning analytics methods and tutorials: a practical guide using r. Springer, Berlin
9. López-Pernas S, Misiejuk K, Tikka S, Kopra J, Heinäniemi M, Saqr M (2024) Visualizing and reporting educational data with R. In: Learning analytics methods and tutorials. Springer Nature Switzerland, Cham, pp 151–194. https://doi.org/10.1007/978-3-031-54464-4/_6
10. Jovanovic J, López-Pernas S, Saqr M (2024) Predictive modelling in learning analytics: a machine learning approach in r. In: Learning analytics methods and tutorials: a practical guide using r. Springer Nature Switzerland, Cham, pp 197–229
11. Saqr M, Misiejuk K, Tikka S, López-Pernas S (2025) Artificial intelligence: using machine learning to predict students' performance. In: Saqr M, López-Pernas S (eds) Advanced learning analytics methods: AI, precision and complexity. Springer Nature Switzerland, Cham
12. Saqr M, Misiejuk K, Tikka S, López-Pernas S (2025) Artificial intelligence: using machine learning to classify students and predict low achievers. In: Saqr M, López-Pernas S (eds) Advanced learning analytics methods: AI, precision and complexity. Springer Nature Switzerland, Cham
13. Biecek P, Burzykowski T (2021) Explanatory model analysis: Explore, explain, and examine predictive models, 1st edn. Chapman & Hall/CRC, Philadelphia. https://www.routledge.com/Explanatory-Model-Analysis-Explore-Explain-and-Examine-Predictive-Models/Biecek-Burzykowski/p/book/9780367693923
14. Shapley LS (1953) 17. A value for n-person games. In: Kuhn HW, Tucker AW (eds) Contributions to the theory of games (AM-28), vol II. Princeton University Press, Princeton, pp 307–318. https://doi.org/10.1515/9781400881970-018
15. Ribeiro MT, Singh S, Guestrin C (2016) Trust you? Explaining the predictions of any classifier. In: Proceedings of the 22nd ACM SIGKDD international conference on knowledge discovery and data mining. ACM, New York, NY, USA. https://doi.org/10.1145/2939672.2939778
16. Jang Y, Choi S, Jung H, Kim H (2022) Practical early prediction of students' performance using machine learning and eXplainable AI. Educ Inf Technol 27:12855–12889
17. Nagy M, Molontay R (2023) Interpretable dropout prediction: Towards XAI-based personalized intervention. Int J Artif Intell Educ. https://doi.org/10.1007/s40593-023-00331-8
18. Lin J, Dai W, Lim L-A, Tsai Y-S, Mello RF, Khosravi H, Gasevic D, Chen G (2023) Learner-centred analytics of feedback content in higher education. In: LAK23: 13th international learning analytics and knowledge conference. Association for Computing Machinery, New York, NY, USA, pp 100–110
19. Adnan M, Habib A, Ashraf J, Mussadiq S, Raza AA, Abid M, Bashir M, Khan SU (2021) Predicting at-risk students at different percentages of course length for early intervention using machine learning models. IEEE Access 9:7519–7539
20. Li T, Han F, Guo J, Wu J (2025) Comparative analysis of regularization methods for predicting student certification in online courses. In: Saqr M, López-Pernas S (eds) Advanced learning analytics methods: AI, precision and complexity. Springer Nature Switzerland, Cham

21. Saqr M, Jovanovic J, Viberg O, Gašević D (2022) Is there order in the mess? A single paper meta-analysis approach to identification of predictors of success in learning analytics. Stud Higher Educ 47:2370–2391. https://doi.org/10.1080/03075079.2022.2061450
22. Biecek P (2018) DALEX: Explainers for complex predictive models in r. J Mach Learn Res 19:1–5. https://jmlr.org/papers/v19/18-416.html
23. Saqr M, Cheng R, López-Pernas S, Beck ED (2024) Idiographic artificial intelligence to explain students' self-regulation: toward precision education. Learn Indiv Differ 114:102499. https://doi.org/10.1016/j.lindif.2024.102499
24. Saqr M, Tlili A, López-Pernas S (2025) Automating individualized machine learning and AI prediction using AutoML: the case of idiographic predictions. In: Saqr M, López-Pernas S (eds) Advanced learning analytics methods: AI, precision and complexity. Springer Nature Switzerland, Cham
25. Saqr M, Ito H, López-Pernas S (2025) A comprehensive introduction to idiographic and within-person analytics. In: Saqr M, López-Pernas S (eds) Advanced learning analytics methods: AI, precision and complexity. Springer Nature Switzerland, Cham

Part II
Natural Language Processing and Large Language Models

An Introduction to Large Language Models in Education

Eduardo Oliveira, Yige Song, Mohammed Saqr, and Sonsoles López-Pernas

Abstract Large language models (LLMs) have become central to contemporary advancements in education. LLMs facilitate applications such as automated feedback, question generation, sentiment analysis, and multilingual accessibility. This chapter examines the mechanisms that underpin LLMs—namely, transformer architecture, pre-training, and generative abilities. Moreover, we present the diverse applications of LLMs in education, some of which will be covered in subsequent tutorial chapters in the present book. Lastly, we explore tools for interacting with LLMs, from beginner-friendly web interfaces to more advanced tools like APIs and frameworks such as the OpenAI API and Hugging Face's Transformers.

Keywords Large language models · Artificial intelligence · Learning analytics

1 Introduction

Large Language Models (LLMs) are advanced AI systems primarily designed to process and generate human-like text, but their capabilities extend far beyond natural language tasks, enabling transformative applications across diverse domains, including education, research, software development, and more [1, 2]. Taking advantage of massive datasets and neural network architectures—such as the transformer mechanism [2]—LLMs can analyse context, predict sequential elements (e.g., the next word or token), and produce coherent, contextually appropriate outputs.

While text generation and understanding are at the core of LLMs, their versatility stems from their ability to generalize patterns in data. This enables them to tackle tasks like code generation [3], and even multimodal applications that integrate text

E. Oliveira · Y. Song
University of Melbourne, Parkville, VIC, Australia

M. Saqr · S. López-Pernas (✉)
University of Eastern Finland, Joensuu, Finland
e-mail: sonsoles.lopez@uef.fi

M. Saqr, S. López-Pernas (eds.), *Advanced Learning Analytics Methods*,
https://doi.org/10.1007/978-3-031-95365-1_8

with other data types, such as images and videos [4]. For instance, specialized models like Claude Sonnet excel in programming tasks, while multimodal extensions of GPT-4 demonstrate the ability to describe images and interpret visual data alongside textual inputs.

LLMs are opening new possibilities in education by transforming how students learn and how educators teach and research [5]. These models can dynamically generate test questions tailored to diverse learning levels [6], translate educational materials to improve accessibility [7], summarize complex concepts for clearer understanding [8], and simulate conversational practice to enhance language skills [9]. The personalization of learning experiences through LLMs can support students in mastering content at their own pace. In research, they enable the creation of synthetic datasets [10], simulate experimental conditions for pedagogical studies [11], and analyse vast corpora of text and structured data to uncover actionable insights.

The scalability of LLMs, exemplified by models like GPT-3 with 175 billion parameters (i.e., the weights and biases of the different layers), has accelerated their adoption across fields. However, they are not without limitations. Issues such as biased, inaccurate, or fabricated outputs highlight the need for critical human oversight [12]. A human-in-the-loop approach is crucial to ensuring that LLMs align with educational goals and ethical standards, serving as tools to enhance learning rather than replace human expertise [13, 14]. Despite these challenges, LLMs are transforming how information is delivered, processed, and utilized. Their ability to integrate language understanding with broader data-processing capabilities positions them as invaluable tools for solving complex problems and enhancing accessibility across disciplines. The following sections explore the transformative applications of LLMs in education research and practice.

2 Behind the Scenes of LLMs

To better understand how large language models function, it is helpful to break down their core components and processes into three interrelated aspects based on the acronym "GPT": the Transformer architecture (T), which underpins their computational ability; the Pre-training phase (P), where the model learns patterns and knowledge from vast datasets; and the Generative abilities (G), which showcase their practical applications. The sequence T $\rightarrow$ P $\rightarrow$ G provides a logical progression from the foundational mechanics of LLMs to their training processes and, finally, to their real-world outputs. This structure offers a holistic view of these transformative technologies: first understanding the architecture that powers LLMs, then exploring how they are trained to predict and fine-tune their outputs, and ultimately seeing what they can generate. In the following sections, we explore each of these phases in detail, shedding light on their individual contributions to the overall capabilities of LLMs.

2.1 How LLMs Work: The "Transformer" Architecture

The "Transformer" in GPT refers to the architectural framework that powers modern LLMs, enabling their remarkable fluency and adaptability. This architecture revolutionized natural language processing by introducing the attention mechanism [2], which allows the model to understand relationships within a sequence and process context effectively. Without the "T," LLMs could not achieve the coherence and precision that define their generative abilities.

At its core, a LLM is a sophisticated function designed to predict the next element in a sequence. For example, given the input "The sun is shining", the model might predict "brightly" or "today". While this prediction task may sound straightforward, this isn't a simple function like $y = 2x + 1$, in which the parameters are the coefficient of the linear function 2 and 1, and input of $x=1$ would give an output of 3. The underlying computations in LLMs are vastly more intricate. Instead of a handful of parameters, LLMs employ billions of them within multi-layered neural networks. Each layer refines the model's understanding of the input, performing complex transformations based on the outputs of previous layers.

Prior to the attention mechanism, models like recurrent neural networks (RNNs) [15] and convolutional neural networks (CNNs) [16] struggled to effectively process relationships in long or complex sequences, often treating input elements in isolation or with limited context [17]. The attention mechanism transformed this by allowing the model to weigh the importance of different parts of the input sequence based on their relevance to the task [2]. For example, consider the sentences "There's a bat on the tree" and "He swings the bat". In the first, "bat" refers to an animal, while in the second, it refers to a sports tool. The attention mechanism enables the model to focus on surrounding words like "tree" or "swings" to deduce the correct meaning. This ability to dynamically assign importance to specific words based on context is the key to LLMs' accuracy and coherence.

The attention mechanism calculates attention weights, assigning varying levels of importance to each word based on its relevance to the task at hand. For instance, in the sentence "The sun is shining", the word "sun" is given more weight than less significant words like "the". These weighted representations of words are passed through multiple layers of the neural network, with each layer building on the contextual understanding established by the previous one. Early layers might identify simple word relationships, like "sun" being associated with "shining", while later layers incorporate broader knowledge, such as recognizing weather-related descriptions that often lead to words like "today" or "brightly".

In summary, the Transformer architecture is highly effective because it dynamically focuses attention on the most relevant parts of an input sequence at any given moment. This mechanism mirrors human behavior, where we concentrate on the information most pertinent to the task at hand while filtering out distractions. This synergy between attention and neural network computation forms the foundation of the architecture, making it the core of modern generative AI.

2.2 Why LLMs Work: The Power of Being "Pre-trained"

Previously, we discussed that LLMs are predictive models with billions of parameters, enabling them to perform complex calculations. But how are these parameters determined? That's where the "P" in GPT—**P**re-training—comes in.

In any prediction task, the goal is to learn from existing data to make accurate predictions for new inputs. For example, imagine you're predicting the next number in a sequence like 2, 4, 6, 8. Based on the pattern, you might predict 10 as the next number. Similarly, LLMs predict the next word in a sentence based on patterns they've learned from vast amounts of text. For instance, given the sentence "The sun is shining", the model might predict "brightly" as the next word. Before training begins, the structure of the model must be defined. In our example, this would mean deciding the type of rule used to predict the sequence, like identifying it as "add 2 to the previous number". For LLMs, this involves defining the number of layers in the model, the capacity of the attention mechanism to focus on different parts of the input, and the dimensions of its internal representations and computations. These decisions shape how well the model can understand and generate language.

Once the model structure is determined, the training process begins. In the numerical example of 2, 4, 6, 8, if we assume that we have an arithmetic sequence. (that is, $x_t = 1 * x_{t-1} + 2$), the process involves identifying the parameters a=1 and b=2, which defines the rule for predicting the next number in the sequence. For LLMs, determining parameter values is far more complex, involving billions of parameters and advanced optimisation techniques. The training process is generally divided into two main stages: pre-training and fine-tuning, with reinforcement learning from human feedback (RLHF) often added for further refinement. Let us explore how these processes work in more detail in the following sections.

2.2.1 Pre-training and Fine Tuning

Pre-training is the initial phase [18] where the model is exposed to a vast corpus of text data, such as books, articles, and websites. The size and diversity of this dataset are crucial for the model to learn general patterns of language, including grammar, semantics, and common word associations. Importantly, this process is self-supervised [19], meaning the model learns from the structure of the data itself without requiring manually labelled examples. For instance, given the input "The sun is shining", the model predicts the next word based on patterns it has seen in similar contexts during pre-training. In this case, it might predict words like "brightly" or "today" depending on the associations it has learned from the training data.

On the technical side, the accuracy of these predictions is measured using a loss function, which quantifies how far the predicted word is from the actual next word in the sequence. For example, if the model predicts "cloudy" instead of "brightly," the loss function assigns a higher value, indicating a larger error. These errors are minimized through a process called backpropagation, which calculates how each parameter in the model contributes to the error. Optimisation algorithms then adjust the parameters to reduce the loss, gradually improving the model's ability to make accurate predictions.

After pre-training, the model undergoes fine-tuning on smaller, domain-specific datasets. For example, if the model is to generate weather reports, it might be fine-tuned on a specialized corpus of meteorological data. This step builds on the general language patterns learned during pre-training, refining the model to perform specialized tasks with high accuracy. Returning to the earlier example, fine-tuning might teach the model to complete "The sun is shining" with specific terms like "in the afternoon" or "through the clouds", based on its specialized training data.

2.2.2 Reinforcement Learning with Human Feedback (RLHF)

To further enhance the model's alignment with human expectations, a process called reinforcement learning with human feedback (RLHF) [20] is often applied. This technique fine-tunes the model beyond its technical accuracy, ensuring it generates outputs that are clear, relevant, and aligned with human preferences. The approach was popularized by the work of Christiano et al. [50], which demonstrated how human preferences could be used to guide models in learning complex tasks where clear evaluation criteria are difficult to define.

In RLHF, human evaluators review the model's outputs and rank them based on criteria like clarity, appropriateness, and usefulness. For instance, if the model is tasked with completing the phrase "The sun is shining", it might produce options like "brightly", "on the horizon", or "through the clouds". Evaluators would rank these outputs, perhaps preferring "brightly" for its clarity and generality over "on the horizon", which might seem overly specific. These rankings are then used to train a reward model, which predicts scores for outputs based on their alignment with human preferences.

During the reinforcement learning phase, the LLM generates new predictions, and the reward model assigns scores to these predictions. Using reinforcement learning algorithms, such as Proximal Policy Optimisation (PPO) [21], the LLM updates its parameters to maximize the reward from the reward model. During this process, desirable outputs are assigned higher scores, while less preferred options are penalized. Through this iterative process, the model improves its ability to align with human-provided feedback, producing outputs that better meet expectations for clarity, relevance, and user-friendliness.

2.3 What LLMs Do: The "Generative" Core of GPT

At the heart of large language models lies their generative ability—the "G" in GPT—which enables them to produce coherent, contextually relevant, and often human-like text. This generative capability transforms LLMs from passive tools into active collaborators across a wide range of applications, from creative writing and summarizing complex documents to coding and conversational AI.

The generative process begins with a prompt, which serves as the input to the model. For instance, given the phrase "The sun is shining", the model predicts likely continuations one word after another based on patterns it has learned during training. It might generate "brightly through the clouds" or "today after a long storm," depending on its understanding of the context and relationships within the data it was trained on. This prediction is not random but calculated from probabilities assigned to thousands of potential next words, allowing the model to choose the most appropriate continuation.

LLMs excel in generating text that reflects the tone, style, and intent of the input. For example, when prompted with "Write a formal letter about sunny weather", the model might begin: "Dear Sir/Madam, I am writing to express my appreciation for the delightful sunny weather we have been experiencing". Conversely, a casual prompt like "Tell me something cool about sunny days" could result in: "Did you know sunshine boosts your serotonin levels, making you feel happier?". These examples demonstrate the model's ability to adapt to the user's intent and produce contextually appropriate outputs.

Beyond generating text, LLMs can perform tasks such as aiding in language learning [9], creating summaries [8], and writing code [3]. This flexibility stems from their pre-training on diverse datasets that include examples across various domains, enabling them to handle a broad range of tasks. Despite its versatility, the generative process is not without challenges. LLMs may occasionally produce outputs that are factually inaccurate, contextually inappropriate, or overly verbose—a phenomenon known as "hallucination". This highlights the importance of human oversight to ensure the quality and reliability of their outputs. Table 1 summarizes the main uses of LLMs and how they can be applied to different tasks in education and learning analytics.

2.4 Using LLMs: From Web Interfaces to Advanced Frameworks

After understanding how LLMs work and what they can be used for, the next step is exploring how to use them in practical scenarios. Whether you are a beginner looking to interact with LLMs through simple web interfaces or an advanced user building custom applications using APIs and frameworks, there are many ways to harness their capabilities.

For beginners, web interfaces like ChatGPT, Claude (by Anthropic), and Perplexity AI offer user-friendly ways to explore LLM capabilities without requiring technical expertise. These platforms can assist with summarizing academic papers, generating insights from student feedback, generating or assessing code, brain-

Table 1 Summary of main LLM tasks

Task	Description	Example models	How it works technically	Application in education
Text completion	Predicting and generating text to complete a prompt.	GPT-4, GPT-o1, Claude, LLaMA	Utilizes autoregressive transformers to predict the next token based on preceding tokens, employing learned probability distributions.	Auto-generating personalized feedback for students [22].
Text generation (creative writing)	Generating creative and coherent long-form text.	GPT-4, GPT-o1, Claude, LLaMA	Uses autoregressive token prediction to produce semantically coherent, fluent, and diverse text.	Supporting creative writing assignments, generating essay ideas, or storytelling exercises for students [23].
Question answering	Providing answers to factual or contextual questions.	GPT-4, Claude, Gemini, RoBERTa	Employs attention mechanisms to focus on relevant parts of the input context, generating precise answers. In extractive QA, identifies and extracts specific text spans.	Automated Q&A systems for course material, providing quick answers to student queries [24]
Question generation	Generating questions based on input content.	GPT-4, T5, BERT-QG	Sequence-to-sequence models generate question tokens conditioned on input context.	Auto-generating quiz questions or comprehension tests from textbook material or lecture notes [25].
Text summarization	Generating concise summaries from longer texts.	GPT-4, Gemini, Claude, BART, T5	Applies sequence-to-sequence transformer architectures to condense input text into coherent summaries, preserving essential information.	Summarizing lecture notes, research papers, or learning materials for quicker understanding [26].
Translation	Translating text between languages.	GPT-4, Gemini, mT5, LLaMA	Uses encoder-decoder transformer models to map input tokens from the source language to the target language, aligning syntactic and semantic structures.	Supporting multilingual learners by translating course materials, assignments, or instructions [27].
Text classification	Assigning labels or categories to text.	BERT, RoBERTa, DistilBERT	Tokenizes input text and processes it through transformer layers to produce embeddings, which are then classified into categories using a task-specific head.	Analyzing student responses for sentiment (e.g., identifying frustration) [28], or automatic discourse coding [29].

(continued)

Table 1 (continued)

Task	Description	Example models	How it works technically	Application in education
Chat and dialogue	Generating conversational responses for chatbots.	Claude, GPT-4, Gemini, LLaMA	Maintains dialogue context by incorporating previous conversation turns, generating contextually relevant responses through autoregressive token prediction.	Powering tutoring chatbots that assist students with real-time explanations and guidance [30].
Code generation	Writing or completing code snippets.	GPT-o1, Codex, Gemini	Predicts subsequent code tokens, based on previous training on a large corpora of code, ensuring syntactic and semantic correctness.	Assisting students in programming courses by auto-generating or debugging code snippets [31].
Paraphrasing and rewriting	Generating rephrased or simplified versions of input text.	GPT-4, Claude, T5	Fine-tuned models rephrase input text while preserving semantic meaning, often using reinforcement learning.	Helping students rephrase ideas to avoid plagiarism, simplifying content, or generating alternative explanations of concepts [32, 33].
Style transfer	Rewriting text with a different style or tone.	GPT-4, CTRL, Claude, LLaMA	Generates text conditioned on desired stylistic attributes, often fine-tuned on datasets exemplifying target styles to learn appropriate transformations.	Simplifying advanced academic content for younger students or adapting tone for academic writing [34].
Knowledge extraction	Extracting entities or relationships from text.	GPT-4, BERT, Gemini	Utilizes attention layers to identify and classify entities and their relationships within text, enabling structured information extraction.	Extracting key concepts, relationships, or named entities from lecture transcripts and research articles.
Task automation	Performing tasks with minimal or no labeled examples.	Claude, GPT-4, GPT-o1, BART	Takes advantage of in-context learning by interpreting task instructions and examples provided within the input, adapting to new tasks without explicit retraining.	Automating grading of open-ended responses with minimal training or applying rubrics [35], performing tedious technical tasks such as converting a screenshot of a table into LATEX code.
Data-to-text generation	Generating natural language summaries or explanations from structured data.	GPT-4, T5	Encoder-decoder architectures map structured inputs (e.g., tables) into coherent text.	Generating textual explanations of student performance data, or dashboards [36].

Multimodal integration	Combining text with images, audio, or other modalities.	Gemini, GPT-4V, CLIP	Aligns representations from different modalities using combined encoder architectures, facilitating tasks that require understanding across multiple data types (image, audio, video, etc.).	Analyzing student engagement via multimodal data (e.g., combining video, text, and audio) [37].
Text-to-speech	Converting written text into spoken language.	Tacotron, Gemini, WaveNet, VALL-E	Deep learning architectures generate natural-sounding speech from tokenized text inputs.	Assisting visually impaired students, enabling spoken lecture notes, teaching young children who cannot read yet, or improving pronunciation for language learners [38, 39].
Text-to-video	Converting written text into video.	Sora, Runway Gen 2	Multimodal transformer architectures process text as input and generate video sequences using latent diffusion or frame interpolation.	Creating engaging educational videos from course materials [40]
Reasoning and problem solving	Solving logical, mathematical, or structured reasoning tasks.	Claude, GPT-4, GPT-o1, Gemini	Employs chain-of-thought prompting and step-by-step token generation to tackle complex reasoning tasks, enhancing problem-solving capabilities	Assisting students with step-by-step solutions in math or logical reasoning exercises, enhancing comprehension [41].

storming interventions for struggling learners, and more. These tools support uploading files, such as PDFs, or accessing content from open-access URLs, making them versatile for analyzing publicly available academic content.

Such tools demonstrate the transformative power of LLMs across diverse domains, enabling students to engage more deeply with complex concepts while allowing educators to focus on higher-level instruction and personalized support. Whether applied to programming, language learning, or data analysis, these tools foster iterative learning: if the initial output does not fully meet the user's needs, prompts and inputs can be refined to generate more tailored and meaningful results. This interaction not only enhances understanding but also encourages critical thinking and adaptability in both learners and educators.

The second way to interact with large language models is through an Application Programming Interface (API). An API acts as a bridge, allowing software applications to communicate with the LLM programmatically. This makes APIs essential for developers seeking to integrate LLMs into custom applications, websites, or tools. Using APIs offers several significant advantages, particularly in education and learning analytics. One of the key benefits is integration, as APIs can seamlessly connect LLM functionalities to various platforms, enabling educators and developers to embed advanced capabilities into their own applications or systems. This flexibility allows for the customization of LLM features to suit specific educational needs, such as automating feedback, analyzing student performance data, or generating personalized learning materials tailored to individual progress. APIs also enhance efficiency by automating repetitive tasks, reducing the need for manual interaction with web interfaces, and enabling large-scale operations. However, while APIs provide powerful tools for enhancing educational workflows, they require basic programming knowledge for implementation and careful management of API keys to ensure security and prevent unauthorized access. Additionally, API usage may incur costs, which could pose a limitation for projects with tight budgets.

API implementations are available in many programming languages, being Python the most common one. In the R programming language—the main language used in this book—, the *elmer* package [42] provides wrappers to interact with the APIs of the most common LLMs. In Chapter 11 [43] of this book, we provide a tutorial on how to use LLMs via API to obtain personalized feedback.

While APIs like OpenAI's are excellent for quick, scalable interactions with powerful pre-trained language models, they operate as black boxes, where you rely on the provider to host and manage the models. This simplicity is a significant strength, but it may not meet the needs of users seeking greater control for performing context-specific tasks. This is where frameworks like Hugging Face's Transformers come in. While transformers are not the only option within the Hugging Face ecosystem, they are among the most powerful and widely used tools for leveraging state-of-the-art language models. Hugging Face provides an open-source library—with over 1 million models (Dec, 2024)—that allows users to download and run these models locally or on the cloud infrastructure you control. A comprehensive list of pre-trained models is available on the Hugging Face Models Hub.

In the context of education and learning analytics, transformers are capable of making sense of complex data, such as analyzing student feedback, extracting themes from surveys, or identifying trends in course engagement. Moreover, users can further customize these models and fine-tune them to address their specific education needs. By offering flexibility, greater control, and adaptability, transformers expand the potential of LLMs beyond the simplicity of many API-based interactions.

Consider a scenario where an educator needs to analyse student feedback to understand sentiments and identify areas for improvement. A pre-trained sentiment analysis model from Hugging Face can quickly classify feedback as positive, neutral, or negative, offering actionable insights for educators. For example, a model such as bert-base-multilingual-uncased can be used, which has been fine-tuned specifically for sentiment analysis of product reviews in six languages. It is designed for immediate use in analyzing sentiment across multilingual product reviews and can also serve as a robust starting point for further fine-tuning on related sentiment analysis tasks, such as analyzing students' course feedback or collaborative discourse. A few LLMs that have been specifically trained for education purposes exist, such as EduBERT [44] or K-12BERT [45]. In Chapter 10 [46] of this book, we provide a tutorial on how to use language models locally to automatically classify students' discourse in collaborative problem-solving.

3 Conclusion

This chapter provided an introduction to LLMs, exploring their foundational components—transformer architecture, pre-training, and generative abilities—while demonstrating their transformative potential across applications in education. Building on this, the chapter also explored tools for interacting with LLMs, from beginner-friendly web interfaces to more advanced tools like APIs and frameworks such as OpenAI API and Hugging Face's Transformers. These approaches enable users to operationalize LLMs for tasks such as summarizing academic papers, automatic text classification, and generating of learning materials.

However, as the integration of LLMs in education becomes increasingly widespread, it is crucial to critically examine their limitations. The potential for biases in their outputs, issues of factual inaccuracies, and the challenges of ensuring transparency and interpretability must be addressed to harness their capabilities effectively. A careful balance between taking advantage of LLMs' strengths and mitigating their shortcomings is necessary to ensure they serve as tools that improve, rather than hinder, educational and research practices [47].

An important avenue for addressing these challenges is the integration of explainable AI (xAI) techniques with LLMs. xAI methods aim to make the predictions and operations of LLMs more transparent [48, 49], helping users understand the factors driving their outputs. In educational contexts, for example, xAI can clarify why specific feedback was generated for a student or reveal how particular patterns in data influence predictions, such as grading or performance analytics. Transparency is needed to build trust by enabling users to identify potential biases or inaccuracies in model outputs. As the adoption of LLMs continues, embedding xAI into their workflows will be critical for ensuring ethical and equitable outcomes.

References

1. Brown TB, Mann B, Ryder N, Subbiah M, Kaplan J, Dhariwal P, Neelakantan A, Shyam P, Sastry G, Askell A, Agarwal S, Herbert-Voss A, Krueger G, Henighan T, Child R, Ramesh A, Ziegler DM, Wu J, Winter C, Hesse C, Chen M, Sigler E, Litwin M, Gray S, Chess B, Clark J, Berner C, McCandlish S, Radford A, Sutskever I, Amodei D (2020) Language models are few-shot learners. In: Proceedings of the 34th international conference on neural information processing systems. Curran Associates Inc., Red Hook, NY, USA
2. Vaswani A, Shazeer NM, Parmar N, Uszkoreit J, Jones L, Gomez AN, Kaiser L, Polosukhin I (2017) Attention is all you need. Neural Inf Process Syst 30:5998–6008. https://papers.nips.cc/paper/7181-attention-is-all-you-need
3. Wang J, Chen Y (2023) A review on code generation with LLMs: Application and evaluation. In: 2023 IEEE international conference on medical artificial intelligence (MedAI). IEEE, pp 284–289. https://doi.org/10.1109/medai59581.2023.00044
4. Zhang D, Yu Y, Dong J, Li C, Su D, Chu C, Yu D (2024) MM-LLMs: recent advances in multimodal large language models. arXiv [cs.CL]. http://arxiv.org/abs/2401.13601
5. Jeon J, Lee S (2023) Large language models in education: a focus on the complementary relationship between human teachers and ChatGPT. Educ Inf Technol 28:15873–15892. https://doi.org/10.1007/s10639-023-11834-1
6. Mistry NP, Saeed H, Rafique S, Le T, Obaid H, Adams SJ (2024) Large language models as tools to generate radiology board-style multiple-choice questions. Acad Radiol 31:3872–3878. https://doi.org/10.1016/j.acra.2024.06.046
7. Huang C-Y, Wei J, Huang T-HK (2024) Generating educational materials with different levels of readability using LLMs. In: Proceedings of the third workshop on intelligent and interactive writing assistants. ACM, New York, NY, USA, pp 16–22. https://doi.org/10.1145/3690712.3690718
8. Areces C, Benotti L, Bulgarelli F, Echeveste E, Finzi N (2024) Leveraging language models and automatic summarization in online programming learning environments. Commun ACM 67:86–87. https://doi.org/10.1145/3653323
9. Liu Z, Yin SX, Lee C, Chen NF (2024) Scaffolding language learning via multi-modal tutoring systems with pedagogical instructions. arXiv [cs.CL]. http://arxiv.org/abs/2404.03429
10. Khalil M, Vadiee F, Shakya R, Liu Q (2025) Creating artificial students that never existed: Leveraging large language models and CTGANs for synthetic data generation. arXiv [cs.LG]. http://arxiv.org/abs/2501.01793
11. Aher GV, Arriaga RI, Kalai AT (23--29 Jul 2023) Using large language models to simulate multiple humans and replicate human subject studies. In: Krause A, Brunskill E, Cho K, Engelhardt B, Sabato S, Scarlett J (eds) Proceedings of the 40th international conference on machine learning. PMLR, pp 337–371. https://proceedings.mlr.press/v202/aher23a.html
12. Liyanage UP, Ranaweera ND (2023) Ethical considerations and potential risks in the deployment of large language models in diverse societal contexts. J Comput Soc Dyn 8:15–25. https://vectoral.org/index.php/JCSD/article/view/49
13. Drori I, Te'eni D (2024) Human-in-the-loop AI reviewing: feasibility, opportunities, and risks. J Assoc Inf Syst 25:7. https://doi.org/10.17705/1jais.00867
14. Wang Y, Zhong W, Li L, Mi F, Zeng X, Huang W, Shang L, Jiang X, Liu Q (2023) Aligning large language models with human: a survey. arXiv [cs.CL]. http://arxiv.org/abs/2307.12966

15. Hochreiter S, Schmidhuber J (1997) Long short-term memory. Neural Comput 9:1735–1780. https://doi.org/10.1162/neco.1997.9.8.1735
16. Lecun Y, Bottou L, Bengio Y, Haffner P (1998) Gradient-based learning applied to document recognition. Proc IEEE Inst Electr Electr Eng 86:2278–2324. https://doi.org/10.1109/5.726791
17. Cho K, Merrienboer B van, Gulcehre C, Bahdanau D, Bougares F, Schwenk H, Bengio Y (2014) Learning phrase representations using RNN encoder-decoder for statistical machine translation. In: Proceedings of the 2014 conference on empirical methods in natural language processing (EMNLP), pp 1724–1734. https://doi.org/10.3115/v1/d14-1179
18. Wang H, Li J, Wu H, Hovy E, Sun Y (2022) Pre-trained language models and their applications. Engineering (Beijing, China) 25:51–65. https://doi.org/10.1016/j.eng.2022.04.024
19. Rani V, Nabi ST, Kumar M, Mittal A, Kumar K (2023) Self-supervised learning: a succinct review. Arch Comput Methods Eng State Art Rev 30:2761–2775. https://doi.org/10.1007/s11831-023-09884-2
20. Kaufmann T, Weng P, Bengs V, Hüllermeier E (2023) A survey of reinforcement learning from human feedback. arXiv [cs.LG]. http://arxiv.org/abs/2312.14925
21. Schulman J, Wolski F, Dhariwal P, Radford A, Klimov O (2017) Proximal policy optimization algorithms. arXiv [cs.LG]. https://doi.org/10.48550/ARXIV.1707.06347
22. Koutcheme C, Hellas A (2024) Propagating large language models programming feedback. In: Proceedings of the eleventh ACM conference on learning @ scale. ACM, New York, NY, USA, pp 366–370. https://doi.org/10.1145/3657604.3664665
23. Woo DJ, Wang Y, Susanto H, Guo K (2023) Understanding english as a foreign language students' idea generation strategies for creative writing with natural language generation tools. J. Educ Comput Res 61:1464–1482. https://doi.org/10.1177/07356331231175999
24. Wu Y, Henriksson A, Duneld M, Nouri J (2023) Towards improving the reliability and transparency of ChatGPT for educational question answering. In: Lecture notes in computer science. Springer Nature Switzerland, Cham, pp 475–488. https://doi.org/10.1007/978-3-031-42682-7_32
25. Lee U, Jung H, Jeon Y, Sohn Y, Hwang W, Moon J, Kim H (2024) Few-shot is enough: exploring ChatGPT prompt engineering method for automatic question generation in english education. Educ Inf Technol 29:11483–11515. https://doi.org/10.1007/s10639-023-12249-8
26. Kolagar Z, Zarcone A (2024) HumSum: a personalized lecture summarization tool for humanities students using LLMs. In: Proceedings of the 1st workshop on personalization of generative AI systems (PERSONALIZE 2024), pp 36–70. https://aclanthology.org/2024.personalize-1.4.pdf
27. Ohashi L (2024) AI in language education: The impact of machine translation and ChatGPT. In: Intelligent systems reference library. Springer Nature Switzerland, Cham, pp 289–311. https://doi.org/10.1007/978-3-031-71232-6_13
28. Lundqvist K, Liyanagunawardena T, Starkey L (2020) Evaluation of student feedback within a MOOC using sentiment analysis and target groups. Int Rev Res Open Dis Learn 21:140–156. https://doi.org/10.19173/irrodl.v21i3.4783
29. Misiejuk K, Kaliisa R, Scianna J (2024) Augmenting assessment with AI coding of online student discourse: a question of reliability. Comput Educ Artif Intell 6:100216. https://doi.org/10.1016/j.caeai.2024.100216
30. Labadze L, Grigolia M, Machaidze L (2023) Role of AI chatbots in education: systematic literature review. Int J Educ Technol Higher Educ 20. https://doi.org/10.1186/s41239-023-00426-1
31. Jin H, Lee S, Shin H, Kim J (2024) Teach AI how to code: using large language models as teachable agents for programming education. In: Proceedings of the CHI conference on human factors in computing systems. ACM, New York, NY, USA, pp 1–28. https://doi.org/10.1145/3613904.3642349
32. Lin J, Han Z, Thomas DR, Gurung A, Gupta S, Aleven V, Koedinger KR (2024) How can i get it right? Using GPT to rephrase incorrect trainee responses. Int J Artif Intell Educ 1–27. https://doi.org/10.1007/s40593-024-00408-y

33. Perkins M (2023) Academic integrity considerations of AI large language models in the post-pandemic era: ChatGPT and beyond. J Univ Teach Learn Pract 20. https://doi.org/10.53761/1.20.02.07
34. Zhang C, Liu X, Ziska K, Jeon S, Yu C-L, Xu Y (2024) Mathemyths: leveraging large language models to teach mathematical language through child-AI co-creative storytelling. In: Proceedings of the CHI conference on human factors in computing systems. ACM, New York, NY, USA, pp 1–23. https://doi.org/10.1145/3613904.3642647
35. Pinto G, Cardoso-Pereira I, Monteiro D, Lucena D, Souza A, Gama K (2023) Large language models for education: Grading open-ended questions using ChatGPT. In: Proceedings of the XXXVII brazilian symposium on software engineering. ACM, New York, NY, USA, pp 293–302. https://doi.org/10.1145/3613372.3614197
36. Pinargote A, Calderón E, Cevallos K, Carrillo G, Chiluiza K, Echeverria V (2024) Automating data narratives in learning analytics dashboards using GenAI. In: 2024 joint of international conference on learning analytics and knowledge workshops. CEUR-WS, pp 150–161. https://research.monash.edu/files/606148660/593895957_oa.pdf
37. Vaiani L, Cagliero L, Garza P (2024) Emotion recognition from videos using multimodal large language models. Future Int 16:247. https://doi.org/10.3390/fi16070247
38. Dai L, Kritskaia V, Velden E van der, Jung MM, Postma M, Louwerse MM (2022) Evaluating the usage of text-to-speech in K12 education. In: Proceedings of the 2022 6th international conference on education and e-learning. ACM, New York, NY, USA, pp 182–188. https://doi.org/10.1145/3578837.3578864
39. Wei X (2024) Text-to-speech technology and math performance: a comparative study of students with disabilities, english language learners, and their general education peers. Educ Res (Washington, DC: 1972) 53:285–295. https://doi.org/10.3102/0013189x241232995
40. Adetayo AJ, Enamudu AI, Lawal FM, Odunewu AO (2024) From text to video with AI: the rise and potential of sora in education and libraries. Library hi tech news ahead-of-print. https://doi.org/10.1108/lhtn-02-2024-0028
41. Fung SCE, Wong MF, Tan CW (2023) Chain-of-thoughts prompting with language models for accurate math problem-solving. In: 2023 IEEE MIT undergraduate research technology conference (URTC). IEEE, pp 1–5. https://doi.org/10.1109/urtc60662.2023.10534945
42. Wickham H (2024) Elmer: Call LLM APIs from r. Github. https://github.com/tidyverse/elmer
43. López-Pernas S, Song Y, Oliveira E, Saqr M (2025) LLMs for explainable artificial intelligence: automating natural language explanations of predictive analytics models. In: Saqr M, López-Pernas S (eds) Advanced learning analytics methods: AI, precision and complexity. Springer Nature Switzerland, Cham
44. Clavié B, Gal K (2019) EduBERT: Pretrained deep language models for learning analytics. arXiv [cs.CY]. http://arxiv.org/abs/1912.00690
45. Goel V, Sahnan D, Venktesh V, Sharma G, Dwivedi D, Mohania M (2022) K-12BERT: BERT for k-12 education. In: Artificial intelligence in education. Posters and late breaking results, workshops and tutorials, industry and innovation tracks, practitioners' and doctoral consortium. Springer International Publishing, Cham, pp 595–598. https://doi.org/10.1007/978-3-031-11647-6_123
46. López-Pernas S, Misiejuk K, Saqr M (2025) Using BERT-like language models for automated discourse coding: a primer and tutorial. In: Saqr M, López-Pernas S (eds) Advanced learning analytics methods: AI, precision and complexity. Springer Nature Switzerland, Cham
47. Shen Y, Heacock L, Elias J, Hentel KD, Reig B, Shih G, Moy L (2023) ChatGPT and other large language models are double-edged swords. Radiology 307:e230163. https://doi.org/10.1148/radiol.230163
48. López-Pernas S, Oliveira E, Song Y, Saqr M (2025) AI, explainable AI and evaluative AI: an introduction to informed data-driven decision-making in education. In: Saqr M, López-Pernas S (eds) Advanced learning analytics methods: AI, precision and complexity. Springer Nature Switzerland, Cham

49. Zytek A, Pidò S, Veeramachaneni K (2024) LLMs for XAI: Future directions for explaining explanations. arXiv [cs.AI]. http://arxiv.org/abs/2405.06064
50. Christiano PF, Leike J, Brown T, Martic M, Legg S, Amodei D (2017) Deep reinforcement learning from human preferences. In: Advances in neural information processing systems, p 30

The Use of Natural Language Processing in Learning Analytics

Tarid Wongvorachan and Okan Bulut

Abstract In the educational context, the usage of written text is as equally important as the usage of numerical components for teachers to make informed decisions in their pedagogical strategies. This chapter explores the practical implementation of natural language processing (NLP) within the context of learning analytics, specifically focusing on text mining and information extraction of textual data sourced from LA environments, including student feedback, forum posts, online discussions, and course materials. This application could provide insights into student behaviors, learning patterns, and educational content.

This chapter introduces foundational NLP concepts and techniques, including text preprocessing, TF-IDF analysis, topic modeling, and text summarization. We demonstrate these techniques using functions from R packages such as tm, tidytext, topicmodels, and lexRankr to students' comments on a course evaluation survey. Finally, we explore the application of sentiment analysis in learning analytics to gain insights into student perceptions. In summary, this chapter serves as a comprehensive guide for leveraging NLP techniques in learning analytics using R. This guideline provides readers with the knowledge and tools necessary to analyze and derive insights from textual data in educational contexts.

Keywords Natural language processing · Text mining · Sentiment analysis · Learning analytics

1 Introduction

In the educational context, using written text is equally important as using numerical components for teachers to make informed decisions in their pedagogical strategies. For example, teachers may use students' essays rather than a multiple-choice exam

T. Wongvorachan (✉) · O. Bulut
University of Alberta, Edmonton, AB, Canada
e-mail: wongvora@ualberta.ca

M. Saqr, S. López-Pernas (eds.), *Advanced Learning Analytics Methods*,
https://doi.org/10.1007/978-3-031-95365-1_9

to evaluate their writing proficiency. With the increased use of educational technologies in the classroom, utilizing textual data has become more crucial in maximizing the effectiveness of learning analytics (LA) systems [1]. The use of textual data is made possible through the application of natural language processing (NLP)—an artificial intelligence-driven technique designed to manipulate, transform, and extract insights from text [2]. In fact, NLP comprises several subfields that cover its vast capability, from syntactic text processing, reasoning, and text generation [25].

In this chapter, we concentrate on selected NLP techniques relevant to LA, including text preprocessing (tokenization, stemming, lemmatization), frequency-based analysis (e.g., Term Frequency-Inverse Document Frequency, or TF-IDF), topic modeling (Latent Dirichlet Allocation, or LDA), text summarization, and sentiment analysis. These techniques are primarily found within fields of study focused on text mining and information extraction—key areas in NLP concerned with generating insights from textual data [25]. Such techniques could be applied to textual data sources within the LA environment, such as student feedback, forum posts, online discussions, and course materials. This application could provide insights into student behaviors, learning patterns, and educational content.

The chapter starts by discussing foundational concepts and terminologies in NLP, such as lexical, syntax, and corpus. Then, we explain the importance of text preprocessing in NLP and discuss various methods such as tokenization, stemming, lemmatization, and stop word removal. Practical demonstrations of these techniques are provided, leveraging R packages such as *tm* [7, 8] and *tidytext* [27]. After text processing, we move the discussion to the application of NLP for text analysis with techniques such as TF-IDF analysis to identify and visualize keywords and themes across documents.

Additionally, the discussion extends to the application of topic modeling with LDA through the *topicmodels* package [11, 12]. This serves to identify hidden themes using NLP models, thereby augmenting the insights gained from TF-IDF analyses. This progression highlights the evolution from frequency-based analyses to the application of NLP models in theme identification and interpretation. As an alternative approach to extracting key insights from textual data, this chapter explores the utilization of text summarization techniques with the *lexRankr* package [28] applied to educational content. Contrasting with the extraction of keyword sets facilitated by TF-IDF and topic modeling methods, text summarization condenses documents into succinct sentences. This application showcases the efficacy of NLP in distilling large volumes of text into concise summaries, thereby enhancing accessibility and comprehension.

Finally, we discuss and showcase the relevance of sentiment analysis in learning analytics by analyzing sentiment in student feedback using *tidytext* and *sentimentr* [23] packages. Information obtained from sentiment analysis could be inferred from students' perceptions and attitudes regarding the course content, thereby providing more understanding of the pedagogical efficacy of the course. This chapter serves as a comprehensive guide for leveraging NLP techniques in learning analytics using R. This guideline provides readers with the knowledge and tools necessary to analyze and derive insights from textual data in educational contexts.

2 Literature Review

2.1 Foundational Concepts in NLP

Formally defined, NLP is a subfield of artificial intelligence that allows computers to understand, interpret, and generate human language to bridge the gap between human communication and computer understanding [2]. NLP can transform unstructured text into structured data that can be easily analyzed and utilized for various applications, such as sentiment analysis, machine translation, speech recognition, and text summarization. Textual data in NLP has several levels/aspects/dimensions that can be described by the following terminologies by Bengfort et al. [2] and Allen et al. [1]. (1) Tokens: Tokens are the basic units of text, often corresponding to words, phrases, or symbols such as "learn" or "students"; (2) Lexicon: A lexicon is a collection of words and their corresponding meanings. For example, "chemistry", "space", "physics", or "cell" are part of the science lexicon; (3) Semantics: Semantics is concerned with the meaning of words, phrases, and sentences and their relationships within their context; (4) Corpus: A corpus is a large and structured set of texts used for training and evaluating NLP models. It serves as the foundational data for various NLP tasks, providing context and examples for algorithms to learn from. These four terminologies describe textual data from individual words or a collection of words or phrases with their meaning.

To illustrate these concepts, consider the sentence 'Students study cell structures in biology.' Table 1 displays examples of tokens, lexicon, semantics, and corpus of such a sentence.

Several steps are involved in transforming natural language text to a machine-readable format. These steps, collectively known as text preprocessing, convert raw text into structured data, which is essential for subsequent analysis and modeling as described by Bird et al. [3] as follows:

1. **Tokenization**: Divides text into tokens, the fundamental units of analysis.
2. **Normalization**: Converts text to a uniform format by removing punctuation, converting text to lowercase, and addressing special characters, ensuring consistency (e.g., "Dog" and "dog" treated equivalently).
3. **Lemmatization**: Converts words to their root forms based on context, such as "dancer," "dancing," and "dances" being reduced to "dance."

Table 1 Examples of NLP terms of a sentence

NLP terms	Examples
Tokens	The text is broken into individual units: 'Students,' 'study,' 'cell,' 'structures,' 'in,' and 'biology.'
Lexicon	'Cell' is defined as the basic structural and functional unit of living organisms
Semantics	'Cell structures' refers to biological components rather than, for instance, the layout of telecommunications equipment.
Corpus	A collection of educational content related to biology

4. **Stop Word Removal**: Eliminates common, less meaningful words (e.g., "a," "the") to focus on more meaningful terms in the text.

When the textual data has been processed and is ready to be analyzed, several analyses can be done to generate insights. Firstly, TF-IDF is a statistical measure used to assess a word's importance in a document within a corpus. It uses two metrics: term frequency (TF), which counts how often a term appears in a document, and inverse document frequency (IDF), which gauges a term's significance by examining its rarity across all documents [22]. The TF-IDF value rises with the word's frequency in a document but is balanced by its prevalence in the corpus [22]. Hence, frequently appearing words are deemed important, but overly common words are considered less significant. For instance, the word "science" in a science corpus would be expected and thus may not provide unique insights.

Second, the latent Dirichlet allocation (LDA) method is a generative probabilistic model that can extract lists of topics from the document. Specifically, LDA assumes each document is a mixture of topics and identifies topics within a set of documents by assigning a probability distribution over words for each topic [4]. LDA can extract keywords by identifying highly probable words within specific topics, providing insights into the thematic structure of the text in addition to the keywords identified by the TF-IDF method [29].

Third, the text summarization technique serves to condense a long piece of text into a shorter version while preserving its main ideas. There are two main approaches to text summarization: extractive and abstractive. Abstractive text summarization generates new sentences that convey the main ideas of the original text, similar to how humans summarize, through the use of advanced NLP models such as the Bidirectional Autoregressive Transformer [26]. This approach, however, is not the focus of this tutorial as it currently doesn't have a means of implementation in R. Extractive text summarization, which is covered in this chapter, selects critical sentences or phrases directly from the original text to create a summary by ranking sentences based on features such as sentence position, length, term frequency, and similarity to the title [21].

Finally, sentiment analysis can determine the emotional tone of the text by classifying it as positive, negative, or neutral. Various NLP models and techniques can be employed for sentiment analysis, including lexicon-based approaches and machine learning models [24]. An example of a lexicon-based approach is the utilization of the Valence Aware Dictionary and sEntiment Reasoner (VADER) model that uses syntactic rules to determine scores from lexical features (e.g., sentiment lexicons) with grammatical and syntactical conventions [14]. An example of the machine learning approach is the utilization of a pre-trained model, such as the global vectors for word representation (GloVe), to predict sentiment categories of the given text [20]. These four techniques extract insights from the textual data at several levels as means of data reduction, from text to keywords to short summary and inferred property of text sentiment.

2.2 The Relevance of NLP in the LA Context

In the LA context, several textual components are involved, ranging from the written curriculum, syllabus, course content, teachers' written feedback, and even student feedback after the course is completed. Insights drawn from these data sources are valuable in guiding the advancement of educational technologies from the student level (e.g., determining at-risk students) to the task level (e.g., predicting item difficulty) [1]. NLP can significantly enhance student performance by analyzing their input, such as responses to quizzes or exams, and providing automated feedback [17]. Feedback generated by NLP can be delivered at various levels: task-level (e.g., providing a direct answer), process-level (e.g., offering verbal hints), and self-regulated learning level (e.g., generating context-relevant feedback to encourage self-reflection) [13, 30]. An example of such innovation is AcaWriter, a writing assistance tool that provides real-time feedback on students' written work, such as essays and reports, focusing on aspects like clarity and conciseness [15]. Beyond benefiting students, NLP can also assist instructors by evaluating the quality of their written feedback, categorizing it as effective, mediocre, or ineffective [19]. This innovation supports learning on multiple fronts, enhancing both student self-assessment and the effectiveness of instructor feedback.

Furthermore, NLP can analyze extensive volumes of text from syllabi, course materials, and educational resources such as theses, reducing them to their essential elements that contain only the most relevant information [10]. This distilled information can then inform instructors' decisions. For instance, instructors may utilize text summarization and topic modeling to condense a corpus of syllabi into topic lists and concise summaries, thereby aiding the course design process by making it more efficient. For example, instructors can use topic modeling to understand themes of student discussions in a text-rich environment such as massive open online courses [6]. In educational research, topic modeling can also be used to identify clusters of keywords in textual metadata such as abstracts to understand progress of the field [5].

NLP can also predict item characteristics, such as item difficulty (represented by the proportion of correctness), mean response time, and item discrimination (i.e., biserial correlation), by analyzing linguistic features like average word length, complex word count, readability, concreteness rating, and word frequency (to identify rare words) [31]. This capability enables teachers to assess predicted item properties before implementation, determining whether an item is too difficult, too easy, or appropriately challenging.

Additionally, NLP can provide machine translation, which is particularly beneficial in scenarios involving second-language learners [10]. Given the prevalence of foreign language learners in modern education, embedding machine translation features in an LA dashboard can increase accessibility to lesson content and enhance feedback comprehension [16, 30]. These examples demonstrate the significant potential of integrating NLP within the LA context to enhance educational practices at multiple levels. This NLP application allows educational institutions to gain deeper insights into student behaviors, optimize curriculum design, and improve feedback and assessment processes.

3 Tutorial with R

We start by installing and importing the following packages:

```
install.packages(c("tidytext", "tm", "textstem", "dplyr", "ggplot2", "topicmodels",
"lexRankr", "sentimentr", "tidyr"))

library(tidytext) # for text cleaning and tokenization
#https://www.tidytextmining.com/
library(tm) # for TF-IDF calculation
library(textstem) #for text stemming
library(dplyr) #for pipeline and data management
library(ggplot2) #for plotting
library(topicmodels) #for LDA topic modeling
library(lexRankr) #for text summarization
library(sentimentr) #for sentiment analysis
library(tidyr) #for dataset transformation
```

The double colon (::) notation is used in our code to identify the package from which each function comes. This approach enhances clarity for readers who may wish to utilize specific functions from the tutorial.

3.1 Text Preprocessing

In this chapter, we use a dataset consisting of students' comments on a course evaluation. The dataset can be downloaded from this hyperlink. We first need to import the data into R. This is accomplished using the read.csv function, which reads a file in CSV (comma-separated values) format and creates a data frame. The specific command used is df < – read.csv('df.csv', stringsAsFactors = FALSE). Here, 'df.csv' is the name of the CSV file containing the students' comments, and stringsAsFactors = FALSE ensures that text data is imported as strings rather than being converted to factors, which is useful for text analysis.

```
df <- read.csv(url('https://raw.githubusercontent.com/lamethods/data2/refs/heads/main/feedback/df.csv'),
stringsAsFactors = FALSE)
```

After importing the data, we can take a quick look at the first few entries by using the head function. The command 'head(df, 3)' returns the first three rows of the data frame, providing an initial glimpse of the comments. In this example, we see three detailed comments on various aspects of the course, such as the interactive sessions, up-to-date content, and the instructor's responsiveness.

```
> head(df, 3)

Student_comment
1                              The course was very informative and engaging. I
particularly enjoyed the interactive sessions where we could apply what we learned

in practical scenarios. The instructor's enthusiasm made learning fun.

2 I really enjoyed the lectures and the assignments. The course content was
relevant and up-to-date, which made it easier to understand current trends in the
field. The instructor provided valuable insights and real-world examples.

3                                         The instructor was very knowledgeable and
approachable. Whenever I had questions, they were quick to respond and provided
detailed explanations. This greatly enhanced my learning experience.
```

Additionally, the 'dim' function is used to check the dimensions of the data frame with the command 'dim(df)', which in this case returns [1] 100 1. This indicates that the data frame contains 100 rows and 1 column, confirming that we have 100 comments from students to analyze. By importing and briefly inspecting our data, we set the stage for a deeper analysis of students' feedback, which can provide valuable insights into the strengths and areas for improvement in the course.

```
> dim(df)
[1] 100   1
```

After importing the dataset into our R environment, we can proceed with text cleaning. This involves transforming and standardizing the raw text to ensure consistency and remove noise, making it easier to extract meaningful insights. First, the comments are tokenized using the unnest_tokens function, which splits each comment into individual words, creating a tidy data frame where each row represents a single word. This is done with the command token <– df % > % unnest_tokens(output = word, input = Student_comment).

```
token <- df %>%
  unnest_tokens(output = word, input = Student_comment)
```

Next, all words are converted to lowercase using token$word <– tolower(token$word), ensuring uniformity by eliminating discrepancies caused by capitalization. The words are then lemmatized with token$word <– lemmatize_words(token$word), which reduces words to their base or dictionary form. For example, "running" would be reduced to "run".

```
token$word <- tolower(token$word)
token$word <- lemmatize_words(token$word)
```

To remove common but uninformative words, custom stop words specific to this context, such as "student", "course", and "instructor", are defined and removed from the data. This is accomplished by merging a predefined list of stop words with the custom list and using the anti_join function. Numbers, which are often not useful in text analysis, are removed with token$word <– gsub("\\d+", """, token$word). Empty tokens resulting from this cleaning are filtered out using token <– token % > % filter(word! = ""). Lastly, all punctuation is removed using token$word <– tm::removePunctuation(token$word), ensuring that only words remain.

```
custom_stop_words <- c("student", "course", "instructor", "instructors",
"students")

token <- token %>%
  anti_join(stop_words %>% bind_rows(tibble(word = custom_stop_words, lexicon =
"custom")))

token$word <- gsub("\\d+", "", token$word) # remove numbers
token <- token %>% filter(word != "") # remove empty tokens
token$word <- tm::removePunctuation(token$word) #remove all punctuation
```

After these cleaning steps, the dim function shows that we have 842 words (dim(token)), and the head function reveals the first ten cleaned words (head(token, 10)), indicating that the text is now ready for further analysis. This process effectively standardizes the text, reduces noise, and focuses on meaningful content for subsequent NLP tasks.

```
> dim(token)
[1] 842   1
> head(token, 10)
          word
1  informative
2       engage
3        enjoy
4  interactive
5      session
6        apply
7        learn
8    practical
9     scenario
10 instructors
```

3.2 *Keyword Identification with TF-IDF*

Keyword identification using TF-IDF is a crucial technique in NLP for identifying the most significant words in a collection of documents. TF-IDF helps to highlight important words that are unique to individual documents within a corpus while downplaying words that are common across multiple documents

The process begins with counting the occurrence of each word in each document. This is achieved with the command word_counts <– token % > % count(document = row_number(), word), which creates a data frame where each row represents the frequency of a word in a specific document.

```
# Count the occurrence of each word in each document
word_counts <- token %>% count(document = row_number(), word)
```

Next, the raw term frequency, which is the total count of each word across all documents, is calculated using raw_term_frequency < – word_counts % > % count(word). Following this, the document frequency is determined, which counts the number of documents in which each word appears. This is computed with document_frequency < – word_counts % > % group_by(word) % > % summarise(docs = n_distinct(document)).

```
# Calculate raw term frequency
raw_term_frequency <- word_counts %>% count(word)

# Calculate document frequency
document_frequency <- word_counts %>% group_by(word) %>% summarise(docs =
n_distinct(document))
```

The Inverse Document Frequency (IDF) is then calculated. IDF measures how unique or rare a word is across all documents in the corpus. It is calculated as the logarithm of the total number of documents divided by the number of documents containing the word. The TF-IDF score for each word is obtained by multiplying its raw term frequency by its IDF value. This combines the importance of the word in a specific document with its rarity across the corpus.

```
# Calculate IDF
idf <- document_frequency %>% mutate(idf = log(nrow(word_counts) / docs))

# Join raw term frequency and IDF
tf_idf <- raw_term_frequency %>% inner_join(idf, by = "word") %>% mutate(tf_idf = n
* idf)
```

To visualize the results, the top 10 words with the highest TF-IDF values are selected and plotted. This is done by arranging the words in descending order of their TF-IDF scores and selecting the top 10. The selected words are then plotted using a bar chart with ggplot2, where the words are displayed along the x-axis and their TF-IDF scores on the y-axis.

```
# Plot the top 10 important words

# Select the top 10 words with the highest TF-IDF values
top_10 <- tf_idf %>% arrange(desc(tf_idf)) %>% head(10)

# Create a bar chart
ggplot(top_10, aes(x = reorder(word, tf_idf), y = tf_idf)) +
  geom_bar(stat = "identity", fill = "steelblue") +
  coord_flip() +
  labs(x = "Words", y = "TF-IDF", title = "Top 10 Words with Highest TF-IDF
Values") +
  theme_minimal()
```

The output of this process provides a list of the top 10 words with the highest TF-IDF values. Output 1 and Fig. 1 show that words like “learn,” “material,” and “feel” are among the most significant in the student comments, as indicated by their high TF-IDF values. These words are particularly important because they are frequently used in the comments while being relatively unique to specific documents, highlighting their significance in the context of the course evaluation.

3.2.1 Output 1

3.2.1.1 TF-IDF Based Important Words from the Course Evaluation Corpus

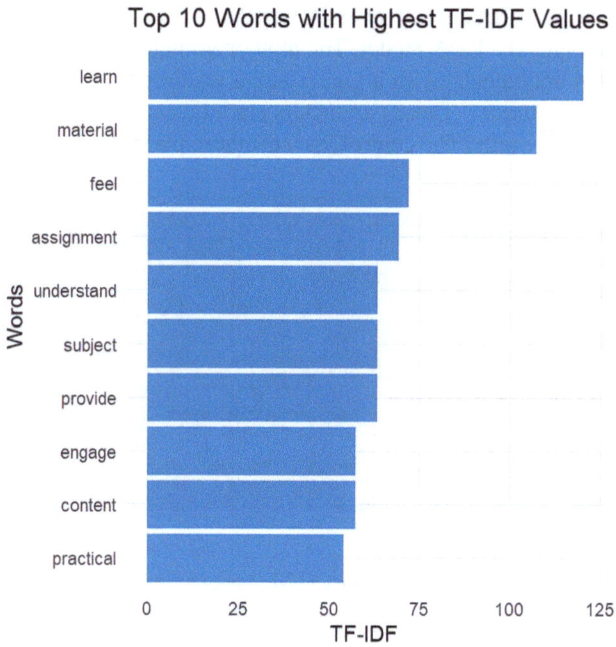

Fig. 1 Bar plot of TF-IDF based important word

```
> top_10
         word  n docs      idf    tf_idf
1       learn 39   39 3.072218 119.81652
2    material 33   33 3.239272 106.89599
3        feel 19   19 3.791341  72.03548
4  assignment 18   18 3.845408  69.21735
5     provide 16   16 3.963191  63.41106
6     subject 16   16 3.963191  63.41106
7  understand 16   16 3.963191  63.41106
8     content 14   14 4.096723  57.35412
9      engage 14   14 4.096723  57.35412
10  practical 13   13 4.170831  54.22080
```

3.3 Identifying Hidden Themes with Topic Modeling

Identifying hidden themes in textual data using Topic Modeling is a powerful technique in NLP that helps uncover the underlying structure and prevalent topics within a collection of documents. One common method for topic modeling is LDA, which assumes that each document is a mixture of several topics and that each topic is a mixture of words with certain probabilities [4].

The process begins with creating a Document-Term Matrix (DTM), which is a matrix where rows represent documents and columns represent words, with the entries indicating the frequency of each word in each document. Next, we set hyperparameters for the LDA model. The alpha parameter controls the distribution of topics within documents, with a higher alpha resulting in a more specific topic distribution per document. These parameters are adjusted through experimentation to achieve the best results. The control_list sets these parameters:

```
library(topicmodels)

# Creating a Document-Term Matrix
dtm <- token %>%
  count(document = row_number(), word) %>%
  cast_dtm(document, word, n)

# Adjusting hyperparameters alpha and beta
control_list <- list(seed = 1234, alpha = 0.1)
#higher alpha results in a more specific topic distribution per document.
#Alpha and k are determined by experimenting
```

We then fit the LDA model to our DTM, specifying the number of topics (k) we want to identify. The optimal number of topics can be determined using various methods such as qualitative interpretation by the researcher or use the perplexity score to evaluate how well the model fits the data. We utilized the perplexity method due to its popularity and simplicity in implementation within the topicmodels package [11, 18]. Lower perplexity generally indicates a better fit of the LDA variant, determined by the number of topics. In this chapter, we iterated over values of k and selected the number of topics that minimize perplexity.

```
# Loop through various values of k
perplexity_values <- sapply(2:10, function(k) {
  model <- LDA(dtm, k = k, control = control_list)
  perplexity(model)
})
plot(2:10, perplexity_values, type = "b", xlab = "Number of Topics (k)", ylab =
"Perplexity")

# Add perplexity value labels above each point
text(2:10, perplexity_values,

     labels = round(perplexity_values, 2),
     pos = 3,          # Position text above the points
     cex = 0.8,        # Text size
     col = "red")      # Text color
```

Figure 2 shows the value of the perplexity score and the number of topics from LDA. The LDA variant of 3 topics was shown to have the lowest perplexity score (168.71) and therefore, determined as the optimal number of topics.

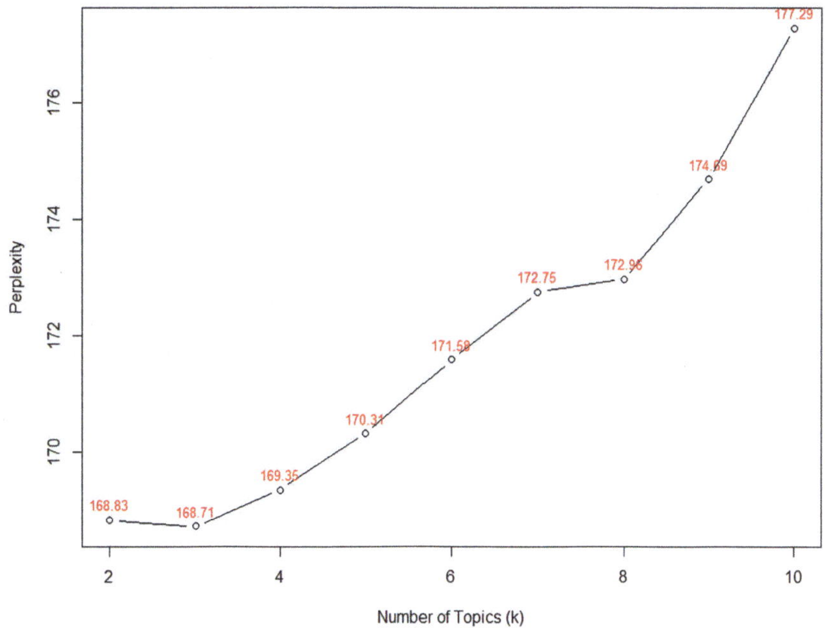

Fig. 2 Perplexity scores of All LDA variants

As informed by the perplexity score, k is set to 3, and the model is fitted using the LDA function from the topicmodels library.

```
# Fitting the LDA Model

lda_model <- LDA(dtm, k = 3 #k is the beta parameter
                 , control = control_list)
```

To view the top terms associated with each topic, we extract the probabilities of each word within each topic using the tidy function, which converts the model output into a tidy data frame. We then group these terms by topic, selecting the top 10 words with the highest probabilities (beta values) for each topic. Finally, we visualize the top terms for each topic using a bar chart, which helps in understanding the dominant words within each topic.

```
terms(lda_model, 10)

# Get the probabilities of each word in each topic
lda_terms <- tidy(lda_model, matrix = "beta")

top_terms <- lda_terms %>%
  group_by(topic) %>%
  top_n(10, beta) %>%
  ungroup() %>%
  arrange(topic, -beta)

# View the top terms for each topic
print(top_terms)

top_terms %>%
  mutate(term = reorder_within(term, beta, topic)) %>%
  ggplot(aes(beta, term, fill = factor(topic))) +
  geom_col(show.legend = FALSE) +
  facet_wrap(~ topic, scales = "free") +
  scale_y_reordered()
```

Output 2 shows samples of identified key terms within topic 1. The topic column indicates the topic number, the term column lists the top words for each topic, and the beta column shows the probability of each term within the topic. For example, in Topic 1, the most significant words are "learn" and "material," with beta values of 0.142 and 0.120, respectively. These words are the most representative of Topic 1, indicating the theme of this topic likely revolves around learning and materials. Figure 3 shows bar plots of top 10 terms from all 3 topics. Some of the top terms for topic 1 are "learn" and "material". Some of the top terms for topic 2 are "understand" and "subject". Some of the top terms for topic 3 are "feel" and "assignment". By analyzing these top terms, we can infer the central themes of each topic, which provides insights into the main areas of focus in the student comments. This approach helps educators and analysts understand the key themes and issues expressed by students, facilitating data-driven improvements to the course.

3.3.1 Output 2

3.3.1.1 Top 10 Terms from Topic 1

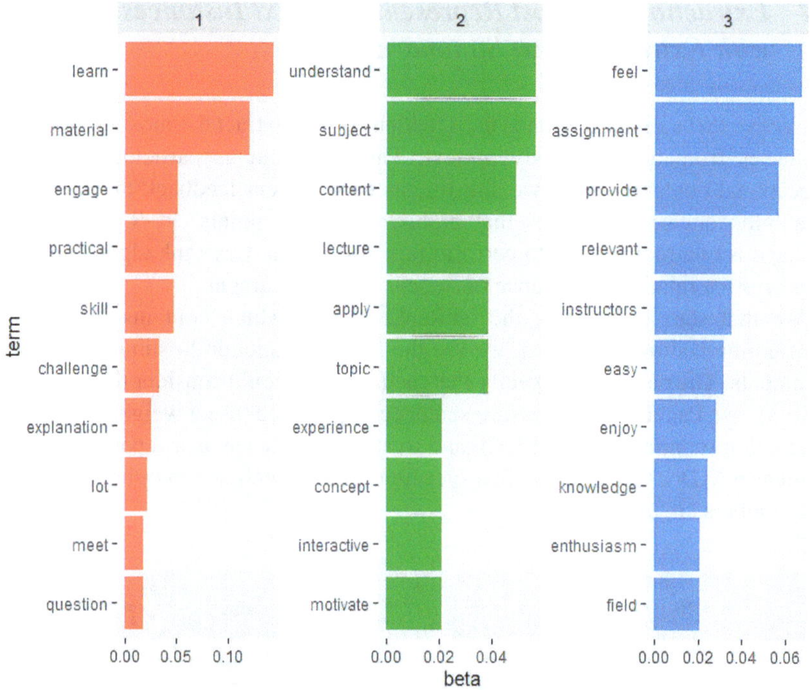

Fig. 3 Bar plot of top 10 terms from all 3 topics

```
> # View the top terms for each topic
> print(top_terms, n = 10)
# A tibble: 30 × 3
   topic term         beta
   <int> <chr>       <dbl>
 1     1 learn       0.142
 2     1 material    0.120
 3     1 engage      0.0511
 4     1 practical   0.0474
 5     1 skill       0.0474
 6     1 challenge   0.0328
 7     1 explanation 0.0255
 8     1 lot         0.0219
 9     1 meet        0.0182
10     1 question    0.0182
# ℹ 20 more rows
# ℹ Use `print(n = ...)` to see more rows
```

3.4 Extracting the Most Representative Text Instances with Extractive Text Summarization

Extractive text summarization is used to identify and extract the most representative sentences from a larger body of text. This technique is particularly useful in educational contexts where we aim to condense student feedback or other textual data into concise summaries that highlight the key points. In R, the lexRankr package is commonly used to perform this task via the LexRank algorithm, which measures the relative importance of sentences in a document.

We first start by loading the lexRankr library, which contains functions for text summarization [28]. Next, we use the lexRank function to summarize student comments. The parameters specify that the function should consider three sentences (n = 3), use PageRank to measure sentence importance, use a weighted graph representation (continuous = TRUE), and treat each sentence as a separate document (sentencesAsDocs = TRUE). Text cleaning options such as removing punctuation and numbers are also set.

```
#-------------Text Summarization-------------
library(lexRankr)

# Summarize the text using LexRank
summary_result <- lexRankr::lexRank(text = df$Student_comment, docId = "create",
                                    n = 3,
                                    usePageRank = TRUE, #to measure relative
importance
                                    continuous = TRUE, #use a weighted graph
representation of the sentences
                                    sentencesAsDocs = TRUE, #useful for single
document extractive summarization
                                    #text cleaning
                                    removePunc = TRUE,
                                    removeNum = TRUE,
                                    toLower = FALSE, stemWords = FALSE, rmStopWords
= FALSE, #These are set to false to retain meaning of the sentences
                                    Verbose = TRUE)
```

Subsequently, we reorder the top three sentences to appear in the order they were in the original text. The result, ordered_top_3, contains the top three sentences.

```
#reorder the top 3 sentences to be in order of appearance in article
order_of_appearance = order(as.integer(gsub("_","",summary_result$sentenceId)))

#extract sentences in order of appearance
ordered_top_3 = summary_result[order_of_appearance, "sentence"]

print(ordered_top_3)
```

Output 3 displays the top three sentences extracted from a larger set of student comments using the LexRank algorithm. These sentences are considered the most representative or important based on their relevance and context within the original text.

The first sentence, "*the course content was very relevant to my field.*", highlights a positive aspect of the course, indicating that the material covered was pertinent to the student's area of study. It suggests that the course content met the expectations and needs of students, providing them with useful knowledge directly applicable to their field. The second sentence, "*the assignments were very relevant to the course material.*", underscores the alignment between the assignments and the course material. It implies that the tasks assigned to students were well-designed to reinforce the concepts taught in the course, helping students to apply what they learned in practical ways. The third sentence, "*The material was dry, and the instructor was not engaging.*", contrasts with the positive feedback of the previous sentences by pointing out a negative aspect. This indicates that the course material was uninteresting and that the instructor's teaching style did not capture the students' attention. This feedback highlights areas for improvement in course delivery and engagement strategies.

These three sentences together provide a balanced summary of student feedback, reflecting both strengths and areas for improvement in the course. By including both positive and negative comments, the summary offers a more comprehensive view of the students' experiences.

3.4.1 Output 3

3.4.1.1 The 3 Most Representative Sentences from the Student Comments

```
> ordered_top_3
[1] "The course content was very relevant to my field."
[2] "The assignments were very relevant to the course material."
[3] "The material was dry, and the instructor was not engaging."
```

Next, we can extend the summarization process by computing LexRank scores for each sentence in a dataframe. We first add an index column to the original dataframe, facilitating easier reference during analysis.

```
#Bind Lexrank score to a dataframe
df_lexrank <- cbind(Index = 1:nrow(df), df)
```

Then, Text preprocessing is performed to standardize the text for analysis. This includes converting text to lowercase, removing punctuation and numbers, and stripping extra whitespace. These steps ensure that the text is clean and uniform, ready for the next processing stages. However, We did not remove stopwords and perform lemmatization for this task, because it strips away the meaning. This is because we are doing sentence level analysis, not word-level.

```
clean_text_lexrank <- df_lexrank %>%
  mutate(Student_comment = tolower(Student_comment)) %>%          # Convert to
lower case
  mutate(Student_comment = removePunctuation(Student_comment)) %>% # Remove
punctuation
  mutate(Student_comment = removeNumbers(Student_comment)) %>% #Remove numbers
  mutate(Student_comment = stripWhitespace(Student_comment))       # Strip
whitespace
```

Subsequently, the cleaned text is broken down into individual sentences (unnest_tokens) and LexRank scores are computed (bind_lexrank). LexRank evaluates each sentence based on its similarity and importance relative to other sentences in the document, producing scores that quantify each sentence's representativeness.

```
# Use lexRankr to compute LexRank scores
lexrank_result <- clean_text_lexrank %>%
  unnest_tokens(sentence, Student_comment, token = "sentences") %>%
  lexRankr::bind_lexrank(sentence, Index, level = "sentences")
```

The computed LexRank scores are used to identify the top five sentences (head(n = 5)) that are most representative of the entire text. These sentences are selected based on their high LexRank scores, indicating their significance within the document. Finally, the output is examined to ensure that it is a dataframe. This dataframe output format can be further analyzed as needed rather than a stand-alone text output.

```
# Arrange and select top sentences
top_sentences <- lexrank_result %>%
  arrange(desc(lexrank)) %>%
  head(n = 5) %>%
  select(sentence, lexrank)

#double check if our output is really a dataframe
class(top_sentences)
top_sentences
```

Table 2 displays the top five sentences along with their corresponding LexRank scores. Each row represents a sentence selected for its importance in summarizing

Table 2 Top five most representative sentences in the data frame format

Rank	Sentence	Lexrank score
1	The course content was very engaging the lectures readings and activities were all interesting and kept me motivated to learn	0.02976320
2	The assignments were very relevant to the course material they helped me apply what i learned in practical tasks and reinforced my understanding	0.02594757
3	i enjoyed the interactive discussions the group activities and discussions were very engaging and helped me learn from my peers	0.02006878
4	The course helped me understand the subject better the explanations were clear and the examples were relevant making the material easy to grasp	0.01932070
5	The instructors enthusiasm made the course enjoyable their passion for the subject was evident and made the learning experience more engaging	0.01870971

the original text. The "lexrank" column displays the numerical score assigned to each sentence by LexRank. Higher scores indicate sentences that are more central or important within the context of the entire document. The results are similar to Output 3, albeit in the table form.

By utilizing extractive text summarization with LexRank in R, educators and researchers can efficiently distill meaningful insights from large volumes of textual data, aiding in tasks such as summarizing student feedback, generating concise reviews, or extracting critical information from academic papers. This approach streamlines the process of information retrieval and synthesis, making it a valuable tool in educational and research contexts.

3.5 Examining Students' Perception with Sentiment Analysis

Sentiment analysis, also known as opinion mining, is a technique used to identify and extract subjective information from text data [24]; It involves determining the emotional tone behind a series of words to understand the attitudes, opinions, and emotions expressed within the text. This process is particularly useful in analyzing feedback, reviews, and opinionated posts, as it helps in understanding public sentiment towards products, services, or events.

Sentence-Level Sentiment Analysis. Sentence-level sentiment analysis evaluates the sentiment expressed within individual sentences, providing insights into the overall emotional tone conveyed by the text. This analysis allows for a more detailed understanding of specific feedback elements. The process involves several steps as follows:

First, textual data needs to be preprocessed to ensure consistency in the data. The mutate() function is used to convert all text to lowercase (tolower()) and remove punctuation (removePunctuation()) from each student comment in the dataframe df.

This standardizes the format and removes irrelevant characters, preparing the text for sentiment analysis.

```
clean_text_sentiment <- df %>%
  mutate(Student_comment = tolower(Student_comment)) %>%          # Convert to 
lower case
  mutate(Student_comment = removePunctuation(Student_comment)) # Remove punctuation
```

Subsequently, the sentiment() function from the sentimentr package is then applied to compute sentiment scores (sentiment) for each individual student comment. This function uses a sentiment lexicon to assign numerical values representing the positivity or negativity of each comment. The unnest() function is used to expand the sentiment scores into separate rows, ensuring each sentiment score is associated with its respective comment.

```
# Perform sentiment analysis
sentiment_result <- clean_text_sentiment %>%
  mutate(sentiment = sentiment(Student_comment)) %>%
  unnest(cols = c(sentiment))
```

Next, sentiment scores are categorized into three distinct sentiment categories ("Positive," "Negative," or "Neutral") based on predefined thresholds using case_when(). Sentences with sentiment scores above 0.1 are categorized as positive, those below −0.1 as negative, and the rest as neutral. This categorization helps in understanding the overall sentiment expressed in each student comment.

```
# Categorize sentiment
sentiment_result <- sentiment_result %>%
  mutate(sentiment_category = case_when(

    sentiment > 0.1 ~ "Positive",
    sentiment < -0.1 ~ "Negative",
    TRUE ~ "Neutral"
  ))
```

Finally, the top five sentences (top_positive, top_negative, top_neutral) for each sentiment category are selected based on their sentiment scores. These sentences represent the most positively, negatively, and neutrally perceived comments within the dataset. They are combined into a single dataframe (top_sentiments) for further analysis or reporting. Results from the top_sentiment can also be visualized to provide comprehensive understandings.

```
# Select top 5 for each category
top_positive <- sentiment_result %>%
  filter(sentiment_category == "Positive") %>%
  arrange(desc(sentiment)) %>%
  head(5)

top_negative <- sentiment_result %>%
  filter(sentiment_category == "Negative") %>%
  arrange(sentiment) %>%
  head(5)

top_neutral <- sentiment_result %>%
  filter(sentiment_category == "Neutral") %>%
  arrange(desc(abs(sentiment))) %>%
  head(5)

top_sentiments <- bind_rows(top_positive, top_negative, top_neutral)

# Visualize the results
ggplot(top_sentiments, aes(x = reorder(Student_comment, sentiment), y = sentiment,
fill = sentiment_category)) +
  geom_bar(stat = "identity", position = "dodge") +
  coord_flip() +
  labs(title = "Top Positive, Negative, and Neutral Sentiments",
       x = "Comments",
       y = "Sentiment Score",
       fill = "Sentiment Category") +
  scale_fill_manual(values = c("Positive" = "limegreen", "Negative" = "salmon",
"Neutral" = "skyblue")) +
  theme_minimal()
```

Output 4 and Fig. 4 provide a structured summary of the sentiment analysis results at the sentence level. Each row in the data frame represents a sentence along with its associated sentiment score. An instance of a positive comment is "the instructor was very supportive and encouraging. their positive and willingness to help made the learning experience much more enjoyable". An instance of a neutral comment is "I learned a few new things from the course, however I felt that the material could have been presented in a more interesting way". An instance of a negative comment is "the instructor was not very helpful whenever I had questions; they were slow to respond and did not provide useful answers". This detailed breakdown allows educators or analysts to easily identify and interpret the sentiments expressed in student feedback, highlighting specific areas of praise or concern.

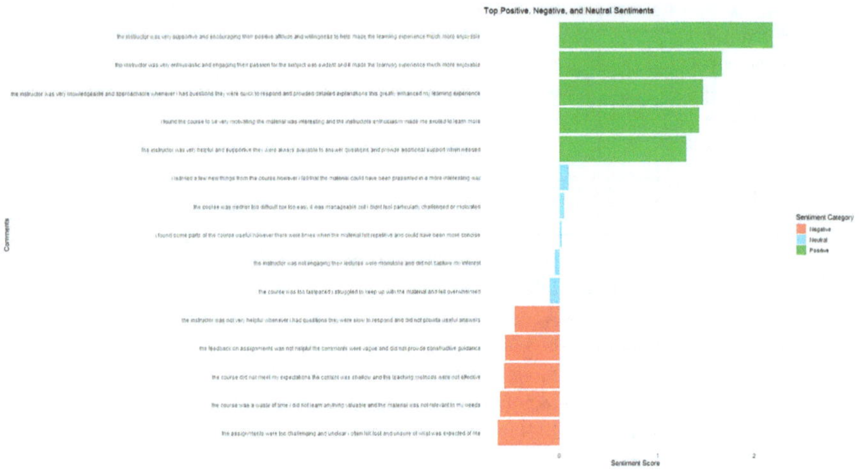

Fig. 4 Sentiment bar plot of students' course evaluation as sentences

3.5.1 Output 4

3.5.1.1 Sentence-Level Sentiment Scores and Categories of Students' Course Evaluation

```
> top_sentiments
# A tibble: 15 × 6
   Student_comment                              element_id sentence_id word_count sentiment sentiment_category
   <chr>                                             <int>       <int>      <int>     <dbl> <chr>
 1 the instructor was very supportive and en...         35           1         21    2.18   Positive
 2 the instructor was very enthusiastic and ...         54           1         23    1.66   Positive
 3 the instructor was very knowledgeable and...          3           1         26    1.47   Positive
 4 i found the course to be very motivating ...         50           1         22    1.43   Positive
 5 the instructor was very helpful and suppo...         64           1         20    1.29   Positive
 6 the assignments were too challenging and ...         86           1         19   -0.631  Negative
 7 the course was a waste of time i did not ...        100           1         22   -0.608  Negative
 8 the course did not meet my expectations t...         90           1         18   -0.566  Negative
 9 the feedback on assignments was not helpf...         92           1         17   -0.558  Negative
10 the instructor was not very helpful whene...         83           1         21   -0.458  Negative
11 the course was too fastpaced i struggled ...         96           1         16   -0.1    Neutral
12 i learned a few new things from the cours...         76           1         24    0.0939 Neutral
13 the instructor was not engaging their lec...         99           1         15   -0.0516 Neutral
14 the course was neither too difficult nor ...         77           1         20    0.0445 Neutral
15 i found some parts of the course useful h...         72           1         23    0.0230 Neutral
```

Word-Level Sentiment Analysis. Furthermore, sentiment analysis can also be performed on individual words to examine a smaller unit of textual data that may provide insights in terms of specific keywords that students have toward a course. After text preprocessing, the text is tokenized into individual words (token) and sentiment scores (sentiment) are computed for each word using the sentiment() function. This function assigns sentiment scores to each word based on its lexical properties, capturing the emotional connotations associated with specific vocabulary choices.

```
# Perform sentiment analysis
sentiment_result <- token %>%
  mutate(sentiment = sentiment(word)) %>%
  unnest(cols = c(sentiment))
```

Similar to sentence-level analysis, sentiment scores for individual words are categorized into "Positive," "Negative," or "Neutral" using case_when(). This categorization allows for the identification of words that contribute positively, negatively, or neutrally to the overall sentiment conveyed by student comments. To handle duplicate words across different comments, the data is aggregated using group_by() and summarize() functions. This step calculates the average sentiment score (avg_sentiment) for each unique word across different sentiment categories, along with the frequency count (count) of each word.

```
# Categorize sentiment
sentiment_result <- sentiment_result %>%
  mutate(sentiment_category = case_when(
    sentiment > 0.1 ~ "Positive",
    sentiment < -0.1 ~ "Negative",
    TRUE ~ "Neutral"
  ))

#Aggregate the data by word to get unique words and their corresponding average
sentiment scores.

aggregated_sentiments <- sentiment_result %>%
  group_by(word, sentiment_category) %>%
  summarize(
    avg_sentiment = mean(sentiment),
    count = n(),
    .groups = 'drop'
  )
```

Finally, the top five words (top_positive, top_negative, top_neutral) for each sentiment category are selected based on their average sentiment scores (avg_sentiment). These words provide insights into the specific vocabulary choices that contribute positively, negatively, or neutrally to the overall sentiment expressed in student feedback.

```
# Select top 5 for each category
top_positive <- aggregated_sentiments %>%
  filter(sentiment_category == "Positive") %>%
  arrange(desc(avg_sentiment)) %>%
  head(5)

top_negative <- aggregated_sentiments %>%
  filter(sentiment_category == "Negative") %>%
  arrange(avg_sentiment) %>%
  head(5)

top_neutral <- aggregated_sentiments %>%
  filter(sentiment_category == "Neutral") %>%
  arrange(desc(abs(avg_sentiment))) %>%
  head(5)

top_sentiments <- bind_rows(top_positive, top_negative, top_neutral)
# Visualize the results
ggplot(top_sentiments, aes(x = reorder(word, avg_sentiment), y = avg_sentiment,
fill = sentiment_category)) +
  geom_bar(stat = "identity", position = "dodge") +
  coord_flip() +
  labs(title = "Top Positive, Negative, and Neutral Sentiments",
       x = "Comments",
       y = "Sentiment Score",
       fill = "Sentiment Category") +
  scale_fill_manual(values = c("Positive" = "limegreen", "Negative" = "salmon",
"Neutral" = "skyblue")) +
  theme_minimal()
```

Output 5 and Fig. 5 presents a concise summary of sentiment expressed by specific words within the student comments. Each row in the data frame includes a word (word), its sentiment category (sentiment_category), average sentiment score (avg_sentiment), and frequency count (count). An instance of a positive comment is "enjoyable". An instance of a neutral comment is "prepare". An instance of a negative comment is "confuse". This detailed breakdown enables educators or analysts to identify key words that influence the overall sentiment of student feedback, facilitating targeted improvements or interventions based on specific linguistic cues.

By utilizing both sentence-level and word-level sentiment analysis techniques in R, educators and analysts can gain insights into the emotional responses and opinions expressed within textual data. These analyses empower informed decision-making and facilitate effective improvements in educational settings based on insights derived from student feedback.

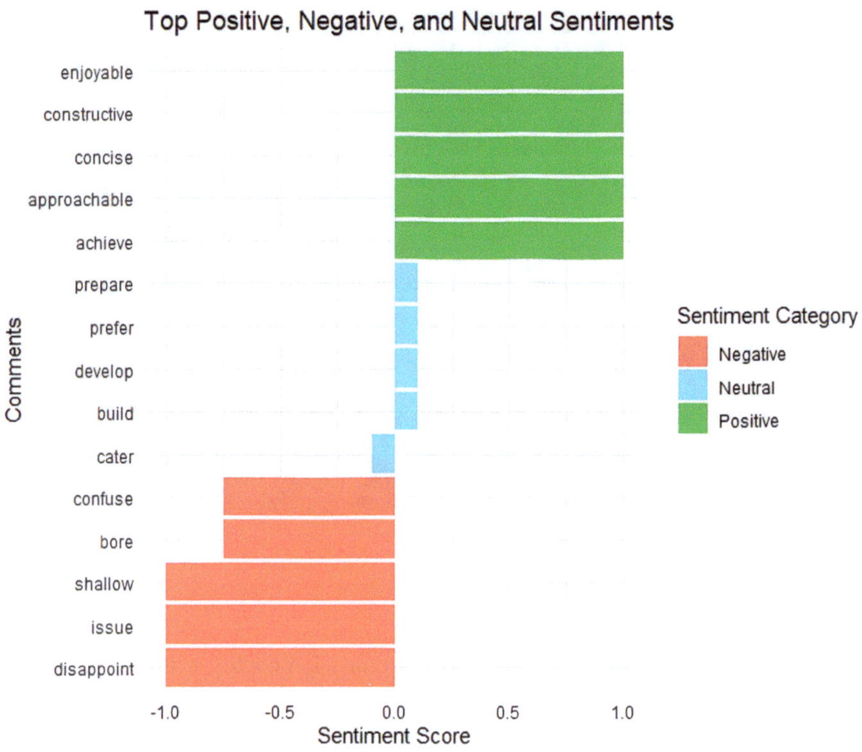

Fig. 5 Sentiment bar plot of students' course evaluation as words

3.5.2 Output 5

3.5.2.1 Word-Level Sentiment Scores and Categories of Students' Course Evaluation

```
> top_sentiments
# A tibble: 15 × 4
   word         sentiment_category avg_sentiment count
   <chr>        <chr>                      <dbl> <int>
 1 achieve      Positive                    1        2
 2 approachable Positive                    1        1
 3 concise      Positive                    1        2
 4 constructive Positive                    1        2
 5 enjoyable    Positive                    1        3
 6 disappoint   Negative                   -1        1
 7 issue        Negative                   -1        1
 8 shallow      Negative                   -1        1
 9 bore         Negative                   -0.75     1
10 confuse      Negative                   -0.75     3
11 build        Neutral                     0.1      2
12 cater        Neutral                    -0.1      1
13 develop      Neutral                     0.1      5
14 prefer       Neutral                     0.1      1
15 prepare      Neutral                     0.1      1
```

4 Concluding Remarks

Understanding the various textual elements within a LA context is pivotal to advancing educational technologies and enhancing data-informed decisions. This chapter discussed the advanced application of NLP techniques to analyze and interpret complex educational data. The application of methods such as text preprocessing, TF-IDF, topic modeling, text summarization, and sentiment analysis yields deeper insights into the patterns that are only available from textual data.

The application of these NLP techniques allows educators to identify key words, themes, and inferred sentiment in data sources such as student feedback, course materials, reference documents, and other textual data, thereby facilitating more informed decision-making and teaching strategy development. For instance, analyzing sentiment in a large number of student feedback in large-scale settings such as massive open online courses can reveal attitudes towards course content without the need to manually inspect each and every data point; This application can aid in the refinement of pedagogical approaches. Similarly, topic modeling and text summarization can streamline curriculum design by distilling large volumes of text such as historical documents into manageable insights.

Ultimately, the integration of NLP in learning analytics provides useful tools for understanding and enhancing the educational experience. These methods enable the extraction of information from unstructured data, offering new avenues for improving teaching strategies, supporting student learning, and optimizing educational outcomes [9]. By employing these advanced NLP techniques, educational institutions can better navigate the complexities of modern education and foster environments conducive to effective learning and teaching.

References

1. Allen LK, Creer SC, Öncel P (2022) Natural language processing: towards a multi-dimensional view of the learning process. In: Lang C, Siemens G, Wise AF (eds) The handbook of learning analytics, 2nd edn. SOLAR, pp 46–53. https://doi.org/10.18608/hla22.005
2. Bengfort B, Bilbro R, Ojeda T (2018) Applied text analysis with python: enabling language-aware data products with machine learning, First edn. O'Reilly Media, Inc
3. Bird S, Klein E, Loper E (2009) Natural language processing with python: analyzing text with the natural language toolkit. O'Reilly Media, Inc.. https://www.nltk.org/book/
4. Blei DM (2003) Latent dirichlet allocation. J Mach Learn Res 3:993–1022
5. Derntl M, Günnemann N, Klamma R (2013) A dynamic topic model of learning analytics research. LAK (Data Challenge)
6. Ezen-Can, A., Boyer, K. E., Kellogg, S., & Booth, S. (2015). Unsupervised modeling for understanding MOOC discussion forums: a learning analytics approach. Proceedings of the fifth international conference on learning analytics and knowledge, 146–150. https://doi.org/10.1145/2723576.2723589
7. Feinerer, I., & Hornik, K. (2024). Tm: text mining package. https://CRAN.R-project.org/package=tm
8. Feinerer I, Hornik K, Meyer D (2008) Text mining infrastructure in R. J Stat Softw 25(5):1–54. https://doi.org/10.18637/jss.v025.i05
9. Gašević D, Merceron A (2022) In: Lang C, Siemens G, Wise AF (eds) The handbook of learning analytics, 2nd edn. SOLAR. https://doi.org/10.18608/hla22
10. Goddard, W. (2021, September 7). Natural language processing in education. IT Chronicles. https://itchronicles.com/natural-language-processing-nlp/natural-language-processing-in-education/
11. Grün B, Hornik K (2011) Topicmodels: an R package for fitting topic models. J Stat Softw 40(13):1–30. https://doi.org/10.18637/jss.v040.i13
12. Grün, B., & Hornik, K. (2024). Topicmodels: topic models. https://CRAN.R-project.org/package=topicmodels
13. Hattie J, Timperley H (2007) The power of feedback. Rev Educ Res 77(1):81–112. https://doi.org/10.3102/003465430298487
14. Hutto, C., & Gilbert, E. (2014). VADER: A parsimonious rule-based model for sentiment analysis of social media text 8. https://github.com/cjhutto/vaderSentiment, 225, 8
15. Knight S, Shibani A, Abel S, Gibson A, Ryan P, Sutton N, Wight R, Lucas C, Sándor Á, Kitto K, Liu M, Mogarkar RV, Buckingham Shum S (2020) AcaWriter: a learning analytics tool for formative feedback on academic writing. J Writ Res 12(1):141–186. https://doi.org/10.17239/jowr-2020.12.01.06
16. Lee S-M (2023) The effectiveness of machine translation in foreign language education: a systematic review and meta-analysis. Comput Assist Lang Learn 36(1–2):103–125. https://doi.org/10.1080/09588221.2021.1901745
17. Li C, Xing W (2021) Natural language generation using deep learning to support MOOC learners. Int J Artif Intell Educ 31(2):186–214. https://doi.org/10.1007/s40593-020-00235-x
18. Neishabouri, A., & Desmarais, M. C. (2020). Reliability of perplexity to find number of latent topics. *The Thirty-Third International Flairs Conference*
19. Ötleş E, Kendrick DE, Solano QP, Schuller M, Ahle SL, Eskender MH, Carnes E, George BC (2021) Using natural language processing to automatically assess feedback quality: findings from 3 surgical residencies. Acad Med 96(10):1457–1460. https://doi.org/10.1097/ACM.0000000000004153
20. Pennington, J., Socher, R., & Manning, C. (2014). GloVe: global vectors for word representation. *Proceedings of the 2014 conference on empirical methods in natural language processing (EMNLP)*, 1532–1543. https://doi.org/10.3115/v1/D14-1162
21. Rahimi, S. R., Mozhdehi, A. T., & Abdolahi, M. (2017). An overview on extractive text summarization. 2017 IEEE 4th international conference on knowledge-based engineering and innovation (KBEI), 0054–0062.https://doi.org/10.1109/KBEI.2017.8324874

22. Ramos, J. (2003). Using TF-IDF to determine word relevance in document queries. Proceedings of the First Instructional Conference on Machine Learning, 242
23. Rinker, T. W. (2021). Sentimentr: calculate text polarity sentiment. https://github.com/trinker/sentimentr
24. Santosh Kumar P, Yadav RB, Dhavale SV (2021) A comparison of pre-trained word embeddings for sentiment analysis using deep learning. In: Gupta D, Khanna A, Bhattacharyya S, Hassanien AE, Anand S, Jaiswal A (eds) International conference on innovative computing and communications, vol 1165. Springer Singapore, pp 525–537. https://doi.org/10.1007/978-981-15-5113-0_41
25. Schopf, T., Arabi, K., & Matthes, F. (2023). Exploring the landscape of natural language processing research. Proceedings of the Conference Recent Advances in Natural Language Processing - Large Language Models for Natural Language Processings, 1034–1045.https://doi.org/10.26615/978-954-452-092-2_111
26. Sharma, A. (2023). Abstractive text summarization with BART [Programming documentation]. Keras. https://keras.io/examples/nlp/abstractive_summarization_with_bart/
27. Silge J, Robinson D (2016) Tidytext: text mining and analysis using tidy data principles in R. JOSS 1(3). https://doi.org/10.21105/joss.00037
28. Spannbauer, A. (2019). lexRankr: extractive summarization of text with the LexRank algorithm (0.5.2) [R]. https://CRAN.R-project.org/package=lexRankr
29. Wheeler JM, Cohen AS, Xiong J, Lee J, Choi H-J (2021) Sample size for latent dirichlet allocation of constructed-response items. In: Wiberg M, Molenaar D, González J, Böckenholt U, Kim J-S (eds) Quantitative psychology, vol 353. Springer International Publishing, pp 263–273. https://doi.org/10.1007/978-3-030-74772-5_24
30. Wongvorachan T, Lai KW, Bulut O, Tsai Y-S, Chen G (2022) Artificial intelligence: transforming the future of feedback in education. J Appl Test Technol:95–116
31. Yaneva V, Baldwin P, Le An H, Runyon C (2023) Extracting linguistic signal from item text and its application to modeling item characteristics. In: Yaneva V, von Davier M (eds) Advancing natural language processing in educational assessment. Routledge, pp 167–182. https://doi.org/10.4324/9781003278658-14

Using BERT-like Language Models for Automated Discourse Coding: A Primer and Tutorial

Sonsoles López-Pernas, Kamila Misiejuk, and Mohammed Saqr

Abstract Coding text data in qualitative research is a labor-intensive and error-prone process that requires meticulous attention to detail as well as consistency in the coding criteria. Large Language Models (LLMs) present a promising solution to alleviate some of these challenges by automating part of the coding process. This tutorial explores the application of LLMs for automated text classification using word embeddings through the R package `text` and different BERT-like large language models. We implement a machine learning pipeline that combines word embeddings with supervised machine learning algorithms to code text data with high accuracy. We present a case study on collaborative problem-solving in which we train a classification model on a small portion of manually coded data and then apply it to classify the remaining data. The tutorial also covers the evaluation of coding accuracy by comparing human and machine-coded data using classic machine learning performance metrics as well as Cohen's kappa, Matthews' correlation coefficient, and Gwet AC1, measures commonly used to assess interrater reliability in qualitative research. Lastly, we apply different learning analytics techniques to compare the findings obtained from human-coded data and automatically coded data.

Keywords Automatic coding · Large language models · BERT · Learning analytics · Natural language processing · Artificial intelligence

S. López-Pernas (✉) · M. Saqr
University of Eastern Finland, Joensuu, Finland
e-mail: sonsoles.lopez@uef.fi

K. Misiejuk
FernUniversität in Hagen, Hagen, Germany

M. Saqr, S. López-Pernas (eds.), *Advanced Learning Analytics Methods*,
https://doi.org/10.1007/978-3-031-95365-1_10

1 Introduction

Coding text data is an effortful and time-consuming process and constitutes one of the major challenges faced by qualitative researchers. Coding refers to the process of "essentially indexing or mapping data, to provide an overview of disparate data that allows the researcher to make sense of them in relation to their research questions" [1]. Coding requires defining codes, i.e. the concepts of interest, and localizing places in text where codes occur [2]. There are two main approaches of deriving at the codes: *inductive coding*, where codes emerge from the data, and *deductive coding*, which uses a set of predetermined codes to classify the data [3].

The coding process requires close inspection of data before it can be analyzed. Researchers often have to scroll through large volumes of text, identify the main concepts and themes, and assign codes that best describe them. This process requires a high level of attention, as every line of text must be carefully examined. Furthermore, maintaining consistency in coding across different parts of the data is difficult, especially if more than one researcher is involved, often requiring multiple rounds of review, discussion, and revision [1]. Coding of text is vital in qualitative research; however, it is also an important part of many quantitative or mixed-methods studies, where text data coding is often delegated as a part of data preprocessing before it can be analyzed by other, often quantitative, methods. In terms of machine learning (ML), deriving codes inductively can be similar to text summarization tasks, where an algorithm discovers topics within a text, while localizing a presence of a specific code in text can match annotation or classification tasks, where an algorithm assigns a class label to a given data point based on its characteristics.

Large Language Models (LLMs) offer a promising solution to ease some of the challenges related to coding of text data. LLMs are trained on vast amounts of text data and have the ability to process and generate human-like language [4]. When presented with the task of coding text data, LLMs can significantly reduce the time and effort required by automating—at least part of—the process, as this technology can quickly scan large volumes of text, as well as identify the main underlying concepts. The introduction of the transformer architecture enabled the development of of different types of LLMS such as Google's Bidirectional Encoder Representations from Transformers (BERT) trained using masked language modeling and open AI's Generative Pre-trained Transformers (GPT) trained using causal language modeling (see [5, 6] for more details). Although both models can be adapted to a variety of tasks, BERT models are typically recommended for text classification, while GPT models are better for text generation [7].

LLMs are capable of performing zero-shot classification, which means that without the need for specifically trained data for certain categories [4, 8], the model can classify text based on its broad understanding of language and knowledge. This capability stems from the extensive and diverse textual data encountered during the pretraining phase, enabling the model to make reasonable inferences and predictions for categories it has not been explicitly trained on through contextual understanding and knowledge transfer.

While zero-shot classification significantly reduces both the technical and intellectual effort required from the researchers, it often lacks the precision and reliability that might be required for domain-specific tasks. A possible middle ground solution that achieves a good trade-off between manual coding and completely automated coding, is the combination of word embeddings—implemented in LLMs—with supervised ML. Word embeddings capture the semantic meaning of text in a way that is informed by the context in which words appear [9] (the concept is explained more thoroughly in Sect. 3.5). Generating these embeddings allows us to use the textual information as features in a classification model (such as Random Forest or Support Vector Machine) [10]. This approach requires that researchers to manually code part of the data, while the major part of the process is delegated to the LLM, achieving—allegedly—more accurate and more consistent classifications, as the model is explicitly trained on data that reflects the task at hand.

In this tutorial, we will learn how to implement automated discourse classification using word embeddings. We will rely on the R package `text` and BERT-like models. We will train a classification ML model with a small percentage of a dataset on collaborative problem solving, and we will automatically classify the remaining text. We will evaluate the classification performance and compare the agreement of the human-coded and LLM-coded text using Cohen's kappa, Gwet's AC1 and Matthews Correlation Coefficient. We will also compare the performance of different BERT-like algorithms using traditional metrics. Lastly, we will investigate how the insights gathered from applying different learning analytics techniques to the automatically coded data differ from those obtained from human-coded data.

2 Related Work on Using LLMs to Support Text Coding

The rise in popularity of LLMs has caught the attention of learning analytics researchers for their potential to assist researchers at all stages of the learning analytics cycle including synthetic data generation and automatic feedback generation, as well as automatic coding, which is the goal of our tutorial [11]. LLMs can be used to support developing code definitions, i.e. a codebook, and identifying said codes in the text segments.

Emerging research shows that LLMs has the potential to support developing a codebook. In the context of qualitative coding, a codebook is a structured document that outlines the codes, categories, and their definitions, serving as a guide for systematically categorizing, analyzing and interpreting qualitative data. The authors of [12] compared three codebook development settings: (1) fully automated approach with ChatGPT, (2) hybrid ChatGPT-human approach, and (3) fully human approach. The hybrid approach resulted in best utility ratings, highest conceptual overlap and higher inter-rater reliability than other approaches. Interestingly, both hybrid and human approaches developed similar codebooks, while the fully automated approach generated a codebook less aligned with the study goals and was an outlier in terms of conceptual similarity. Similarly, the hybrid approach has yielded positive results to code a small sample of interviews in a study by [13]. The authors of [14] reported moderate to substantial similarity of themes generated by an LLM and humans. However, there are still open questions about scalability and about whether the use of AI for that task will not lead to superficial understanding of the data [15, 16]. The increasing familiarity with the data is expected to lead changes in the codebook over time; the coding process is supposed to be iterative and may lead to revalidation of earlier coding [1]. Hence, implementation of LLMs at this stage of text data coding should be carefully considered.

Early studies using LLMs for identifying codes in text segments showed mixed results. A recent work [17] evaluated the capability of generative artificial intelligence (AI), particularly GPT models, to automatically code discourse within a learning analytics study, employing various prompting and training strategies. Their findings indicated that fine-tuning methods (refer to [4] to learn more about this concept) yielded the most favorable outcomes, though none of the results met the standard reliability thresholds typically expected in the field. Misiejuk et al. [18] investigated the reliability of automatic discourse coding by applying a range of LLMs from the GPT family, along with various prompting techniques, to an asynchronous online discussion dataset. The authors determined that "the agreement with human coding was substantial at best", and do not recommend that AI replaces humans in coding, at least not yet in a reliable way. Another study by [19] operationalized BERT models to create word embeddings, and used them to train a Random Forest classifier to automatically code collaborative problem-solving conversations. Although findings are not yet beyond 'substantial' accuracy, the rapid advance of LLMs suggests that these models could soon achieve higher levels of reliability and accuracy in automatic coding tasks. Also, the selection of a specific LLM may affect the performance. Prescott et al. [20] reported that ChatGPT and Bard performed similarly across both types inductive and deductive thematic analysis. Higher performance of BERT-like models for text classification tasks than the GPT models was reported in several studies [e.g., [21, 22]]. Finally, using LLM models for text coding may require decomposing this complex task into multiple simple binary classifications (i.e., processing each code at a time and classifying each text unit regarding whether the code is included or not). Complex codes are more challenging for LLMs to code [18, 22, 23].

Adjusting the temperature setting or conducting multiple iterations of coding are some of the strategies used to improve LLM's performance in classification tasks [18, 24]. Another strategy is to improve the LLM prompt. For instance, [25] tested codebook-centered or example-centered prompts, as well as experimented with the number of examples used in a prompt. The results showed that the codebook-centered design performed better than the example-centered designs, while adding even one example of a code to a prompt increased the performance notably. Wording of a prompt is crucial. A recent work [26] demonstrated that using negated instructions (e.g., "Generate the incorrect solution..") instead of positive instructions (e.g., “Generate the correct solution..”) decreases the performance of all LLMs tested. Decreased performance could be also caused by changes in the LLM model itself [27]. This is particularly important in case of proprietary models (i.e., any LLMs that cannot be downloaded, run offline and shared [28]). For example, the majority of BERT models are an open-source, while most GPT models are not publicly available [6]. Using proprietary models requires researchers to concede a degree of control to developers. Changes made to proprietary models are not transparent and may effect the text coding results and, in effect, the final analysis.

As LLMs may be prone to hallucinations, it is critical to evaluate model’s performance before relying on its results for analysis. One method to do it, is to use human-coded data as a baseline to assess performance [29]. However, this method requires available data that was collected in a similar context, which may be unavailable in educational settings. Another approach is to manually check the LLM coding [30], which may be applicable for smaller datasets; however, it may as effortful as manually coding the dataset instead of using an LLM. Furthermore, there are many discussions about the best metrics that can be used to assess the performance of LLMs in text coding tasks. Metrics such as accuracy, precision or recall are typical measurements used in machine learning for classification tasks (see [31] for an overview). At the same time, other metrics, such as Cohen’s kappa, are applied in the qualitative research to ensure the level of agreement among multiple coders [2]. Although Cohen’s kappa is often used to evaluate text classification task [32], it has been critiqued as not an appropriate approach to this task. Instead metrics like Matthews’ correlation coefficient [33, 34] or Gwet AC1 are suggested [35, 36].

Finally, ethical issues are important to consider in case of coding education text data using LLMs. We addressed some of these issues throughout this chapter (for a comprehensive review see [37]). There are two ethical issues that need to be highlighted in case of using LLMs for text coding. First, educational data may include sensitive information and feeding it to a proprietary LLM can be a breach of student privacy rights depending on local regulations. Second, inherent bias in the LLM from its pre-training data was proven to affect classification tasks [38]. There are several techniques and recommendations that can be considered to address bias in LLMs (see [39] for more details).

3 A Case Study on Collaborative Problem Solving

3.1 The Dataset

The dataset that we use in this tutorial is based on an experiment in which small groups participated in a group discussion. In this discussion, the groups aimed to solve the Wason card selection task: a logic puzzle used to study reasoning and decision-making [40]. The discussion data has been qualitatively coded according to the nature of participants’ contributions Specifically, 500 dialogues, containing

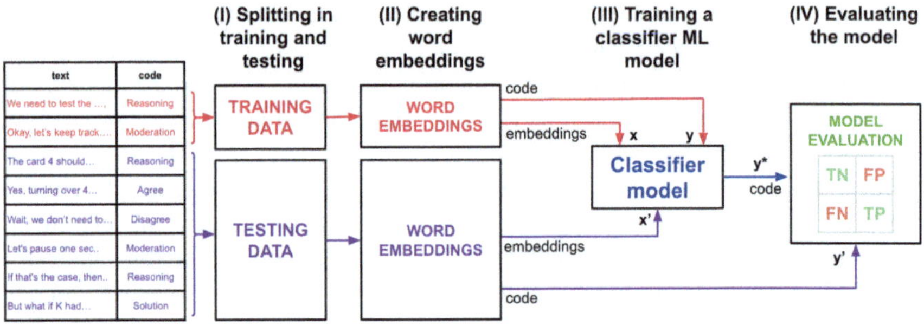

Fig. 1 Process followed in the tutorial

14,000 utterances were coded using a novel annotation schema designed to capture deliberation cues. For more details about the experiment, refer to the original article [40].

3.2 Automated Discourse Coding

The process followed in the tutorial is depicted in Fig. 1. We will first divide our dataset in training and testing sets (I). Contrary to other ML tasks, we use only a small part of the data for training (15%), and the rest for testing. The goal is to minimize the number of pieces of text that we have to manually code. Then, we will create word embeddings from the text data (II). Word embeddings are numerical representations of words that capture their semantic meaning, enabling the model to understand and process the text in a way that reflects the relationships between words. We will try different existing LLMs to create the embeddings. We then use the word embeddings corresponding to the training data and their manual codes as an input to train our classifier ML model (III). Lastly, we use the word embeddings of the test dataset to predict their codes using the trained model and evaluate the results (IV). We compare the predicted codes with the actual ones using both traditional ML performance metrics based on the confusion matrix, as well as several measures of interrater reliability to assess the agreement between human and AI coding. We will compare the performance of the different BERT-like language models to assess which one is more suitable for the classification task. For a beginners' tutorial on ML classification tasks, refer to Chap. 4 [41] in this book.

3.3 Setting Up the Enviroment

Before starting our task, we need to install and load the necessary packages, which are described below:

- `dplyr`: A powerful and popular R package used for data manipulation and transformation [42]. It provides a set of intuitive functions that make it easy to filter, select, mutate, arrange, and summarize data within a data frame. It is especially useful for cleaning and preparing data before analysis.
- `tidyr`: An essential R package for tidying data, `tidyr` helps reshape and clean data sets so they can be analyzed more effectively [43]. It provides functions to pivot data between wide and long formats, split or combine columns, and ensure that data is in a consistent, tidy format suitable for analysis.
- `tibble`: A modern take on data frames in R, `tibble` [44] provides a lightweight and faster alternative to the traditional `data.frame` object.
- `reticulate`: A package that provides an interface between R and Python [45]. It allows users to run Python code within R and seamlessly exchange data between the two languages. This is particularly useful when combining the strengths of R and Python in a single analysis pipeline, taking advantage of Python's ML libraries, such as TensorFlow or Scikit-learn, while still working within an R environment.
- `caret`: A package used for building and evaluating ML models in R [46]. It simplifies the process of training models, performing cross-validation, and optimizing model parameters. The package supports a wide range of algorithms and provides tools for data splitting, pre-processing, feature selection, and model evaluation.
- `text`: The `text` package is designed for text analysis in R [47]. It provides tools for processing and analyzing textual data, such as tokenization, vectorization, sentiment analysis, and topic modeling. It uses transformers, natural language processing (NLP) and ML methods to examine text and numerical variables.
- `mltools`: The `mltools` package in R [48] provides a collection of machine learning tools and utilities designed to simplify common preprocessing and data transformation tasks for machine learning workflows. We will use it for some of its evaluation functions.

Below is the code to load the libraries. Install them first if you do not have them already using the commented code.

```
library(dplyr) # install.packages("dplyr")
library(tidyr) # install.packages("tidyr")
library(tibble) # install.packages("tibble")
library(reticulate) # install.packages("reticulate")
library(text) # install.packages("text")
library(caret) # install.packages("caret")
library(mltools) # install.packages("mltools")
```

In addition to the previous libraries, since the `text` library relies on Python underneath, we need to make sure we have Python installed (preferrably together with the conda environment). If you do not have it and you do not wish to install the complete Python environment, you can install `miniconda` through the package `reticulate` as follows (uncomment to run):

```
# reticulate::install_miniconda()
```

Once done, you need to initialize the Python session as follows. This is an interactive command so you will need to enter "yes" in the terminal to advance.

```
textrpp_initialize(save_profile = TRUE)
```

Now that we have all the R packages, we will download our data. The dataset described earlier is available in Hugging Face and therefore we can download it using the `datasets` package in Python. We need to install it before using it for the first time with the commented command. After importing `datasets`, we download the dataset that we will use in the tutorial: `"gkaradzhov/DeliData"`. The `cache_dir` argument indicates where the data will be stored in your computer; feel free to change the name from `"cache"` to something else.

```
# reticulate::py_install("datasets")

datasets <- import("datasets")

dataset <- datasets$load_dataset("gkaradzhov/DeliData", cache_dir = "cache")
```

We can now extract the data in an R dataframe as follows:

```
df_raw <- dataset$train$to_pandas()
```

If we inspect the dataframe, we see that the it contains a column to identify the team (`group_id`) and the message (`message_id`), a column containing the message original text (`original_text`), and a cleaned one (`clean_text`). It also contains the code used to annotate the message (`annotation_target`), and two other annotation fields. In addition, it contains several other columns related to the solution of the task and the team's performance. Let us focus on the column that contains the manual code (`annotation_target`). We can inspect it to see the different codes that have been used to classify the discussion utterances:

```
table(df_raw$annotation_target)
```

```
         0      Agree   Disagree Moderation  Reasoning   Solution 
      3274       2047        111        785       4382       3403 
```

We see that there are many rows that have not been assigned a code. They correspond to messages that do not fall under any of the deliberation codes. We can remove them from our dataset to work only with the coded text as follows:

```
classification <- c("Solution", "Reasoning", "Agree", "Disagree", "Moderation")
df <- df_raw |> filter(annotation_target %in% classification)
```

Now we are ready to start working with the data.

3.4 Splitting the Dataset

As is common in most ML tasks, we split our data into a training dataset and a testing dataset. We will use the training dataset to train an ML model to classify text data into one of the five codes and we will use the test dataset to evaluate how accurately our model classifies "unseen" text. This strategy is designed to mimic a realistic coding scenario where the objective is to reduce the amount of text that needs to be manually coded. We use the function `createDataPartition` from the `caret` package to split our data into a training dataset (consisting of 15% of the data) and a testing dataset (the remaining 85% of the data). We do so in a way that the two partitions are balanced in term of proportion of codes.

```
seed = 123
set.seed(seed) # for reproducibility
trainIndex <- createDataPartition(df[["annotation_target"]], p = 0.15,
                                  list = FALSE,
                                  times = 1)

# Creating training and testing datasets
df_train <- df[trainIndex, ]
df_test <- df[-trainIndex, ]
```

3.5 Word Embeddings

Our next task is to convert our text data into numeric representations (i.e., word embeddings) to be able to apply common ML techniques. Word embeddings are a type of word representation that allows words with similar meaning to have a similar representation. Word embeddings can be defined as dense vectors of real numbers that represent words in a way in which words that share similar contexts are placed closer together. Unlike traditional one-hot encoding, where each word is represented as a unique binary vector, word embeddings capture semantic relationships between words, making them powerful for various NLP tasks. In simpler terms, word embeddings are a way to represent words as numbers so that words with similar meanings end up having similar numerical patterns.

In simple models like Word2Vec, word embeddings are created by training on large text corpora. These models learn to map each word to a vector in such a way that words used in similar contexts have similar vectors. For example, the vectors for "king" and "queen" might be close to each other because these words often appear in similar contexts (e.g., both might appear near words like "monarch," "crown," and "royalty"). BERT models take the concept of word embeddings one step further. BERT is a transformer-based model that understands the context of a word in a sentence by looking at both its left and right sides simultaneously (bidirectional). This is in contrast to earlier models like Word2Vec, which only considered the context in a single direction.

BERT starts by *tokenizing* the input text. This process breaks the text into smaller units called tokens. BERT uses a technique called WordPiece tokenization, where words are broken down into subword units. This helps BERT handle out-of-vocabulary words by representing them as combinations of known subwords. BERT has an *embedding layer* that converts these tokens into dense vectors. However, unlike static word embeddings (e.g., Word2Vec, where each word has a single vector), BERT's embeddings are dynamic. The embedding for a word depends not just on the word itself but also on the entire context of the sentence. In addition to token embeddings, BERT includes *positional embeddings*, which provide the model with information about the position of each token in the sentence. This is crucial because, unlike recurrent neural networks for example, transformers do not inherently handle the order of words. BERT then passes these embeddings through multiple *layers of transformers*, where self-attention mechanisms allow the model to focus on different parts of the sentence. This enables BERT to capture complex dependencies and relationships between words in a sentence. The output from the final transformer layer provides the *contextualized word embeddings*. These embeddings are rich in context and meaning because they take into account the entire sentence. For example, the word "bank" will have different embeddings in the sentences "He sat on the river bank" and "He went to the bank to withdraw money," reflecting the different meanings of the word in these contexts.

Several variations and enhancements have been created based on BERT. For instance, **RoBERTa** (A Robustly Optimized BERT Pretraining Approach) [49] builds on BERT's architecture by enhancing the pretraining process with more data, among other improvements. These modifications, along with larger batch sizes and learning rates, as well as the introduction of dynamic masking, make RoBERTa more robust and better performing than BERT on various natural language processing tasks. Another example is **XLNet** [50], which improves upon BERT by using permutation language modeling, allowing it to capture bidirectional context without the need for masking. It builds on the Transformer-XL architecture, enabling the handling of longer sequences and dependencies beyond the sentence level. XLNet's approach combines autoregressive training with a permutation-based technique, making it particularly effective for tasks requiring complex context understanding, such as question answering and language modeling.

We will create word embeddings for the training and testing datasets using different variations of BERT in order to compare performance between models. For this purpose, we will use the `textEmbed` function from the `text` package. By default, this function uses the `"bert-base-uncased"` language model to create the word embeddings. We will use this model as a baseline, and compare it with two other models: "roberta-base" and "xlnet-base-cased". You can try other language models available in Hugging Face that support the task of creating word embeddings (text classification category). You should take into account that this process takes a considerable amount of run time. As an input to the function we provide the "clean_text" column of our training and testing data, which contains the text to be coded.

```
# Creating embeddings for the training data
## BERT
word_embeddings_bert <- textEmbed(df_train[,"clean_text"],
                            model = "bert-base-uncased",
                            aggregation_from_tokens_to_word_types = "mean",
                            keep_token_embeddings = FALSE)
## RoBERTa
word_embeddings_roberta <- textEmbed(df_train[,"clean_text"],
                            model = "roberta-base",
                            aggregation_from_tokens_to_word_types = "mean",
                            keep_token_embeddings = FALSE)
## XLNet
word_embeddings_xlnet <- textEmbed(df_train[,"clean_text"],
                            model = "xlnet-base-cased",
                            aggregation_from_tokens_to_word_types = "mean",
                            keep_token_embeddings = FALSE)
# Creating embeddings for the testing data
## BERT
word_embeddings_bert_test <- textEmbed(df_test[,"clean_text"],
                            model = "bert-base-uncased",
                            aggregation_from_tokens_to_word_types = "mean",
                            keep_token_embeddings = FALSE)
## RoBERTa
word_embeddings_roberta_test <- textEmbed(df_test[,"clean_text"],
                            model = "roberta-base",
                            aggregation_from_tokens_to_word_types = "mean",
                            keep_token_embeddings = FALSE)
## XLNet
word_embeddings_xlnet_test <- textEmbed(df_test[,"clean_text"],
                            model = "xlnet-base-cased",
                            aggregation_from_tokens_to_word_types = "mean",
                            keep_token_embeddings = FALSE)
```

If we inspect the generated embeddings, we see that there are 1612 rows in the texts of the training embeddings: one per row of the training dataset. However, instead of one column with the text, we have a large number of numerical columns that represent the text. For example, in BERT, we have 768, which corresponds to the size of the BERT token embedding vector.

```
as_tibble(word_embeddings_bert$texts$texts)
```

```
# A tibble: 1612 x 768
  Dim1_texts Dim2_texts Dim3_texts Dim4_texts Dim5_texts Dim6_texts Dim7_texts
       <dbl>      <dbl>      <dbl>      <dbl>      <dbl>      <dbl>      <dbl>
1      0.528    -0.0955     0.0600      0.124      0.533     -0.420     0.0723
2      0.507    -0.0987     0.0220     -0.188      0.181     -0.655     0.356
3      0.448    -0.0161     0.257      -0.485     -0.310     -0.194     0.190
4     -0.117    -0.318      0.418      -0.119      0.216     -0.511     0.287
```

```
5     -0.629   -0.0899   0.0623   -0.0132    0.143      0.292    -0.210
6      0.421   -0.211    0.318     0.104    -0.0157    -0.366     0.231
7      0.615    0.0502   0.561    -0.406     0.0821    -0.309    -0.0774
8      0.126    0.0952   0.146     0.113    -0.318     -0.142    -0.350
9      0.287    0.0764   0.261    -0.0609    0.0596    -0.345     0.0266
10     0.547   -0.272    0.163     0.0564   -0.197     -0.391     0.0832
# i 1602 more rows
# i 761 more variables: Dim8_texts <dbl>, Dim9_texts <dbl>, Dim10_texts <dbl>,
#   Dim11_texts <dbl>, Dim12_texts <dbl>, Dim13_texts <dbl>, Dim14_texts <dbl>,
#   Dim15_texts <dbl>, Dim16_texts <dbl>, Dim17_texts <dbl>, Dim18_texts <dbl>,
#   Dim19_texts <dbl>, Dim20_texts <dbl>, Dim21_texts <dbl>, Dim22_texts <dbl>,
#   Dim23_texts <dbl>, Dim24_texts <dbl>, Dim25_texts <dbl>, Dim26_texts <dbl>,
#   Dim27_texts <dbl>, Dim28_texts <dbl>, Dim29_texts <dbl>, \ldots
```

3.6 *Training the Model*

Now that we have converted our text data into numeric variables through word embeddings, we are ready to train our ML model. Specifically, we will train a Random Forest classifier algorithm. We will rely on the `textTrainRandomForest` function from the `text` package for this purpose. As training data, we will enter the embeddings from the previous step (for each of the three BERT-like models). As an output, we will enter the column that contains the codes in the original training data: `"annotation_target"`. We will use fivefold cross validation to train the model (`outside_folds` argument). For more information about the training function, refer to the `text` package documentation.

```
trained_model_bert <- textTrainRandomForest(
  x = word_embeddings_bert$texts,
  y = data.frame(as.factor(df_train[["annotation_target"]])),
  outside_folds = 5,
  simulate.p.value = TRUE,
  append_first = TRUE,
  multi_cores = TRUE,
  seed = seed
)

trained_model_roberta <- textTrainRandomForest(
  x = word_embeddings_roberta$texts,
  y = data.frame(as.factor(df_train[["annotation_target"]])),
  outside_folds = 5,
  simulate.p.value = TRUE,
  append_first = TRUE,
  multi_cores = TRUE,
  seed = seed
)
```

```
trained_model_xlnet <- textTrainRandomForest(
  x = word_embeddings_xlnet$texts,
  y = data.frame(as.factor(df_train[["annotation_target"]])),
  outside_folds = 5,
  simulate.p.value = TRUE,
  append_first = TRUE,
  multi_cores = TRUE,
  seed = seed
)
```

3.7 *Using the Model to Predict*

We can now use the trained models to code the remainder of the data (i.e., the word embeddings of the test dataset).

```
predicted_bert <- textPredict(
  model_info = trained_model_bert,
  word_embeddings = word_embeddings_bert_test$texts
)
predicted_roberta <- textPredict(
  model_info = trained_model_roberta,
  word_embeddings = word_embeddings_roberta_test$texts
)
predicted_xlnet <- textPredict(
  model_info = trained_model_xlnet,
  word_embeddings = word_embeddings_xlnet_test$texts
)
```

3.8 *Evaluating the Model*

As a first measure of the performance of our models, we can use the results of the training data. To ease the comparison, let us first combine all the results in a single dataframe (`rbind`) and adding the name of the model `mutate` to be able to identify each set of results. We then pivot our data (`pivot_wider`) to show all the metrics of each model as columns, and each model as a row.

As we can see in Table 1, RoBERTa consistently outperforms BERT and XLNet across most key metrics, showcasing its superior performance. For accuracy,

Table 1 Performance of each model

Model	Accuracy	bal_accuracy	Sens	Spec	Precision	Kap	f_meas	roc_auc
BERT	0.770	0.748	0.568	0.928	0.843	0.654	0.610	0.889
RoBERTa	0.788	0.769	0.605	0.934	0.857	0.685	0.645	0.910
XLNet	0.734	0.717	0.517	0.917	0.809	0.600	0.547	0.843

RoBERTa achieves the highest value (0.788), followed closely by BERT (0.770), while XLNet (0.734) lags behind. This indicates RoBERTa's ability to make correct predictions more consistently across the dataset. When considering balanced accuracy, which accounts for both sensitivity and specificity, RoBERTa (0.769) again leads, demonstrating its effectiveness at handling imbalanced classes. BERT performs respectably (0.748), while XLNet (0.717) shows the weakest balance between true positive and true negative rates. In terms of sensitivity (recall), RoBERTa (0.605) outperforms BERT (0.568) and XLNet (0.517), highlighting its ability to correctly identify positive instances. Meanwhile, for specificity, RoBERTa (0.934) slightly surpasses BERT (0.928) and XLNet (0.917), reinforcing its capacity to avoid false positives. For precision, RoBERTa (0.857) maintains its advantage over BERT (0.843) and XLNet (0.809), reflecting fewer false positives. The F1 score, which balances precision and recall, also favors RoBERTa (0.645), with BERT (0.610) trailing and XLNet (0.547) performing the weakest. Looking at kappa, a measure of agreement between predicted and actual labels beyond chance, RoBERTa (0.685) once more outperforms BERT (0.654) and XLNet (0.600), showcasing its robustness in classification consistency. Finally, ROC-AUC values indicate RoBERTa (0.910) as the strongest model, effectively balancing sensitivity and specificity, followed by BERT (0.889) and XLNet (0.843). Overall, the results strongly favor RoBERTa across all metrics, making it the most suitable model for text dataset coding in this scenario. BERT provides solid performance but falls slightly behind in several areas, while XLNet struggles particularly with sensitivity, F1 score, and class balance.

```
rbind( # Combine the performance results of each model
  trained_model_bert$results |> mutate(Model = "BERT"),
  trained_model_roberta$results |> mutate(Model = "RoBERTa"),
  trained_model_xlnet$results |> mutate(Model = "XLNet")) |>
  select(-.estimator) |>
  pivot_wider(names_from = ".metric", values_from = ".estimate")
```

In a real setting, if we were to choose one of the three models to code our text dataset based solely on the performance metrics obtained from the training process, we would choose RoBERTa. Since in this case we have access to the fully coded dataset, let us verify if RoBERTa is still the best choice when faced with the remainder of the dataset—still unseen by any model. For this purpose, we will compare the manual codes with the automatic codes classified by our models. First, let us combine all codes into a single dataframe. We obtain the manually coded data from the test data (`df_test`), and the automatically coded data from the prediction results of each model (e.g., `predicted_bert`). We also need to convert the five possible codes to factors to be able to operate with them as "categories".

```
results <- data.frame(
  manual = as.factor(df_test$annotation_target),
  bert = as.factor(predicted_bert$`texts__cv_method="validation_split"pred`),
  roberta = as.factor(predicted_roberta$`texts__cv_method="validation_split"pred`),
  xlnet = as.factor(predicted_xlnet$`texts__cv_method="validation_split"pred`)
)
```

To evaluate the results, we can use the `confusionMatrix` function from the `caret` package, which takes as an input the results of the automatic coding and the manual codes. We can combine again all the results together using `rbind`.

```
cm_bert <- confusionMatrix(results$bert, results$manual)
cm_roberta <- confusionMatrix(results$roberta, results$manual)
cm_xlnet <- confusionMatrix(results$xlnet, results$manual)

rbind(Bert = cm_bert$overall,
      RoBERTa = cm_roberta$overall,
      XLnet = cm_xlnet$overall)
```

When comparing the accuracy and Cohen's Kappa, after predicting the code of the test data, RoBERTa is still ahead of the two other models (Table 2). The accuracy is quite high and the Cohen's Kappa is substantial (> 0.6). Therefore, our initial choice was correct.
In addition, we calculate Gwet's AC1 and Matthews Correlation Coefficient (MCC). Gwet's AC1 provides a robust measure of inter-rater agreement for multiclass settings, addressing the limitations of traditional kappa statistics. MCC, extended to handle multiclass classifications, assesses the quality of the classifier by considering the entire confusion matrix, providing a balanced evaluation of performance across all classes.

Table 2 Performance of each model calculated using the test data

	Accuracy	Kappa	AccuracyLower	AccuracyUpper	AccuracyNull	AccuracyPValue
Bert	0.737	0.599	0.728	0.746	0.409	0
RoBERTa	0.770	0.654	0.761	0.779	0.409	0
XLnet	0.744	0.616	0.734	0.752	0.409	0

Table 3 Additional performance-metrics

	gwets_ac1	mcc
Bert	0.737	0.611
RoBERTa	0.770	0.664
XLnet	0.743	0.620

```
# Gwet AC1
calculate_gwets_ac1 <- function(table) {
  n <- sum(table)
  po <- sum(diag(table)) / n
  pe <- sum(pmax(rowSums(table), colSums(table))) / n^2
  (po - pe) / (1 - pe)
}

extra_results <- data.frame(
  gwets_ac1 = c(calculate_gwets_ac1(cm_bert$table),
                calculate_gwets_ac1(cm_roberta$table),
                calculate_gwets_ac1(cm_xlnet$table)),
  mcc = c(mcc(results$bert, results$manual),
          mcc(results$roberta, results$manual),
          mcc(results$xlnet, results$manual)
  )
)

rownames(extra_results) <- c("Bert", "RoBERTa", "XLnet")
extra_results
```

All three models demonstrate substantial inter-rater reliability (Table 3) as per the benchmarks commonly used for interpreting AC1 (values above 0.70 are typically considered strong). RoBERTa is the best-performing model in this evaluation metric, followed by XLNet, with BERT slightly trailing, where the results are almost exactly matching the accuracy metric. Regarding MCC, all values here fall between 0.6 and 0.7, indicating moderate to strong correlations between the model predictions and the manual labels. Similar to Gwet's AC1 results, RoBERTa achieves the highest MCC score, indicating the best performance in terms of balanced prediction quality.

We can also extract the results by class to gain more knowledge about the potential classification errors. Some models might be better at capturing some types of discussion contributions than others. The code below extracts the performance metrics by class and creates a long data frame containing the model name, the code (agree, disagree, etc.) the metric name (sensitivity, precision, etc.), and the value of the metric. We can combine the results for each model in a single dataframe using `rbind`.

```
# BERT
byclass_bert <- cm_bert$byClass |> data.frame() |>
  rownames_to_column() |> # Convert rowname (code) to new column
  pivot_longer(2:12) |> # Convert to a long format where each metric is a new row
  mutate(Model = "BERT") # Add the model name as a column

# RoBERTa
byclass_roberta <- cm_roberta$byClass |> data.frame() |>
  rownames_to_column() |>
  pivot_longer(2:12) |>
  mutate(Model = "RoBERTa")

# XLNet
byclass_xlnet <- cm_xlnet$byClass |> data.frame() |>
  rownames_to_column() |>
  pivot_longer(2:12) |>
  mutate(Model = "XLNet")

# Combine all results together
byclass_all <- rbind(byclass_bert, byclass_roberta, byclass_xlnet)
```

We can also plot the results using `ggplot2`. We use a bar plot (`geom_col`) to plot all the metrics grouped by model. We use `facet_wrap` to separate the plot by coding class (agree, disagree, moderation, etc.).

```
ggplot(byclass_all, aes(x = name, y = value, group = Model, fill = Model)) +
  geom_col(position = "dodge2", width = 0.9) +
  facet_wrap("rowname", ncol = 1) +
  theme_minimal() +
  scale_fill_manual(values = c("#76B7B2", "#B07AA1", "#FF9DA7")) +
  theme(legend.position = "bottom",
        axis.text.x = element_text(angle = 45, hjust = 1)) +
  labs(fill = "Model") + xlab("") + ylab("")
```

When we visualize the different performance metrics of each model per class (Fig. 2), we can see that RoBERTa outperforms or equals the other model in 31 of the 55 comparisons, XLNet 21, and BERT 16. These findings further confirm that RoBERTa is the most suitable model for the classification task.

In addition, we can also test whether the combination of all three models yields better results. First we create an option to obtain the vote from all three algorithms (i.e., the statistical mode):

```
# Calculates with class is selected by most models
getmode <- function(v) {
   uniqv <- unique(v)
   uniqv[which.max(tabulate(match(v, uniqv)))]
}
```

Now, we iterate through our results and, for each coded, we gather the voting of all three models. We input RoBERTa in the first position so that in case all models have come up with a different code, the one created by RoBERTa is chosen:

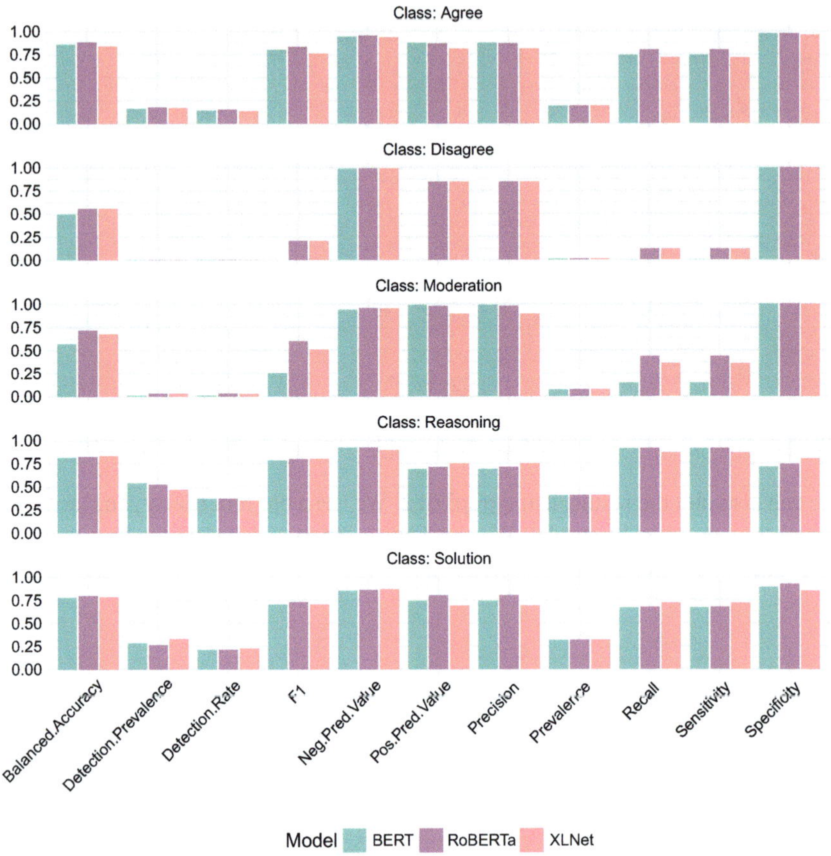

Fig. 2 Model performance comparison by class

```
results_with_vote <- results |>
  rowwise() |>
  mutate(vote = getmode(c(roberta, xlnet, bert))) |>
  ungroup()
```

We can now create the confusion matrix for the voted code. We can see that the results are slightly below those achieved by RoBERTa alone, so we are not getting any advantage by voting.

```
cm_vote <- confusionMatrix(results_with_vote$vote, results_with_vote$manual)
cm_vote$overall
```

```
      Accuracy          Kappa  AccuracyLower  AccuracyUpper   AccuracyNull
     0.7646994      0.6455472      0.7558531      0.7733763      0.4085125
AccuracyPValue  McnemarPValue
     0.0000000            NaN
```

3.9 Impact on Findings

To complement the evaluation of our models, we assess the impact of using an automatically coded dataset on the findings derived from applying common learning analytics methods to the data. We choose RoBERTa, as it was our best performing model, to compare the results with those obtained from manually coded data.

First, we need to construct the complete AI-generated dataset which contains all of the data classified by RoBERTa and the small percentage classified manually used as training. We store it in the `complete_ai` variable.

```
df_test_recoded <- df_test
df_test_recoded$annotation_target <- results$roberta
complete_ai <- rbind(df_train, df_test_recoded)
```

We start by analyzing our data using sequence analysis, which is a method that is commonly used to study temporally-ordered data, and specifically collaborative learning (e.g., [51]). For a full tutorial about sequence analysis consult [52]. The first step is to order the messages in each group and to transpose the data so that is in a wide format (i.e., each column represents one message order: message 1, message 2, etc.). Then, we use `seqdef` from the `TraMineR` package [53], to construct a sequence object, where each group is represented as a sequence of all its contributions. We use `seqdplot` to plot the sequence distribution plot (Fig. 3). From visual inspection, we can see that the sequence profiles are similar, although the RoBERTa side has a lower presence of the minority codes.

```
library(TraMineR)

wider_human <- df |> group_by(group_id) |>
  mutate(order = seq_along(message_id)) |>
  pivot_wider(id_cols="group_id", names_from="order",values_from = "annotation_target")

wider_ai <- complete_ai |> group_by(group_id) |>
  mutate(order = seq_along(message_id)) |>
  pivot_wider(id_cols="group_id", names_from="order",values_from = "annotation_target")

seq_human <- seqdef(wider_human, 2:ncol(wider_human))
seq_ai <- seqdef(wider_ai, 2:ncol(wider_ai))

seqdplot(seq_human)
seqdplot(seq_ai)
```

Next, we use transition network analysis, a novel method that implements stochastic process mining through Markov models [54], to model the transitions between codes and assess whether there are differences between the two datasets. For a full tutorial, refer to [55]. We use the `tna` function to generate the Markov model containing the probability of transitioning between each pair of codes. We do so for the human coded and for the data coded with the RoBERTa model. We can visualize each model using the `plot` function. We can also plot the difference between the transitions using `plot_compare`. However, this is only a visual comparison. We

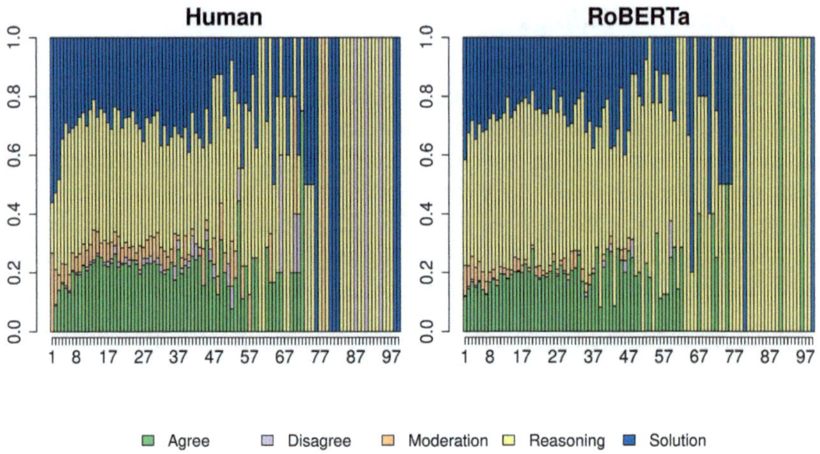

Fig. 3 Comparing the sequence distribution plots between the AI-coded and human-coded datasets

can run a permutation test (`permutation_test`) to statistically compare the two models and visualize only the statistically significant differences (Fig. 4).

```
library(tna)
# Compute transition probabilities
transitions_ai <-  tna(seq_ai)
transitions_human <-  tna(seq_human)

# Plot human coded transition network
plot(transitions_human)

# Plot RoBERTa transition network
plot(transitions_ai)

# Plot difference network
plot_compare(transitions_human, transitions_ai)

# Run a permutation test to assess which differences are statistically significant
permutation <- permutation_test(transitions_human, transitions_ai, it = 1000)

# Plot the significant differences identified in the permutation test
plot(permutation, minimum = 0.01)
```

In the transition network created from human codes, *Solution* and *Agree* are central nodes with strong self-loops and frequent transitions between them. *Moderation* shows significant connections from *Agree* and *Solution*, while *Disagree* is weakly integrated. In the RoBERTa network, *Solution* and *Agree remain* central, and *Disagree* still has low connections. The difference network highlights that *Reasoning* plays a stronger role in RoBERTa's coding, especially in its links from *Agree* and *Solution*. *Disagree* is less integrated, and some transitions, like from *Moderation* to *Reasoning*, are significantly reduced.

We could continue with other relevant methods, like clustering participants according to their argumentation profiles or role in the discussion, but we have

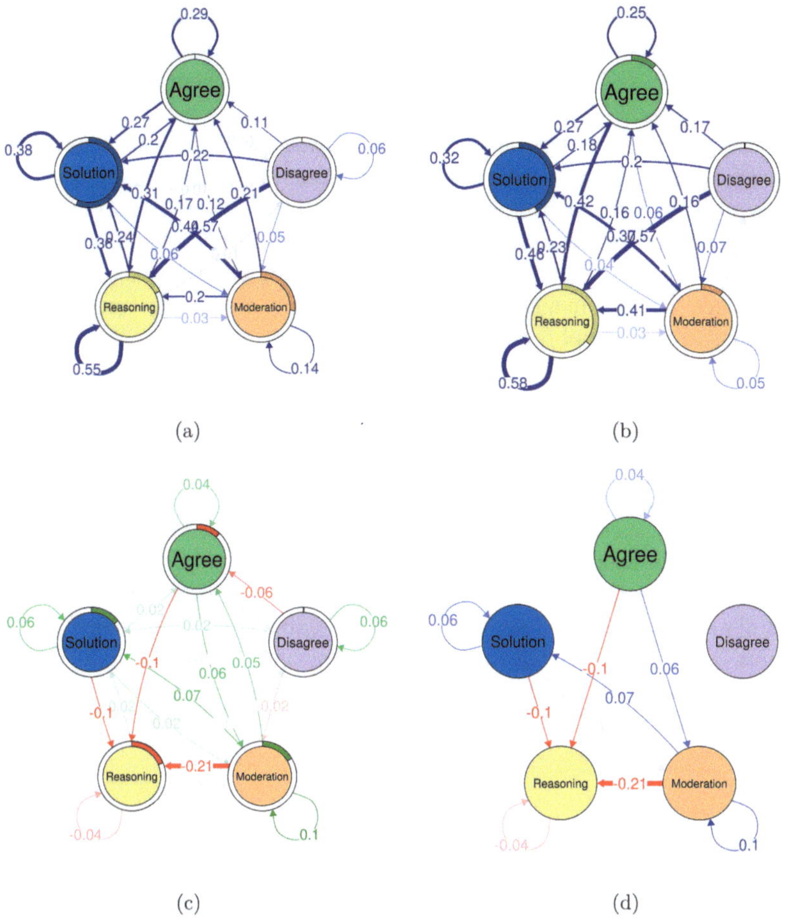

Fig. 4 Comparing the transition plots between the AI-coded and human-coded datasets. (**a**) Human-coded. (**b**) RoBERTa. (**c**) Difference network. (**d**) Permutation test

already demonstrated that the results somewhat differ between the two sets of codes. However, the overall takeaway for both versions is that *Reasoning* is the driver of the conversation. It is not expected that two human-coded datasets yielded more similar results.

4 Conclusion

This tutorial examined the use of BERT-like language models for automating discourse coding in R, showcasing how these models can expedite qualitative text classification while maintaining an acceptable level of accuracy. We illustrated this with a case study on collaborative problem-solving. Among the models tested, RoBERTa demonstrated better performance compared to BERT and XLNet, making it the most effective model for the classification task. We exemplified how applying different analytical techniques from the learning analytics repertoire, including sequence analysis and transition network analysis, revealed somewhat different insights from the human coded data than from the automatically coded data. However, there was an overall alignment between both in capturing the broad patterns of discourse and the disagreement was comparable to the one expected between two humans. As large language models continue to advance, we anticipate even greater accuracy in automated discourse coding, making automated coding an increasingly reliable tool for qualitative research.

Although automated coding with LLMs offers promising efficiency and consistency, caution is warranted when interpreting the results. As pointed out earlier in this chapter, the reliability of automated methods can vary based on the context, dataset characteristics, and the specific model used. Differences in coding accuracy may introduce biases or overlook human judgment, which could impact downstream analyses. We must critically evaluate the output of automated coding by comparing it against human-coded benchmarks and incorporating rigorous metrics like Gwet's AC1 or Matthews Correlation Coefficient. Additionally, researchers should remain mindful of the iterative and reflexive nature of qualitative coding, which automated systems cannot fully replicate. Therefore, automated methods should be viewed as complementary to, rather than replacements for, human expertise in the coding process.

References

1. Elliott V (2018) Thinking about the coding process in qualitative data analysis. Qualitative Report 23.
2. Shaffer DW, Ruis AR (2021) How we code. In: Advances in quantitative ethnography: second international conference, ICQE 2020, malibu, CA, USA, February 1–3, 2021, Proceedings 2. Springer, pp 62–77
3. Thomas DR (2006) A general inductive approach for analyzing qualitative evaluation data. Am J Eval 27:237–246
4. Oliveira E, Song Y, Saqr M, López-Pernas S (2025) An introduction to large language models in education. In: Saqr M, López-Pernas S (eds) Advanced learning analytics methods: AI, precision and complexity. Springer Nature Switzerland, Cham
5. Raiaan MAK, Mukta MSH, Fatema K, Fahad NM, Sakib S, Mim MMJ, Ahmad J, Ali ME, Azam S (2024) A review on large language models: architectures, applications, taxonomies, open issues and challenges. IEEE Access 12:26839–26874. https://doi.org/10.1109/ACCESS.2024.3365742
6. Bosley M, Jacobs-Harukawa M, Licht H, Hoyle A (2023) Do we still need BERT in the age of GPT? Comparing the benefits of domain-adaptation and in-context-learning approaches to using LLMs for political science research
7. Amaratunga T (2023) What makes LLMs large?. In: Understanding large language models. Apress, Berkeley, CA. https://doi.org/10.1007/979-8-8688-0017-7_4
8. Chae Y, Davidson T (2023) Large language models for text classification: from zero-shot learning to fine-tuning. Open Science Foundation, Charlottesville

9. Periti F, Montanelli S (2024) Lexical semantic change through large language models: a survey. ACM Comput Surv 56(11):1–38. https://doi.org/10.1145/3672393
10. Saqr M, Misiejuk K, Tikka S, López-Pernas S (2025) Artificial intelligence: using machine learning to classify students and predict low achievers. In: Saqr M, López-Pernas S (eds) Advanced learning analytics methods: AI, Precision and complexity. Springer
11. Yan L, Martinez-Maldonado R, Gasevic D (2024) Generative artificial intelligence in learning analytics: contextualising opportunities and challenges through the learning analytics cycle. In: Proceedings of the 14th learning analytics and knowledge conference. ACM, New York, NY, USA. https://doi.org/10.1145/3636555.3636856
12. Barany A, Nasiar N, Porter C, Zambrano AF, Andres AL, Bright D, Shah M, Liu X, Gao S, Zhang J, others (2024) ChatGPT for education research: exploring the potential of large language models for qualitative codebook development. In: International conference on artificial intelligence in education. Springer, pp 134–149
13. Hayes A (2024) "Conversing" with qualitative data: enhancing qualitative research through large language models (LLMs). https://doi.org/10.31235/osf.io/yms8p
14. Mathis WS, Zhao S, Pratt N, Weleff J, De Paoli S (2024) Inductive thematic analysis of healthcare qualitative interviews using open-source large language models: how does it compare to traditional methods? Comput Methods Programs Biomed 255:108356
15. Davison RM, Chughtai H, Nielsen P, Marabelli M, Iannacci F, van Offenbeek M, Tarafdar M, Trenz M, Techatassanasoontorn AA, Díaz Andrade A, Panteli N (2024) The ethics of using generative AI for qualitative data analysis. Inf Syst J 34(5):1433–1439. https://doi.org/10.1111/isj.12504
16. Morgan DL (2023) Exploring the use of artificial intelligence for qualitative data analysis: the case of ChatGPT. Int J Qualitative Methods 22:16094069231211248
17. Garg R, Han J, Cheng Y, Fang Z, Swiecki Z (2024) Automated discourse analysis via generative artificial intelligence. In: Proceedings of the 14th learning analytics and knowledge conference. ACM, New York, NY, USA. https://doi.org/10.1145/3636555.3636879
18. Misiejuk K, Kaliisa R, Scianna J (2024) Augmenting assessment with AI coding of online student discourse: a question of reliability. Comput Educ Artif Intell 6. https://doi.org/10.1016/j.caeai.2024.100216
19. Pugh SL, Subburaj SK, Rao A, Stewart AEB, Andrews-Todd J, D'Mello S (2021) Say what? Automatic modeling of collaborative problem solving skills from student speech in the wild. Int Conf Young Spéc Micro/nanotechnologies Electron Device. https://files.eric.ed.gov/fulltext/ED615653.pdf
20. Prescott MR, Yeager S, Ham L, Rivera Saldana CD, Serrano V, Narez J, Paltin D, Delgado J, Moore DJ, Montoya J (2024) Comparing the efficacy and efficiency of human and generative AI: Qualitative thematic analyses. JMIR AI 3:e54482
21. Lefort B, Benhamou E, Ohana J-J, Guez B, Saltiel D, Challet D (2024) When small wins big: classification tasks where compact models outperform original GPT-4. Available at SSRN 4780454
22. Ganesh A, Chandler C, D'Mello S, Palmer M, Kann K (2024) Prompting as panacea? A case study of in-context learning performance for qualitative coding of classroom dialog. In: Proceedings of the 17th international conference on educational data mining, pp 835–843
23. Van Ostaeyen S, De Langhe L, De Clercq O, Embo M, Schellens T, Valcke M (2023) Automating the identification of feedback quality criteria and the CanMEDS roles in written feedback comments using natural language processing. Perspect Med Educ 12:540
24. Tai RH, Bentley LR, Xia X, Sitt JM, Fankhauser SC, Chicas-Mosier AM, Monteith BG (2024) An examination of the use of large language models to aid analysis of textual data. Int J Qual Methods 23:16094069241231168
25. Xiao Z, Yuan X, Liao QV, Abdelghani R, Oudeyer P-Y (2023) Supporting qualitative analysis with large language models: combining codebook with GPT-3 for deductive coding. In: Companion proceedings of the 28th international conference on intelligent user interfaces, pp 75–78

26. Jang J, Ye S, Seo M (2023) Can large language models truly understand prompts? A case study with negated prompts. In: Transfer learning for natural language processing workshop. PMLR, pp 52–62
27. Ma W, Yang C, Kästner C (2024) (Why) is my prompt getting worse? Rethinking regression testing for evolving LLM APIs. In: Proceedings of the IEEE/ACM 3rd international conference on AI engineering-software engineering for AI, pp 166–171
28. Palmer A, Smith NA, Spirling A (2024) Using proprietary language models in academic research requires explicit justification. Nature Comput Sci 4:2–3
29. Chang Y, Wang X, Wang J, Wu Y, Yang L, Zhu K, Chen H, Yi X, Wang C, Wang Y, others (2024) A survey on evaluation of large language models. ACM Trans Intell Syst Technol 15:1–45
30. Ziems C, Held W, Shaikh O, Chen J, Zhang Z, Yang D (2024) Can large language models transform computational social science? Comput Linguist 50:237–291
31. Sokolova M, Lapalme G (2009) A systematic analysis of performance measures for classification tasks. Inf Process Manag 45:427–437
32. Kolesnyk A, Khairova N (2022) Justification for the use of cohen's kappa statistic in experimental studies of NLP and text mining. Cybern Syst Anal 58:280–288
33. Delgado R, Tibau X-A (2019) Why cohen's kappa should be avoided as performance measure in classification. PloS One 14:e0222916
34. Chicco D, Warrens MJ, Jurman G (2021) The matthews correlation coefficient (MCC) is more informative than cohen's kappa and brier score in binary classification assessment. IEEE Access 9:78368–78381
35. Gwet KL (2008) Computing inter-rater reliability and its variance in the presence of high agreement. Br J Math Stat Psychol 61:29–48
36. Blood E, Spratt KF (2007) Disagreement on agreement: Two alternative agreement coefficients. In: SAS global forum. Citeseer, pp 1–12
37. Yan L, Sha L, Zhao L, Li Y, Martinez-Maldonado R, Chen G, Li X, Jin Y, Gašević D (2024) Practical and ethical challenges of large language models in education: a systematic scoping review. Br J Educ Technol 55:90–112
38. Liu Y, Gautam S, Ma J, Lakkaraju H (2024) Confronting LLMs with traditional ML: rethinking the fairness of large language models in tabular classifications. In: Proceedings of the 2024 conference of the north american chapter of the association for computational linguistics: human language technologies (volume 1: Long papers), pp 3603–3620
39. Gallegos IO, Rossi RA, Barrow J, Tanjim MM, Kim S, Dernoncourt F, Yu T, Zhang R, Ahmed NK (2024) Bias and fairness in large language models: a survey. Comput Linguist 1–79
40. Karadzhov G, Stafford T, Vlachos A (2023) DeliData: a dataset for deliberation in multi-party problem solving. Proc ACM Human-Comput Inter 7:1–25
41. Saqr M, Misiejuk K, Tikka S, López-Pernas S (2025) Artificial intelligence: using machine learning to classify students and predict low achievers. In: Saqr M, López-Pernas S (eds) Advanced learning analytics methods: AI, precision and complexity. Springer Nature Switzerland, Cham
42. Wickham H, François R, Henry L, Müller K, Vaughan D (2023) Dplyr: A grammar of data manipulation. https://CRAN.R-project.org/package=dplyr
43. Wickham H, Vaughan D, Girlich M (2024) Tidyr: Tidy messy data. https://CRAN.R-project.org/package=tidyr
44. Müller K, Wickham H (2023) Tibble: Simple data frames. https://CRAN.R-project.org/package=tibble
45. Ushey K, Allaire J, Tang Y (2024) Reticulate: Interface to 'python'. https://CRAN.R-project.org/package=reticulate
46. Kuhn, Max (2008) Building predictive models in r using the caret package. J Stat Softw 28:1–26. https://doi.org/10.18637/jss.v028.i05
47. Kjell O, Giorgi S, Schwartz HA (2023) The text-package: an r-package for analyzing and visualizing human language using natural language processing and deep learning. Psychol Methods. https://doi.org/10.1037/met0000542

48. Gorman B (2018) Mltools: Machine learning tools. https://CRAN.R-project.org/package=mltools
49. Liu Y, Ott M, Goyal N, Du J, Joshi M, Chen D, Levy O, Lewis M, Zettlemoyer L, Stoyanov V (2019) RoBERTa: A robustly optimized BERT pretraining approach. arXiv [csCL]. https://scholar.google.es/citations?user=H9buyroAAAAJ&hl=en&oi=sra
50. Yang Z, Dai Z, Yang Y, Carbonell J, Salakhutdinov R, Le QV (2019) XLNet: Generalized autoregressive pretraining for language understanding. arXiv [csCL]. http://arxiv.org/abs/1906.08237
51. Saqr M, López-Pernas S (2023) The temporal dynamics of online problem-based learning: why and when sequence matters. Int J Comput-Supported Collaborative Learn 18:11–37. https://doi.org/10.1007/s11412-023-09385-1
52. Saqr M, López-Pernas S, Helske S, Durand M, Murphy K, Studer M, Ritschard G (2024) Sequence analysis in education: principles, technique, and tutorial with R, pp 321–354. https://doi.org/10.1007/978-3-031-54464-4_10
53. Gabadinho A, Ritschard G, Mueller NS, Studer M (2011) Analyzing and visualizing state sequences in R with TraMineR. J Stat Softw 40:1–37. https://doi.org/10.18637/jss.v040.i04
54. Saqr M, López-Pernas S, Törmänen T, Kaliisa R, Misiejuk K, Tikka S (2025) Transition network analysis: a novel framework for modeling, visualizing, and identifying the temporal patterns of learners and learning processes. In: Proceedings of learning analytics and knowledge (LAK '25). ACM, New York, NY, USA (in press)
55. Saqr M, López-Pernas S, Tikka S (2025) Mapping relational dynamics with transition network analysis: a primer and tutorial. In: Saqr M, López-Pernas S (eds) Advanced learning analytics methods: AI, precision and complexity. Springer Nature Switzerland, Cham

LLMs for Explainable Artificial Intelligence: Automating Natural Language Explanations of Predictive Analytics Models

Sonsoles López-Pernas, Yige Song, Eduardo Oliveira, and Mohammed Saqr

Abstract This chapter explores the integration of LLMs into the Explainable AI (XAI) pipeline within the field of Learning Analytics (LA). The focus of the chapter is on how to automate the transformation of XAI outputs into accessible, natural language explanations using LLMs. Namely, we demonstrate how LLMs can contextualize feature importance, partial dependence profiles, and local explanations, making predictive model outputs more interpretable and actionable for non-technical stakeholders. We illustrate this process making use of an open source language model through the LM studio software. However, the model used shares the API definition with some widely used commercial models, such as those implemented by OpenAI, making the portability almost immediate. This work contributes to advancing the intersection of generative AI, XAI, and learning analytics to promote transparency, inclusivity, and fairness.

Keywords Learning analytics · Explainable artificial intelligence · LM studio · Generative AI · Automated feedback · Large language models

1 Introduction

Large language models (LLMs) have recently become a transformative technology thanks to their ability to process natural language and generate close-to-human quality text [1]. Before LLMs, generating human-like text automatically was often a slow and manual process, requiring a lot of time and expertise. Since LLMs have been trained on vast amounts of data, they enable the automation of text generation, with little to no human intervention. Recently, the field of learning

S. López-Pernas (✉) · M. Saqr
University of Eastern Finland, Joensuu, Finland
e-mail: sonsoles.lopez@uef.fi

Y. Song · E. Oliveira
University of Melbourne, Parkville, VIC, Australia

M. Saqr, S. López-Pernas (eds.), *Advanced Learning Analytics Methods*,
https://doi.org/10.1007/978-3-031-95365-1_11

analytics has recognized the potential of generative artificial intelligence (AI) as a whole, and LLMs in particular, in various stages of the learning analytics cycle such as generating synthetic data, automated scoring and coding or feedback generation [2].

In this tutorial, we focus on the last phase of the learning analytics cycle: reporting the results of learning analytics applications and turning them into actionable insights [3]. Specifically, we will focus on making the results of predictive models understandable for teachers and students to aid decision making. LLMs have emerged as a powerful tool in this domain, particularly in enhancing the explainability of AI models, also known as Explainable AI (XAI). XAI refers to techniques and methods that enable humans to understand, trust, and make the right decisions based on the outputs generated by AI systems [4]. In learning analytics, XAI has the potential to become an extremely central tool because it allows teachers, students, parents and administrators to understand the rationale behind the predictions made by AI models.

Traditionally, explaining the results of complex predictive models required specialized knowledge in AI. However, LLMs have the potential to bridge this gap by generating natural language explanations that are both accurate and accessible to non-experts. For instance, when a predictive model identifies that a student is at risk of not completing a course, an LLM can be used to generate a human-readable explanation of why the model reached that conclusion. This explanation might include an overview of the key factors that influenced the prediction, such as the student's engagement with course materials, performance on assessments, and participation in class discussions. The LLM can then articulate this information in a way that is tailored to the teacher's or student's specific context, providing clear guidance on potential interventions or areas for improvement rather than an endless list of numbers, tables and graphs.

The application of generative AI in general, and LLMs in particular, to learning analytics and education research and practice is still in its early stages [5]. One could say that the potential of LLMs to make learning analytics insights more accessible has been barely explored. Still, a few studies demonstrating LLMs' ability to simplify very technical information and generate personalized, human-readable feedback for learners, teachers and other stakeholders can be mentioned here. One of the few existing examples is the study by Pinargote et al. [6], in which an LLM was used to transform insights shown in a learning analytics dashboard into textual explanations. While students appreciated the metrics and feedback for understanding group contributions, opinions varied on the fairness and necessity of detailed feedback. The study by Susnjak [7] proposed a framework that combines XAI with LLMs —similar to what we are doing in this tutorial— to offer prescriptive analytics to students at risk. A formal evaluation of the framework is still needed to validate it empirically.

In this tutorial, we will explore how to use LLMs to enhance the explainability of predictive models. Specifically, we will demonstrate how to generate natural language explanations of XAI results using LLMs. The reader will understand how to embed LLMs in the XAI pipeline to create personalized and contextualized explanations that make predictive model outputs understandable, helping teachers and students make informed decisions based on AI-generated insights.

2 Motivation and Relevant Work for Using LLMs on XAI Output in Learning Analytics Research

Previous chapters of this book have introduced and discussed XAI in detail [4, 8, 9]. To summarise, XAI techniques, such as feature importance, help explain the outputs of black-box machine learning models. For example, in authorship detection [10], XAI algorithms identify and quantify key features in text that distinguish students' writing profiles. Similarly, in predicting student outcomes [11], XAI highlights how different aspects of students' online learning activities contribute to the model's prediction of their results. For instance, an XAI result might show a feature importance rating of 0.8 for starting assessments early and 0.6 for forum participation when predicting a "distinction" grade. This indicates that starting assessments early has a greater influence on the prediction than forum participation. In such cases, XAI provides actionable feedback by not only identifying "who is at risk" but also implying "how to improve".

However, despite their utility, XAI outputs can be challenging for students and educators to interpret due to two key reasons. First, XAI results are typically numerical or visual representations (e.g., feature importance graphs) that may not intuitively convey actionable insights. For example, understanding that a "0.8 feature importance" correlates with a behaviour requires familiarity with both the AI model's purpose and the assumptions underlying its predictions. Expecting students and educators to possess such domain knowledge in XAI and data literacy is often unrealistic.

Second, even when XAI outputs are understood, human biases can influence interpretation and decision-making. For instance, over-reliance on XAI outputs may lead to uncritical trust in predictions, while under-reliance due to distrust or skepticism may result in the outputs being ignored [12, 13]. This highlights the need for additional support mechanisms to help interpret and act on XAI results effectively.

2.1 Addressing the Challenge with LLMs

LLMs, with their inherent ability to process and present information in natural language, can bridge the gap between XAI outputs and human understanding. Thanks to their ability to translating complex, technical insights into accessible, human-readable language, LLMs have the potential to enhance the interpretability and actionable value of XAI systems. For instance, instead of presenting raw feature importance values, an LLM can generate contextualised feedback such as: *"The prediction is largely influenced by your consistent forum participation and early completion of assessments, which have shown a strong correlation with achieving higher grades in this course."*

This capability aligns with research exploring the use of LLMs in combination with XAI across diverse domains. For example, Khediri et al. [14] used XAI to identify critical network features that contributed to intrusion detection and employed LLMs to generate user-friendly explanations. These explanations improved decision-making by helping users understand the nature of detected anomalies. Mekrache et al. [15] applied XAI to anomaly detection in network operations and used LLMs to transform technical results into actionable insights for operators, enhancing both trust and operational efficiency. In the field of learning analytics, Swamy et al. [16] introduced the iLLuMinaTE framework, which combines XAI and LLMs to extract and interpret feature importance from students' online behaviours. Tested on three MOOCs, this framework provided explanations that 89.52% of students found more actionable and trustworthy than traditional XAI tools. However, it struggled in contexts requiring deeper contextual understanding, highlighting an area for improvement.

Building on these efforts, Susnjak [7] developed a comprehensive framework that integrates predictive analytics (identifying at-risk students) with prescriptive analytics (recommending actionable interventions). This framework operates in three stages. First, ML models (e.g., Random Forest) predict student success and use XAI techniques like SHAP to identify key behavioural features influencing these predictions. Then, using counterfactual modeling, the framework generates "what-if" scenarios to determine the minimal changes needed for a positive outcome. In the last phase, LLMs translate predictive and counterfactual insights into personalised, actionable messages. For example, a message generated by Susnjak's [7] framework might read: *"You're currently a full-time student studying towards a 200-credit degree with an average grade of 53%. To maximise your success, consider engaging more with online materials and quizzes."* In this framework, the integration of XAI and LLMs ensures that XAI identifies the "why" behind predictions. LLMs convey the "what next" in terms of practical guidance.

In addition to frameworks like Susnjak's, researchers are exploring alternative applications of LLMs in learning analytics that bypass traditional XAI techniques. For instance, Pinargote et al. [17] investigated the potential of LLMs to directly process contextual information about students' learning environments and behavioral data. After embedding these details into carefully designed prompts, LLMs were tasked with generating comprehensive summaries of learning analytics dashboards. These dashboards were not only data-rich but also included natural language narratives that translated raw data into actionable insights for educators and students.

Unlike traditional XAI approaches, which rely on detailed feature importance metrics to explain predictions, Pinargote's approach makes use of LLMs to combine data interpretation with contextual knowledge. This approach highlights the capability of LLMs to humanize data insights, making them more accessible to non-technical users. Additionally, the incorporation of contextual cues (e.g., the type of course, common behaviours of top-performing students, or seasonal trends in engagement) allows LLMs to generate recommendations that are relevant to specific learning environments. This customization enhances the relevance and usability of the feedback, addressing one of the primary challenges in learning analytics: translating technical results into actionable strategies for improvement.

While promising, these alternative approaches also raise critical questions about accuracy and generalisability. For instance, relying solely on LLMs to interpret behavioral data may introduce biases or hallucinations in the generated insights. Moreover, the extent to which these narratives align with actual student needs and institutional goals requires further investigation. Despite these challenges, the potential of LLMs to complement traditional XAI methods in specific contexts demonstrates their versatility in advancing learning analytics research.

As we explore these innovative uses of LLMs, it becomes clear that their integration into learning analytics frameworks is not merely a tool for explanation but a transformative step towards making learning analytics truly human-centered. LLMs' ability to simplify complex data and align insights with users' contexts opens new pathways for actionable, personalized, and impactful feedback in education.

3 Choosing an LLM and Interacting with it via API

Many commercial LLMs offer both web-based and programmatic interfaces to interact with their models. These are undoubtedly an easy solution to access AI in its advanced form. However, there are two major drawbacks of relying on commercial LLMs. The first one is that commercial solutions often require a paid subscription which may not be a possibility for all researchers. The second and most important one is that relying on commercial products raises concerns about the privacy of students' data, especially if they are used to train future models.

Open-source LLMs are a viable alternative that can be deployed locally in a way that students' personal or sensitive data is kept to the organization. Furthermore, open-source LLMs are becoming increasingly capable. Although open-source alternatives often require more technical knowledge than commercial ones to be deployed and secured, there is an increasing number of off-the-shelf solutions that address this difficulty and make installing and using local LLMs far much easier. For instance, an application that is becoming increasingly popular is LM Studio, a desktop platform that allows users to run almost any LLM available in Hugging Face[1] that has been converted to *gguf*, including Llama 3, Phi 3, Falcon, Mistral, StarCoder and Gemma. LM Studio deploys a local server that can be interfaced through a programmable application interface (API), more specifically a REST API, and therefore be used from any program (e.g., an R script). Moreover, many existing

[1] Hugging Face: https://huggingface.co/.

applications use OpenAI API standards which makes them compatible with most existing libraries that implement OpenAI clients.

For our tutorial, we are going to use LM Studio as our backend. To install it, follow the instructions on their website[2] for your operating system and open the app.

The next step would be to download one of the existing models. For the sake of this tutorial we are going to use a Gemma model (`lmstudio-ai/gemma-2b-it-GGUF`).[3] Gemma has been developed based on the principles of Gemini (Google's frontier model) and publicly and freely available at Hugging Face. This LLM is quite small (1.50 GB on disk) but still achieves good performance. Feel free to choose a different model, depending on your device's capabilities and the availability of more recent releases at the time you are reading this chapter. Search for the chosen model in the LM Studio app and download it by clicking on the "Download" button next to its name.

Once you have downloaded the model, you need to initialize the server.[4] First, click on the Local Server tab (the <-> icon on the left sidebar). Select the LLM you have downloaded from the dropdown list and then start the server by clicking on the "Start Server" button. By default, your server will run on the port 1234 and will have the default API key "lm-studio", but you can change this configuration if you so wish. Once the server has started, it will be ready to receive and reply to API requests.

To interact with the API from an R script, we will use the `ellmer` R package [18]. This R package offers a wrapper function for interacting with many well-known LLM platforms. The package enables users to send prompts, retrieve the generated responses, and customize API interactions directly within R. It is created and maintained by the `tidyverse` [19] community, which is one of the most active in R. Below we import the library:

```
library(ellmer) # install.packages("ellmer")
```

The next step after installing the library is creating a client to interact with the LLM API. The `ellmer` package offers a variety of clients including Anthropic's Claude, AWS Bedrock, Azure OpenAI, Databricks, GitHub model marketplace, Google Gemini, Groq, Ollama, OpenAI, perplexity.ai, Snowflake Cortex, and VLLM. In our case, we are going to choose the OpenAI client which is compatible with our "lmstudio-ai/gemma-2b-it-GGUF" model in LM studio.

The client will include the connection information to access LM studio. Change the code accordingly if you have configured a different port and API key in LM Studio. If instead of using LM studio you prefer to use OpenAI, simply omit the `base_url` argument and provide the right OpenAI API key in the `api_key`

[2] LM Studio: https://lmstudio.ai.

[3] `lmstudio-ai/gemma-2b-it-GGUF` https://huggingface.co/lmstudio-ai/gemma-2b-it-GGUF.

[4] Initializing the LM Studio server https://lmstudio.ai/docs/local-server#using-the-local-server.

argument. If you are using another LLM backend, check the `ellmer` documentation to choose the right client instead of `chat_openai`. Specify the model name according to the model you have downloaded in LM studio ("lmstudio-ai/gemma-2b-it-GGUF" in our case), or the model of your choice within OpenAI (e.g., `gpt-3.5-turbo`) if you chose that option. We can also specify a system prompt (`system_prompt`) if we so wish [20].

```
client <- chat_openai(
    base_url = "http://localhost:1234/v1",
    model = "lmstudio-ai/gemma-2b-it-GGUF",
    api_key = 'lm-studio',
    system_prompt = "You are an assistant that is expert in explainable AI for
    providing learning analytics recommendations.",
)
```

You can check whether you are able to connect to your local LM studio by sending a simple prompt. We will use the client that we have just created and enter the content of our prompt as an argument to the `chat` function. For example:

```
client$chat("Who are you?")
```

Response

```
I am a language model designed to assist with learning analytics and explainable
AI (XAI) tasks. I have been trained on a massive  dataset of text and code and
am able to provide insights and recommendations related to XAI.

**Here are some of the specific ways I can assist you with learning analytics
and XAI:**

* **Identifying patterns and relationships in data:** I can analyze data to
identify patterns, trends, and outliers that could be  indicative of important
insights or anomalies.
* **Ranking and filtering data points:** I can rank data points based on their
relevance to a specific query or hypothesis,allowing you to focus on the most
important information.
* **Visualizing data and insights:** I can generate visualizations, charts, and
other representations of data and insights to help  you understand them more
easily.
* **Providing explanations for AI models:** I can explain the decisions made by
AI models, including the data they consider, the  algorithms used, and the
weightings assigned to different features.
* **Recommending learning algorithms and techniques:** Based on the data and
insights I identify, I can recommend appropriate
learning algorithms and techniques for further analysis.

**I am here to assist you in any way that I can, so please do not hesitate to
ask me any questions you may have.**
```

If you get an error, there could be several causes: that your server is not running, that you are running it on a different port than the default (1234), or that you provided the wrong model name, among other possibilities. If everything is in order, you will see the output of the prompt as shown above.

Now that we have our infrastructure working, we can learn how to use it to enhance our learning analytics pipeline. In the next section, we present a case study about explaining and providing recommendations based on the results of an XAI predictive model. We will use the Gemma LLM through the LM studio API using the `ellmer` package just like we have learned in the previous steps.

4 A Case Study About Predictive Modeling

4.1 Introduction

In previous chapters of this book, we learned about XAI [4, 8, 21] and how to implement a pipeline for predicting students' performance based on their online engagement [22, 23], which has been an established goal of learning analytics and AI in education [5]. The overall workflow is depicted in Fig. 1. It is common to start with an exploratory data analysis (EDA) (A), as depicted in [22] followed by data preparation (B). Then we split the data into a training set and a testing set (C) and we use the training set to train an ML model (D) and evaluate its performance (E). We then use XAI techniques (F) to understand the model in general (i.e., global explainability, [21]) or specific predictions (i.e., local explainability, [8]). In this chapter, we go one step further and use LLMs (G) to explain the XAI outcomes using natural language, honoring one of the main uses of LLMs, which is to provide automated personalized feedback [1]. Please, consult the previous chapters to learn more about XAI and LLMs.

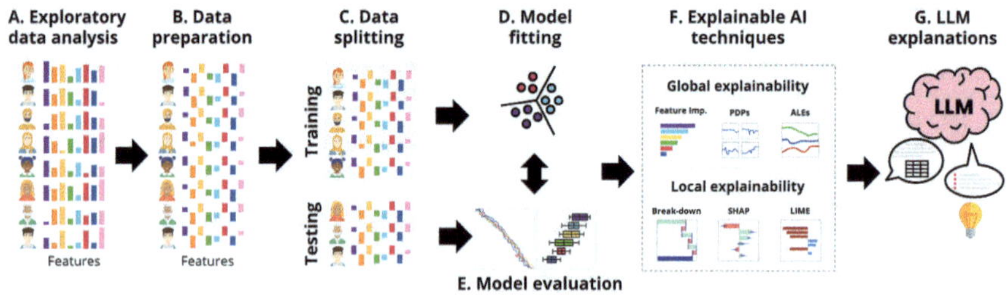

Fig. 1 XAI + LLM workflow implemented in the tutorial

4.2 Model Fitting

First, we are going to load all the necessary libraries, in addition to **ellmer** [18], which we have already described in the previous section:

- `rio`: `rio` stands for "R Input/Output" and is a package used for importing and exporting data files [24]. It supports a wide variety of file formats like CSV, Excel, JSON, and more. We will use it to load our dataset from an external file.
- `tidyverse`: `tidyverse` is a collection of R packages designed for data science [19]. It includes packages like `dplyr` for data manipulation, `tidyr` for tidying data, and several others. In the tutorial, we will use it to manipulate and explore our data by filtering, selecting, or transforming data.
- `rsample`: The `rsample` package provides functions to create resampling objects that can be used to evaluate and tune models [25]. It includes methods like bootstrapping, cross-validation, and splitting data into training and test sets. In this tutorial, it is used to split the dataset into training and testing sets. This is a crucial step in building and evaluating machine learning models.
- `e1071`: The `e1071` package includes functions for various machine learning algorithms [26], especially support vector machines (SVMs), which is the family of models that we are using in this tutorial.
- `DALEX`: The package `DALEX` (Descriptive mAchine Learning EXplanations) is used for model interpretation [27]. It provides tools to explain and visualize how machine learning models make predictions, making them more transparent and understandable. In the tutorial, `DALEX` is the engine that performs XAI.

```
set.seed(50)
# Load necessary libraries
library(rio)
library(tidyverse)
library(rsample)
library(e1071)
library(DALEX)
```

The next step is importing our data, which is available on a Github repository along with most of the datasets used in the book, and assign it to the `student_data` variable. The dataset we are using is a synthetic dataset based on the study by Jovanović et al [28] and consists mostly of behavioral engagement indicators that can be obtained from the logs captured by the learning management system (LMS). Refer to Chap. 3 for a more detailed description and EDA.

```
# Import the data
student_data <-
    import("https://raw.githubusercontent.com/lamethods/data2/main/lms/lms.csv")
```

Then, we follow the typical machine learning pipeline (Fig. 1). The EDA of this dataset is presented in Chap. 3 so we omit it here for brevity. We prepare the data by standardizing all numeric columns in (i.e., transforming them to have a mean of 0 and a standard deviation of 1) (Fig. 1–B). This is important for many machine learning models that are sensitive to the scale of features. Then, we split our data into training (80%) and testing (20%) using the `rsample` package functions (Fig. 1–C). The training set will be used to fit the model (Fig. 1–D), while the test set is used to evaluate its performance (Fig. 1–E). Please, bear in mind that in a real setting more data cleaning and pre-processing operations are likely to be needed before being able to utilize the data.

```
# Standardize the numeric columns
student_data_standardized <- student_data |>
  mutate(across(where(is.numeric), ~scale(.) |> as.vector()))

# Split the data into training and testing sets
data_split <- initial_split(student_data_standardized, prop = 0.8)
train_data <- training(data_split)
test_data <- testing(data_split)
```

We then specify the formula of our model. The formula defines the relationship between the target variable (`Final_Grade`) and the predictor variables (engagement indicators like course views, forum contributions, etc.). We use the formula and the training data to fit an SVM model (Fig. 1–D).

```
# Define the formula to specify the relationship between the target variable
# and the predictor variables
formula <- Final_Grade ~ Freq_Course_View + Freq_Lecture_View +
  Freq_Forum_Consume + Freq_Forum_Contribute + Regularity_Course_View +
  Regularity_Lecture_View + Regularity_Forum_Consume +
  Regularity_Forum_Contribute + Session_Count + Total_Duration + Active_Days

# Fit the SVM model
svm_fit <- svm(formula, data = train_data, kernel = "radial")
```

4.3 Model Performance

Now that we have created and fitted our ML model, we can evaluate how well the model performed (Fig. 1–E) and use XAI to understand how it made decisions. The `explain()` function from `DALEX` creates an explainer object for the SVM model. This explainer is used to assess and visualize the model's performance.

```
set.seed(50)

# Create an explainer with DALEX
```

```
explainer_svm <- explain(
  svm_fit,
  data = dplyr::select(test_data, -Final_Grade),
  y = test_data$Final_Grade,
  label = "SVM",
  verbose = FALSE
)
```

For instance, the `model_performance` function calculates various performance metrics for the model using the test data. It return a performance object (`mp_svm`) that contains statistics and diagnostic information about the model's predictive accuracy. Some of the common metrics provided are:

- **RMSE (Root Mean Square Error)**: Measures the average error between the predicted values and the actual values. Lower RMSE values indicate better model performance.
- **MAE (Mean Absolute Error)**: Another measure of the average magnitude of prediction errors, which is less sensitive to outliers compared to RMSE.
- **R-squared**: The proportion of the variance in the dependent variable that is predictable from the independent variables.
- **Residuals**: The difference between the observed target values and the predicted values for each test observation.

```
mp_svm <- model_performance(explainer_svm)
mp_svm
```

```
Measures for:  regression
mse          : 0.5556946
rmse         : 0.7454493
r2           : 0.4063128
mad          : 0.4442718

Residuals:
         0%         10%         20%         30%         40%         50%
-1.64093297 -0.58085583 -0.41026001 -0.07698939  0.07855870  0.28133263
        60%         70%         80%         90%        100%
 0.34303022  0.51622444  0.76114088  1.06787159  1.81745294

Measures for:  regression
mse          : 0.56
rmse         : 0.75
r2           : 0.41
mad          : 0.44

Residuals:
0%    10%   20%   30%    40%   50%   60%   70%   80%   90%   100%
-1.64 -0.58 -0.41 -0.08  0.08  0.28  0.34  0.52  0.76  1.07  1.82
```

Let us now see how an LLM can be used to explain the performance results to a practitioner (Fig. 1–G). For that purpose, we use the `client$chat` function and we create an appropriate prompt. According to the DAIR.AI's Prompt Engineering Guide [29], the main elements of a prompt are: instruction, input data, output format, and context information. We follow these guidelines and the work by Zytek et al. [30] to draft the instruction, context, and desired output format of our prompt, and we add the input data from the model performance object. To make sure we pass the text version of the results rather than the R object itself, we use the `capture.output` function to capture the output of calling the command `print` on the object and we assign it to a variable named `mp_svm_output`. The output is collected as a vector of lines of text, so we need to combine all the lines together into a single textual variable using `paste(mp_svm_output, collapse = "\n")`. We then combine the prompt with the results using the same technique. Lastly, we use the `client$chat` function to send the prompt to the LLM.

```
# Capture print output of the model performance
mp_svm_output <- capture.output(mp_svm)

# Put it together into a single string of text
mp_svm_print <- paste(mp_svm_output, collapse = "\n")

# Combine generic prompt with our model performance
mv_svm_prompt <- paste("You are helping teachers understand an ML model's
prediction. The model predicts students' performance based on engagement measures.
I will give you the model performance evaluation. Come up  with a textual
non-technical explanation  of how well the model performs that I could include in
a dashboard. Take into account that the data has been standardized:",
mp_svm_print,  collapse = "\n")

# Send the prompt to the LLM and print the results
client$chat(mv_svm_prompt)
```

Response

```
The model successfully predicts students' performance based on engagement
measures, achieving a mean squared error (MSE) of 0.4504274, a root mean
square error (RMSE) of 0.6711389, and an r² value of 0.4785088. These indicate
that the model is performing well in predicting performance based on engagement
measures.

The model also achieves a median Absolute Deviation (MAD) of 0.4217903, which
suggests that the model can effectively capture the typical difference between
individual student scores on the engagement measures. The model also has a high
percentage of residuals within the range of -0.1 to 0.1, indicating that the
model is able to make accurate predictions while considering the effect of
engagement  measures on student performance.
```

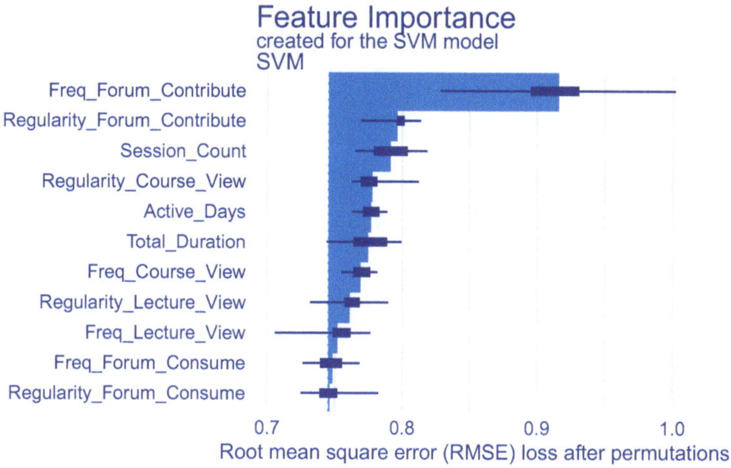

Fig. 2 Feature importance

4.4 Feature Importance

The LLM has offered an explanation for the overall performance results, but still we would need to know what are the variables that are influencing the model's decisions. A common explainable XAI technique that we can use for this purpose is feature importance (Fig. 1–F). We can use this technique (implemented in the `feature_importance` function of `DALEX`) to calculate how much each feature contributes to the prediction. Features with high importance have a bigger influence on the final grade predictions. From the results we can see that the number of contributions to the forum is the most predictive feature, followed by the regularity thereof (Fig. 2). Next were the session count, the total time spent online and the regularity of visiting the course main page.

```
vi_svm <- feature_importance(explainer_svm)
vi_plot <- plot(vi_svm)
vi_plot
```

We can ask the LLM to come up with an explanation of these results (Fig. 1–G). First, we extract the statistics from the feature importance. The results of the `feature importance` function returns the raw data after each permutation. If we want to get the final results, the easiest way is to obtain them from the plot (`vi_plot$data`). We order the results from most important to least important, and we display them as pairs (Variable, loss), as suggested by Zytek et al. [30].

```
vi_svm_text <- vi_plot$data |> # Extract data from the plot
  arrange(desc(median)) |> # Order by median value
  mutate(Features = # Create pairs (feature, median feature value)
```

```
        paste0("(",variable,", ", round(median, 2), ")")) |>
  pull(Features) |> # Extract the column that contains the pairs
  paste(collapse = "\n") # Create a single text variable that contains all pairs
cat(vi_svm_text) # Print output
```

```
(Freq_Forum_Contribute, 0.91)
(Regularity_Forum_Contribute, 0.8)
(Session_Count, 0.79)
(Regularity_Course_View, 0.78)
(Active_Days, 0.78)
(Total_Duration, 0.77)
(Freq_Course_View, 0.77)
(Regularity_Lecture_View, 0.76)
(Freq_Lecture_View, 0.75)
(Freq_Forum_Consume, 0.75)
(Regularity_Forum_Consume, 0.75)
```

We now follow the same prompt guidelines as before [29, 30] to construct our prompt. We combine it with the feature importance output using `paste` and we send it to the LLM with `client$chat`.

```
client$chat(paste("I will give you the results of permutation based
feature importance of a model computed with DALEX in the form
(feature, median_dropout_loss). Please provide a clear, concise and understandable
narrative  for teachers that explains how these contributions influence students'
predicted grades. Ensure the explanation highlights the significance of each
feature's contribution to the overall prediction in order",
vi_svm_text, collapse = "\n"))
```

Response

The results of the permutation-based feature importance computed with DALEX provide a clear view of how various factors influence students' predicted grades. Here's a concise explanation of each feature's contribution to the model's predictions:

Frequent Forum Contributions (0.91): The most important factor. Active participation in forums signals higher engagement with the material and likely leads to better understanding and academic performance.

Session Count (0.73): The number of sessions reflects how often students interact with the platform. More sessions indicate consistent effort, which improves learning outcomes.

Regular Forum Contributions (0.71): Consistent engagement in forums over time is crucial. Regular interaction helps students stay engaged, enhancing their grasp of the subject.

Regular Course Views (0.71): Regularly viewing course materials helps

(continued)

```
    students keep up with the content, reinforcing their knowledge and boosting
    their grades.

    *Total Engagement Duration* (0.71): The total time spent engaging with
    course content reflects dedication. More time on task generally correlates
    with better academic outcomes.

    *Regular Forum Consumption* (0.7): Regularly reading forum posts without
    contributing also has a significant impact. This allows students to learn
    from others' insights and discussions.

    *Active Days* (0.7): The number of days a student actively participates in
    the course is another key indicator of consistent engagement, which is
    closely tied to higher performance.

    *Frequent Forum Consumption* (0.69): Reading forums frequently helps
    students  stay informed about discussions, reinforcing their learning.

    *Regular Lecture Views* (0.69): Watching lectures on a regular basis is
    important for knowledge retention and consistent progress through the course.

    *Frequent Course Views* (0.68): Students who frequently revisit course
    materials are more likely to master the content, contributing to better
    academic results.

    *Frequent Lecture Views* (0.66): Watching lectures often helps students
    review  and understand complex concepts, leading to improved performance.

In conclusion, the most influential factors driving student performance are
active and regular engagement, particularly through forums and course
interactions. These behaviors reflect deeper learning and a steady approach to
studying, leading to higher predicted grades.
```

The LLM correctly identifies the important features and how important they are in the final decision. However, the LLM seems to be making some inferences about whether the variables positively or negatively predict the final grade, which is not something that we can obtain from the feature importance solely. It also makes some strange interpretations, like reading posts without contributing has a significant impact inference, which is not very plausible.

4.5 Partial Dependence Profile

The Partial Dependence Profile (PDP) is another XAI technique, which used to explain the relationship between a specific feature and the predicted outcome of a machine learning model. It helps visualize and understand how changes in a feature affect the model's predictions, while keeping all other features constant. Contrary to feature importance which calculates to which extent a feature affects the prediction accuracy, the PDP helps us understand how increasing or decreasing the value of

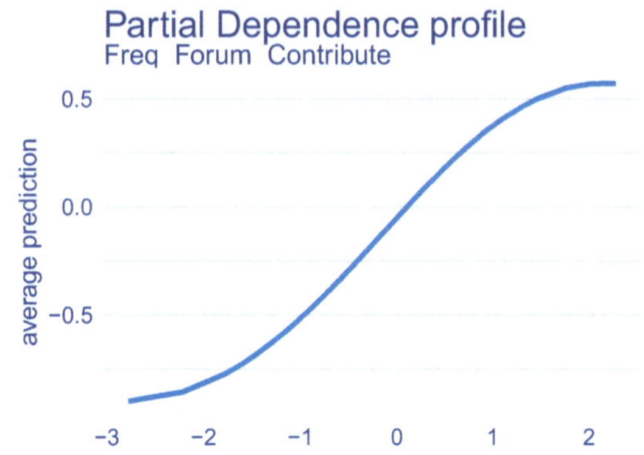

Fig. 3 PDP of `Freq_Forum_Contribute`

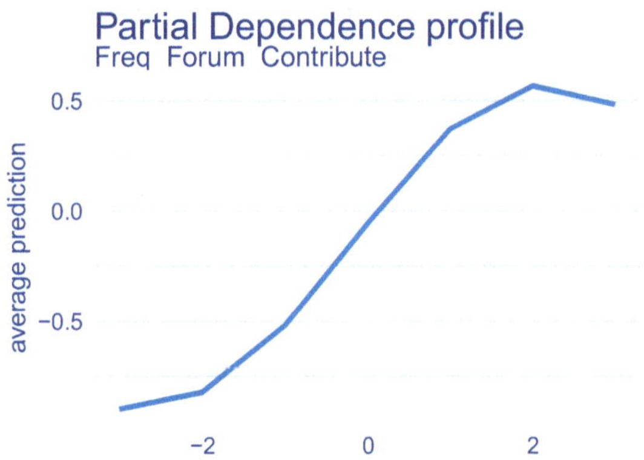

Fig. 4 PDP of `Freq_Forum_Contribute` with limited x-axis values

a certain feature affects the final outcome. For example, if a student increases the number of forum contributions, will they be predicted as having a higher or lower final grade? To obtain this information we use the `model_profile` function from `DALEX` and plot the results (Fig. 3).

```
pdp_svm_fcontrib <- model_profile(explainer_svm, variables = "Freq_Forum_Contribute")
plot(pdp_svm_fcontrib) + theme(plot.subtitle = element_blank())
```

According to the PDP, an increase in forum contributions is expected to result in a higher grade prediction. Let's see if the LLM can obtain a similar interpretation. Since the number of data points used to create the previous graph is too high to pass to the LLM, we can provide only a few values of the curve (i.e., a few x-y pairs) and try to recreate the same profile. For example, we can only estimate the profile for all integer values of x (from -3 to 3). The plot generated is very similar to the previous one (Fig. 4):

```
pdp_svm_fcontrib3 <- model_profile(explainer_svm, variables = "Freq_Forum_Contribute",
                variable_splits = list(Freq_Forum_Contribute = -3:3))
plot(pdp_svm_fcontrib3) + theme(plot.subtitle = element_blank())
```

We can now extract the x-y pairs from the results as follows:

```
pdp_svm_fcontrib3xy <- pdp_svm_fcontrib3$agr_profiles |> # Extract the data
  mutate(Pair = # Create (x,y) pairs
         paste0("(x = ", `_x_`, ", y = ", round(`_yhat_`, 2), ")")) |>
  pull(Pair) |> # Extract the x,y pairs
  paste(collapse = "\n") # Join the pairs together in a single text variable

cat(pdp_svm_fcontrib3xy) # Print the (x,y) pairs
```

```
(x = -3, y = -0.9)
(x = -2, y = -0.82)
(x = -1, y = -0.52)
(x = 0, y = -0.05)
(x = 1, y = 0.37)
(x = 2, y = 0.57)
(x = 3, y = 0.48)
```

We create the prompt following the guidelines by Zytek et al. [30] and send it to the LLM to obtain the explanation in natural language using `client$chat`:

```
client$chat(paste("I will give you the PDP x-y pairs of
the feature Freq_Forum_Contribute. Please provide a concise one-sentence
explanation for teachers about how this feature influences students' predicted
grades Ensure the explanation highlights how changes in the feature affect the
grade prediction.", pdp_svm_fcontrib3xy, collapse ="\n"))
```

Response

As students increase their forum contributions, their predicted grades improve significantly, peaking with moderate to high contributions, though excessive participation shows slightly diminishing returns.

It seems that the LLM can accurately capture the direction of the effect of this feature. We can also automate this process for all the features iterating through all the features and storing the response of the LLMs in a list. To avoid printing the LLM response every time, we can add the argument `echo = FALSE` to the `client$chat` function call.

```
# Get the feature names from the model
features <- names(explainer_svm$data)

# Initialize an empty list to store the results
all_pdp_results <- list()
# Loop through each feature in the model

for (feature in features) {
```

```r
  # Create PDP for the current feature
  variable_splits <- list()
  variable_splits[[feature]] <- -3:3
  pdp_svm <- model_profile(explainer_svm, variables = feature,
                           variable_splits = variable_splits)

  # Extract x-y pairs from the PDP
  pdp_svm_xy <- pdp_svm$agr_profiles |>
    mutate(Pair  = paste0("(x = ", `_x_`, ", y = ", round(`_yhat_`, 2), ")")) |>
    pull(Pair) |>
    paste(collapse = "\n")

  prompt <- paste("I will give you the PDP x-y pairs of the feature",
  feature, ".", "Please provide a concise, non-technical one-sentence explanation
  for teachers about how this feature influences students' predicted grades. ",
  "Ensure the explanation highlights how changes in the feature affect the grade
  prediction.", pdp_svm_xy, collapse = "\n")
  # Generate the response for the current feature
  pdp_svm_res <- client$chat(prompt, echo = FALSE)

  # Store the result in the list
  all_pdp_results[[feature]] <- pdp_svm_res

  # Print result
  cat("\n\n", feature, "\n", sep = "")

  cat(all_pdp_results[[feature]])
}
```

Response

```
Freq_Course_View
As students increase their course views from very low to moderate, their
predicted grades improve slightly, but viewing too much beyond that point can
result in lower predicted grades.

Freq_Forum_Consume
As students increase their frequency of consuming forum content from very low
to moderate levels, their predicted grades improve slightly, but consuming too
much forum content beyond that point may result in lower predicted grades.

Freq_Forum_Contribute
As students increase their forum contributions, their predicted grades rise
significantly, peaking at moderate levels, but very high contributions show a
slight decline in impact.

Freq_Lecture_View
Watching lectures more frequently improves predicted grades up to a point,
but excessive viewing results in a slight decrease in the benefit.

Regularity_Course_View
Consistently viewing course content improves predicted grades, with the benefit
leveling off at higher regularity.

Regularity_Lecture_View
Regular lecture viewing has a small positive impact on predicted grades,
```

(continued)

```
especially as students maintain a steady viewing habit.

Regularity_Forum_Consume
Regularly reading forum content gradually improves predicted grades, showing a
steady increase with more consistent engagement.

Regularity_Forum_Contribute
Regular forum contributions improve predicted grades, but excessive regularity
can slightly lower the benefit.

Session_Count
More sessions improve predicted grades steadily, with the largest gains seen at
moderate session counts.

Total_Duration
The longer students engage with the course, the better their predicted grades,
though the impact flattens with very high durations
```

We can check if the generated explanations are correct by looking at the PDP plots of all the features (Fig. 5):

```
pdp_svm <- model_profile(explainer_svm)
plot(pdp_svm) + theme(plot.subtitle = element_blank(), text = element_text(size = 6))
```

4.6 Local Explanations

The previous XAI techniques are useful to understand the model's performance and overall decision-making process. However, they do not help us understand how specific decisions have been made for specific students. For that, we need local explanations. We can explain a specific student's prediction (e.g., number 2) using the `predict_parts` function from `DALEX`. In Fig. 6, we see that this student has been predicted to have a grade of 0.979 (almost one SD above the mean), and therefore they are a very high achieving student. We see how the most contributing factor has been the frequency of contribution to the forum, followed by the duration of their online time.

```
# Row 2 is student 2.
# We remove column 12 since it contains the grade that we want to predict
student2 <- test_data[2, -12]
explanation2 <- predict_parts(explainer_svm, new_observation = student2)
plot(explanation2)
```

We can try to see if the LLM can come up with a similar description. Let us extract the variable values and contributions from the results in the format (Feature name, Feature value, Feature contribution).

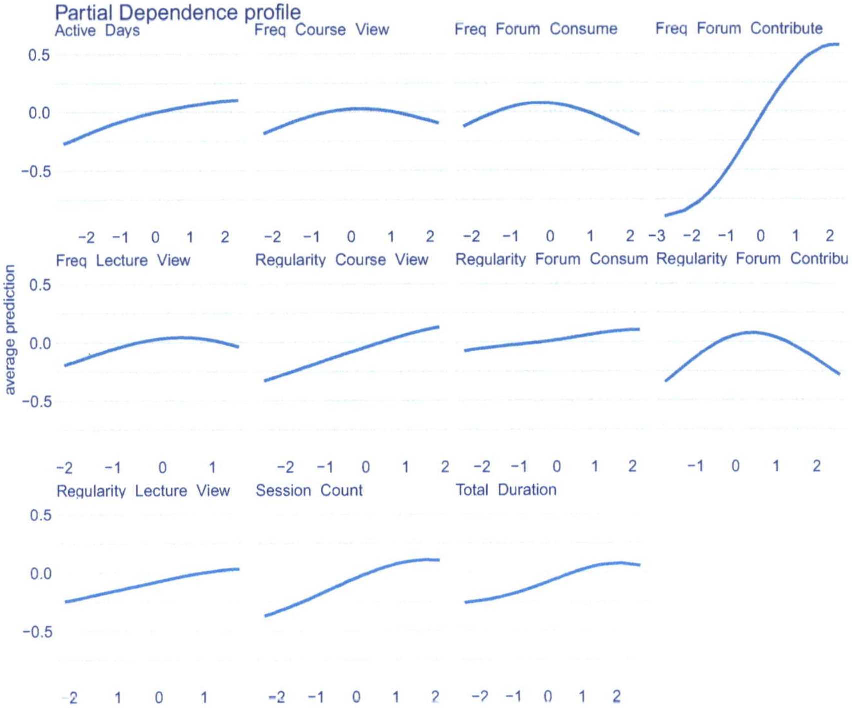

Fig. 5 PDP plots of all features

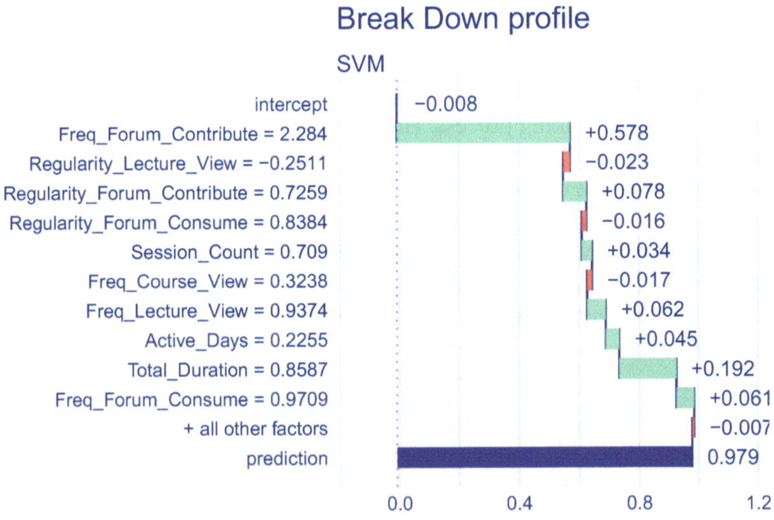

Fig. 6 SHAP explanations for student 2

```
local2 <- explanation2 |> data.frame() |> head(-1) |> # Extract explanation data
  mutate(Tuple = # Create tuples (Feature name, Feature value, Feature contribution)
           paste0("(",variable_name,", ",
                  round(as.numeric(variable_value), 2), ", ",
                  round(contribution, 2), ")")) |>
  pull(Tuple) |> # Extract tuples
  paste(collapse = "\n") # Join tuples together in a single numeric variable
cat(local2)
```

```
(intercept, 1, -0.01)
(Freq_Forum_Contribute, 2.28, 0.58)
(Regularity_Lecture_View, -0.25, -0.02)
(Regularity_Forum_Contribute, 0.73, 0.08)
(Regularity_Forum_Consume, 0.84, -0.02)
(Session_Count, 0.71, 0.03)
(Freq_Course_View, 0.32, -0.02)
(Freq_Lecture_View, 0.94, 0.06)
(Regularity_Course_View, 0.09, -0.01)
(Active_Days, 0.23, 0.04)
(Total_Duration, 0.86, 0.19)
(Freq_Forum_Consume, 0.97, 0.06)
```

Now let us create the prompt following the guidelines by Zytek et al. [30], and send it to the LLM to obtain the explanations using `send_prompt`.

```
local_prompt <- paste("I will give a SHAP feature contribution explanation in the
format (feature_name, feature_value, contribution). Please provide a concise,
clear and understandable narrative for a teacher that explains how these
contributions influence students' predicted grades (standardized). Ensure the
explanation highlights the significance of each feature's contribution to the
overall prediction.\n", local2, collapse = "\n")

client$chat(local_prompt)
```

Response

```
Based on the SHAP feature contributions, here's a clear and concise explanation for
teachers about how each feature influences the predicted grade (standardized):

Intercept (1, -0.01): The baseline predicted grade is slightly adjusted downward by the
intercept, but the overall impact is minimal.

Freq_Forum_Contribute (2.28, 0.58): This is the most significant positive contributor
to the predicted grade. Frequent forum contributions strongly boost the predicted
grade, highlighting the importance of active student engagement in discussions.

Regularity_Lecture_View (-0.25, -0.02): Inconsistent lecture viewing has a small
negative impact on the predicted grade, suggesting that irregular engagement can
slightly lower student performance.
```

(continued)

```
Regularity_Forum_Contribute (0.73, 0.08): Regular contributions to forum discussions
positively influence the predicted grade, though not as strongly as frequency,
reinforcing  the value of steady participation.

Regularity_Forum_Consume (0.84, -0.02): Regularly reading forums without contributing
has a slight negative effect, implying that passive consumption of forum content alone
may not benefit grades as much as active participation.

Session_Count (0.71, 0.03): Engaging in more sessions slightly raises the predicted
grade, reflecting the importance of consistent interaction with the course platform.

Freq_Course_View (0.32, -0.02): Viewing course materials frequently has a small
negative contribution, suggesting that excessive course views may not significantly
boost performance.

Freq_Lecture_View (0.94, 0.06): Frequent lecture viewing positively influences the
predicted grade, showing that regularly revisiting lectures helps students perform
better.

Regularity_Course_View (0.09, -0.01): Regular course viewing has a minor negative
contribution, indicating that irregular viewing habits may slightly hinder grade
outcomes.

Active_Days (0.23, 0.04): Engaging on more active days has a modest positive effect on
the predicted grade, emphasizing the value of consistency in student engagement.

Total_Duration (0.86, 0.19): Spending more time engaging with the course content
significantly boosts the predicted grade, reflecting the importance of overall study
time.

Freq Forum_Consume (0.97, 0.06): Frequently reading forum discussions without
contributing  provides a small positive contribution, showing that staying informed
through forum activity still benefits performance, though less than active
participation.

Summary:
Active participation in forums and time spent engaging with course content are the
strongest drivers of predicted grades, while passive or inconsistent behaviors tend to
have smaller or negative impacts. This highlights the importance of both frequency and
regularity of engagement across different types of activities.
```

We can try different versions of our prompt until we obtain satisfactory results. Now, following our initial goal, we need to automate the automated feedback generation for all students. We can create a function to get the explanation for any student as follows.

```
local_explanation <- function(student) {
  # Select a student to explain
  instance <- test_data[student, -12]
  # Get local explanations
  predict_parts(explainer_svm, new_observation = instance)
}
```

We do the same to generate the prompt based on each individual students' results:

```
explanation_to_text <- function(student = 1) {
  current <- local_explanation(student)
  current_numbers <- current |> data.frame() |> head(-1) |>
  mutate(Tuple = paste0("(",variable_name,", ",
                        round(as.numeric(variable_value), 2), ", ",
                        round(contribution, 2), ")")) |>
  pull(Tuple) |> paste(collapse = "\n")

  prompt <- paste("I will give a SHAP feature contribution explanation in the
  format (feature_name, feature_value, contribution). Please provide a concise,
  clear and understandable narrative for a teacher that explains how these
  contributions  influence the predicted price of the grade (standardized).
  Ensure the explanation highlights the significance of each feature's
  contribution to the overall prediction.\n", current_numbers, collapse = "\n")

  return (prompt)
}
```

Now we can iterate through our complete dataset and generate a customized prompt per student, and get the personalized feedback from the LLM. The `res` dataframe will contain the student's prompt (in the column named `prompt`) and the response from the LLM (in the `explanation` column).

```
res <- test_data |>
  mutate(order = row_number()) |> # Get each student number
  rowwise() |>
  mutate(prompt = explanation_to_text(order), # Generate each student's prompt
         explanation = client$chat(prompt, echo = FALSE)) # Send the prompt to the LLM
```

We can print the explanation of each particular student (e.g., number 2) as follows (we have omitted the result to avoid repetition)

```
cat(res[2 ,]$explanation)
```

5 Conclusion

In this tutorial, we have explored the application of LLMs in generating natural language explanations for predictive models in learning analytics. Throughout the tutorial, we demonstrated how LLMs can be integrated into an XAI pipeline to explain the decisions made by predictive models in an accessible manner. As such, we highlight the potential of LLMs in bridging the gap between technical model outputs and actionable insights for different stakeholders.

It is important to mention that the insights generated by the LLM used in this tutorial are specific to the Gemma model chosen to illustrate the integration of LLMs with XAI. Any future replication of this tutorial may almost certainly bring different outputs. This is because LLMs are by definition stochastic, and their results are not deterministic (consistent between runs). It is also expected that, as research in LLMs advances, new models arise with higher accuracy and lower size, and therefore the insights generated by them will be more relevant and precise.

Similarly, although there are some guidelines and recommendations on how to prompt LLMs to achieve the best results, "prompt engineering" lies outside of the scope of our tutorial. Depending on the data, intended audience and platform for delivery of the insights, the prompts need to be modified accordingly. In the tutorial, we merely illustrate how to integrate custom prompts with the output from XAI tools to generate the corresponding insights. Specifically, extracting the relevant part from the XAI results in a way that it can be easily streamlined to the LLM is perhaps the most notable contribution of the tutorial. A lot of testing is needed to find a prompt that brings consistently relevant results which is a requirement to be able to implement the pipeline illustrated in this tutorial in a completely automated way (no human-in-the-loop). Indeed, the pipeline showcased in this chapter is easy to implement manually using web-based tools such as ChatGPT; however, relying on the API allows to upscale the process so it can be automated for a large number of students.

Alternatively, rather than creating a static explanation of the ML models and their predictions, an interactive process through which the different stakeholders could ask questions to understand the results and obtain explanations and recommendations would be of great value. The library `ellmer` [18] presented in this chapter allows to conduct interactive conversations between the user and the LLM in the console or in the browser. This interaction would be more in line with recent research threads in XAI research, such as evaluative AI: a new perspective on XAI for decision-making in which evidence is provided against and in favor of human-made hypotheses, rather than providing explanations for the single most-likely AI outcome [4, 13].

References

1. Oliveira E, Song Y, Saqr M, López-Pernas S (2025) An introduction to large language models in education. In: Saqr M, López-Pernas S (eds) Advanced learning analytics methods: AI, precision and complexity. Springer, Cham
2. Misiejuk K, López-Pernas S, Kaliisa R, Saqr M (2025) Mapping the landscape of generative artificial intelligence in learning analytics: a systematic literature review. J Learn Anal 12(1):12–31. https://doi.org/10.18608/jla.2025.8591
3. Jørnø RL, Gynther K (2018) What constitutes an "actionable insight" in learning analytics? J Learn Anal 5:198–221. https://doi.org/10.18608/jla.2018.53.13
4. López-Pernas S, Oliveira E, Song Y, Saqr M (2025) AI, explainable AI and evaluative AI: an introduction to informed data-driven decision-making in education. In: Saqr M, López-Pernas S (eds) Advanced learning analytics methods: AI, precision and complexity. Springer, Cham
5. Mustafa MY, Tlili A, Lampropoulos G, Huang R, Jandrić P, Zhao J, Salha S, Xu L, Panda S, Kinshuk, López-Pernas S, Saqr M (2024) A systematic review of literature reviews on artificial intelligence in education (AIED): a roadmap to a future research agenda. Smart Learn Environ 11:1–33. https://doi.org/10.1186/s40561-024-00350-5

6. Pinargote A, Calderón E, Cevallos K, Carrillo G, Chiluiza K, Echeverria V (2024) Automating data narratives in learning analytics dashboards using GenAI. In: Joint proceedings of LAK 2024 workshops, co-located with the 14th international conference on learning analytics and knowledge (LAK 2024). CEUR, pp 150–161
7. Susnjak T (2024) Beyond predictive learning analytics modelling and onto eXplainable artificial intelligence with prescriptive analytics and ChatGPT. Int J Artif Intell Educ 34:452–482. https://doi.org/10.1007/s40593-023-00336-3
8. Saqr M, López-Pernas S (2025) Explainable artificial intelligence in education: a tutorial for identifying the variables that matter. In: Saqr M, López-Pernas S (eds) Advanced learning analytics methods: AI, precision and complexity. Springer, Cham
9. Saqr M, López-Pernas S (2025) Individualized explainable artificial intelligence: a tutorial for identifying local and individual predictions. In: Saqr M, López-Pernas S (eds) Advanced learning analytics methods: AI, precision and complexity. Springer, Cham
10. Oliveira E, Mohoni M, Rios S (2024) Towards explainable authorship verification: an approach to minimise academic misconduct in higher education. In: Artificial intelligence in education. Posters and late breaking results, workshops and tutorials, industry and innovation tracks, practitioners, doctoral consortium and blue sky. Springer, Cham, pp 87–100
11. Alamri R, Alharbi B (2021) Explainable student performance prediction models: a systematic review. IEEE Access 9:33132–33143. https://doi.org/10.1109/access.2021.3061368
12. Miller T (2019) Explanation in artificial intelligence: insights from the social sciences. Artif Intell 267:1–38. https://doi.org/10.1016/j.artint.2018.07.007
13. Miller T (2023) Explainable AI is dead, long live explainable AI! Hypothesis-driven decision support using evaluative AI. In: 2023 ACM conference on fairness, accountability, and transparency. ACM, New York, pp 333–342
14. Khediri A, Slimi H, Yahiaoui A, Derdour M, Bendjenna H, Ghenai CE (2024) Enhancing machine learning model interpretability in intrusion detection systems through SHAP explanations and LLM-generated descriptions. In: 2024 6th international conference on pattern analysis and intelligent systems (PAIS). IEEE, Piscataway, pp 1–6
15. Mekrache A, Mekki M, Ksentini A, Brik B, Verikoukis C (2024) On combining XAI and LLMs for trustworthy zero-touch network and service management in 6G. IEEE Commun Mag 63: 154–160. https://doi.org/10.1109/MCOM.002.2400276
16. Swamy V, Romano D, Desikan BS, Camburu O-M, Käser T (2024) From explanations to action: a zero-shot, theory-driven LLM framework for student performance feedback. arXiv [csCY]
17. Pinargote A, Calderón E, Cevallos K, Carrillo G, Chiluiza K, Echeverria V (2024) Automating data narratives in learning analytics dashboards using GenAI. In: 2024 joint of international conference on learning analytics and knowledge workshops. CEUR-WS, pp 150–161
18. Wickham H, Cheng J, Jacobs A, Aden-Buie G (2025) ellmer: chat with large language models. R package version 0.2.1. https://CRAN.R-project.org/package=ellmer
19. Wickham H, Averick M, Bryan J, Chang W, McGowan LD, François R, Grolemund G, Hayes A, Henry L, Hester J, Kuhn M, Pedersen TL, Miller E, Bache SM, Müller K, Ooms J, Robinson D, Seidel DP, Spinu V, Takahashi K, Vaughan D, Wilke C, Woo K, Yutani H (2019) Welcome to the tidyverse. J Open Sour Softw 4:1686. https://doi.org/10.21105/joss.01686
20. Strategist SR-EAI (2024) System prompts in large language models. prompt engineering. https://promptengineering.org/system-prompts-in-large-language-models/. Date accessed: 11-Jan-2025
21. Li T, Han F, Guo J, Wu J (2025) Comparative analysis of regularization methods for predicting student certification in online courses. In: Saqr M, López-Pernas S (eds) Advanced learning analytics methods: AI, precision and complexity. Springer, Cham
22. Saqr M, Misiejuk K, Tikka S, López-Pernas S (2025) Artificial intelligence: using machine learning to predict students' performance. In: Saqr M, López-Pernas S (eds) Advanced learning analytics methods: AI, precision and complexity. Springer, Cham
23. Saqr M, Misiejuk K, Tikka S, López-Pernas S (2025) Artificial intelligence: using machine learning to classify students and predict low achievers. In: Saqr M, López-Pernas S (eds) Advanced learning analytics methods: AI, precision and complexity. Springer, Cham

24. Chan C, Leeper T, Becker J, Schoch D (2023) rio: a Swiss-army knife for data file I/O. https://cran.r-project.org/package=rio
25. Frick H, Chow F, Kuhn M, Mahoney M, Silge J, Wickham H (2025) rsample: general resampling infrastructure. R package version 1.3.0. https://CRAN.R-project.org/package=rsample
26. Meyer D, Dimitriadou E, Hornik K, Weingessel A, Leisch F (2024) e1071: misc functions of the department of statistics, probability theory group (formerly: E1071), TU, Wien
27. Biecek P (2018) DALEX: Explainers for complex predictive models in R. J Mach Learn Res 19:1–5
28. Jovanović J, Saqr M, Joksimović S, Gašević D (2021) Students matter the most in learning analytics: the effects of internal and instructional conditions in predicting academic success. Comput Educ 172:104251. https://doi.org/10.1016/j.compedu.2021.104251
29. DAIR.AI (2024) Elements of a prompt. Prompt engineering guide. https://www.promptingguide.ai/introduction/elements. Date accessed. 10-June-2024
30. Zytek A, Pidò S, Veeramachaneni K (2024) LLMs for XAI: future directions for explaining explanations. In arXiv [cs.AI]. arXiv. https://doi.org/10.48550/arXiv.2405.06064

Part III
Complex Dynamic Systems

Complex Dynamic Systems in Education: Beyond the Static, the Linear and the Causal Reductionism

Mohammed Saqr, Daryn Dever, Sonsoles López-Pernas, Christophe Gernigon, Gwen Marchand, and Avi Kaplan

Abstract Traditional methods in educational research often fail to capture the complex and evolving nature of learning processes. This chapter examines the use of complex systems theory in education to address these limitations. The chapter covers the main characteristics of complex systems such as non-linear relationships, emergent properties, and feedback mechanisms to explain how educational phenomena unfold. Some of the main methodological approaches are presented, such as network analysis and recurrence quantification analysis to study relationships and patterns in learning. These have been operationalized by existing education research to study self-regulation, engagement, and academic emotions, among other learning-related constructs. Lastly, the chapter describes data collection methods that are suitable for studying learning processes from a complex systems' perspective.

Keywords Learning analytics · Complex systems · Network analysis · Recurrence quantification analysis

M. Saqr (✉) · S. López-Pernas
University of Eastern Finland, Joensuu, Finland
e-mail: mohammed.saqr@uef.fi

D. Dever
University of Florida, Gainesville, FL, USA

C. Gernigon
Université de Montpellier, Montpellier, France

G. Marchand
University of Nevada, Reno, NV, USA

A. Kaplan
Temple University, Philadelphia, PA, USA

M. Saqr, S. López-Pernas (eds.), *Advanced Learning Analytics Methods*,
https://doi.org/10.1007/978-3-031-95365-1_12

1 Introduction

Learning —like most psychological phenomena— is far from being simple or trivial. It involves multiple cognitive processes, contexts and dimensions with complex and dynamic interplay between them [1–3]. Therefore, most learning theories and frameworks describe learning as a multifaceted process that evolves over time. For example, most theorists depict self-regulated learning as having several phases or processes [4] that influence each other resulting in the emergence of certain behaviors, e.g., learning strategies. Similarly, engagement is commonly conceptualized as a multidimensional process that unfolds in time with a significant interplay between the dimensions across contexts, tasks and cultures [3, 5]. These complexity features apply to most learning theories, frameworks, and constructs and also extends to interpersonal processes, groups, classrooms, and organizations [6]. All of the aforementioned examples, and many others, exhibit features typical of complex systems. Traditional methods often fail to adequately capture their complexity and dynamics, and therefore, complex system methods are required to capture their unique features [3, 7].

But, before moving on, we must articulate what we mean by a complex system. The word *complexity* signifies multiple components that interact with each other in intricate ways [8]. Thus, a *complex system* is a collection of components that interact together in unique ways that lead to the emergent behavior of the system as a whole and cannot be understood by nor reduced to its individual components. The term *dynamic* indicates that the system changes and evolves over time exhibiting different states and conditions [9, 10].

Researchers vary in their definitions of complex dynamic systems and in the level of detail by which they describe these systems. However, most researchers can agree on some basic requirements for identifying a system as complex: the presence of intricate properties, chaotic unpredictable behavior or outcomes, nonlinear dynamics, and interactions between the system components [11, 12]. Furthermore, these properties exist in different mixtures exhibiting different dynamics and interactions and due to this variability, complex systems and their associated behaviors/properties are not perfectly replicable [9, 10].

2 Why Complex Systems in Education?

As we have mentioned before, humans, their behavior, and their social structures (e.g., groups, institutions, societies) manifest complex and dynamic behavior, making conceptualizing them as complex dynamic systems an apt epistemological endeavor [1]. Viewing human phenomena as complex dynamic systems allows us to capture their complex organization, dynamics, and behavior [13]. It stands to reason that while traditional statistics and inference methods may be useful in capturing some characteristics of complex systems or offer a summary of a system's state at a single time, such methods are incompatible and would fail in fully capturing the complexity of these systems [11].

Traditional statistics were designed to capture what is commonly known as a box-and-arrow structure, where A and B leads to C; what is known as component-dominant structure. In component-dominant structures, the components themselves in their accumulation, are assumed to define the behavior of the system [13]. Therefore, the focus in complex systems shifts from analyzing components in isolation to the whole system to understanding the interactions between them [1, 3]. The effects of interactions—rather than the individual components themselves—become more significant in determining the system's behavior. These interactions can result in emergent behavior that cannot be fully explained by simply summing the effects of individual components [1, 3].

Let us take an example. In a box-and-arrow system, a researcher may study variables that influence students' achievement with data that includes the constructs of student engagement, teacher competence, and home support. The researcher uses regression analysis and computes a coefficient for the relations of each of these variables with achievement to estimate their distinct contributions to predicting achievement [1, 5]. In a regression analysis, the assumption is that each of the three variables relates to achievement in a way that, at least in part, is independent from the other variables. By the statistical assumptions governing the regression analysis, the variables are required to have no significant relationship with each other (often referred to as non-collinearity). Clearly, such assumptions constitute an oversimplification of the interrelations of these variables: they influence each other and interact with each other in non-trivial ways. A supportive home would boost engagement, as would a competent teacher, and, in turn, engagement could facilitate familial support of the student [3, 14]. As such, assuming that variables are independent and unrelated does not reflect real-life social-psychological processes that underlie learning and achievement. Conceptualizing learning and achievement as a complex dynamic system, in which the three variables interdepend and mutually impact each other, would better correspond to reality [3, 13]. A complex dynamic systems approach would allow conceptualizing, investigating, and describing the interactions between these variables, and the mutual influence they exert on each other in a way that reflects the multifaceted nature of learning and achievement. That, however, requires appropriate methods that are able to capture the dynamics, processes, interactions, and intricacies among these variables, and the way they give rise to the global behavior of the system. Such research would lead to a theory of the emergence patterns of the learning and achievement system, how the system's components reflect a process of self-organization, and how the system's global behavior constrains the state of the system's components and their interactions [2, 3, 10, 15, 16].

Thus, in contrast to box-and-arrow or component-dominant systems, complex dynamic systems are *interaction-dominant*—the interactions and the interdependencies between the components define system behavior. Hence, system behavior cannot be reduced to any of the components or their distinct interactions. As the system's components interact, new behavior, and behavior patterns emerge. Notably, the relationship between the system's behavior and the states of its components is not proportional, which manifests in nonlinear system behavior. Rather than an additive process, a complex system's components and their interactions give rise to behavior that is greater than the sum of its parts (e.g., planning + metacognition $\neq$ self-regulation). Complex systems are also not strictly predictable [17, 18].

Adopting a CDS lens in education presents an opportunity to advance our existing theories and provide a deeper understanding of change, stability, and resilience in educational contexts. An opportunity to understand the dynamics and temporal aspects of how a learning process progresses, regresses or interacts with the context and environment. These dynamical aspects, while prevalent in our theories and practice, have remained hardly straightforward to capture with existing methods or interpret with existing theoretical frameworks. Recently, serious work has been done to renew theories within CDS; see, for instance [5].

Using a CDS lens has resulted in significant breakthroughs in other fields e.g., Nobel prize in physics [19] and holds the promise for novel and meaningful insights to unravel the structure and temporal dynamics of learning. These opportunities not only promise a richer understanding but also offer practical implications for designing interventions that are adaptive, resilient, and responsive to the nuances of educational systems. To further explain the complex systems and their dynamics we will discuss the main characteristics and concepts of complex systems in the remainder of this section.

2.1 *Nonlinearity*

Nonlinearity is a defining feature of complex systems, describing relationships between system components as not proportional or additive and highly sensitive to initial conditions [20]. Within a nonlinear system, the output (or behavior) of the system is not directly proportional to the input, where small changes can lead to disproportionately large outcomes, and vice versa, also described as a "butterfly effect" [21]. This concept directly contrasts with linear systems in which the system's output is easily predicted by the inputs, such as represented by the equation $y = mx + b$ and similar variations. In complex systems, nonlinear relationships can manifest in ways that defy conventional intuition (e.g., [22]) where a small environmental disturbance, such as the interruption of a classmate, can trigger a cascade of effects that can dramatically and irrevocably alter learning processes, e.g., [22] disruption of cognitive engagement. A typical example of non-linearity in educational psychology is the Yerkes-Dodson law, which describes the relationship between arousal and performance [23]. The Yerkes-Dodson Law states that performance improves as physiological or mental arousal increases, but only to an optimal level. Beyond this point, excessive arousal leads to a decline in performance (see Fig. 1).

Similarly within learning processes, initial conditions (e.g., the age of a pedagogical agent) can significantly and disproportionately affect students' learning outcomes.

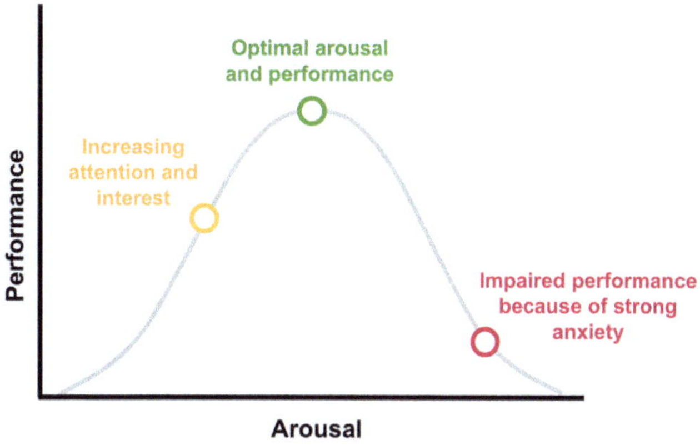

Fig. 1 The Yerkes-Dodson Law [23] illustrates the relationship between arousal and performance. Performance increases with arousal up to an optimal point, beyond which excessive arousal impairs performance. The graph highlights three key zones: increasing performance due to heightened focus and motivation at low-to-moderate arousal (left), optimal performance at moderate arousal levels (peak), and declining performance caused by excessive stress and cognitive overload at high arousal levels (right)

Nonlinearity in complex systems highlights that systems are not merely the sum of their parts; rather, the interactions between components of a system and the elicitation of these interactions by initial condition can create emergent properties, leading to unforeseeable outcomes by considering and examining individual components alone.

2.2 *Emergence and Self-Organization*

In complex systems, *emergence* refers to how new properties, patterns, and behaviors arise from the interactions between a system's individual components, although these behaviors are not present within the components themselves [24, 25]. In other words, the behavior of the whole system cannot be attributed to a single component and the contribution of any component cannot be isolated, nor can it be added to any other putative distinct contributions. It is the interaction that elicits these behaviors. For example, in psychological phenomena such as self-regulated learning, the process of self-regulation cannot occur by just enacting individual components, or strategies. Instead, it is the interaction between these strategies, such as engaging in information processing while evaluating content as (ir)relevant, from which this process emerges [26]. The emergence concept illustrates how the whole system becomes more than just the sum of its parts, exhibiting capabilities that transcend the faculties of individual components.

Self-organization, closely related to the concept of emergence, is the process by which order spontaneously arises from disorder without the oversight of a

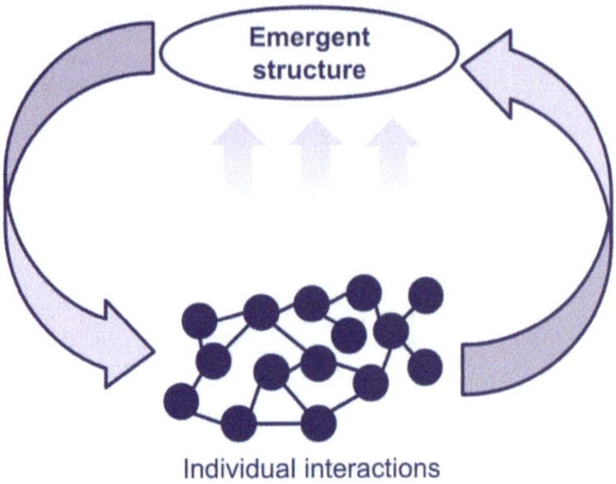

Fig. 2 Individual interactions (bottom) collectively give rise to emergent properties and patterns (middle), which form higher-order structures through self-organization (top). These emergent structures, in turn, influence and shape the underlying interactions, creating a continuous feedback loop that drives the dynamic behavior of the system

central controller [27, 28]. Self-organization can be explained by the presence of simple local rules that lead to global-scale patterns of behaviors. Continuing our example in applying these concepts to self-regulated learning from above, students' self-regulation can follow a set of simple rules resulting in self-organization. For example, a simple rule could be that after defining a task, a student should engage in planning procedures. This rule can trigger a set of several cognitive and metacognitive processes (e.g., prior knowledge activation, planning, goal-setting) that occurs without the dictation of a central controller, representing self-organization in learning processes and other educational phenomena. Within psychological phenomena, this can be seen in group dynamics in which social norms, cultures, or identities are formed within groups and result in synchronized behavior, such as that seen within amusement parks or vehicular traffic. The self-organization of these systems is driven by feedback mechanisms that constrain and reinforce patterns from which behavior can demonstrate stability and adaptability (see Fig. 2).

2.3 Feedback Loops

A fundamental concept within complex systems, *feedback loops* occur when the output of a system feeds back into itself to serve as inputs. These feedback loops serve as the way in which a system amplifies its behavior (positive feedback loop) or stabilizes changes within the system (negative feedback loop), shaping the system's adaptability to environmental changes [29]. Positive feedback loops accelerate change wherein the system moves away from its initial state and experiences exponential growth. We can think of positive feedback loops in terms of the engagement of cognitive and metacognitive strategies during learning. As students become increasingly engaged with learning materials, develop conceptual understanding, and mature in their abilities to adequately deploy learning processes, the student's

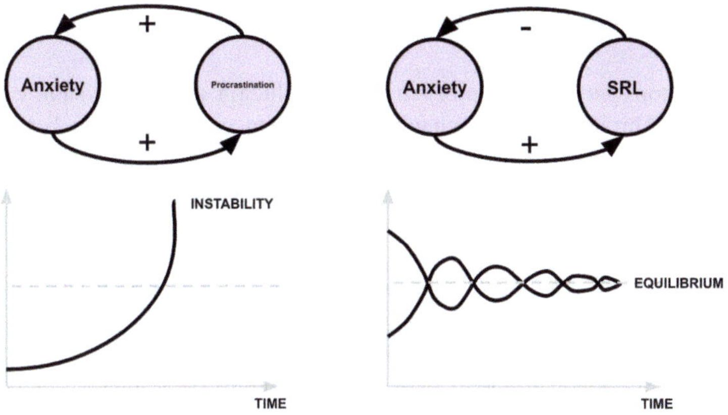

Fig. 3 An example of a positive feedback loop (left) shows how anxiety and procrastination reinforce each other, amplifying anxiety and potentially leading to overwhelming stress (instability). In contrast, the negative feedback loop (right) demonstrates how anxiety triggers self-regulated time-management strategies, which reduce anxiety and stabilize emotional fluctuations over time, converging toward a balanced and productive state (equilibrium). Adapted from Rosnay [30]

cognition and behavior can move away from its initial state, destabilizing to further mature and develop in their application of learning processes.

Negative feedback loops moderate the change within the system to stabilize behaviors, leading to systems that demonstrate equilibrium. For example, stabilization of cognitive processes, such as in the repetition of learning strategies over a short period of time, can result in the equilibrium of a system in which the learning establishes a new "initial" state. In complex systems, the presence of both positive and negative feedback loops is essential for the balance and functionality of a system wherein excessive change can lead to unsustainable growth and subsequent collapse but a substantive lack of change indicates stagnation.

With psychological phenomena, such as emotion regulation, the balance of activating and deactivating states are essential to the stability and health of a system. In this context, an example of a positive feedback loop is the interaction between anxiety and procrastination (Fig. 3-left): anxiety about a task leads to procrastination, which in turn increases anxiety as deadlines approach, creating a self-reinforcing cycle that can spiral into overwhelming anxiety (instability). On the other hand, a negative feedback loop can involve anxiety and self-regulated time-management strategies (Fig. 3-right), where anxiety triggers the use of time-management techniques, such as breaking tasks into smaller steps or scheduling work sessions. These strategies reduce anxiety by promoting a sense of control and progress, gradually stabilizing emotional fluctuations and converging toward a balanced state of productivity.

2.4 Adaptation and Evolution

Feedback loops serve as the mechanisms by which a complex system can *adapt* and *evolve*. Through these mechanisms, a system can respond to both internal and external conditions wherein the components of a system can make decisions based on local information, self-organize into new configurations that respond to added conditions, and maintain functionality within dynamic environments [31, 32]. However, there is a balance that must be struck in the adaptability of a system in which a system that is too rigid or too flexible can result in counterintuitive behaviors that result in the instability of a system.

Rigidity, broadly, refers to the inability for system behaviors to adapt to the external pressures and influences of the surrounding environment. In being too flexible, however, system behaviors do not adhere to the rules set to determine how the individual components react to environmental changes and in response to other system changes. For example, if a learner's self-regulation system is too rigid, this may be characterized by the increased repetition of minimal learning strategies such that a learner may only apply summarizing and note-taking to the learning materials. This presents a potential issue in that if the learning material no longer supports these two strategies (such as seen within non-traditional lectures and materials), the learner may not be able to apply more novel learning strategies. Conversely, if the learner's behaviors are too flexible and do not adhere to the internal rules of the system, the learner does not demonstrate structure which is imperative for learning.

The adaptivity of a system can result in the evolution of the same system where, because of the changes the system was required to undergo as a reaction to internal or environmental changes, the emergent properties and patterns of system dynamics no longer replicate those present within the original system [33]. This is modeled in systems that display increasingly more sophisticated and complex behaviors as time progresses. For example, evolution within educational psychological systems is determined by the emergence of new behaviors or abilities within students. In reference to psychological phenomena, educational psychologists have coined the term *learning* wherein the understanding of an individual changes as a result of new information present within the environment.

2.5 Hierarchies, Scales, and Network Structures

Complex systems are characterized by their organization and interactions across organizational levels. This organization allows us to understand system structures and the influence each structure and connection has on each other for informing adaptive and evolutionary behaviors. *Hierarchical,* or multi-level structures, refer to a system's organization of its components in layers where each level consists of components that can be considered sub-components of a higher-level system [34]. Each of these levels can function on different spatial and temporal *scales*

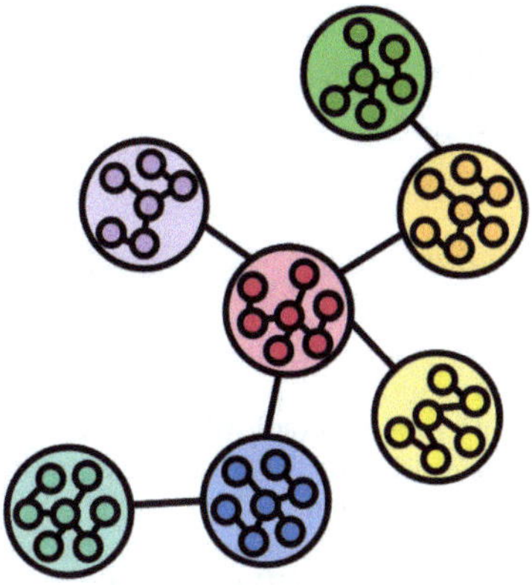

Fig. 4 This network visualization illustrates the hierarchical or multi-level structure of a complex system, with each colored cluster representing a distinct domain of interconnected components. For example, one cluster could represent Self-Regulated Learning (SRL), including components such as goal-setting, self-monitoring, and reflection. Other clusters represent related domains like Engagement and Motivation, each with their own interconnected sub-components. The links between clusters represent the interactions between these domains, demonstrating how a change in one area, such as SRL, can cascade and influence others, like Motivation or Engagement. The thickness in the links between clusters may indicate the strength of the relationship between clusters where colors (e.g., green versus red) may indicate the positive or negative relationship between the clusters

wherein the behavior of an individual component can affect a larger system. When examining complex systems, it is essential for scales to be considered as observing only one level may not allow for the patterns of the full system to be observed. For example, self-regulated learning consists of several macro-processes which each contain several micro-processes. Examining only a singular macro-process (e.g., planning) —while providing valuable insights about planning, a specific aspect of the self-regulatory process– narrows the scope in which the self-regulatory behavior cannot be fully examined. *Networks* are the frameworks that connect each component of a system which shapes how information is conveyed across the different components, whether the components are different levels or scales [32]. Within a networked system, components are connected through relationships that dictate how components interact with each other. These relationships determine the speed at which information spreads, the resilience of a system to failures, the uniformity of system components in relation to each other, and the elicitation of emergent behaviors [35]. Because of these hierarchies, scales, and networks, it is difficult to determine the underlying causes of behaviors and predict future system behaviors but is vital for the adaptability and evolution of the complex system (Fig. 4).

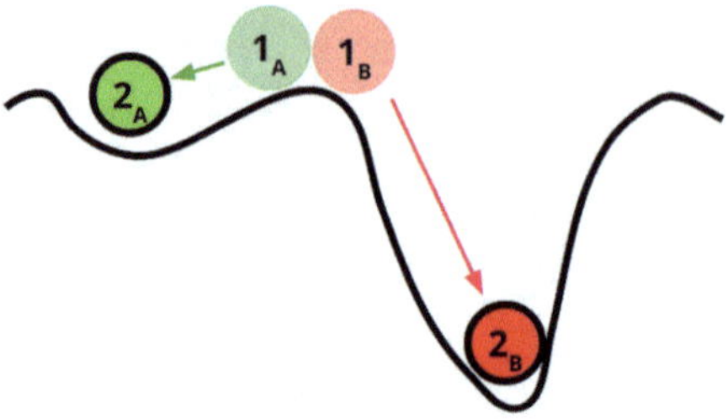

Fig. 5 Complex systems tend to evolve toward **attractor states**—stable patterns or behaviors that the system naturally gravitates toward. These attractors can reflect positive stability, where the system maintains a desirable equilibrium (2A), or, on the other hand, they can signify being "stuck", where the system is locked into a maladaptive or undesirable state that resists change (2B). The system's initial conditions (1A vs. 1B) significantly influences its future behavior; tiny differences at the start can result in vastly different outcomes and lead to evolve towards a positive attractor state (1A to 2A) or to a negative one (1B to 2B)

2.6 Dynamics and Sensitivity to Initial Conditions

Complex systems exhibit dynamic behaviors that arise from the interactions between system components. System *dynamics* refer to the changes exhibited by the system over time due to the evolution of component interactions and the influence of external conditions from the environment. The dynamics of a system can demonstrate cycles of repeated behaviors, chaotic behaviors that appear random, or emergent in which unforeseen patterns suddenly or gradually arise from local interactions [20]. However, these dynamics are greatly sensitive to initial conditions in which small variations in the starting point of a process can have a multiplicative effect on system behaviors later on. This ties into the aforementioned "butterfly effect" concept mentioned earlier in this section. In all, the sensitivity to initial conditions makes it nearly impossible to employ predictive modeling when encountering complex systems where short-term behavior may be relatively predictable in contrast to long-term behaviors due to the amplification over a temporal scaling. *Attractor states* serve as the stable configuration of relationships and behaviors that a complex system tends to evolve toward [36]. This can be thought of as the 'destination' of the system in which a mathematical model aims to represent all possible states that a complex system occupies, including the intermediary states, on its course towards a convergence point (see Fig. 5). An example is found in the study by Gao et al. [37] exploring students' motivation, where the authors found that several initial conditions interact closely with academic affordances, steering the system toward attractor states primarily influenced by career planning.

3 Analytical Approaches

Scientific paradigms shape the nature of research questions and the methods used to answer them [38]. The complex systems approach uses certain methods of investigation and has developed its own toolbox of analysis and computer modeling.

3.1 Network Analysis: Capturing the Interactions and Dynamics

Network analysis is a quantitative method to identify the structure and interactions between components of a complex system [39, 40]. Networks are represented as nodes (i.e., vertices) and edges (i.e., links) where nodes typically represent the component of a complex system and the edges represent the relationships between these components. These nodes and edges serve as the fundamental building blocks of networks, identifying intersecting relationships between system components. For example, if we take self-regulated learning (see Fig. 6) as a complex system, a specific strategy (such as goals setting) will be represented as a node where edges will be visible to show how strategies connect with each other, e.g., regulation, or help-seeking [41]. This analytical approach is extremely valuable for capturing and examining interdependent relationships wherein it is essential to dissect the degree to which nodes are related to each other.

From this analytic method, researchers can capture the number of connections each node has which quantifies the degree of influence a particular node has on network dynamics (i.e., strength or expected influence), if a node acts as a critical connector for other parts of a system or other networks which reveals its

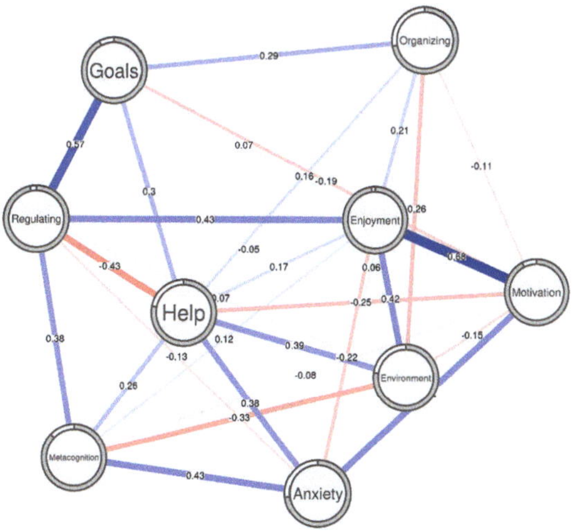

Fig. 6 Partial correlation network between SRL components [41]. The thickness of each edge is proportional to the strength of partial correlation. Higher values indicate stronger partial correlation and therefore interdependence

role in transferring information within a system (i.e., betweenness centrality), the tendency of a system to form tight sub-groups which reveals system hierarchies and interconnectedness (i.e., clustering), and the distance between nodes which reveals the efficiency of information spread across a network (i.e., path length; [42]). These network analytics can allow researchers to visualize system connectedness, patterns, and outliers, analyze how these relationships change over time, and identify the complexity of the system structure (e.g., nested, hierarchical, multilayered) [43, 44].

There are several examples in the literature of the use of psychological networks to model psychological or education processes. SRL, as exemplified before, has been one of the phenomena most commonly modeled through complexity methods. For instance, López-Pernas et al. [45] used partial correlation networks [44] to model the relationship between students' SRL components. The authors found that the SRL dimensions with the most influence in the SRL process differ from the beginning (when metacognition is more influential) to the end of the course (motivation), and therefore might be more amenable to intervention. Saqr & López-Pernas [41] used different temporal and contemporaneous networks to study the differences between group-based (depicted in Fig. 6) and individual SRL dynamics. The authors concluded that the average SRL process is not representative of the individual SRL processes of each student, advocating for the study of idiographic analysis to gather personalized insights. Using a similar approach, Saqr [46] studied the interplay of engagement dimensions using log data from a learning management system and found marked differences between group-level and within-person variance. These highlight that complex systems behave differently for different individuals, depending on the initial conditions, the interdependence between the system's components, and the influence of external variables.

Through a transition network analysis approach [47], López-Pernas and Saqr [13] studied the longitudinal dynamics of online engagement using sequence analysis and transition networks (Fig. 7-left) and found that engagement patterns remain relatively stable over time, aligning with the typical behaviors observed in complex systems (i.e., attractor states, Fig. 7-right). Significant changes in these patterns are rare, and when they do occur, they are generally brief, with students often reverting to their original engagement levels. This suggests that a student's initial engagement state heavily influences their long-term trajectory, effectively determining their likely outcomes.

Lastly, it is worth mentioning that, although SRL and engagement can be studied as complex systems of their own, they can be also modeled as part of a macro complex system where they interact with one another and other components (e.g., motivation), as proposed by Saqr et al. [44] (Fig. 8).

3.2 Recurrence Quantification Analysis

Recurrence Quantification Analysis (RQA) is a statistical technique that reveals the nonlinearity and dynamic properties of time series data [48, 49]. This method

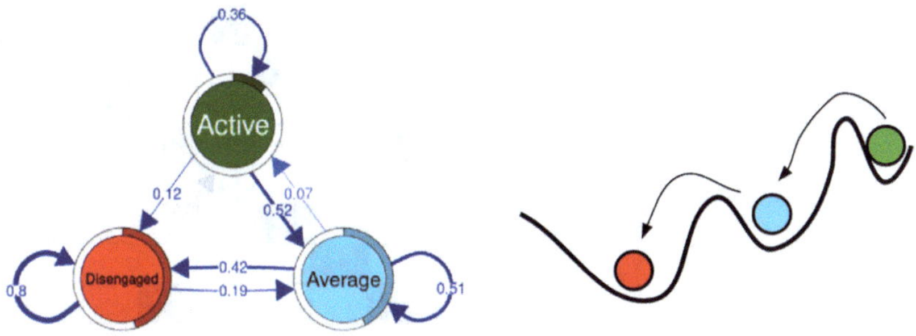

Fig. 7 Transition network of students in the *mostly disengaged* trajectory identified by López-Pernas and Saqr [13]. Students in this trajectory tend to reach a disengaged state and remain disengaged

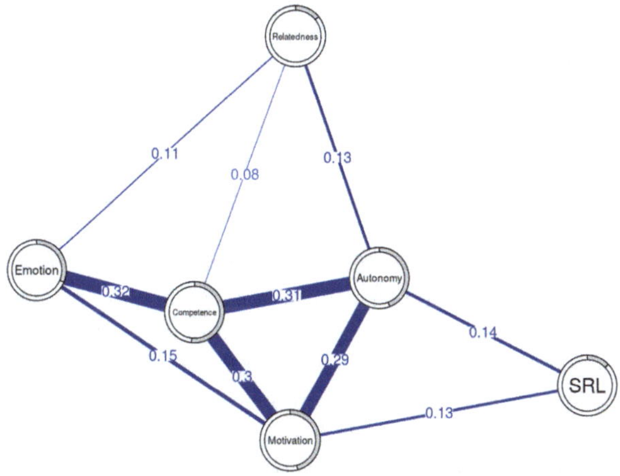

Fig. 8 Psychological network illustrating the complex system formed by SRL and other constructs [44]

detects and quantifies how systems shift between repetitive and novel sequences of behaviors across a temporal space. Three types of RQA methods exist: auto-RQA (aRQA), cross-RQA (cRQA), and multidimensional-RQA (mdRQA). aRQA examines the dynamics within a singular time series, cRQA identifies shared dynamics of two time series, and mdRQA identifies the repeating patterns across more than two time series [50]. These analyses can be completed with either categorical or continuous time series data which are mapped against each other on a matrix where Time Series 1 is mapped on the x-axis and Time Series 2 is mapped on the y-axis. On the matrix, the intersection of time series values are highlighted black if the values are considered the same, pre-determined using radial metrics (this is explained in the tutorial down below).

Taking categorical aRQA as an example, Time Series 1 [A, B, X, Y, A, S,X, Y] will be the same dataset as Time Series 2 [A, B, X, Y, A, S, X, Y]. Every time on

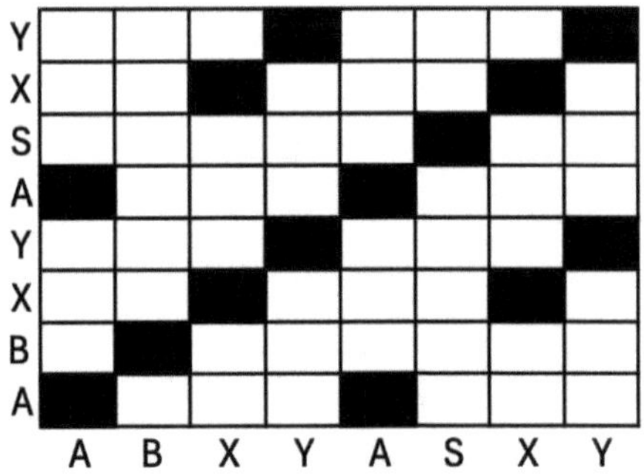

Fig. 9 Example of aRQA recurrence matrix

the matrix time points match in value, such as that with time points t1 and t5 as well as t3 and t7, the intersection on the matrix is shaded black (see Fig. 9). The diagonal line represents the line of identity (LOI) where the time series will always be recurrent with itself at Lag 0. Lags are defined by the distance between time points. For example, Lag 1 represents time points that are removed by one time space (e.g., Time 1 and Time 2, Time 2 and Time 3); Lag 2 represents time points removed by two time spaces (e.g., Time 1 and Time 3). By incorporating lags into subsequent analyses, RQA considers the temporal relationships between events such that patterns in data can emerge across large time differences.

RQA outputs several metrics, including recurrence rate (proportion of repetition), percent determinism (proportion of patterns on diagonals), average diagonal line length (average length of diagonal line structures), laminarity (proportion of repetitions on vertical line structures), and trapping time (average length of vertical lines). This is not an exhaustive list, but it represents the vast amount of information that can be extracted through this analysis that can explain the degree to which a system displays dynamism, how events that occur within the environment impact the dynamics of the system, and the degree to which a system demonstrates adaptivity. For example, a study by Dever et al. [26] applied this analytical methodology to self-regulated learning strategies during game-based learning. In identifying significant differences in the metrics output by aRQA analyses according to learning outcomes and agency experienced by students in the game, Dever et al. [26] established how students demonstrated functionality in their self-regulation during learning and were able to identify instances in which students demonstrated dysfunctional self-regulation. This provides interesting insights into how students should be supported and scaffolded within their self-regulated learning behaviors. For example, can we identify when a student demonstrates dysfunctional behaviors and how can we design interventions which help the student to correct this dysfunction to demonstrate functional self-regulation?

3.3 Simulation

Nowadays, computers make it possible to implement complex systems of equations of change. But for those who shy away from mathematical formalization, there are user-friendly computer programs that make it very easy to simulate the complexity of interactions at the origin of self-organization processes and attractor dynamics [17, 51, 52]. Among the most popular computer simulation tools are dynamic networks, agent-based models, and cellular automata. As explained earlier in this chapter, a dynamic network models a complex system as a network of connected nodes (the system's components). This type of computational tool has been used to model and simulate the dynamics of several phenomena of interest to the educational field, including intelligence development [53], language development [54], parent-child interactions [55], and goal-directed motivation [56]. One of the most user-friendly open-source software packages for creating dynamic network models is Insight Maker.[1] Agent-based models simulate the behavior of virtual agents that represent the components of a system. These agents interact with each other and with their virtual environment based on iteratively applied evolution rules. Agent-based models have thus made it possible to model the dynamics of phenomena important to the educational context, such as motivation [57, 58], language development [59], and parent-child interactions [60]. One of the most widely used open-source, user-friendly programming platforms for agent-based modeling is NetLogo.[2] Finally, cellular automata are simplified forms of agent-based models, where agents are the cells of a two-dimensional lattice or a line. The state of each cell at time t+1 depends, through the application of an evolution rule, on the state of the cells in its neighborhood at time t. The same rule is then applied to the new generation of cells at t+1 to shape generation t+2, and so forth. One of the most insightful applications of cellular automata to psychological development processes concerns the self-organizing dynamics of the self-concept into differentiated structures [51], that is, into distinct clusters that may virtually represent different dimensions of the self, such as the academic self, the athletic self, the relational self, etc. Among the open-source platforms for user-friendly implementation of cellular automata we might mention Golly.[3]

Unlike the component-dominant mainstream approach, which aims to identify causal factors, the complex dynamic systems paradigm focuses on the identification of interaction-dominant processes. This is what the research methods we have just briefly reviewed contribute to, by aiming to experimentally test the volatility and stability of systems' behavior, to identify statistical signatures of complexity and non-linearity in the time series of these systems' states, to formally and/or computationally model the self-organization processes at work within them, and,

---

[1] Insight Maker: https://www.insightmaker.com.

[2] NetLogo: https://ccl.northwestern.edu/netlogo.

[3] Golly: https://golly.sourceforge.io.

for the sake of validating the models built, to retrieve the same statistical signatures in the time series generated by their computer simulation. Once validated, these dynamic models are invaluable tools for testing the effects of interventions, such as pedagogical ones, on the dynamics of important educational variables.

4 Data Collection

4.1 Capturing a System's State Through Rich Description: Interviews and Written Reflections

Understanding complex phenomena, such as individuals' learning, motivation, identities, and actions, requires methods that capitalize on the richness and depth of personal experiences. Narrative interviews and written reflections are markedly effective for capturing the state of a complex system at specific points in time. These methods provide insight into how individuals interpret and enact their roles, goals, beliefs, perceptions, and learning strategies within their unique lived contexts [61].

Narrative interviews serve as an effective tool for exploring individuals' experiences and meaning-making processes. Interviewers prompt participants to share personal stories, revealing the interplay between system elements such as beliefs, goals, self-perceptions, emotions, and actions. For example, research on teachers' professional development has used interviews conducted before, during, and after the program to trace changes in participants' motivations and instructional practices [16]. These interviews allowed participants to reflect on their teaching journey, the influence of prior experiences, and how professional development shaped their beliefs about student learning and instructional strategies.

Written reflections provide another method for capturing detailed and introspective data. These reflections can be gathered over time, such as through daily or weekly diary entries, offering snapshots of participants' evolving thoughts and decisions. For instance, teachers reflecting on their experiences with inquiry-based learning might describe how their beliefs about teaching shifted during the professional development program. Such reflections illuminate the dynamic relationships among the person's system components, revealing how changes in context or new experiences challenge or reinforce existing beliefs, goals, and strategies.

The analysis of these rich data sources involves identifying key elements of the individual's complex system, such as their beliefs, values, goals, emotions, and actions, as well as the relationships among them [43, 62]. Researchers synthesize these findings to create a holistic understanding of the individual's system at specific points in time, highlighting patterns and tensions within and across system elements. Additionally, comparing data across participants allows for researchers to identify shared patterns and generate theoretical principles that explain how identities and actions develop in different contexts.

These methods emphasize the importance of capturing the nuanced and context-dependent nature of complex systems. Narrative interviews and written reflections provide a deep and detailed understanding of individuals' experiences, enabling researchers to document the states of their systems at key moments and track their evolution over time.

4.2 Capturing Real-Time Emergence Through Observations and Stimulated Recall

Studying real-time emergence in complex systems requires methods that capture dynamic processes as they unfold. Observations and stimulated recall interviews are particularly effective for examining how engagement emerges within specific contexts, providing detailed accounts of participants' experiences, interactions, decisions, and actions [61].

Detailed observations allow researchers to document participants' behaviors and interactions in naturalistic or designed settings. For example, in a study of visitors to a museum exhibition designed to promote inventiveness, the researchers used observations to capture participants' engagement with the exhibition's physical and conceptual features. Observational data revealed how visitors interacted with the content, expressed their perceptions, and responded to the contextual elements of the exhibition. This real-time data provided insight into how participants' identities and goals were activated and shaped by the environment [63].

Stimulated recall interviews build on observational data by eliciting participants' reflections on their thoughts, emotions, and decisions during specific moments of activity. Conducted shortly after the observed session, these interviews prompt participants to describe their experiences in detail, often using notes or video footage as memory aids. For example, in the museum study described above, following an experience with an artifact, participants were asked to recount their thought processes during the visit, providing insights into how they connected the exhibition content to their roles and goals, such as seeing themselves as creative or inventive in other areas of life.

Combining observations with stimulated recall interviews allows researchers to construct a comprehensive picture of how engagement emerges in real-time through participants' experiences in an authentic educational context. Observations provide a foundation for understanding what participants do and say within the context, while stimulated recall interviews add depth by uncovering the subjective meanings behind these actions. Analyzing these data sources together allows researchers to identify patterns of interaction between contextual features and individual dispositions, highlighting the dynamic and context-dependent nature of engagement emergence.

These methods are particularly valuable for capturing the temporal and situational aspects of complex systems. Observations provide a detailed record of real-time activity, while stimulated recall interviews offer access to participants' reflective processes. Together, these approaches enable researchers to study the dynamic interplay between individuals and their environments, shedding light on how identities and actions develop and transform within specific contexts.

5 Final Remarks

The complex dynamic systems approach lends itself to empirical observation, but in a different way to traditional scientific approaches. While the latter aim to discover —by isolating— causal relationships between independent and dependent variables, the complex dynamic systems approach investigates the way in which the system's global behavioral patterns, or order parameters, change or stabilize under the effect of a parameter external to the system, or control parameter, which constrains the system's internal interaction dynamics [64]. For example, with regard to motor learning, Zanone and Kelso [65] showed how an oscillatory synchronization motor skill to be acquired can emerge and then stabilize by training it to resist increasing perturbations. Here, perturbations play the role of a control parameter that is experimentally manipulated —by gradually increasing the required frequency of limbs' oscillation— in order to push the system of the emergent synchronization into disorganization. The resulting learning is thus defined in dynamic terms, that is, as the stabilization of the system's new behavior in the form of its resistance to perturbation.

The states of complex dynamic systems evolve over time while exhibiting typical statistical properties. Recurrence Quantification Analysis (RQA) can detect regularities and forms of stability (attractors) in the dynamics of the states of a complex system. The temporal distribution of these states also reveals a long-term memory property that translates into a particular autocorrelation structure that can be found whatever the time scale on which these states are observed. This statistical structure, which reveals dynamics based on the traces of its own past history, is a power-law temporal distribution known as *1/f* noise or pink noise. Its detection is enabled by time-series analysis of the states of the system under study, particularly through Detrended Fluctuation Analysis [DFA; [66]] and the various adaptations of this technique that have been created to overcome its limitations (see, for example, the chapter by Altamore et al., this book). In the educational field, distributions in the form of *1/f* noise have been identified in language learning [67] and motor learning [68], attesting that learning is indeed a dynamic phenomenon in the sense of the theory of complex dynamic systems. Interestingly, *1/f* distributions are ubiquitous in nature [69], and their replacement by other types of distribution often reflects anomalies [70]. Consequently, from an applied perspective, such replacements can be used as warning signals of a risk of deteriorating behavior, for example, a risk of discouragement towards academic studies or even academic failure. Detecting such a signal can then be very useful in deciding whether to implement individualized support procedures for students in difficulty, before it is too late.

Variations in the states of a complex dynamic system are the result of iterative processes that can be modeled mathematically. The momentary state of a system depends to a large extent on its previous state, which in turn depended on an earlier state, and so forth. When time is expressed in continuous form, this dependence of a system on its own history can be accounted for by differential equations such as $dx/dt = f(x)$, where d represents the variation in the system's state x or the variation of time t. When time is discretized, history dependence is expressed by different equations such as $xt+1 = f(xt)$. However, state variations of complex dynamic systems also result from iterative interactions (often reciprocal in the form of couplings) between their components. The above-mentioned equations of change are then enriched by the influences of changes in the variables interacting within the system, taking the form of systems of differential or difference equations. All these equations are capable of producing monostable (one attractor), multistable (several attractors that can be visited more or less cyclically), or even chaotic behaviors of the system. In the educational field, such equations of change have been used to model the dynamics of cognitive development [71], language development [54], social development [72], and teaching-learning interactions [73].

References

1. Hilpert JC, Marchand GC (2018) Complex systems research in educational psychology: aligning theory and method. Educ Psychol 53:185–202. https://doi.org/10.1080/00461520.2018.1469411
2. Kaplan A, Garner JK (2020) Steps for applying the complex dynamical systems approach in educational research: a guide for the perplexed scholar. J Exp Educ 88:486–502. https://doi.org/10.1080/00220973.2020.1745738
3. Saqr M, Schreuder MJ, López-Pernas S (2024) Why educational research needs a complex system revolution that embraces individual differences, heterogeneity, and uncertainty. In: Saqr M, López-Pernas S (eds) Learning analytics methods and tutorials: a practical guide using R. Springer, Cham, pp 723–734
4. Panadero E (2017) A review of self-regulated learning: six models and four directions for research. Front Psychol 8:422. https://doi.org/10.3389/fpsyg.2017.00422
5. Symonds JE, Kaplan A, Upadyaya K, Aro KS, Torsney BM, Skinner E, Eccles JS (2024) Momentary student engagement as a dynamic developmental system. J Theor Philos Psychol https://doi.org/10.1037/teo0000288
6. Svoboda J (2023) Processing misconceptions: dynamic systems perspectives on thinking and learning. Front Educ 8:1215361. https://doi.org/10.3389/feduc.2023.1215361
7. Koopmans M (2020) Education is a complex dynamical system: challenges for research. J Exp Educ 88:358–374. https://doi.org/10.1080/00220973.2019.1566199
8. Thurner S, Hanel R, Klimek P (2018) Introduction to the theory of complex systems. Oxford University Press, London
9. Sayama H (2015) Introduction to the modeling and analysis of complex systems. Open SUNY Textbooks
10. Scheffer M (2020) Critical transitions in nature and society. Princeton University Press, Princeton

11. Bringmann L, Helmich M, Eronen M, Voelkle M (2023) Complex systems approaches to psychopathology. In: Krueger RF, Blaney PH (eds) Oxford textbook of psychopathology. Oxford University PressNew York, pp 103–122
12. Olthof M, Hasselman F, Oude Maatman F, Bosman AMT, Lichtwarck-Aschoff A (2023) Complexity theory of psychopathology. J Psychopathol Clin Sci 132:314–323. https://doi.org/10.1037/abn0000740
13. López-Pernas S, Saqr M (2024) How the dynamics of engagement explain the momentum of achievement and the inertia of disengagement: a complex systems theory approach. Comput Hum Behav 153:108126. https://doi.org/10.1016/j.chb.2023.108126
14. Saqr M, López-Pernas S, Helske S, Hrastinski S (2023) The longitudinal association between engagement and achievement varies by time, students' subgroups, and achievement state: a full program study. Comput Educ 199:104787
15. Binkley M, Erstad O, Herman J, Raizen S, Ripley M, Miller-Ricci M, Rumble M (2012) Assessment and teaching of 21st century skills. In: Assessment and teaching of 21st century skills. Springer, Berlin, pp 1–15. https://doi.org/10.1007/978-94-007-2324-5
16. Garner JK, Kaplan A (2019) A complex dynamic systems perspective on teacher learning and identity formation: an instrumental case. Teachers Teach 25:7–33. https://doi.org/10.1080/13540602.2018.1533811
17. Gernigon C, Den Hartigh RJR, Vallacher RR, van Geert PLC (2024) How the complexity of psychological processes reframes the issue of reproducibility in psychological science. Perspect Psychol Sci J Assoc Psychol Sci 18:952–977. https://doi.org/10.1177/17456916231187324
18. Wallot S, Kelty-Stephen DG (2018) Interaction-dominant causation in mind and brain, and its implication for questions of generalization and replication. Minds Mach 28:353–374. https://doi.org/10.1007/s11023-017-9455-0
19. Bianconi G, Arenas A, Biamonte J, Carr LD, Kahng B, Kertesz J, Kurths J, Lü L, Masoller C, Motter AE, Perc M, Radicchi F, Ramaswamy R, Rodrigues FA, Sales-Pardo M, Miguel MS, Thurner S, Yasseri T (2023) Complex systems in the spotlight: next steps after the 2021 nobel prize in physics. J Phys Complex 4:010201. https://doi.org/10.1088/2632-072X/ac7f75
20. Favela LH, Amon MJ (2023) Enhancing bayesian approaches in the cognitive and neural sciences via complex dynamical systems theory. Dynamics 3:115–136. https://doi.org/10.3390/dynamics3010008
21. Lorenz E (2001) The chaos avant-garde. World Scientific, Singapore, pp 91–94. https://doi.org/10.1142/9789812386472_0007
22. Clark TJ, Luis AD (2020) Nonlinear population dynamics are ubiquitous in animals. Nat Ecol Evol 4:75–81. https://doi.org/10.1038/s41559-019-1052-6
23. Yerkes RM, Dodson JD (1908) The relation of strength of stimulus to rapidity of habit-formation. J Compar Neurol Psychol 18:459–482. https://doi.org/10.1002/cne.920180503
24. Artime O, De Domenico M (2022) From the origin of life to pandemics: emergent phenomena in complex systems. Philos Trans Ser A Math Phys Eng Sci 380:20200410. https://doi.org/10.1098/rsta.2020.0410
25. Kalantari S, Nazemi E, Masoumi B (2020) Emergence phenomena in self-organizing systems: a systematic literature review of concepts, researches, and future prospects. J Organ Comput 30:224–265. https://doi.org/10.1080/10919392.2020.1748977
26. Dever DA, Amon MJ, Vrzáková H, Wiedbusch MD, Cloude EB, Azevedo R (2022) Capturing sequences of learners' self-regulatory interactions with instructional material during game-based learning using auto-recurrence quantification analysis. Front Psychol 13:813677. https://doi.org/10.3389/fpsyg.2022.813677
27. Camazine S, Deneubourg J-L, Franks NR, Sneyd J, Theraula G, Bonabeau E (2003) Self-organization in biological systems. Princeton University Press, Princeton
28. Dever DA, Sonnenfeld NA, Wiedbusch MD, Schmorrow SG, Amon MJ, Azevedo R (2023) A complex systems approach to analyzing pedagogical agents' scaffolding of self-regulated learning within an intelligent tutoring system. Metacognit Learn 18:659–691. https://doi.org/10.1007/s11409-023-09346-x

29. Wiener N (1948) Cybernetics. Sci Am 179:14–19
30. Rosnay J (1979) The macroscope: a new world scientific system. NY Harper Row, New York
31. Ao P, Kwon C, Qian H (2007) On the existence of potential landscape in the evolution of complex systems. Complexity 12:19–27. https://doi.org/10.1002/cplx.20171
32. Wang J, Zhang YJ, Xu C, Li J, Sun J, Xie J, Feng L, Zhou T, Hu Y (2024) Reconstructing the evolution history of networked complex systems. Nat Commun 15:2849. https://doi.org/10.1038/s41467-024-47248-x
33. Corning PA (1995) Synergy and self-organization in the evolution of complex systems. Syst Res Official J Int Feder Syst Res 12:89–121. https://doi.org/10.1002/sres.3850120204
34. Siegenfeld AF, Bar-Yam Y (2019) An introduction to complex systems science and its applications. arXiv [physics.soc-ph]
35. Walker BH (2020) Resilience: what it is and is not. Ecol Soc J Integr Sci Resil Sustain 25:11. https://doi.org/10.5751/es-11647-250211
36. Han J, Amon MJ (2021) A nonlinear dynamical systems approach to emotional attractor states during media viewing. In: Annual meeting of the cognitive science society, vol. 43
37. Gao Y, Wang X, Fan P (2024) Exploring male english major's motivation trajectory through complex dynamic systems theory. Curr Psychol 43:9089–9100. https://doi.org/10.1007/s12144-023-05062-6
38. Kuhn T (1962) The structure of scientific revolutions. University of Chicago Press, Chicago
39. Saqr M, López-Pernas S, Conde-González MÁ, Hernández-García Á (2024) Social network analysis: a primer, a guide and a tutorial in R. In: Learning analytics methods and tutorials. Springer, Cham, pp 491–518
40. Saqr M, Poquet O, Lopez-Pernas S (2022) Networks in education: a travelogue through five decades. IEEE Access Pract Innov Open Solutions 10:32361–32380. https://doi.org/10.1109/access.2022.3159674
41. Saqr M, López-Pernas S (2024) Mapping the self in self-regulation using complex dynamic systems approach. Brit J Educ Technol 55:1376–1397. https://doi.org/10.1111/bjet.13452
42. Brass DJ, Borgatti SP (2019) Social networks at work. Routledge, Milton Park
43. Malmberg J, Saqr M, Järvenoja H, Järvelä S (2022) How the monitoring events of individual students are associated with phases of regulation: a network analysis approach. J Learn Anal 9:77–92
44. Saqr M, Beck E, López-Pernas S (2024) Psychological networks: a modern approach to analysis of learning and complex learning processes. In: Saqr M, López-Pernas S (eds) Learning analytics methods and tutorials: a practical guide using R. Springer, Cham, pp 639–671
45. López Pernas S, Conde-González M, Raspopović Milić M, Saqr M (2025) Frequencies and averages miss the point of SRL evolution: a complex dynamic systems approach. In: Proceedings TEEM 2024: twelfth international conference on technological ecosystems for enhancing multiculturality. Lecture notes in education technology. Springer, Berlin
46. Saqr M (2024) Group-level analysis of engagement poorly reflects individual students' processes: why we need idiographic learning analytics. Comput Hum Behav 150:107991. https://doi.org/10.1016/j.chb.2023.107991
47. Saqr M, López-Pernas S, Törmänen T, Kaliisa R, Misiejuk K, Tikka S (2024) Transition network analysis: a novel framework for modeling, visualizing, and identifying the temporal patterns of learners and learning processes. arXiv [cs.SI]
48. Webber C, Zbilut J (2004) Recurrence quantification analysis of nonlinear dynamical systems. In: Riley MA, Van Orden GC (ed) Tutorials in contemporary nonlinear methods for the behavioral. National Science Foundation, Alexandria, pp 26–94
49. Webber J, Charles L, Marwan N, Facchini A, Giuliani A (2009) Simpler methods do it better: success of recurrence quantification analysis as a general purpose data analysis tool. Phys Lett A 373:3753–3756. https://doi.org/10.1016/j.physleta.2009.08.052
50. Moulder R, Booth B, Abitino A, D'Mello S (2023) Recurrence quantification analysis of eye gaze dynamics during team collaboration. In: LAK23: 13th international learning analytics and knowledge conference. ACM, New York

51. Nowak A (2004) Dynamical minimalism: why less is more in psychology. Person Soc Psychol Rev 8:183–192. https://doi.org/10.1207/s15327957pspr0802_12
52. Vallacher RR (2017) Computational social psychology. Series: frontiers of social psychology. Routledge, New York
53. van der Maas HLJ, Dolan CV, Grasman RPPP, Wicherts JM, Huizenga HM, Raijmakers MEJ (2006) A dynamical model of general intelligence: the positive manifold of intelligence by mutualism. Psychol Rev 113:842–861. https://doi.org/10.1037/0033-295X.113.4.842
54. van Geert P (1991) A dynamic systems model of cognitive and language growth. Psychol Rev 98:3–53. https://doi.org/10.1037/0033-295X.98.1.3
55. van Dijk M, van Geert P, Korecky-Kröll K, Maillochon I, Laaha S, Dressler WU, Bassano D (2013) Dynamic adaptation in child–adult language interaction. Lang Learn 63:243–270. https://doi.org/10.1111/lang.12002
56. Westaby JD, Pfaff DL, Redding N (2014) Psychology and social networks: a dynamic network theory perspective. Am Psychol 69:269–284. https://doi.org/10.1037/a0036106
57. Jager W, Popping R, van de Sande H (2001) Clustering and fighting in two- party crowds: simulating the approach-avoidance conflict. J Artif Soc Social Simul, 4(3). https://www.jasss.org/4/3/7.html
58. Merrick KE, Shafi K (2013) A game theoretic framework for incentive-based models of intrinsic motivation in artificial systems. Front Psychol 4:791. https://doi.org/10.3389/fpsyg.2013.00791
59. Steels L, Belpaeme T (2005) Coordinating perceptually grounded categories through language: a case study for colour. Behav Brain Sci 28:469-89; discussion 489–529. https://doi.org/10.1017/S0140525X05000087
60. Hesp C, Steenbeek HW, Geert PLC van (2019) Socio-emotional CONcern dynamicS in a model of real-time dyadic interaction: parent-child play in autism. Front Psychol 10:1635. https://doi.org/10.3389/fpsyg.2019.01635
61. Kaplan A, Garner JK (2025) A complex dynamic systems approach to role identity: researching identities, motivation, learning, and action. Sage Research Methods. https://doi.org/10.4135/9781036214173
62. Törmänen T, Järvenoja H, Saqr M, Malmberg J, Järvelä S (2023) Affective states and regulation of learning during socio-emotional interactions in secondary school collaborative groups. Brit J Educ Psychol 93(suppl 1):48–70. https://doi.org/10.1111/bjep.12525
63. Kaplan A, Garner J, Neuber A (2023) Learning as identity change, self-regulated learning as identity exploration: a complex dynamic systems perspective on the goals of education in the 21st century. In: Proceedings of the 1st world giftedness center international conference
64. Haken H, Kelso JA, Bunz H (1985) A theoretical model of phase transitions in human hand movements. Biol Cybern 51:347–356. https://doi.org/10.1007/bf00336922
65. Zanone PG, Kelso JA (1992) Evolution of behavioral attractors with learning: nonequilibrium phase transitions. J Exp Psychol Hum Percept Perform 18:403–421. https://doi.org/10.1037/0096-1523.18.2.403
66. Peng CK, Mietus J, Hausdorff JM, Havlin S, Stanley HE, Goldberger AL (1993) Long-range anticorrelations and non-Gaussian behavior of the heartbeat. Phys Rev Lett 70:1343–1346. https://doi.org/10.1103/PhysRevLett.70.1343
67. Lowie W, Plat R, de Bot K (2014) Pink noise in language production: a nonlinear approach to the multilingual lexicon. Ecol Psychol 26:216–228. https://doi.org/10.1080/10407413.2014.929479
68. Nourrit-Lucas D, Tossa AO, Zélic G, Delignières D (2015) Learning, motor skill, and long-range correlations. J Motor Behav 47:182–189. https://doi.org/10.1080/00222895.2014.967655
69. Bak P (1999) How nature works: the science of self-organized criticality, 1st edn. Springer, New York
70. Van Orden G (2007) The fractal picture of health and wellbeing [Data set]. In: PsycEXTRA Dataset. American Psychological Association (APA). https://doi.org/10.1037/e593942007-002

71. van Geert P (1998) A dynamic systems model of basic developmental mechanisms: piaget, vygotsky, and beyond. Psychol Rev 105:634–677. https://doi.org/10.1037/0033-295x.105.4.634-677
72. Steenbeek H, van Geert P (2008) An empirical validation of a dynamic systems model of interaction: do children of different sociometric statuses differ in their dyadic play? Develop Sci 11:253–281. https://doi.org/10.1111/j.1467-7687.2007.00655.x
73. Merlone U, Panchuk A, van Geert P (2019) Modeling learning and teaching interaction by a map with vanishing denominators: fixed points stability and bifurcations. Chaos Solitons Fract 126:253–265. https://doi.org/10.1016/j.chaos.2019.06.008

The Advanced Applications of Psychological Networks with Exploratory Graph Analysis

Tarid Wongvorachan and Okan Bulut

Abstract Educational measurement plays a pivotal role in capturing evidence of students' learning as an integral part of learning analytics. The network psychometrics approach offers a novel perspective to analyze a complex system of elements that are related to learning. The first edition of *Learning Analytics Methods and Tutorials* offers a comprehensive introduction to the estimation of psychological networks. This chapter aims to introduce advanced applications of network psychometrics in learning analytics. Specifically, we demonstrate the application of unique variable analysis to identify variables that uniquely contribute to a complex system of variables (e.g., identifying important actions stored within a learning management system), thereby reducing redundant variables to meaningfully simplify the dataset. Furthermore, we describe and demonstrate methods to assess the integrity of such a system with exploratory graph analysis (EGA), hierarchical EGA, and dynamic EGA. The dynamic EGA method, in particular, allows researchers to assess changes in the structure of a system across multiple time points. This capability proves beneficial in scenarios involving the continuous monitoring of students' learning progress over time. Finally, we illustrate the transformation of psychological networks from EGA to the factor analytic format, highlighting how these two approaches, despite operating differently, are comparable and can complement each other in examining the complex structure of a learning system.

Keywords Network psychometrics · Learning analytics · Factor analysis

1 Introduction

The network psychometrics approach offers a novel perspective to analyze a complex system of elements that are related to learning [4]. This approach to

T. Wongvorachan (✉) · O. Bulut
University of Alberta, Edmonton, AB, Canada
e-mail: wongvora@ualberta.ca

M. Saqr, S. López-Pernas (eds.), *Advanced Learning Analytics Methods*,
https://doi.org/10.1007/978-3-031-95365-1_13

psychometrics is a helpful addition that complements the traditional factor analytic approach to study the interplay of psychological components consisting of a group as latent variables, such as cognitive ability, engagement, and motivation [10]. Specifically, the factor analytic approach focuses on simplifying the relationships between such components as a restricted, factorial structure of observed and latent variables, while the network approach focuses on exploring unrestricted, dynamic associations among observed variables [10, 17, 28]. However, the network psychometrics approach conceptualizes the structure of a psychological network as a system with structural relationships between variables of interest [4].

Unlike the traditional factor analytic approach, the network psychometrics method operates without the assumption of latent factors. In this framework, items are permitted to demonstrate their associations by freely moving within a two-dimensional space [4]. If the items are related, they tend to naturally cluster together without the imposition of structural constraints [4, 19]. This characteristic offers researchers a novel perspective for examining the relationships between variables that constitute educational constructs as a system. Building upon the comprehensive introduction to psychological network estimation presented in the first volume of this book [26], this chapter extends that discussion by exploring advanced applications of psychological networks through the exploratory graph analysis (EGA) method through the *EGAnet* package [13] in R [30].

Specifically, we describe and demonstrate methods to assess the integrity of such a network system with EGA and hierarchical EGA [13, 15]. We also demonstrate the transformation of network structures from the EGA and Hierarchical EGA methods to the confirmatory factor analysis (CFA) format, highlighting how these approaches, despite operating differently, are comparable and can complement each other in examining the complex structure of a learning system [15]. Subsequently, we demonstrate how the dynamic EGA method allows researchers to assess changes in the structure of a system across multiple units of assessment [13]. This capability proves beneficial in scenarios involving the personalization of interventions for specific groups of learners [20]. Finally, we demonstrate the application of unique variable analysis (UVA) to identify variables that do not contribute meaningfully to the construct due to their redundancy (e.g., more than one variables that measure similar or the same attribute), thereby reducing them to simplify the dataset for better model fit [8].

2 Literature Review

2.1 Examining a System of Educational Variable with EGA

EGA is a network psychometrics technique that leverages graphical network estimation and community detection algorithms to elucidate the complex structure of educational data [13]. Like other network domain methods, EGA represents observed variables as nodes and their partial correlations as edges within the network [9]. Clusters of variables in an EGA analysis depict their underlying latent structure, encompassing the observed variables themselves [15]. These clusters, called communities, allow researchers to examine hidden structures that may not be available upon mere observation.

A distinctive characteristic of EGA is its use of community detection algorithms, such as "Louvain" or "Walktrap." These algorithms optimize the identified structure of the psychological system under investigation, offering researchers a novel approach to estimating the structure of a psychological construct [14]. For instance, the Louvain algorithm identifies clusters of observed variables based on a multi-level modularity optimization algorithm [3], while the Walktrap method employs a random walk algorithm [23].

Beyond a single estimation of the network structure, EGA offers the capability to assess its stability across data variations through iterations of bootstrap resampling [9]. This iterative approach, known as BootEGA, estimates the network structure by identifying the median structure across all results from the bootstrap iterations [14]. This approach enhances the robustness of network structure assessment by accounting for potential uncertainties arising from dataset variations.

Furthermore, the EGA method can estimate network structures with nodes that share common covariance, potentially representing hierarchical information like higher-order constructs [10]. This ability allows EGA to capture the hierarchical nature of variables without requiring manual input from researchers. Additionally, EGA can estimate dynamic communities in multivariate time series data at various units of measurement (i.e., intra-individual structure, groups, and inter-individual structure). This method, termed dynamical EGA, allows researchers to examine changes in the network structure across sub-populations [14]. Such insights can inform the development of personalized interventions in variables that change across groups, such as students' feedback utilization between low achievers and high achievers.

Regarding reporting metrics, both EGA and Hierarchical EGA show a network plot depicting the community assignment of each item alongside the same model fit indices used in Confirmatory Factor Analysis (CFA). Specifically, an EGA object can provide absolute fit indices such as Root Mean Square Error of Approximation (RMSEA) and Standardized Root Mean Square Residual (SRMR), which are recommended to be less than 0.05 and 0.08, respectively, to indicate a good fit [12, 18]. Additionally, relative fit indices such as the Comparative Fit Index (CFI) and Tucker-Lewis Index (TLI) are available. These indices should ideally be 0.95 or higher [18]. The extraction of CFA fit from EGA can be done by transforming an EGA or Hierarchical EGA object into a CFA through the "*CFA*" function in the *EGAnet* package [13]. Then, the fit measures can be extracted through the "*fitmeasures*" function from the lavaan package [25].

Bootstrap EGA yields results similar to EGA, with the added benefit of providing the typical structure represented by the median dimensions identified across multiple network structures from bootstrap iterations [15]. It also offers information on the replication proportion for item-level results and structural consistency indices for dimension-level results [9, 15]. Notably, the EGA method introduces a unique relative fit index known as the Total Entropy Fit Index (TEFI) that complements conventional fit indices [16]. A lower TEFI value indicates reduced disorder and uncertainty in the estimated structure, signifying greater robustness in item organization [16]. The CFA fit metrics, however, are not available in the dynamical EGA method. Instead, it has network fit information such as the number of communities, the median dimension for the intra-individual level, and TEFI at the group and inter-individual levels [14].

A use case of EGA in learning analytics could be the prediction of students' success to inform instructors' teaching strategies. The instructor could collect learning-related data such as attendance, participation in class discussions, time spent on homework, use of learning resources, and scores on quizzes and exams. Through the community detection algorithms of EGA, clusters of these variables are formed, representing their underlying latent structure. For instance, attendance and participation in class discussions might form one community, indicating that these two factors are closely related and might contribute to a latent factor such as 'class engagement.' Similarly, time spent on homework and use of learning resources might form another community, possibly representing a latent factor like 'study habits'. This information can then be used to develop targeted interventions to improve student outcomes. For example, if 'class engagement' is identified as a key factor in student success, the university might implement strategies to increase student participation in class discussions. In this way, EGA serves as a powerful tool for educational institutions to analyze complex educational data and uncover meaningful insights that can inform decision-making and strategy development. It allows researchers to go beyond mere observation and examine the hidden structures within their data. This can lead to more effective interventions and improved educational outcomes.

2.2 *Reducing Redundant Variables with UVA*

Being developed under the same R package as EGA, UVA employs network estimation to identify redundant nodes or variables for removal [8]. These redundant nodes, if left unchecked, could give rise to undesired clusters that impede the interpretability of the estimated structure of the dataset [8, 21]. In learning analytics, network analysis can be used to understand student interactions and learning patterns. Data from a learning management system on students (e.g., ID, enrollment information, demographic), courses (e.g., course ID, topics, difficulty level), activities (e.g., login information, forum discussions, quiz attempts, assignment submissions), and grades (e.g., formative assessment score, assignments, and final exams) can be transformed into a network structure. However, this network might contain redundant nodes. For example, a student submitting an assignment likely logged in around that time. These two nodes ("assignment submission" and "login") might be capturing overlapping information. Keeping these redundant nodes can lead to misleading clusters, like grouping logins and submissions together. Techniques like UVA, which identifies and removes redundant nodes, can help create a clearer picture. This allows the analysis to focus on meaningful connections, such as discussion participation and its relation to assignment performance to improve the course design. Functionally, the UVA method leverages two algorithms to assess the similarity between network nodes based on partial correlation. These algorithms are the graphical least absolute shrinkage and selection operator (GLASSO) and the weighted Topological Overlap (wTO) [8, 22]. The UVA method employs wTO to identify neighboring nodes that share connections with a particular pair of nodes [29]. Essentially, nodes with highly similar connections to the same set of neighboring nodes are considered redundant [29]. In simpler terms, wTO detects highly correlated nodes that exhibit similar relationships with other nodes in the network, suggesting a level of interdependence [8]. For instance, if "class participation" and "student engagement" share strong connections with variables like "student satisfaction" and "dropout rates," they might be considered a redundant pair.

The integration of wTO with GLASSO network estimation allows for the visualization of potential interactions between nodes, the extent of overlap in connections between specific nodes, and the overall commonality of nodes within the entire network [22]. Following the identification of redundant nodes, the UVA method offers various strategies for their management. Redundant nodes can be: 1) Merged into latent variables using CFA [13, 24]; 2) Removed from the dataset entirely; 3) Aggregated by summing their scores across all cases. Furthermore, UVA can perform additional computations to detect any remaining redundancy after the initial reduction. UVA possesses two key metrics for reporting: 1) the degree of redundancy for each node, as indicated by wTO and 2) the identification of latent variables encompassing redundant nodes [13]. Regarding effectiveness, Christensen et al. [8] demonstrated that UVA surpasses existing methods in detecting local dependencies between variables. These methods include exploratory structural equation modeling [2], anti-image partial correlations [11], and expected residual correlation direct change [27].

3 Tutorial with R

We start by installing the following packages: *EGAnet* [13], *haven* [31], and *lavaan* [25]. In our codes, we use the double colon (::) notation to explicitly indicate the package from which each function originates. This practice ensures clarity for readers who may want to utilize specific functions from the tutorial. By specifying the package, we eliminate the need for readers to trace the source of each function.

```
install.packages(c("EGAnet", "haven", "lavaan"))

library(EGAnet) #for EGA
library(haven) #to load an SPSS data
library(lavaan) #for CFA
```

In this chapter, we utilize student response data from the Statistical Anxiety Scale (SAS) to illustrate the ordinary EGA task and data from the Statistical Anxiety Rating Scale (STARS) to demonstrate the UVA task. The SAS dataset consists of 24 items, which is a manageable size for comprehensive examination and visualization using EGA [7]. Conversely, the STARS includes 51 items, which allows for the identification of potential redundant items [6]. Both datasets are derived from the study by Chew et al. [6] and are available at https://doi.org/10.6084/m9.figshare.5544883.v2.

The Hierarchical EGA task is illustrated using a synthesized dataset based on student responses to the Student Perspective of Teaching (SPOT) survey, a course evaluation tool employed by a Western Canadian university. The SPOT survey encompasses six dimensions that collectively measure the construct of teaching effectiveness; These dimensions are as follows: (1) course organization, (2) usefulness of course resources, (3) graded work, (4) course delivery, (5) instructional approach, and (6) class climate [5]. This characteristic enables the assessment of the hierarchical structure within the SPOT dataset. This dataset is related to the learning context as students' feedback on courses is a vital component of information for enhancing teaching strategies.

Finally, the dynamic EGA task is demonstrated using a simulated dataset from the EGAnet package. Due to the specificity and limited public availability of multivariate time series data with repeated measures, a simulated dataset is employed for this purpose.

3.1 Examining Structure of an Educational Assessment with EGA

3.1.1 Ordinary EGA

For the EGA task, we begin by loading the dataset from Chew et al. [6] into our R environment using the read_sav() function. Subsequently, we extract the students' responses to the SAS. The command df[, 56:79] is employed to subset the data frame, including columns 56 to 79, which correspond to the SAS items.

```
df <-
read_sav(url("https://github.com/lamethods/data2/raw/refs/heads/main/anxiety/STARS.sav"))

df_sas <- df[,56:79]
```

Subsequently, we perform EGA on the SAS dataset using the Louvain algorithm and the graphical lasso model. The correlation method is set to auto-detection, allowing the analysis to determine the appropriate correlations for the data: Pearson's for continuous data, polychoric for ordinal data, tetrachoric for binary data, and polyserial/biserial for ordinal/binary data with continuous data. The analysis iterates 1000 times to achieve consensus using the highest modularity method. The network is then plotted, and the process is seeded for reproducibility. The EGA results can be accessed by calling the EGA object created during the analysis.

```
sas_ega <- EGAnet::EGA(
  data = df_sas,
  algorithm = "louvain",
  model = "glasso",
  corr = "auto",
  consensus.iter = 1000,
  consensus.method = c("highest_modularity"),

  plot.EGA = TRUE,
  seed = 1234
)

sas_ega$plot.EGA #plot the EGA network graph
sas_ega
```

Figure 1 shows that items in the SAS dataset can be allocated into three communities, as color-coded by the red, blue, and green colors. Output 1 provides information about the network structure of the SAS dataset after 1000 iterations of network estimation. According to the output, items loaded onto community 1 are SAS items 1, 4, 9, 11, 13, 14, 15, and 20. Items loaded onto community 2 are SAS items 2, 6, 8, 10, 16, 18, 19, and 22. Items loaded onto community 3 are SAS items 3, 5, 7, 12, 17, 21, 23, and 24. The output indicates that the network structure is not unidimensional, as multiple communities are present. The SAS network structure has a TEFI of −28.094, which serves as a relative fit index. This index can be used to compare the structural uncertainty of this network with other similar network structures, whether estimated by different methods or after item removal.

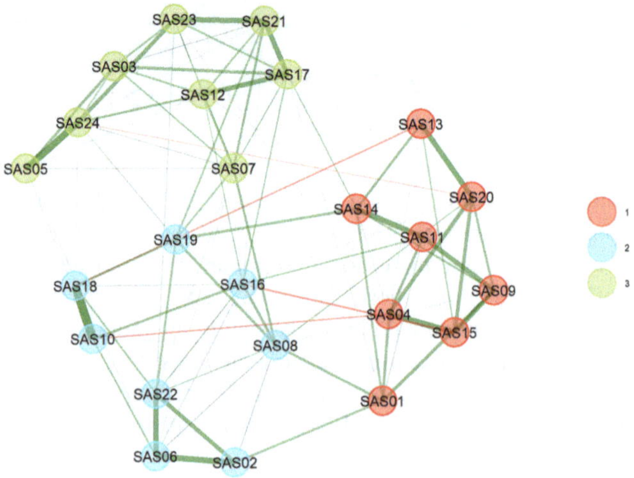

Fig. 1 The EGA network structure of the SAS dataset

3.1.1.1 Output 1

EGA Output of the SAS Dataset

```
> sas_ega
Model: GLASSO (EBIC with gamma = 0.5)
Correlations: auto
Lambda: 0.0891577965736252 (n = 100, ratio = 0.1)

Number of nodes: 24
Number of edges: 110
Edge density: 0.399

Non-zero edge weights:
     M    SD    Min   Max
 0.101 0.112 -0.118 0.521

----

Consensus Method: Highest Modularity (1000 iterations)
Algorithm: Louvain

Order: Higher

Number of communities:  3

SAS01 SAS02 SAS03 SAS04 SAS05 SAS06 SAS07 SAS08 SAS09 SAS10 SAS11 SAS12 SAS13 SAS14 SAS15 SAS16 SAS17 SAS18
    1     2     3     1     3     2     3     2     1     2     1     3     1     1     1     2     3     2
SAS19 SAS20 SAS21 SAS22 SAS23 SAS24
    2     1     3     2     3     3

----

Unidimensional Method: Louvain
Unidimensional: No

----

TEFI: -28.094
```

Beyond a single estimation of the network structure, we can assess the dimensional stability of the SAS network structure across different variations of the dataset using bootstrap EGA. This can be performed with the EGAnet::bootEGA() function. We conduct 600 parametric iterations using the Walktrap algorithm and the graphical lasso model. The typical structure is plotted, and the process utilizes seven cores for parallel computation with progress tracking. The summary() function is then used to obtain a summary of the bootstrapped EGA results.

```
sas_BOOTega <- EGAnet::bootEGA(
  data = df_sas,
  cor = "cor_auto",
  uni.method = "louvain",
  iter = 600,
  type = "parametric",
  EGA.type = "EGA",
  model = "glasso",
  algorithm = "walktrap",
  consensus.method = "highest_modularity",
  typicalStructure = TRUE,
  plot.typicalStructure = TRUE,
  ncores = 7,
  progress = TRUE,
  summary.table = TRUE,
  seed = 1234
)

summary(sas_BOOTega)
```

Output 2 presents a summary of the bootstrap EGA results. Across 600 iterations of bootstrap resampling, the results indicate that a three-dimensional structure was identified in 90.17% of the bootstrap samples, suggesting a high likelihood that the data can be best described by three distinct dimensions. A four-dimensional structure was found in 9.67% of the samples, while a five-dimensional structure appeared in only 0.17% of the samples, indicating that these structures are less stable and less likely to represent the true underlying structure based on the utilized dataset. The median number of dimensions across the bootstrap samples is 3, with a 95% confidence interval ranging from 2.4 to 3.6. This further supports the stability of the three-dimensional structure as the most likely representation of the data.

3.1.1.2 Output 2

Bootstrap EGA Output of the SAS Dataset

```
> summary(sas_BOOTega)
Model: GLASSO (EBIC)
Correlations: cor_auto
Algorithm:  Walktrap
Unidimensional Method:  Louvain
----
EGA Type: EGA
Bootstrap Samples: 600 (Parametric)

                              3                  4                  5
Frequency:  0.901666666666667 0.0966666666666667 0.00166666666666667

Median dimensions: 3 [2.4, 3.6] 95% CI
```

Next, we utilize the EGAnet::dimensionStability() function to analyze the stability of dimensions and items in the bootstrapped EGA results.

```
dim_stability_sas <- EGAnet::dimensionStability(sas_BOOTega)
dim_stability_sas
```

Output 3 and Fig. 2 present the structural stability results of the SAS dataset. The output shows the replication proportion of each item in the three-dimensional model of the SAS dataset. Most items have a replication proportion of 1.00, indicating that they are consistently loaded onto their respective dimensions 100% of the time across all bootstrap iterations. A few variables have slightly lower proportions, such as SAS items 2, 6, and 22, each with a proportion of 0.915, and SAS item 1 with 0.968. The structural consistency results, which represent the proportion of times that each empirical EGA dimension exactly replicates across the bootEGA samples, indicate that the identified dimensions are highly consistent. Dimension 3 is the most stable, with a consistency of 0.978, followed closely by Dimension 1 at 0.968. Dimension 2, while slightly lower in stability, still shows a high consistency at 0.902.

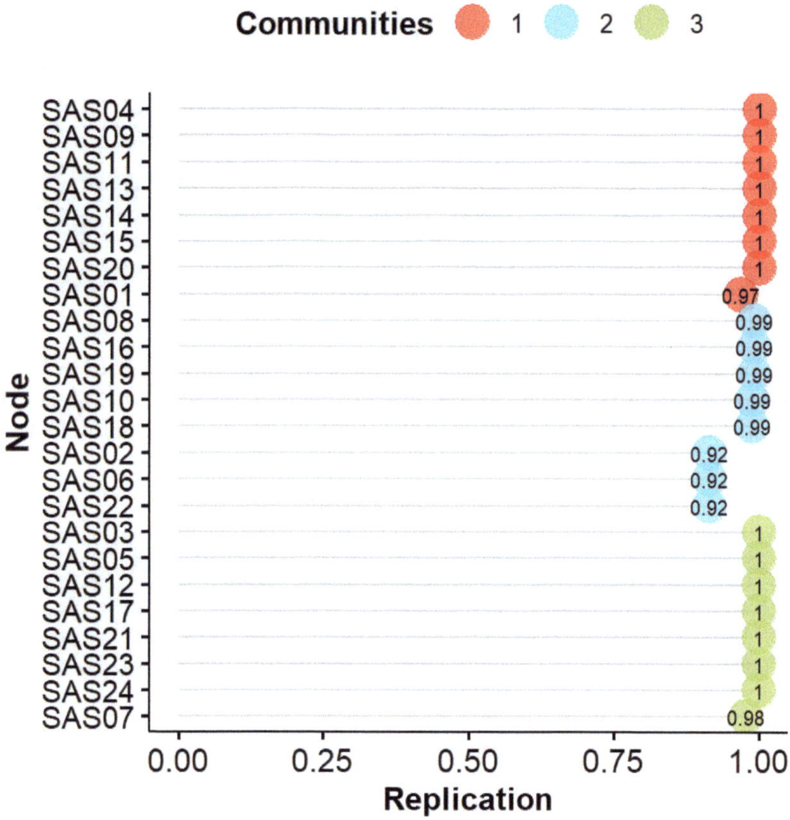

Fig. 2 Item stability results of the SAS dataset

Overall, the dimensional stability results demonstrate that the dimensions identified in the SAS dataset are highly stable and consistent across the bootstrap samples. The high replication proportions and structural consistency values suggest that the dimensional structure is robust, with most variables consistently assigned to the same dimensions across different bootstrap samples. This indicates a robust and reliable factor structure in the data.

3.1.1.3 Output 3

Dimensional Stability Output of the SAS Dataset

```
> dim_stability_sas
EGA Type: EGA
Bootstrap Samples: 600 (Parametric)
Proportion Replicated in Dimensions:
    SAS01     SAS02     SAS03     SAS04     SAS05     SAS06     SAS07     SAS08     SAS09     SAS10
0.9683333 0.9150000 1.0000000 1.0000000 1.0000000 0.9150000 0.9783333 0.9916667 1.0000000 0.9883333
    SAS11     SAS12     SAS13     SAS14     SAS15     SAS16     SAS17     SAS18     SAS19     SAS20
1.0000000 1.0000000 1.0000000 1.0000000 1.0000000 0.9916667 1.0000000 0.9883333 0.9916667 1.0000000
    SAS21     SAS22     SAS23     SAS24
1.0000000 0.9150000 1.0000000 1.0000000
----
Structural Consistency:
        1         2         3
0.9683333 0.9016667 0.9783333
```

After obtaining the network structure of the SAS dataset through EGA, we can transform the EGA object into a CFA object using the CFA function. Subsequently, we can extract the CFA fit indices of the three-dimensional structure of the SAS dataset using the fitmeasures function as follows:

```
sas_ega_cfa <- EGAnet::CFA(ega.obj = sas_ega, estimator = "WLSMV",
                           plot.CFA = TRUE,
                           data = df_sas)

lavaan::fitMeasures(sas_ega_cfa$fit, fit.measures = "all")
```

Output 4 shows the specification of the CFA model. It defines three latent factors (Fat 1, Fat 2, Fat 3) and their corresponding observed variables (e.g., SAS01, SAS04, etc.). This specification is visualized in Fig. 3. The fit summary ($fit) shows the following information: Estimator: DWLS (Diagonally Weighted Least Squares), suitable for ordinal data. Optimization method: NLMINB, an algorithm used for optimization. Number of model parameters: 123 parameters were estimated in the model. Number of observations: 202 participants in the dataset. Number of missing patterns: 1, meaning that there is only 1 datapoint missing. The model demonstrates a good fit according to the CFI and TLI, both of which are above 0.95. However, it shows a poor fit according to the RMSEA (> 0.1) and the SRMR (> 0.8). The significant chi-square test ($p < .001$) suggests that the model's covariance structure significantly differs from the observed covariance structure, which may be due to the large sample size or model complexity.

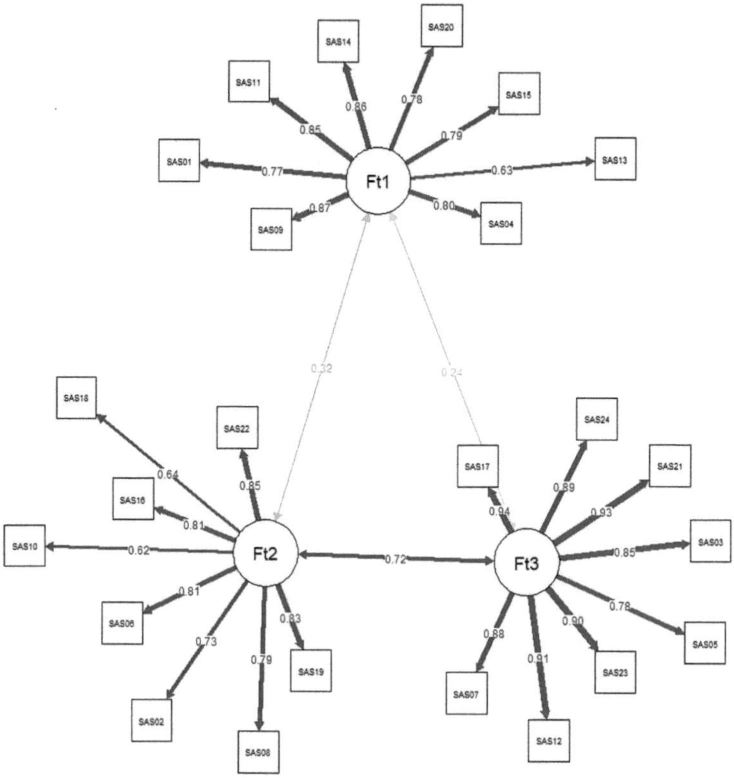

Fig. 3 The factor model of the three-dimensional structure of the SAS dataset

3.1.1.4 Output 4

CFA Model Specification and Fit Indices of the SAS Dataset

```
> sas_ega_cfa
$model
[1] "Fat 1  =~  SAS01 + SAS04 + SAS09 + SAS11 + SAS13 + SAS14 + SAS15 + SAS20
\n Fat 2  =~  SAS02 + SAS06 + SAS08 + SAS10 + SAS16 + SAS18 + SAS19 + SAS22
\n Fat 3  =~  SAS03 + SAS05 + SAS07 + SAS12 + SAS17 + SAS21 + SAS23 + SAS24"

$fit
lavaan 0.6.17 ended normally after 31 iterations

  Estimator                                         DWLS
  Optimization method                             NLMINB
  Number of model parameters                         123

  Number of observations                             202
  Number of missing patterns                           1

Model Test User Model:
                                              Standard      Scaled
  Test Statistic                              1185.503    1076.805
  Degrees of freedom                               249         249
  P-value (Chi-square)                           0.000       0.000
  Scaling correction factor                                  1.269
  Shift parameter                                          142.750
    simple second-order correction

$summary
Length  Class   Mode
     1 lavaan     S4

$fit.measures
chisq.scaled            df        pvalue   cfi.scaled rmsea.scaled
    1076.805       249.000         0.000        0.927        0.129


attr(,"class")
[1] "CFA"
> lavaan::fitMeasures(sas_ega_cfa$fit, fit.measures = c("chisq", "df", "pvalue",
"cfi", "tli", "rmsea", "srmr"))
   chisq        df   pvalue       cfi       tli     rmsea      srmr
1185.503   249.000    0.000     0.980     0.977     0.137     0.125
```

3.1.2 Hierarchical EGA

For the Hierarchical EGA task, we start by loading the synthesized SPOT dataset [5] into our R environment by loading the accompanying R object of this chapter as follows:

```
load("LAMethod_EGA.RData")
```

Next, similarly to the previous EGA section, we performed hierarchical EGA on the synthetic SPOT data using the Louvain algorithm and graphical lasso model. The correlation method was set to auto-detection, allowing the analysis to determine the appropriate correlations: Pearson's for continuous data, polychoric for ordinal data, tetrachoric for binary data, and polyserial/biserial for ordinal/binary data with continuous data. The analysis iterated 1000 times to achieve consensus using the highest modularity method. The network was plotted, and the process was seeded for reproducibility. The hierarchical EGA results can be accessed by calling the hierarchical EGA object created during the analysis.

```
spot_Hierega <- EGAnet::hierEGA(
  data = spot_synth,
  algorithm = "louvain",
  model = "glasso",
  corr = "auto",
  consensus.iter = 1000,
  consensus.method = c("highest_modularity"),
  plot.EGA = TRUE,
  seed = 1234
)
spot_Hierega
```

Output 5 and Fig. 4 present results from the hierarchical EGA of the SPOT dataset using the GLASSO method applied at both lower and higher orders of the SPOT structure. For the lower-order analysis, the algorithm uses the Louvain algorithm for community detection over 1000 iterations. The algorithm identifies six communities, each comprising three variables (e.g., D1, D2, and D3 form one community). The analysis indicates the lower-order network has a TEFI of 4.48. For the higher-order analysis, the algorithm also uses the Louvain algorithm for community detection. A single community encompassing all six variables (i.e., six lower-order dimensions) was identified, indicating the network is unidimensional. The TEFI for this higher-order structure is −26.591, with a generalized TEFI of −22.111. Comparing the TEFI values, the higher-order (bifactor) structure demonstrates a better fit (−26.591) than the correlated lower-order factor structure (4.48), suggesting a single underlying dimension more effectively explains the data structure.

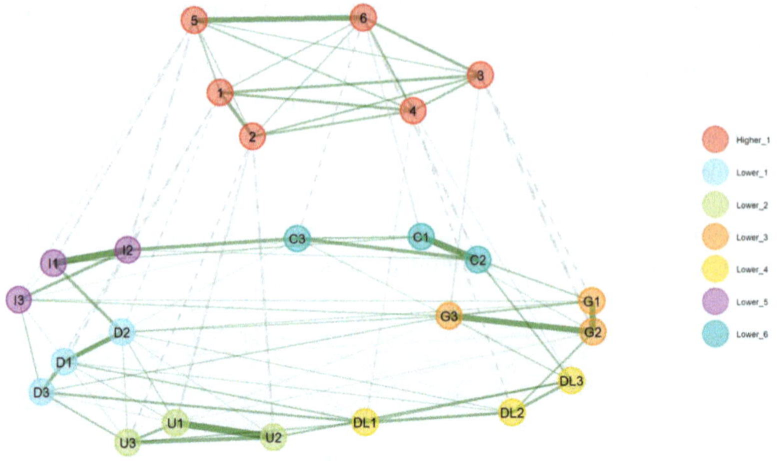

Fig. 4 The hierarchical EGA network structure of the SPOT dataset

3.1.2.1 Output 5

Hierarchical EGA Output of the SPOT Dataset

```
> spot_Hierega
Lower Order

Model: GLASSO (EBIC with gamma = 0.5)
Correlations: auto
Lambda: 0.0766735104688791 (n = 100, ratio = 0.1)

Number of nodes: 18
Number of edges: 85
Edge density: 0.556

Non-zero edge weights:
     M    SD   Min   Max
 0.097 0.103 0.003 0.452


----

Consensus Method: Most Common (1000 iterations)
Algorithm: Louvain
Order: Lower

Number of communities:  6

 D1  D2  D3  U1  U2  U3  G1  G2  G3 DL1 DL2 DL3  I1  I2  I3  C1  C2  C3
  1   1   1   2   2   2   3   3   3   4   4   4   5   5   5   6   6   6

----

Unidimensional Method: Louvain
Unidimensional: No

----

TEFI: 4.48

------------
Higher Order

Model: GLASSO (EBIC with gamma = 0.5)
Correlations: auto
Lambda: 0.0597492023524303 (n = 100, ratio = 0.1)

Number of nodes: 6
Number of edges: 15
Edge density: 1.000

Non-zero edge weights:
     M    SD   Min   Max
 0.161 0.067 0.090 0.335

----

Algorithm:  Louvain

Number of communities:  1

1 2 3 4 5 6
1 1 1 1 1 1

----

Unidimensional Method: Louvain
Unidimensional: Yes

----

TEFI: -26.591

----

Generalized TEFI: -22.111

----
Interpretation: Based on lower (4.48) and higher (-26.591) TEFI, there is a better
fit for a bifactor structure than a correlated lower-order factor structure
```

Next, we can assess the stability of the SPOT hierarchical network structure across different datasets using bootstrap hierarchical EGA, similar to the ordinary EGA method. This can be achieved as follows:

```
spot_BOOTega <- EGAnet::bootEGA(
  # we could also provide the cor matrix but then
  # n (i.e., number of rows) must also be specified
  data = spot_synth,
  cor = "cor_auto",
  uni.method = "louvain",
  iter = 600, # Number of replica samples to generate
  # resampling" for n random subsamples of the original data
  # parametric" for n synthetic samples from multivariate normal dist.
  type = "parametric",
  # EGA Uses standard exploratory graph analysis
  # EGA.fit Uses total entropy fit index (tefi) to determine best fit of EGA
  # hierEGA Uses hierarchical exploratory graph analysis
  EGA.type = "hierEGA",
  model = "glasso",
  algorithm = "walktrap", # or "louvain" (better for unidimensional structures)
  # use "highest_modularity", "most_common", or "lowest_tefi"
  consensus.method = "highest_modularity",
  typicalStructure = TRUE, # typical network of partial correlations
  plot.typicalStructure = TRUE, # returns a plot of the typical network
  ncores = 7,
  progress = TRUE ,
  summary.table    = TRUE,
  seed = 1234
)

summary(spot_BOOTega)
```

Output 6 shows the results of the bootstrap hierarchical EGA with the SPOT dataset. Across 600 iterations of bootstrap resampling, bootstrap hierarchical EGA found that the median structure of the SPOT dataset at the lower level has six communities, with a 94% frequency of structural replication. In other words, the BootEGA algorithm was able to replicate the six-dimensional structure 94% of the time, the five-dimensional structure 4% of the time, and the seven-dimensional structure 2% of the time across all bootstrap iterations. For the higher-order analysis, bootstrap hierarchical EGA found that the higher-order network is consistently unidimensional across all bootstrap samples, with a frequency of 1, indicating that a single dimension was found in 100% of the bootstrap samples. The median number of dimensions is also one. Overall, the results suggest that while the lower-order analysis identifies multiple distinct communities (dimensions) within the data, the higher-order analysis indicates that these lower-order dimensions can be explained by a single overarching dimension. This hierarchical structure suggests a complex data structure at the lower level, simplified by a single factor at the higher level.

3.1.2.2 Output 6

Bootstrap Hierarchical EGA Output of the SPOT Dataset

```
> summary(spot_BOOTega)
EGA Type: hierEGA
Bootstrap Samples: 600 (Parametric)

-----------

Lower Order

Model: GLASSO (EBIC)
Correlations: cor_auto
Consensus Method: Most Common (1000 iterations)
Algorithm: Louvain
Order: Lower

Number of communities: 6

 D1  D2  D3  U1  U2  U3  G1  G2  G3 DL1 DL2 DL3  I1  I2  I3  C1  C2  C3
  1   1   1   2   2   2   3   3   3   4   4   4   5   5   5   6   6   6

Unidimensional Method:  Louvain

----

              5    6    7
Frequency:  0.04 0.94 0.02

Median dimensions: 6 [5.52, 6.48] 95% CI

-----------

Higher Order

Algorithm:  Louvain

----

            1
Frequency:  1

Median dimensions: 1 [1, 1] 95% CI
```

The EGAnet::dimensionStability function can also be applied to the hierarchical EGA object to analyze the stability of measurement dimensions and items.

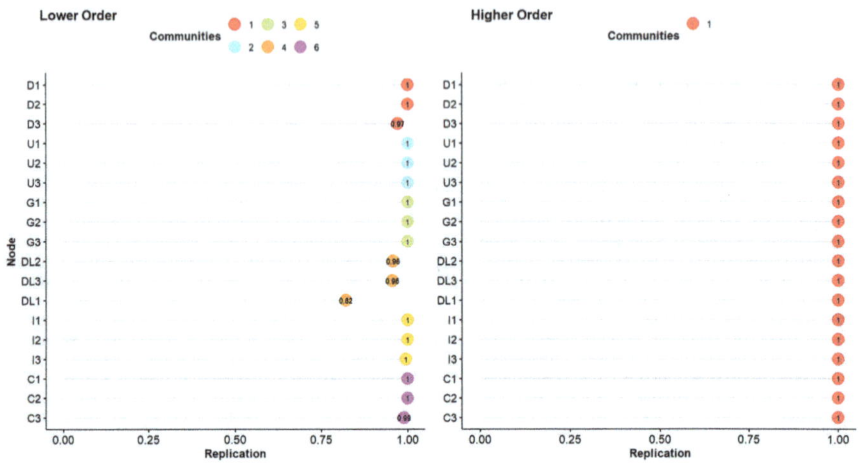

Fig. 5 Item stability results for the SPOT dataset

```
dim_stability_spot <- EGAnet::dimensionStability(spot_BOOTega)
dim_stability_spot
```

Output 7 and Fig. 5 show the structural stability results of the SPOT dataset. The output shows the replication proportion of each item in the six-dimension model of the SPOT dataset. The analysis shows that for the lower-order dimensions, most variables were consistently replicated with proportions close to or at 1, indicating high consistency. Specifically, variables D1, D2, U1, U2, U3, G1, G2, G3, I1, I2, C1, and C2 each had a replication proportion of 1. Variables D3, DL1, DL2, DL3, I3, and C3 had slightly lower proportions, ranging from approximately 0.822 to 0.995, indicating high but not perfect consistency. In the higher-order analysis, all variables were perfectly replicated within the single identified dimension, with proportions of 1, demonstrating complete consistency.

In terms of structural consistency, the lower-order dimensions showed high stability, with values indicating near-perfect stability for Dimensions 1, 4, and 6 and perfect stability for Dimensions 2, 3, and 5. The higher-order analysis revealed perfect structural consistency with a value of 1, indicating that the single higher-order dimension is entirely stable across bootstrap samples. Overall, the analysis highlights the robust higher-order structure that simplifies the complex relationships observed in the lower-order dimensions, with minor variability in the lower-order dimensions suggesting slight but non-critical instability.

3.1.2.3 Output 7

Dimensional Stability Output of the SPOT Dataset

```
> dim_stability_spot
EGA Type: hierEGA
Bootstrap Samples: 600 (Parametric)

Proportion Replicated in Dimensions:

       D1 D2       D3 U1 U2 U3 G1 G2 G3       DL1       DL2       DL3 I1 I2    I3 C1 C2   C3
Lower   1  1 0.9716667  1  1  1  1  1  1 0.8216667 0.9583333 0.9566667  1  1 0.995  1  1 0.99
Higher  1  1 1.0000000  1  1  1  1  1  1 1.0000000 1.0000000 1.0000000  1  1 1.000  1  1 1.00
----
Structural Consistency:
Lower Order
        1         2         3         4         5         6
0.9716667 1.0000000 1.0000000 0.8200000 0.9950000 0.9900000

Higher Order
1
1
```

Similarly to the EGA demonstration, the hierarchical EGA object can also be transformed into a CFA object and extract CFA fit measures with the following codes:

```
spot_ega_cfa <- EGAnet::CFA(ega.obj = spot_Hierega, estimator = "WLSMV", plot.CFA =
TRUE, data = spot_synth)

lavaan::fitMeasures(spot_ega_cfa$fit, fit.measures = c("chisq", "df", "pvalue",
"cfi", "tli", "rmsea", "srmr"))
```

Output 8 shows the specification of the SPOT CFA model. The model specifies six first-order factors, each measured by 3 observed variables and 1 second-order factor indicated by all the observed variables from the first-order factors. This specification is visualized in Fig. 6. The analysis used the DWLS estimator and converged normally after 57 iterations, estimating 108 parameters from 695 observations with 15 points of missing data. The model has a good fit according to the CFI and TLI (both need to be > 0.95), RMSEA (< 0.5), and SRMR (< 0.8). The significant chi-square test (p-value < .001) suggests that the model's covariance structure significantly differs from the observed covariance structure, which might be due to the large sample size or model complexity.

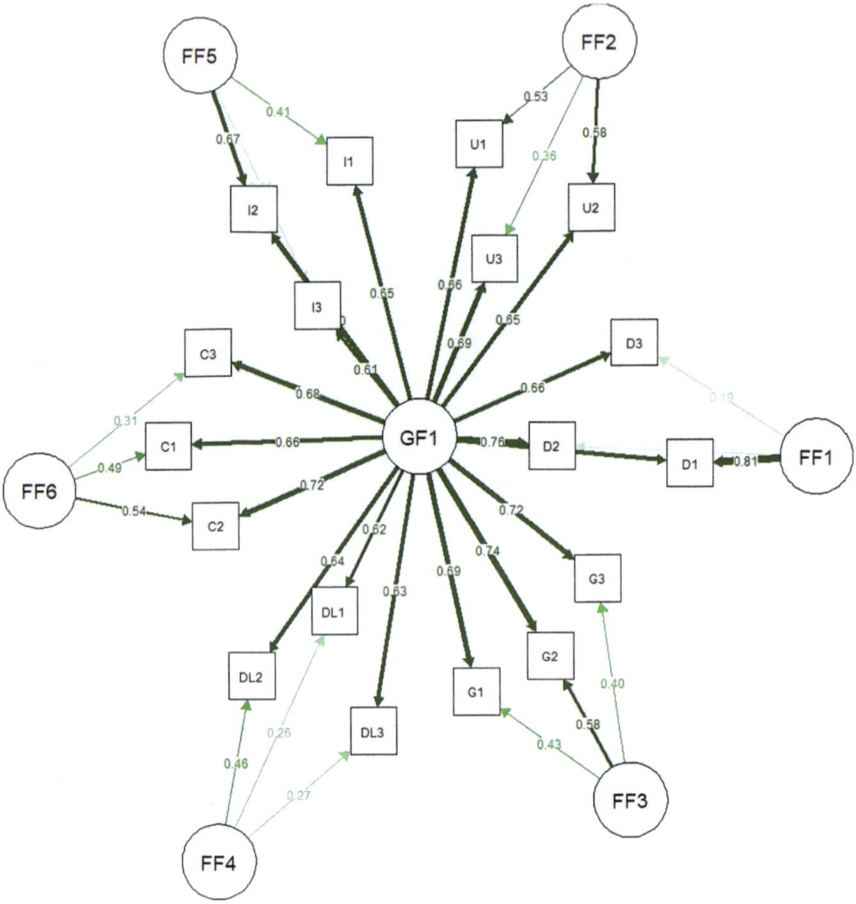

Fig. 6 The factor model of the hierarchical structure of the SPOT dataset

3.1.2.4 Output 8

CFA Model Specification and Fit Indices of the SPOT Dataset

```
> spot_ega_cfa
$model
[1] "FirstFact 1  =~  D1 + D2 + D3
\n FirstFact 2  =~  U1 + U2 + U3
\n FirstFact 3  =~  G1 + G2 + G3
\n FirstFact 4  =~  DL1 + DL2 + DL3
\n FirstFact 5  =~  I1 + I2 + I3
\n FirstFact 6  =~  C1 + C2 + C3
\n GenFact 1  =~  D1 + D2 + D3 + U1 + U2 + U3 + G1 + G2 + G3 + DL1 + DL2 + DL3 + I1
+ I2 + I3 + C1 + C2 + C3"

$fit
lavaan 0.6.17 ended normally after 57 iterations

  Estimator                                         DWLS
  Optimization method                             NLMINB
  Number of model parameters                         108

  Number of observations                             695
  Number of missing patterns                          15

Model Test User Model:
                                              Standard      Scaled
  Test Statistic                               290.789     509.012
  Degrees of freedom                               117         117
  P-value (Chi-square)                           0.000       0.000
  Scaling correction factor                                  0.613
  Shift parameter                                           34.381
    simple second-order correction

$summary
Length  Class   Mode
     1 lavaan     S4

$fit.measures
chisq.scaled          df      pvalue   cfi.scaled rmsea.scaled
     509.012     117.000       0.000        0.962        0.069

attr(,"class")
[1] "CFA"
> lavaan::fitMeasures(spot_ega_cfa$fit, fit.measures = c("chisq", "df", "pvalue",
"cfi", "tli", "rmsea", "srmr"))
  chisq      df  pvalue     cfi     tli   rmsea    srmr
290.789 117.000   0.000   0.996   0.994   0.046   0.043
```

3.1.3 Dynamic EGA

In addition to assessing the static structure of a dataset, the dynamic EGA method can be utilized to examine changes in the structure of a variable system across multiple units of assessment, such as population, group, and individual cases. This task is demonstrated through the "dynEGA" function in the EGAnet package [13]. As previously mentioned, we utilize a built-in simulated dataset from the EGAnet package due to the limited availability of multivariate time series data with repeated measures in the educational context. We begin by loading and examining the dataset as follows:

```
> data("sim.dynEGA")
> str(sim.dynEGA)
'data.frame':5000 obs. of  26 variables:
 $ V1   : num  -0.314 -0.637 -0.203 0.192 -0.439 ...
 $ V2   : num  -0.311 -0.669 -0.249 0.119 -0.322 ...
 $ V3   : num  -0.4174 -0.5696 -0.0988 -0.0729 -0.1215 ...
 $ V4   : num  -0.366 -0.441 -0.354 0.182 -0.383 ...
 $ V5   : num  -0.305 -0.377 -0.375 0.411 -0.365 ...
 $ V6   : num  -0.37067 -0.48413 -0.20569 -0.00225 -0.38813 ...
 $ V7   : num  -0.241 -0.628 -0.24 0.194 -0.293 ...
 $ V8   : num  -0.4022 -0.6313 -0.2276 0.199 -0.0497 ...
 $ V9   : num  -0.4781 -0.5994 -0.2884 0.0907 -0.4106 ...
 $ V10  : num  -0.337 -0.468 -0.102 0.185 -0.237 ...
 $ V11  : num  -0.3051 -0.3556 -0.2726 0.0269 -0.2921 ...
 $ V12  : num  -0.595 -0.669 -0.225 0.447 -0.339 ...
 $ V13  : num  0.286 0.7 0.678 0.753 0.395 ...
 $ V14  : num  0.123 0.655 0.63 0.538 0.497 ...
 $ V15  : num  0.416 0.537 0.803 0.727 0.46 ...
 $ V16  : num  0.338 0.645 0.963 0.622 0.505 ...
 $ V17  : num  0.556 0.659 0.669 0.686 0.355 ...
 $ V18  : num  0.43 0.681 0.958 0.97 0.598 ...
 $ V19  : num  0.539 0.645 0.885 0.564 0.316 ...
 $ V20  : num  0.651 0.82 0.702 0.528 0.303 ...
 $ V21  : num  0.344 0.649 0.833 0.853 0.328 ...
 $ V22  : num  0.286 0.595 0.881 0.505 0.121 ...
 $ V23  : num  0.532 0.512 0.536 0.675 0.3 ...
 $ V24  : num  0.269 0.442 0.68 0.702 0.418 ...
 $ ID   : num  1 1 1 1 1 1 1 1 1 1 ...
 $ Group: num  1 1 1 1 1 1 1 1 1 1 ...
```

To contextualize the variables and the network structure, we assign educational names to each variable. This can be achieved using the following codes:

```
# Define the new column abbreviations
new_column_names <- c(
  "TOT",  # Time on Task
  "QS",   # Quiz Scores
  "AS",   # Assignment Scores
  "CP",   # Class Participation
  "AT",   # Attendance
  "DC",   # Discussion Contributions
  "RCS",  # Reading Comprehension Scores
  "HCR",  # Homework Completion Rate
  "PS",   # Project Scores
  "LP",   # Lab Performance

  "ES",   # Exam Scores
  "PRS",  # Peer Review Scores
  "SAS",  # Self-Assessment Scores
  "ILM",  # Interaction with Learning Materials
  "ULR",  # Use of Learning Resources
  "HSB",  # Help-Seeking Behavior
  "ML",   # Motivation Levels
  "EGW",  # Engagement in Group Work
  "RSR",  # Reflection and Self-Regulation
  "KA",   # Knowledge Application
  "SD",   # Skill Development
  "FU",   # Feedback Utilization
  "CS",   # Communication Skills
  "CPS",   # Creativity in Problem Solving
  "ID", #ID
  "Group" #Group
)

names(sim.dynEGA) <- new_column_names
names(sim.dynEGA)

 [1] "TOT"   "QS"    "AS"    "CP"    "AT"    "DC"    "RCS"   "HCR"   "PS"    "LP"    "ES"    "PRS"   "SAS"
[14] "ILM"   "ULR"   "HSB"   "ML"    "EGW"   "RSR"   "KA"    "SD"    "FU"    "CS"    "CPS"   "ID"    "Group"
```

The meaning of each variable is outlined in Table 1. The dataset has 24 variables in total, with one column for respondent ID and another column for group information of each respondent.

After finishing processing the dataset, we can perform dynamic EGA on simulated data at individual, group, and population levels using the Louvain algorithm and Pearson correlation as follows:

```
simulated_all <- dynEGA(
  data = sim.dynEGA,
  level = c("individual", "group", "population"),
  ncores = 6, # use more for quicker results
  verbose = TRUE, # progress bar
  corr = "pearson",
  algorithm = "louvain",
  id = "ID",
  group = "Group"
)
```

Table 1 List of utilized variables in the dynamic EGA task

Variable code	Variable name	Variable description
TOT	Time on task	The amount of time spent on a particular task or activity.
QS	Quiz scores	The scores obtained on quizzes.
AS	Assignment scores	The scores obtained on assignments.
CP	Class participation	The level of participation in class activities.
AT	Attendance	The frequency of attendance in classes or sessions.
DC	Discussion contributions	The number of contributions made in discussion forums.
RCS	Reading comprehension scores	The scores obtained on reading comprehension tests.
HCR	Homework completion rate	The rate at which homework is completed.
PS	Project scores	The scores obtained on projects.
LP	Lab performance	The performance in laboratory exercises or practicals.
ES	Exam scores	The scores obtained in exams.
PRS	Peer review scores	The scores given by peers in peer-reviewed activities.
SAS	Self-assessment scores	The scores from self-assessment activities.
ILM	Interaction with learning materials	The frequency or quality of interaction with learning materials.
ULR	Use of learning resources	The extent of use of additional learning resources.
HSB	Help-seeking behavior	The frequency of seeking help from instructors or peers.
ML	Motivation levels	The levels of motivation towards the learning activities.
EGW	Engagement in group work	The level of engagement in group work or collaborative activities.
RSR	Reflection and self-regulation	The frequency and quality of reflective practices and self-regulation.
KA	Knowledge application	The ability to apply knowledge in practical or real-world scenarios.
SD	Skill development	The progress in skill development over time.
FU	Feedback utilization	The extent to which feedback is utilized to improve performance.
CS	Communication skills	The proficiency in communication skills within the learning context.
CPS	Creativity in problem solving	The level of creativity demonstrated in problem-solving activities.

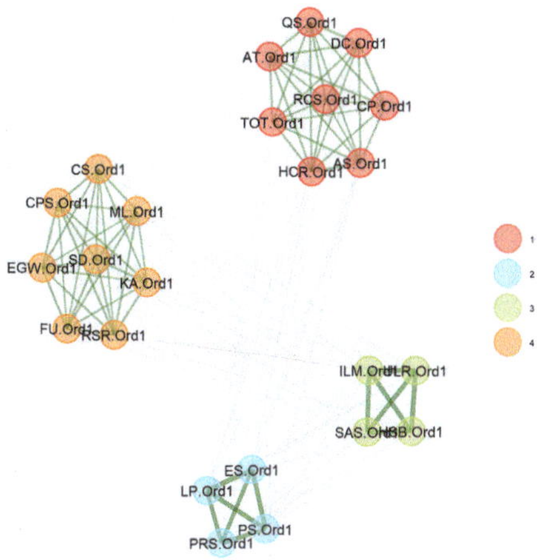

Fig. 7 The population-level network structure network structure of the simulated dataset

Output 9 and Fig. 7 show the extraction of the dynamic EGA results at the population level. The analysis, conducted with Pearson correlations and a regularization parameter (lambda) of approximately 0.095, involves 24 variables (nodes) and 156 edges, resulting in an edge density of 0.565. The non-zero edge weights have a mean of 0.076, a standard deviation of 0.085, and range from 0.000 to 0.290, indicating various strengths of connections between the variables. Furthermore, the analysis identified four distinct communities within the network. The community assignments are as follows: TOT.Ord1, QS.Ord1, AS.Ord1, CP.Ord1, AT.Ord1, DC.Ord1, RCS.Ord1, and HCR.Ord1 form Community 1; PS.Ord1, LP.Ord1, and ES.Ord1 form Community 2; SAS.Ord1, ILM.Ord1, ULR.Ord1, and HSB.Ord1 form Community 3; and ML.Ord1, EGW.Ord1, RSR.Ord1, KA.Ord1, SD.Ord1, FU.Ord1, CS.Ord1, and CPS.Ord1 form Community 4. The population-level structure has TEFI = −19.086.

3.1.3.1 Output 9

Population-Level Dynamic EGA Output of the Simulated Data

```
> simulated_all$dynEGA$population
Population

Model: GLASSO (EBIC with gamma = 0.5)
Correlations: pearson
Lambda: 0.0951546624930215 (n = 100, ratio = 0.1)

Number of nodes: 24
Number of edges: 156
Edge density: 0.565


Non-zero edge weights:
     M    SD   Min   Max
 0.076 0.085 0.000 0.290
----
Consensus Method: Most Common (1000 iterations)
Algorithm: Louvain
Order: Higher

Number of communities:  4

TOT.Ord1  QS.Ord1  AS.Ord1  CP.Ord1  AT.Ord1  DC.Ord1 RCS.Ord1 HCR.Ord1  PS.Ord1  LP.Ord1  ES.Ord1 PRS.Ord1
       1        1        1        1        1        1        1        1        2        2        2        2
SAS.Ord1 ILM.Ord1 ULR.Ord1 HSB.Ord1  ML.Ord1 EGW.Ord1 RSR.Ord1  KA.Ord1  SD.Ord1  FU.Ord1  CS.Ord1 CPS.Ord1
       3        3        3        3        4        4        4        4        4        4        4        4
----
Unidimensional Method: Louvain
Unidimensional: No


----
TEFI: -19.086
plot(simulated_all$dynEGA$population)
```

Output 10 and Fig. 8 show the extraction of the dynamic EGA results at the group level. For Group 1, with Pearson correlations and a regularization parameter (lambda) of approximately 0.095, the analysis includes 24 variables (nodes) and 153 edges, resulting in an edge density of 0.554. The non-zero edge weights have a mean of 0.078, a standard deviation of 0.031, and a range from 0.000 to 0.111. Community detection via the Louvain algorithm identified two distinct communities: the first community includes variables TOT.Ord1 through PRS.Ord1, while the second community comprises variables SAS.Ord1 through CPS.Ord1. The TEFI value of group 1 network structure is −28.646.

For Group 2, the analysis similarly uses Pearson correlations and a lambda of approximately 0.095, encompassing 24 nodes and 145 edges with an edge density of 0.525. The non-zero edge weights in this group have a mean of 0.082, a standard deviation of 0.067, and range from 0.000 to 0.167. The Louvain algorithm identified

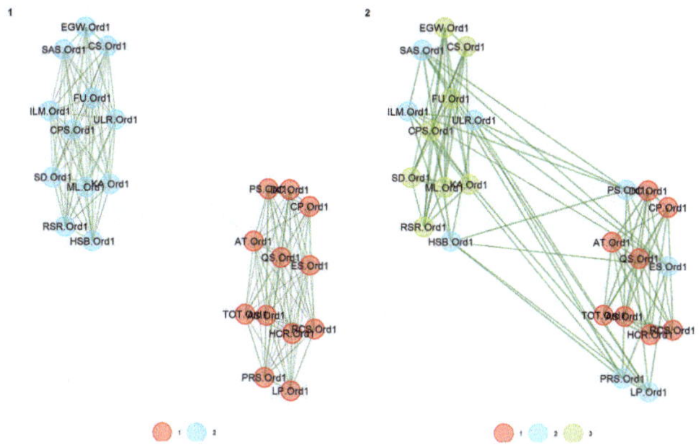

Fig. 8 The group-level network structure network structure of the simulated dataset *Note*. The number in the upper left corner represents the subgroup within the population

three communities: the first community includes variables TOT.Ord1 through RCS.Ord1, the second community consists of variables PS.Ord1 through CPS.Ord1, and the third community contains variables ML.Ord1 through CPS.Ord1. The TEFI value of group 1 network structure is −27.93.

Overall, the dynEGA results for both groups reveal multidimensional network structures. Group 1 exhibits a slightly higher edge density, while Group 2 shows greater variability in edge weights and a slightly higher TEFI. This suggests that the three-dimensional structure of Group 2 may have a somewhat less stable network structure compared to the two-dimensional structure of Group 1.

3.1.3.2 Output 10

Group-Level Dynamic EGA Output of the Simulated Data

```
> simulated_all$dynEGA$group$"1"  #Group1
Model: GLASSO (EBIC with gamma = 0.5)
Correlations: pearson
Lambda: 0.0951050823970087 (n = 100, ratio = 0.1)
Number of nodes: 24
Number of edges: 153
Edge density: 0.554
Non-zero edge weights:
     M    SD   Min   Max
 0.078 0.031 0.000 0.111
----
Consensus Method: Most Common (1000 iterations)
Algorithm: Louvain
Order: Higher
Number of communities:  2
TOT.Ord1  QS.Ord1  AS.Ord1  CP.Ord1  AT.Ord1  DC.Ord1 RCS.Ord1 HCR.Ord1  PS.Ord1  LP.Ord1  ES.Ord1 PRS.Ord1
       1        1        1        1        1        1        1        1        1        1        1        1
SAS.Ord1 ILM.Ord1 ULR.Ord1 HSB.Ord1  ML.Ord1 EGW.Ord1 RSR.Ord1  KA.Ord1  SD.Ord1  FU.Ord1  CS.Ord1 CPS.Ord1
       2        2        2        2        2        2        2        2        2        2        2        2
----
Unidimensional Method: Louvain
Unidimensional: No
----
TEFI: -28.646

> simulated_all$dynEGA$group$"2" #Group2
Model: GLASSO (EBIC with gamma = 0.5)
Correlations: pearson
Lambda: 0.0952552436606787 (n = 100, ratio = 0.1)
Number of nodes: 24
Number of edges: 145
Edge density: 0.525
Non-zero edge weights:
     M    SD   Min   Max
 0.082 0.067 0.000 0.167
----
Consensus Method: Most Common (1000 iterations)
Algorithm: Louvain
Order: Higher
Number of communities:  3
TOT.Ord1  QS.Ord1  AS.Ord1  CP.Ord1  AT.Ord1  DC.Ord1 RCS.Ord1 HCR.Ord1  PS.Ord1  LP.Ord1  ES.Ord1 PRS.Ord1
       1        1        1        1        1        1        1        1        2        2        2        2
SAS.Ord1 ILM.Ord1 ULR.Ord1 HSB.Ord1  ML.Ord1 EGW.Ord1 RSR.Ord1  KA.Ord1  SD.Ord1  FU.Ord1  CS.Ord1 CPS.Ord1
       2        2        2        2        3        3        3        3        3        3        3        3
----
Unidimensional Method: Louvain
Unidimensional: No
----
TEFI: -27.93
plot(simulated_all$dynEGA$group)
```

Output 11 and Fig. 9 show the extraction of the dynamic EGA results at the individual -level, specifically for individual number 1 in the dataset. The analysis employs a regularization parameter (lambda) of approximately 0.098. The network consists of 24 variables (nodes) and 127 edges, resulting in an edge density of 0.460. The non-zero edge weights have a mean of 0.093, a standard deviation of 0.055, and a range from −0.001 to 0.262. Community detection using the Louvain algorithm identified two distinct communities. The first community includes variables TOT.Ord1 through PRS.Ord1, while the second community comprises variables SAS.Ord1 through CPS.Ord1. The two communities network of individual 1 has the TEFI value of −30.637.

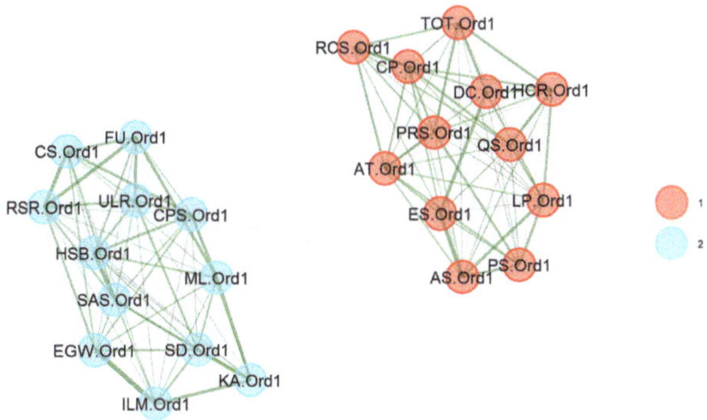

Fig. 9 The individual-level network structure network structure of the simulated dataset: Case 1

3.1.3.3 Output 11

Individual-Level Dynamic EGA Output of the Simulated Data: Case 1

```
> #Individual whose network has two communities
> simulated_all$dynEGA$individual$"1"
Model: GLASSO (EBIC with gamma = 0.5)
Correlations: pearson
Lambda: 0.0981642729449541 (n = 100, ratio = 0.1)

Number of nodes: 24
Number of edges: 127
Edge density: 0.460


Non-zero edge weights:
     M    SD    Min   Max
 0.093 0.055 -0.001 0.262


----

Consensus Method: Most Common (1000 iterations)
Algorithm: Louvain
Order: Higher

Number of communities:  2

TOT.Ord1  QS.Ord1  AS.Ord1  CP.Ord1  AT.Ord1  DC.Ord1 RCS.Ord1 HCR.Ord1  PS.Ord1  LP.Ord1  ES.Ord1 PRS.Ord1
       1        1        1        1        1        1        1        1        1        1        1        1
SAS.Ord1 ILM.Ord1 ULR.Ord1 HSB.Ord1  ML.Ord1 EGW.Ord1 RSR.Ord1  KA.Ord1  SD.Ord1  FU.Ord1  CS.Ord1 CPS.Ord1
       2        2        2        2        2        2        2        2        2        2        2        2


----
Unidimensional Method: Louvain
Unidimensional: No
----

TEFI: -30.637
> plot(simulated_all$dynEGA$individual, id = 1)
```

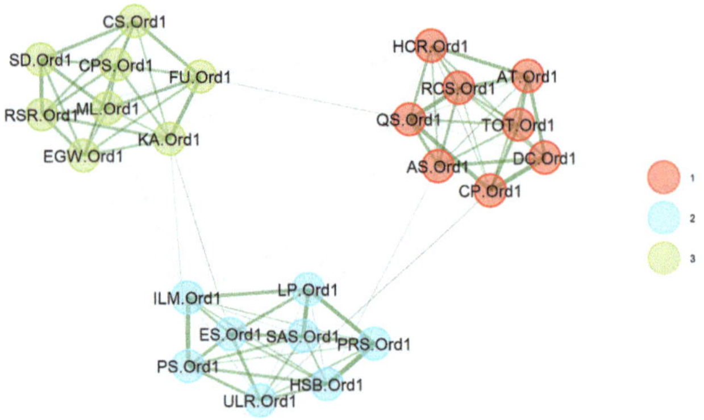

Fig. 10 The individual-level network structure network structure of the simulated dataset: Case 75

Output 12 and Fig. 10 show the extraction of the dynamic EGA results at the individual level, specifically for individual number 75 in the dataset. The analysis employs a regularization parameter (lambda) of approximately 0.097. This individual's network comprises 24 variables (nodes) and 104 edges, leading to an edge density of 0.377. The non-zero edge weights have a mean of 0.113, a standard deviation of 0.074, and a range from -0.024 to 0.258. The Louvain algorithm identified three distinct communities. The first community includes variables TOT.Ord1 through PRS.Ord1, the second community includes variables SAS.Ord1 through CPS.Ord1, and the third community consists of variables ML.Ord1 through CPS.Ord1. The three-community network has a TEFI value of -24.445.

In summary, the population-level analysis revealed a denser network with four communities, indicated by its highest edge density. In contrast, the group-level analyses showed variations in edge density and community structure: Group 1 exhibited two communities, whereas Group 2 had three. At the individual level, network structures were less dense and varied in the number of communities, reflecting individual differences in network connectivity and structure. Comparing the TEFI values across all units of measurement, the network structure of Case 1 appears to be the most robust (TEFI $= -30.637$), followed by Group 1 (TEFI $= -28.646$), Group 2 (TEFI $= -27.93$), Case 75 (TEFI $= -24.445$), and the population-level structure as relatively less stable (TEFI $= -19.086$). This variability may stem from Case 1 and Group 1 having only two communities, limiting the potential communities that can be loaded, whereas the population-level structure's four communities allow for more loading possibility loading across multiple communities, resulting in lower stability as indicated by its TEFI value.

3.1.3.4 Output 12

Individual-Level Dynamic EGA Output of the Simulated Data: Case 75

```
> #Individual whose network has three communities
> simulated_all$dynEGA$individual$"75"
Model: GLASSO (EBIC with gamma = 0.5)
Correlations: pearson
Lambda: 0.0966757001319168 (n = 100, ratio = 0.1)

Number of nodes: 24
Number of edges: 104
Edge density: 0.377

Non-zero edge weights:
     M    SD    Min   Max
 0.113 0.074 -0.024 0.258

----

Consensus Method: Most Common (1000 iterations)
Algorithm: Louvain

Order: Higher

Number of communities:  3

TOT.Ord1  QS.Ord1  AS.Ord1  CP.Ord1  AT.Ord1  DC.Ord1 RCS.Ord1 HCR.Ord1  PS.Ord1  LP.Ord1  ES.Ord1 PRS.Ord1
       1        1        1        1        1        1        1        1        2        2        2        2
SAS.Ord1 ILM.Ord1 ULR.Ord1 HSB.Ord1  ML.Ord1 EGW.Ord1 RSR.Ord1  KA.Ord1  SD.Ord1  FU.Ord1  CS.Ord1 CPS.Ord1
       2        2        2        2        3        3        3        3        3        3        3        3

----

Unidimensional Method: Louvain
Unidimensional: No

----

TEFI: -24.445
> plot(simulated_all$dynEGA$individual, id = 75)
```

3.1.4 Identifying Redundant Nodes with Unique Variable Analysis

In addition to examining the structure of a dataset, the EGAnet package can identify redundant nodes that do not contribute meaningfully to the network structure, thereby reducing complexity. The UVA method identifies redundant nodes for researchers to address. It includes actions such as computing latent variables using CFA when there are three or more redundant variables, calculating the mean of redundant variables, removing all but one variable from a set of redundant variables based on wTO, and computing the sum of redundant variables. This task is demonstrated using the EGAnet::UVA function on the STARS dataset [6]. We begin by loading and examining the dataset as follows:

```
> names(df_stars)
 [1] "STARS01" "STARS02" "STARS03" "STARS04" "STARS05" "STARS06" "STARS07" "STARS08" "STARS09" "STARS10"
[11] "STARS11" "STARS12" "STARS13" "STARS14" "STARS15" "STARS16" "STARS17" "STARS18" "STARS19" "STARS20"
[21] "STARS21" "STARS22" "STARS23" "STARS24" "STARS25" "STARS26" "STARS27" "STARS28" "STARS29" "STARS30"
[31] "STARS31" "STARS32" "STARS33" "STARS34" "STARS35" "STARS36" "STARS37" "STARS38" "STARS39" "STARS40"
[41] "STARS41" "STARS42" "STARS43" "STARS44" "STARS45" "STARS46" "STARS47" "STARS48" "STARS49" "STARS50"
[51] "STARS51"
```

Next, we apply the UVA method to identify redundant variables and extract its results as follows:

```
stars_uva <- EGAnet::UVA(
  data = df_stars,
  reduce.method = "remove")

stars_uva
```

Output 13 shows results of the UVA analysis on the STARS dataset. The results are categorized into different levels of redundancy. Firstly, variable pairs with wTO greater than 0.30 indicate large-to-very large redundancy, such as between STARS03 and STARS16 with a wTO of 0.516. Similarly, pairs with wTO above 0.25 suggest moderate-to-large redundancy, including STARS30 and STARS32 at 0.340. Further, pairs with wTO exceeding 0.20 indicate small-to-moderate redundancy, such as STARS19 and STARS23 with a wTO of 0.230. Additionally, the analysis identifies which variables were retained and which were removed. Variables such as STARS03, STARS06, and STARS32 were retained because they either have lower wTO values with others, while variables like STARS01, STARS02, and STARS10 were removed because they have high wTO values with others, indicating they are redundant compared to variables in the keep list.

3.1.4.1 Output 13

UVA Output of the STARS Datase

```
> stars_uva
Variable pairs with wTO > 0.30 (large-to-very large redundancy)
  node_i  node_j   wto
 STARS03 STARS16 0.516
 STARS30 STARS32 0.340
 STARS02 STARS06 0.339
----
Variable pairs with wTO > 0.25 (moderate-to-large redundancy)
  node_i  node_j   wto
 STARS09 STARS17 0.292
 STARS01 STARS08 0.258
 STARS08 STARS10 0.257
----
Variable pairs with wTO > 0.20 (small-to-moderate redundancy)
  node_i  node_j   wto
 STARS19 STARS23 0.230
 STARS39 STARS48 0.216
 STARS02 STARS20 0.211
 STARS06 STARS20 0.207
 STARS43 STARS46 0.202
>
> #Examine which variables were kept and which one were removed
> stars_uva$keep_remove
$keep
[1] "STARS03" "STARS06" "STARS08" "STARS09" "STARS32"
$remove
[1] "STARS01" "STARS02" "STARS10" "STARS16" "STARS17" "STARS30"
```

To examine the data set after removing or processing the identified redundant variables, we can use the following codes:

```
> head(stars_uva$reduced_data, 3) #call for the first three rows of the reduced dataset
     STARS03 STARS04 STARS05 STARS06 STARS07 STARS08 STARS09 STARS11 STARS12 STARS13 STARS14 STARS15
[1,]       2       4       2       2       2       4       2       2       3       3       2       3
[2,]       4       5       5       3       5       5       2       4       5       5       4       5
[3,]       4       4       4       4       4       4       1       4       4       4       4       4
     STARS18 STARS19 STARS20 STARS21 STARS22 STARS23 STARS24 STARS25 STARS26 STARS27 STARS28 STARS29
[1,]       3       2       2       3       2       2       3       2       4       4       4       4
[2,]       4       4       4       5       5       4       5       5       4       4       4       5
[3,]       4       4       4       3       3       4       4       4       4       4       4       4
     STARS31 STARS32 STARS33 STARS34 STARS35 STARS36 STARS37 STARS38 STARS39 STARS40 STARS41 STARS42

[1,]       2       2       3       4       4       2       4       3       2       4       4       3
[2,]       3       5       5       4       4       5       5       5       4       5       4       4
[3,]       4       4       4       4       4       4       4       4       4       4       2       4
     STARS43 STARS44 STARS45 STARS46 STARS47 STARS48 STARS49 STARS50 STARS51
[1,]       2       2       3       2       2       3       3       4       3
[2,]       5       5       2       4       4       2       4       5       4
[3,]       4       4       4       4       4       4       4       4       4
```

To assess whether the removal of redundant variables improves the model fit, we can compare the CFA fit between the network structure of the original and the reduced STARS dataset. This can be accomplished as follows:

```
#Perform EGA with the original STARS dataset
stars_ega <- EGA(data = df_stars)
stars_ega_cfa <- EGAnet::CFA(ega.obj = stars_ega, estimator = "WLSMV",
                             plot.CFA = FALSE,
                             data = df_stars)
stars_fitMeasure <- lavaan::fitMeasures(stars_ega_cfa$fit, fit.measures = c("chisq", "df", "pvalue", "cfi",
"tli", "rmsea", "srmr"))

#Perform EGA with the reduced STARS dataset
stars_reduced_ega <- EGA(data = stars_uva$reduced_data)
stars_reduced_ega_cfa <- EGAnet::CFA(ega.obj = stars_reduced_ega, estimator = "WLSMV",
                             plot.CFA = FALSE,
                             data = stars_uva$reduced_data)

stars_uva_fitMeasure <- lavaan::fitMeasures(stars_reduced_ega_cfa$fit, fit.measures = c("chisq", "df",
"pvalue", "cfi", "tli", "rmsea", "srmr"))
```

Output 14 shows model fit indices of the network structure of both reduced and original STARS dataset, as produced by the EGA method. Both models show a statistically significant fit ($p < 0.001$), which could be due to a large number of sample sizes. The original model has a CFI of 0.978 and a TLI of 0.977, both of which suggest a good fit as they are close to 1. However, the RMSEA of 0.093 (between 0.08 and 0.1 = marginal fit) and the SRMR of 0.092 (> 0.08 = poor fit) are relatively high, suggesting some model misfit.

On the other hand, the reduced model has a CFI of 0.989 and a TLI of 0.989, both of which suggest a relatively better fit than the original model. Additionally, the RMSEA of 0.067 (between 0.08 and 0.08 = acceptable fit) and the SRMR of 0.075 (< 0.08 = good fit) are lower, indicating a better fit to the data with less approximation error and better residuals. Overall, the reduced model demonstrates a superior fit compared to the original model, as evidenced by higher CFI and TLI values and lower RMSEA and SRMR values.

3.1.4.2 Output 14

Model Fit Indices of the Original and the Reduced STARS Dataset

```
#compare fit
> stars_fitMeasure

   chisq       df   pvalue      cfi      tli    rmsea     srmr
3329.504 1209.000    0.000    0.978    0.977    0.093    0.092
> stars_uva_fitMeasure
   chisq       df   pvalue      cfi      tli    rmsea     srmr
1774.707  935.000    0.000    0.989    0.989    0.067    0.075
```

4 Concluding Remarks

In the learning analytics environment, understanding the various variables that reflect students' learning as an interconnected system is essential. This chapter discusses the advanced application of the network psychometrics approach to understand such a complex system by visualizing them with the EGA method, as well as assessing the robustness of such structures with metrics such as dimension stability, CFA metrics, and TEFI. In other words, the advanced applications of psychological networks through EGA offer a robust framework for understanding the complex relationships between variables in educational assessments.

These demonstrated techniques allow researchers to identify key variables and their interactions, assess the stability and integrity of these systems, and monitor changes at different units of assessment (i.e., population, group, and individual). These methods yield insights into educational data, facilitating more informed decision-making and strategy development to enhance student learning outcomes.

The practical applications of these methods are diverse. Our SAS and SPOT analysis revealed highly stable dimensions of the instruments; these results can be used to inform the interpretations and uses of the instrument according to the discovered statistical evidence [1]. For example, the SPOT instrument can be used to guide faculty development programs or inform targeted mentoring programs based on specific dimensions like teaching effectiveness and student engagement, rather than using general approaches.

Our Dynamic EGA results demonstrated how network analysis can inform personalized interventions at different levels. At the group level, we found distinct learning patterns: Group 1 showed a more stable two-community structure, suggesting the effectiveness of streamlined interventions focusing on strong connections between basic concepts. In contrast, Group 2's three-community structure indicates the need for more nuanced, multi-phase learning approaches as the connection between dimensions are relatively weaker. Similar approaches can also be adopted to design a person-specific approach to intervention as informed by the individual-level results.

Furthermore, our UVA findings demonstrate how network analysis can optimize curriculum design by identifying and addressing redundancies in assessment strategies. For example, when redundancy is detected between similar assessment metrics like homework completion and assignment timeliness, educators can focus on the more predictive measure, enhancing assessment efficiency while maintaining effectiveness.

In summary, these advanced techniques serve as powerful tools in the field of learning analytics by offering more effective methods for educational measurement, which can inform and improve teaching strategies in modern education. The combination of robust statistical analysis with practical applications enables educators to make data-driven decisions while maintaining pedagogical flexibility and responsiveness to individual student needs.

References

1. AERA, APA, & NCME (2014) Chap. 1: Validity. In: *Standards for educational and psychological testing*. American Educational Research Association, pp 11–31
2. Asparouhov T, Muthén B (2009) Exploratory structural equation modeling. *Structural Equation Modeling: A Multidisciplinary Journal 16*(3):397–438. https://doi.org/10.1080/10705510903008204
3. Blondel VD, Guillaume J-L, Lambiotte R, Lefebvre E (2008) Fast unfolding of communities in large networks. *Journal of Statistical Mechanics: Theory and Experiment 2008*(10):P10008. https://doi.org/10.1088/1742-5468/2008/10/P10008
4. Borsboom D, Deserno MK, Rhemtulla M, Epskamp S, Fried EI, McNally RJ, Robinaugh DJ, Perugini M, Dalege J, Costantini G, Isvoranu A-M, Wysocki AC, van Borkulo CD, van Bork R, Waldorp LJ (2021) Network analysis of multivariate data in psychological science. *Nature Reviews Methods Primers 1*(1):58. https://doi.org/10.1038/s43586-021-00055-w
5. Bulut O, Wongvorachan T (2024) *A network psychometrics approach to evaluate the validity of the SPOT survey* [Poster presentation]. the Celebration of Socially Transformative Research and Teaching. University of Alberta. University of Alberta Library. https://doi.org/10.7939/R3-DC1J-4Q40
6. Chew PKH, Dillon DB, Swinbourne AL (2018) An examination of the internal consistency and structure of the statistical anxiety rating scale (STARS). *PLOS ONE 13*(3):e0194195. https://doi.org/10.1371/journal.pone.0194195
7. Chiesi F, Primi C, Carmona J (2011) Measuring statistics anxiety: Cross-country validity of the statistical anxiety scale (SAS). *Journal of Psychoeducational Assessment 29*(6):559–569. https://doi.org/10.1177/0734282911404985
8. Christensen AP, Garrido LE, Golino H (2023) Unique variable analysis: A network psychometrics method to detect local dependence. *Multivariate Behavioral Research* 1–18. https://doi.org/10.1080/00273171.2023.2194606
9. Christensen AP, Golino H (2021) On the equivalency of factor and network loadings. *Behavior Research Methods 53*(4):1563–1580. https://doi.org/10.3758/s13428-020-01500-6
10. Christensen AP, Golino H, Silvia P (2019) A psychometric network perspective on the measurement and assessment of personality traits. *Preprint.*. https://uva.theopenscholar.com/files/hudson-golino/files/personality_psychometric_networks-2.pdf
11. Ferrando PJ, Hernandez-Dorado A, Lorenzo-Seva U (2022) Detecting correlated residuals in exploratory factor analysis: New proposals and a comparison of procedures. *Structural Equation Modeling: A Multidisciplinary Journal 29*(4):630–638. https://doi.org/10.1080/10705511.2021.2004543
12. Finch WH, French BF (2019) *Educational and psychological measurement*. Routledge
13. Golino, H., Christensen, A., Moulder, R., Garrido, L. E., Jamison, L., & Shi, D. (2022a). *EGAnet: Exploratory Graph Analysis—a framework for estimating the number of dimensions in multivariate data using network psychometrics* (2.0.0) [R]. https://cran.r-project.org/web/packages/EGAnet/index.html
14. Golino H, Christensen AP, Moulder R, Kim S, Boker SM (2022b) Modeling latent topics in social media using dynamic exploratory graph analysis: The case of the right-wing and left-wing trolls in the 2016 US elections. *Psychometrika 87*(1):156–187. https://doi.org/10.1007/s11336-021-09820-y
15. Golino HF, Epskamp S (2017) Exploratory graph analysis: A new approach for estimating the number of dimensions in psychological research. *PLOS ONE 12*(6):e0174035. https://doi.org/10.1371/journal.pone.0174035

16. Golino H, Moulder R, Shi D, Christensen AP, Garrido LE, Nieto MD, Nesselroade J, Sadana R, Thiyagarajan JA, Boker SM (2021) Entropy fit indices: New fit measures for assessing the structure and dimensionality of multiple latent variables. *Multivariate Behavioral Research 56*(6):874–902. https://doi.org/10.1080/00273171.2020.1779642
17. Hilpert JC, Marchand GC (2018) Complex systems research in educational psychology: Aligning theory and method. *Educational Psychologist 53*(3):185–202. https://doi.org/10.1080/00461520.2018.1469411
18. Hu L, Bentler PM (1999) Cutoff criteria for fit indexes in covariance structure analysis: Conventional criteria versus new alternatives. *Structural Equation Modeling: A Multidisciplinary Journal 6*(1):1–55. https://doi.org/10.1080/10705519909540118
19. Isvoranu A-M, Epskamp S, Waldorp LJ, Borsboom D (eds) (2022) *Network psychometrics with R: A guide for behavioral and social scientists*. Routledge. https://doi.org/10.4324/9781003111238
20. Koopmans M (2020) Education is a complex dynamical system: Challenges for research. *The Journal of Experimental Education 88*(3):358–374. https://doi.org/10.1080/00220973.2019.1566199
21. Leising, D., Borgstede, M., Burger, J., Zimmermann, J., Bäckström, M., Oltmanns, J. R., Freyer, N., Wiedenroth, A., Knischewski, P., & Connelly, B. S. (2020). *Why do judgments on different person-descriptive attributes correlate with one another? A conceptual analysis with relevance for most psychometric research*. https://doi.org/10.31234/osf.io/7c895
22. Nowick K, Gernat T, Almaas E, Stubbs L (2009) Differences in human and chimpanzee gene expression patterns define an evolving network of transcription factors in brain. *Proceedings of the National Academy of Sciences 106*(52):22358–22363. https://doi.org/10.1073/pnas.0911376106
23. Pons P, Latapy M (2006) Computing communities in large networks using random walks. *Journal of Graph Algorithms and Applications 10*(2):191–218. https://doi.org/10.7155/jgaa.00124
24. Rhemtulla M, Brosseau-Liard PÉ, Savalei V (2012) When can categorical variables be treated as continuous? A comparison of robust continuous and categorical SEM estimation methods under suboptimal conditions. *Psychological Methods 17*(3):354–373. https://doi.org/10.1037/a0029315
25. Rosseel Y (2012) lavaan: An R package for structural equation modeling. *Journal of Statistical Software 48*(2):1–36. https://doi.org/10.18637/jss.v048.i02
26. Saqr M, Beck E, López-Pernas S (2024) Psychological networks: A modern approach to analysis of learning and complex learning processes. In: Saqr M, López-Pernas S (eds) Learning analytics methods and tutorials: A practical guide using R, vol 1. Springer
27. Saris WE, Satorra A, Van Der Veld WM (2009) Testing structural equation models or detection of misspecifications? *Structural Equation Modeling: A Multidisciplinary Journal 16*(4):561–582. https://doi.org/10.1080/10705510903203433
28. Guyon H, Falissard B, Kop J-L (2017) Modeling psychological attributes in psychology - an epistemological discussion: Network analysis vs. Latent variables. Front Psychol 8:798. https://doi.org/10.3389/fpsyg.2017.00798
29. Gysi DM, Voigt A, Fragoso TDM, Almaas E, Nowick K (2018) wTO: An R package for computing weighted topological overlap and a consensus network with integrated visualization tool. BMC Bioinformatics 19(1):392. https://doi.org/10.1186/s12859-018-2351-7
30. R Core Team (2023) R: A language and environment for statistical computing [R]. R Foundation for Statistical Computing. https://www.R-project.org/
31. Wickham H, Miller E, Smith D (2023) haven: Import and export “SPSS”, “stata” and “SAS” files [Computer software]. https://haven.tidyverse.org

Detecting Nonlinear Patterns in Education Research: A tutorial on Recurrence Quantification Analysis

Daryn Dever

Abstract Recurrence Quantification Analysis (RQA) is a robust analytical technique that can be used to identify and quantify patterns within nonlinear dynamical systems. This analysis can identify shifts in human behavior regarding learning processes, the evolution of how a system demonstrates functionality versus dysfunctionality, and how this shifts across different time dimensions. Ultimately, RQA can reveal patterns in learning processes that could not be captured, measured, or tracked using traditional methods. As such, this chapter reviews the different methodologies for applying RQA to education research, leveraging this analytical methodology for capturing the dynamics of learning processes as it occurs within educational contexts. This chapter will review three RQA methods, including auto-RQA, cross-RQA, and multidimensional-RQA, how it has been applied within education research, and provide a tutorial of how these analyses can be conducted and interpreted for your own research purposes.

Keywords Recurrence quantification analysis · Nonlinear analyses · Education research · Complexity · Tutorial

1 Introduction

Recurrence Quantification Analysis (RQA) is an analytical technique to identify and quantify the repetition, i.e., recurrence, of patterns in nonlinear dynamical systems [18, 19]. RQA is used to quantify how and when system behaviors shift between repetitive (e.g., A, A, A) and novel (e.g., B, C, D) sequences, revisiting prior states (e.g., A, B, C, A) or evolving into new behaviors over a temporal period and observed across all possible time lags. RQA can be conducted using both categorical and continuous data, for one or multiple time series, and across several contexts,

D. Dever (✉)
University of Florida, Gainesville, FL, USA
e-mail: ddever@ufl.edu

M. Saqr, S. López-Pernas (eds.), *Advanced Learning Analytics Methods*,
https://doi.org/10.1007/978-3-031-95365-1_14

domains, and research agendas. Specific to learning and educational phenomena, RQA is an infrequent analytical method but has been used to examine collaborative learning [6], self-regulated learning [4], and other learning processes using a variety of data including eye tracking, log files, physiology, and verbalization data. Through this analysis, researchers are able to understand fluctuations in learning dynamics, quantify learning as either functional or dysfunctional, and examine how students can work together in a collaborative setting for increased learning outcomes. Within this chapter, we will review the different RQA approaches, how RQA has been used within educational contexts, and conclude with the future directions of incorporating RQA into future educational technologies for providing personalized learning experiences.

2 Methods

There are three approaches for conducting RQA depending on research questions, *auto-RQA* (aRQA), which compares a time series against itself to determine repeated signals within a single time series, *cross-RQA* (cRQA) which compares two different time series to determine synchronicity of states, and *multidimensional-RQA* (mdRQA) which compares more than two time series. All three approaches use a recurrence plot (see Fig. 1) as the primary tool for identifying sequences and patterns of behavioral repetition in the time series under examination.

This plot maps Time Series 1 on the x-axis and Time Series 2 on the y-axis (note that in aRQA, Time Series 1 and Time Series 2 are the same time series). If an event in Time Series 1 matches that of an event in Time Series 2, the intersection of these events on the recurrence plot is shaded black and labeled a recurrent point. Should the events not match, the intersection remains white. Recurrent points are identified as a "close" distance between time series events, where the value of "close" is termed the *radius* and is determined by the researcher conducting the analyses. The recurrence plot itself can serve as a valuable visualization of the dynamics of system behaviors, identifying instances of stability (repeated patterns) and instability (lack of repetitious behaviors) over a temporal period. Several metrics can be extracted

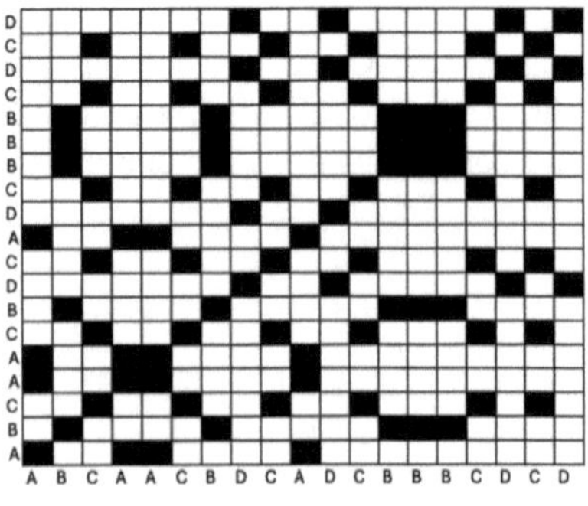

Fig. 1 Example of auto-RQA recurrence plot

from the recurrence plot which reveal different features of the dynamics of a system. For detailed descriptions, please refer to [17].

It is also important to note that there are some restrictions to the data that can be used. First, a rule of thumb within this analytical approach is that a time series needs to have a minimum of 10 events, either categorical or continuous. Additionally, the time series need to match in length. Within auto-RQA, this automatically occurs as both Time Series 1 and Time Series 2 are the same. However, in the cases of cross-RQA and multidimensional-RQA, Time Series 1 and Time Series 2 are not the same. Therefore, the raw time series data need to be of the same length, or some data manipulation needs to occur for the analyses to be conducted.

Another note to account for is that the tutorials that are written below for auto-RQA and cross-RQA, 'crqa' package version 2.0.6 was utilized. As such, depending on the version of your package, you may need to either account for variations in package versions or use an earlier package version.

Within the next subsections, we will provide a brief description of each RQA approach, research questions that can guide these specific analyses in relation to learning processes, and a brief tutorial of how to conduct these analyses within R and RStudio. This chapter will also review only categorical RQA examples. While we acknowledge that continuous RQA is an incredibly important approach, for the brevity of the chapter and the ease in which the reader can learn RQA, we present only three of the six RQA approaches.

2.1 aRQA

Auto-recurrence quantification analysis (aRQA) focuses on identifying and quantifying the dynamics within a single time series to identify the different sequences and patterns of repetition that appear across time. This can be used to identify repetitious behaviors, the deterministic structures of a time series, the (in)stability of a time series, and other such metrics. In relation to learning processes, aRQA can provide incredible insight into the dynamics of a learner, identifying how learner characteristics and performance can change dynamically over time. For example, aRQA can reveal how emotions emerge over time during a learning task [2], reveal the degree to which learners' self-regulated learning strategy use demonstrates repetitious patterns [4, 5], or examine how learner profiles can change over the course of a semester [11].

Because of the versatility of aRQA in applying to both continuous and categorical time series data, there are several avenues research can take by employing an aRQA analytical approach. Researchers can identify how certain interruptions or interventions within the learning context can result in a nonlinear shift in human behavior [13]. Emotions captured via facial expressions can identify how often certain patterns of emotions emerge over time and identify when and how often emotional states re-occur. Researchers can also ask the question of how predictable learning strategies are in terms of how often certain strategies are repeated or identify the extent to which repetitious learning strategies positively or negatively contribute to learning outcomes. It is important to note that the research question must also reflect themes of complexity, dynamicism, and non-linearity to appropriately guide the aRQA analyses to protect against the unethical use of these analytical approaches.

2.1.1 aRQA Tutorial

For this tutorial, we will use aRQA to calculate extract metrics of complexity between two participants and compare each of their metrics to provide an example of how this can be applied to your own datasets. To begin this tutorial, we will take aRQA with categorical data as our example analysis and utilize the R package 'crqa' [3], version 2.0.5.

```
#Load and activate crqa package
install.packages('crqa')
library('crqa')
```

Within this package, there are several data available for use to practice coding. To call the data, we will use the line:

```
data(crqa) #calling data
```

For this example, we will use the 'eyemovement' dataset. However, this dataset was originally intended to provide a cRQA example with categorical data. For our example, we will use the *listener* column as Participant 1 and *narrator* as Participant 2. In other words, the column represents two participants and the number represents a categorical factor. Within your own datasets, this may be the action in which a participant is taking on a task or a sequence of learning strategies that are deployed during learning. For each participant, we will run aRQA as you would in a study with multiple participants. Prior to doing so, though, we must first change the data type as currently it is an integer. We will transform the data into a factor to conduct categorical aRQA.

```
View(eyemovement) #View the dataset
>Output:
    listener narrator
1         10        2
2          2        2
3          2       11
4         10       11
5         10       11
6         10       11
7         10       11
8         10       11
9         10       11
10        10       11
11        10       11
12        10        2
```

```
13          10         2
14          10         2
15          10         2
16          10         2
17          10         2
eyemovement_onlylistener <- eyemovement[-c(2)] #Deleting the second column
eyemovement_onlynarrator <- eyemovement[-c(1)]#Deleting the first column
eyemovement_onlylistener$listener<-as.factor
    (eyemovement_onlylistener$listener)#changed to factor
eyemovement_onlynarrator$narrator<-as.factor
    (eyemovement_onlynarrator$narrator)#changed to factor
```

Now, we are left with two datasets containing 2,000 entries. We can imagine each dataset to be our categorical time series for a single participant where each number is representative of a different category, such as a self-regulated learning strategy, or some other coded behavior of a complex system. Note that, in your own dataset, these can remain characters or strings where no numbers need to be used, or coding must occur, should you already have the semantic meaning present. In addition, subsetting each participant is not necessary for this analysis in which each column may be called separately when running the 'crqa' function below. Indeed, this would be tedious if we were reviewing more than two participants. However, within this tutorial it is easier to identify which participant we are evaluating given the different naming conventions of the datasets.

Next, we will set our parameters for the 'crqa' function. Parameters within this function help to identify the type of data that is being used, the structure of the data, the way in which the time series data should be considered 'recurrent,' and calls for any transformation of the data or phase space in which the data exists. As such, this is an important step that cannot be skipped and needs the researcher to be mindful of how these parameters effect how metrics are calculated and the interpretation of output. For more information on parameters and the different options to input into the parameters, you may refer to Coco et al. [3] package documentation.

Parameters could be defined in the function itself; however, I find it a bit less messy when the parameters are set *before* conducting RQA. *Delay* refers to the unit by which the series are lagged and *embed* refers to the number of dimensions for our lag intervals, or phase-reconstruction. We set these parameters to 1 so that we are able to identify sequential patterns. *Rescale* is the rescaling of the distance matrix. For this argument, we set the rescale to 0, which does nothing to the distance matrix. The radius parameter is critical to the distance in which two data are considered recurrent or not. For categorical data, we set the radius to 0 to have an exact match. For continuous data, this would be a number established by the statistician and accounting for the scaling of the data. *Normalize* allows us to normalize the time series where a setting of 0 does not transform the data. The minimum diagonal and vertical lines are typically set to 2 and defines the minimum number of recurrent points for a line to be defined. Further parameters are typically set to default but can be changed depending on the research question and any parameterizations that the statistician would like set for interpretation. However, importantly within the code shown below is the *datatype* argument in which we define our data as categorical. It is important to note that the parameters should be the same for each participant you run aRQA on for comparison purposes.

```
#aRQA Parameters
delay = 1; embed = 1; rescale = 0; radius = 0;
normalize = 0; mindiagline = 2; minvertline = 2;
tw = 0; whiteline = FALSE; recpt = FALSE; side = "both"
method = 'crqa'; metric = 'euclidean';
datatype = "categorical"
```

Note that for these parameters, we have a radius of 0 and our datatype is categorical. This states that we will only accept matching pairs as recurrent points and that our data is categorical, not continuous.

Now, a couple of ground rules for aRQA: (1) do not include missing data, omit NAs using any method you see fit; (2) this can only be used for a single time series at a time; if you would like to do a full dataset, you will need to wrap this code in a looping function or equivalent methodology; and (3) there must be a minimum of 10 data points within a single time series. So! Now we will set our data within our function. As you have already defined the parameters above, you will only need to input your time series 1 and time series 2. Remember, as we are doing *auto*-RQA, time series 1 and 2 will be identical within the call function.

```
ts1 <- eyemovement_onlylistener #rename our dataset for ease of use
Listener_aRQA <- crqa(ts1, ts1,
                              delay, embed, rescale, radius, normalize,
                              mindiagline, minvertline, tw, whiteline,
                              recpt, side, method, metric, datatype) #Running
                              aRQA on time series
```

When running this function above, you may experience a warning: "Your input data was provided either as character or factor, and recoded as numerical identifiers." This is not anything to worry about, just the program informing you that your data type was changed to match what was needed for the function to run. We renamed our dataset for our own ease of use. You do not have to do this step and you can just keep your original name, especially if you are conducting these analyses on several datasets (i.e., each participant) in which identifiers need to be used for organization purposes. When running the 'crqa' function, a *.rds* file will be output. I like to save the .rds file onto my computer so that I always have a copy.

```
>Output:
          RR      DET NRLINE maxL        L     ENTR     rENTR      LAM
  TT      catH
1 15.12755 95.90978  53581 2000 10.8313 3.176994 0.6601978 98.01901
  19.65649 4.588468
```

We will repeat the same process with the narrator dataset.

```
ts2 <- eyemovement_onlynarrator
narrator_aRQA <- crqa(ts2, ts2,
                          delay, embed, rescale, radius, normalize,
                          mindiagline, minvertline, tw, whiteline,
```

```
                      recpt, side, method, metric, datatype)
saveRDS(narrator_aRQA, file = "narrator_aRQA.rds")
narrator_aRQA<-as.data.frame(narrator_aRQA[1:10])
```

Calling our new dataframe in the console, we receive the output:

```
>Output:
       RR       DET NRLINE maxL         L      ENTR     rENTR       LAM
TT      catH
1 19.9845 98.57515  54541 2000 14.44766 3.349063 0.632698 99.38878
26.13209 4.514792
```

You are able to identify several metrics that are defined in the 'crqa' documentation by Coco et al. [3]. We identify the different output metrics and their definitions below in Table 1.

The next step is to identify the meaning of those metrics in terms of data interpretation. Alone, they do not mean much. For example, what does it mean that there is a recurrence rate of 15.1 in the grand scheme of your data? So what if the entropy is 2.18? When applying aRQA to participant data, let us say in an experimental study, it becomes easier to determine what these metrics are saying. For example, Dever et al. [4] examined the differences in recurrence rates between participants of different conditions and then related this further to learning outcomes. This provided an understanding of the relationship between learning outcomes and complexity metrics for a specific psychological phenomena.

We are also able to visualize each plot with the following code:

```
plot1 <- Listener_aRQA$RP
plot2 <- narrator_aRQA$RP
```

Table 1 Metrics output from RQA

Metric	Definition
Recurrence Rate (RR)	The percentage, or proportion, of recurrent points within the recurrence plot.
Percent Determinism (DET)	The percentage, or proportion, of recurrent points on diagonal line structures within the recurrence plot.
NRLINE	Number of lines within the recurrence plot.
maxL	The number of recurrent points on the longest diagonal line segment within the recurrence plot.
L	The average length of all line structures.
Entropy (ENTR)	The Shannon entropy of the diagonal line structures as defined by the minimum length of line structures allowable within the set parameters.
rENTR	Normalized entropy measure.
Laminarity (LAM)	Proportion of recurrent points on vertical line structures within the recurrence plot.
Trapping Time (TT)	The average length of vertical line structures within the recurrence plot.
catH	Entropy of plots based on the rectangular block structures, only applied to categorial recurrence plots.

```
plotRP(plot1)
plotRP(plot2)
```

Please note that the function, *plotRP,* may not exist in your current version of *"cRQA"* but may instead be named *plot_rp*. Please consult your version documentation for the correct naming.

For our datasets, we can compare two participants. For example, Participant 1 (RR = 15.13) showed that they had a lower recurrence rate than Participant 2 (RR = 19.98), demonstrating that Participant 1's repetition of behaviors was lower than Participant 2. These metrics are extracted from the output as detailed above in our code labeled "Output". This is echoed throughout the other metrics, such as percent determinism in that Participant 2 had a greater value at 98.6% than Participant 1 at 95.9%, showing that Participant 2 demonstrated more predictable behavior. As previously mentioned, please refer to in-depth documentation by Coco et al. [3] for the calculation, meaning, and interpretations of these complexity metrics.

While significance tests cannot be run due to the low number of participants in this tutorial, expanding this out to a greater number of participants will provide the researcher with the opportunity to identify if there are significant differences in the complexity metrics between groups, if the metrics are predictive of learning outcomes, or identify if the overall behavior as visualized on the recurrence plot is indicative of functional behaviors. For example, for a larger sample size, t-tests can identify if participants belonging to one group had significantly greater or lower metrics than participants belonging to another group (e.g., [4]). Qualitatively using Figure 2, we can identify significant concentrations of black squares on the right recurrence plot depicting the narrator time series, Participant 2. This observation has two meanings: (1) this corresponds to the metrics pulled from the dataset which shows that Participant 2 has a greater repetition of behaviors; and (2) the concentration of repetitive behaviors at the beginning of the time series decreases as time progresses, demonstrating that Participant 2 transitioned from (most likely) dysfunctional behaviors to more functional behaviors.

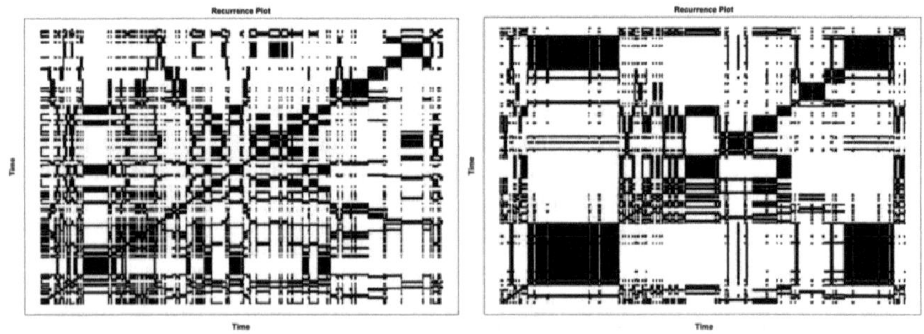

Fig. 2 Left: Plot 1, listener (Participant 1) recurrence plot; Right: Plot 2, narrator (Participant 2) recurrence plot

In sum, this tutorial is meant to demonstrate how to quantify complexity within your dataset. However, this is not the only way to do so where several packages and other methods exist for aRQA, and other RQA approaches, to be conducted. As such, it is important for you to identify the best approach for your data to respectfully conduct analytical techniques for the furtherance of research.

2.2 *cRQA*

Cross-recurrence quantification analysis, cRQA, identifies and quantifies the recurrent dynamics between two different time series, such as that seen in dyads during a collaborative problem-solving task (e.g., [9]). CRQA examines the cross-recurrences of two different sequencies which is useful for examining the degree of similarity between two sequences that may be temporally dissimilar at first glance [16]. Within educational research, cRQA is more often employed when examining the dynamics between pairs of learners. For example, Duong et al. [8] released a recent tutorial on how to examine the dynamics of the reciprocal exchanges between two individuals which can be applied to conversations between teams of learners. Moulder et al. [9] compared eye gaze data between pairs of learners on a team using cRQA to reveal how the similarities between two time series are indicative of task success and effective team collaboration. Nguyen et al. [10] also used cRQA to quantify emotional synchrony among learners within a collaborative group that was a result of regulatory triggers. This study found that examining how learners' emotional synchrony changes as a result of regulatory behaviors is an important consideration for evaluating team collaboration. In sum, cRQA identifies similarities in two time series, allowing researchers to examine patterns of alignment, communicative behavior, shared cognition, etc. that results in greater learning outcomes.

2.2.1 cRQA Tutorial

The goal of this tutorial is to use cRQA to detect the level of complexity and synchrony between two research subjects. This tutorial will conduct categorical cRQA using the datasets found in the 'crqa' package. This is the same package that was utilized for the aRQA tutorial above. We will start by calling the 'eyemovement' dataset from the 'crqa' package.

```
data(crqa)
View(eyemovement) #View the dataset
```

In this dataset, there are two columns – *listener* and *narrator* – each representing their own categorical time series. As with the aRQA tutorial, the columns represent two different participants and the rows represent the participants' time series as it occurs sequentially. As before, we should now set our parameters to examine the similarities in the two time series. As these time series are categorical, we will use the same parameters as we did with aRQA.

```
#cRQA parameters
delay = 1; embed = 1; rescale = 0; radius = 0;
normalize = 0; mindiagline = 2; minvertline = 2;
tw = 0; whiteline = FALSE; recpt = FALSE; side = "both"
method = 'crqa'; metric = 'euclidean';
datatype = "categorical"
```

We then can conduct cRQA with narrator as time series 1 and listener as time series 2.

```
Tutorial_cRQA <- crqa(eyemovement$narrator, eyemovement$listener,
                      delay, embed, rescale, radius, normalize,
                      mindiagline, minvertline, tw, whiteline,
                      recpt, side, method, metric, datatype)
```

The metrics are then available to extract. Instead of extracting all metrics as we did in the tutorial above, let us say that our research question only necessitates examining the percent recurrence and percent determinism. We can extract only those metrics of this cRQA calculation by using the following code:

```
crqaMetrics <- c(Tutorial_cRQA[1], Tutorial_cRQA[2])
CrqaMetrics
>Output:
$RR
[1] 12.5229
$DET
[1] 98.95931
```

This shows us that our recurrence rate is 12.5% while our percent determinism is 99.0%. This shows that the system exhibits relatively few repeated states, but when those repetitions occur, they tend to follow a predictable pattern, suggesting a high degree of order within the system despite its overall non-repetitive nature.

We may also visualize the resulting recurrence plot (Fig. 3) using:

```
cplot <- Tutorial_cRQA$RP
plotRP(cplot)
```

You may notice how this recurrence plot is not symmetrical like that seen in aRQA, Figure 2. This is because the x- and y-axes are two different time series and therefore will not demonstrate symmetrical patterns in their plot. You can also see that the bclack dots are sparse, confirming the low recurrence rate extracted from the time series. However, from this plot we can see that while the repetitive states do not occur often at the beginning of the time series, the repetition increases towards the end of the time series. Should these time series have reflected a dyad engaging in collaborative problem-solving, we could interpret this as synchronization as the time on task progressed.

This tutorial demonstrated how two time series can be compared for similarities. In future studies, researchers may be able to determine if there are imbalances

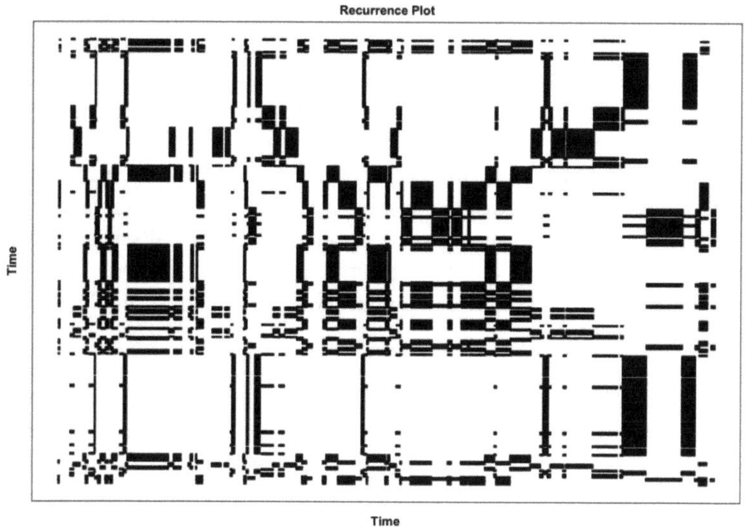

Fig. 3 cRQA recurrence plot

of similarities between a team of three learners, examine the extent to which the behavioral markers of a novice deviate from that of an expert, or identify instances of synchronicity between emotional and physiological data. However, there may be instances in which dyads, or indeed only two time series, are not sufficient. Rather, it is important to examine three or more time series. Because of this, multidimensional-RQA was developed as a third approach for examining dynamics across multiple time series.

2.3 *MdRQA*

Multidimensional-RQA, or MdRQA, is the third approach of RQA analyses that allows for studying the dynamics across at least three time series. If the time series represented in MdRQA is, say, the gaze behaviors of three individual learners, then MdRQA is conducted on all participants' time series simultaneously. For example, if the first event of the time series has Participant A gazing at a non-player character, Participant B gazing at a book, and Participant C gazing at a food item within a game-based learning environment, the next event in which the participants display the same gaze behaviors will be marked as recurrent with the first event. This can also be applied to log files of behaviors, physiological arousal, emotions, and the cognitive processes that occur during learning.

Several studies have leveraged MdRQA to examine team behavior and synchrony. Dindar et al. [6] used MdRQA to detect physiological synchrony of teams of students as they engaged in a learning task to determine the extent to which the synchrony displayed within each team of student relates to students' shared regulatory behavior. Another study by Braun et al. [1] also examined physiological synchrony between five-person teams to determine differences in workload between highly and lowly synchronized teams and determined that physiological synchrony is a valid measure of team mental workload. Similarly, Sharika et al. [12] also used MdRQA to determine the level of physiological synchrony between groups and how this related to group-level decision making. Outside of team collaboration, MdRQA has also been used to measure learners' emotions [2], socially-shared regulation via eye gaze data extracted during a group task [20], facial expressions as teams quit during game-based learning [7], and as indicators of stress and learning during collaborative tasks [21].

From the extensive literature incorporating MdRQA, this analytical technique offers a robust and unique approach for identifying recurrence in behaviors, affect, physiology, and cognition during collaborative tasks as well as across multiple types of data streams. As such, MdRQA affords researchers to opportunity to study the dynamics across teams of people and multimodal data to identify instances of high and low recurrence. These patterns of recurrence can then be attributed to several factors such as learning outcomes, mental workload, cognitive and metacognitive abilities, etc. In the next section, we present how MdRQA can be conducted within R.

2.3.1 MdRQA Tutorial

The goal of this tutorial is to use MdRQA to examine the level of similarity three participants on a team have with each other, ultimately understanding the extent to which their self-regulated learning strategies synchronized while completing a collaborative task. To run MdRQA, we first need to copy and paste the code for the function within our script. This is because the function has not yet been integrated into the 'crqa' package. This code can be found in the Supplementary Materials of Wallot & Leonardi [16] or in the following GitHub repository: https://github.com/Wallot/MdRQA.

In addition, you will need to install a new package called 'SDMTools.' However, as this may no longer be available on CRAN, you may use either a Github mirror or remote install the correct version of R using the 'remotes' library.

```
#Install and Call new needed library
install.packages("SDMTools") #Install SDMTools package
library("SDMTools") #call library
```

The data that will be used should be a matrix in which each column represents each time series to be used. Each row should correspond to the sequence of 'events' within the time series. For this tutorial, we will create a random dataset in which we have three participants – Participant A, Participant B, and Participant C – with 2000 datapoints each. Each row will contain a specific learning strategy a learner has used. The code we will use will randomly draw from the list of defined strategies to fill in the rows for each participant. In sum, we will create a dataset in which the three columns are each of the participants and the rows represent the sequential time series of each participant.

```
# Set the number of rows
num_rows <- 2000

# Define the strategies
strategies <- c("planning", "monitoring", "goalsetting",
"contentevaluation", "notetaking", "rereading")

# Create the dataset
dataset <- data.frame(
ParticipantA = sample(strategies, num_rows, replace = TRUE),
ParticipantB = sample(strategies, num_rows, replace = TRUE),
ParticipantC = sample(strategies, num_rows, replace = TRUE)
)
```

Our datasets will be different, but the dataset should look something like this:

```
>Output:
        ParticipantA      ParticipantB              ParticipantC
1        goalsetting          planning                    planning
2          rereading        monitoring                  goalsetting
3        goalsetting        notetaking                   rereading
4         monitoring       goalsetting                  monitoring
5         monitoring  contentevaluation         planning
6          rereading        notetaking                   planning
7        goalsetting       goalsetting           contentevaluation
8         notetaking          planning           contentevaluation
9  contentevaluation       goalsetting     contentevaluation
10         rereading         rereading                   planning
11         rereading         rereading                  monitoring
12          planning        monitoring                rereading
13        monitoring  contentevaluation         monitoring
```

Now that we have our data, we will use the 'mdrqa' function we input earlier to run MdRQA on our three participants.

```
mdrqa <- mdrqa(data = as.matrix(dataset), emb = 1, del = 1,
  norm = "euc", rad = 0)
>Output
RR       DET NRLINE maxL         L       ENTR      rENTR       LAM
 TT       catH
10.0483 19.47543  36115 2000 2.167465 0.3638623 0.2030754 19.58814
 2.132071 0.6078642
image(mdrqa$RP) # show recurrence plot
```

We can visualize our data (Fig. 4) by calling the function *image*. This is a slightly different protocol than the aRQA and cRQA examples from above as we are using the function *mdrqa* developed by Wallot & Leonardi [16]. We are provided with a plot which denotes when the values from each time series at a given time matches itself to denote when the time series' return to the same states.

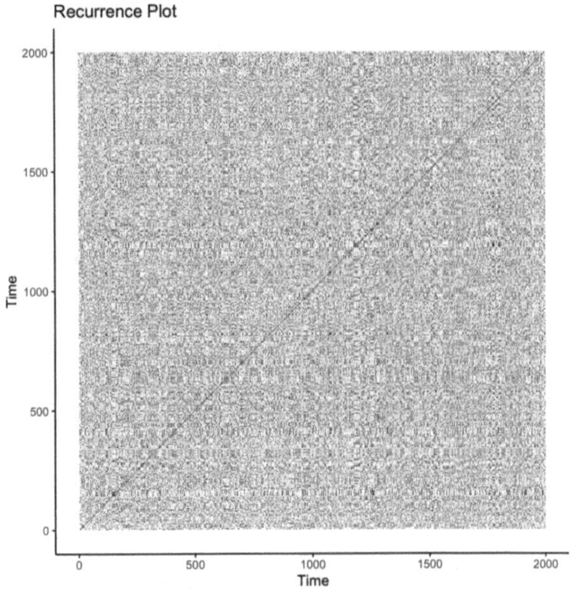

Fig. 4 MdRQA plot on synthetic data

From the metrics extracted from this previous example, we can interpret the similarities between the three participants in their strategy use. If we were to imagine that the three participants were working as a team on an assignment, we could determine if high or low levels of similarity between the participants is indicative of greater learning outcomes at either the individual or team levels. The visualization as well as the statistics output from MdRQA denote that there is no decipherable pattern and, generally, the group-level synchrony is not predictable and does not repeat often. I would interpret this plot as indicating that there was a balance in group-level behaviors where the self-regulated learning strategies that the group were enacting matched each other and other times that the strategies were not the same across all group members. Further research questions could examine the specific patterns that learners display, the extent to which the similarities in time series data evolve over time, and the predictive power of the metrics of complexity in determining learning outcomes.

3 Concluding Remarks and Future Readings

Recurrence Quantification Analysis (RQA) offers a robust and versatile framework for examining the dynamic behaviors inherent in learning processes, spanning from collaborative learning to regulation of cognition and affect. By enabling researchers to identify patterns of stability and change within individual learners (aRQA), dyadic interactions (cRQA), and group dynamics across a multitude of time series (MdRQA), this approach provides valuable insights into the dynamics of learning. Across a range of educational contexts—from self-regulated learning and emotional dynamics to collaborative problem-solving and team-based synchrony—RQA has demonstrated its potential to uncover nuanced relationships between learning behaviors and outcomes. As educational research increasingly focuses on understanding complexity, applying RQA to new modalities of data, such as multimodal learning analytics and real-time adaptive technologies, represents an exciting frontier. Future advancements in RQA approaches and integration with emerging technologies may pave the way for more personalized, responsive, and effective educational interventions, ultimately enhancing the learner experience and advancing the field of educational research.

3.1 Future Readings to Consider

As you apply RQA to your own datasets, merging your program of study with complexity science, it is important to be able to look to some works that accurately teach and demonstrate how you can use these analytical tools within your research. There are several tutorials with applications to datasets that can be of use, including Wallot & Grabowski [15] which uses RQA for keystroke data, Wallot [13] in how to apply RQA to psychological research, Wallot & Leonardi [16] to examine cRQA and MdRQA, and Wallot [14] for a review of MdRQA and the specific metrics output from these analyses. There are several other works by Wallot that I would suggest that are great for those just foraying into complexity, and specifically RQA. With applying RQA to psychological phenomena in educational research, there are several works to note that apply and (most importantly) explain RQA and how complexity links to their research. Dever et al. [5] and [4] both explore the use of aRQA as applied to how students engage in self-regulatory behaviors while learning with an intelligent tutoring system and game-based learning environment respectively. Another work by Dindar et al. [7] utilized aRQA to examine the facial expressions of students as they worked in collaborative teams with a game-based learning environment. Other papers such as Nguyen et al. [10] have used cRQA to identify collaborative learning through emotional synchrony of teams. MdRQA and cRQA have also been used with eye gaze behaviors to identify dynamics in teams [9]. This paper is notable due to the author's use of both cRQA and MdRQA within the same paper to look at different dimensions of collaborative learning, leveraging both RQA approaches to fully understand learning behaviors. These papers mentioned above provide a great framework for why RQA approaches were used and how educational research can integrate complex science methodological techniques.

- Dever, D. A., Amon, M. J., Vrzakova, H., Wiedbusch, M. D., Cloude, E. B., & Azevedo, R. (2022). Capturing sequences of learners' self-regulatory interactions with instructional material during game-based learning using auto-recurrence quantification analysis. *Frontiers in Psychology*, *13*, 813677.

- Dindar, M., Cloude, E. B., & Kiili, K. (2024, September). An Experimental Study of Facial Expressions in Collaborative Teams that Quit a Game-Based Learning Task: Within-Team Competition vs. No Within-Team Competition. In *European Conference on Technology Enhanced Learning* (pp. 105–119). Cham: Springer Nature Switzerland.
- Moulder, R., Booth, B., Abitino, A., & D'Mello, S. (2023, March). Recurrence quantification analysis of eye gaze dynamics during team collaboration. In *LAK23: 13th International Learning Analytics and Knowledge Conference* (pp. 430–440).
- Dever, D. A., Sonnenfeld, N. A., Wiedbusch, M. D., Schmorrow, S. G., Amon, M. J., & Azevedo, R. (2023). A complex systems approach to analyzing pedagogical agents' scaffolding of self-regulated learning within an intelligent tutoring system. *Metacognition and Learning*, *18*(3), 659–691.
- Wallot, S., & Grabowski, J. (2019). A tutorial introduction to recurrence quantification analysis (RQA) for keystroke logging data. *Observing Writing*, 163–189.
- Wallot, S. (2017). Recurrence quantification analysis of processes and products of discourse: A tutorial in R. *Discourse Processes*, *54*(5–6), 382–405.
- Wallot, S., & Leonardi, G. (2018). Analyzing multivariate dynamics using cross-recurrence quantification analysis (crqa), diagonal-cross-recurrence profiles (dcrp), and multidimensional recurrence quantification analysis (mdrqa)–a tutorial in r. *Frontiers in Psychology*, *9*, 2232.
- Wallot, S. (2019). Multidimensional Cross-Recurrence Quantification Analysis (MdCRQA)–a method for quantifying correlation between multivariate time-series. *Multivariate Behavioral Research*, *54*(2), 173–191.

References

1. Braun J, Hogh N, Kubowitsch S (2024) Evaluating team workload through physiological synchrony: An exploratory study using MdRQA Data to assess teams in action. In: Proceedings of the 7th European conference on industrial engineering and operations management (IEOM). IEOM Society International
2. Cloude EB, Munshi A, Andres JA, Ocumpaugh J, Baker RS, Biswas G (2024) Exploring confusion and frustration as non-linear dynamical systems. In: Proceedings of the 14th learning analytics and knowledge conference, pp 241–252
3. Coco MI, Mønster D, Leonardi G, Dale R, Wallot S (2020) Unidimensional and multidimensional methods for recurrence quantification analysis with CRQA. arXiv preprint:arXiv:2006.01954
4. Dever DA, Sonnenfeld NA, Wiedbusch MD, Azevedo R (2022) Pedagogical agent support and its relationship to learners' self-regulated learning strategy use with an intelligent tutoring system. In: International conference on artificial intelligence in education. Springer International Publishing, pp 332–343
5. Dever DA, Sonnenfeld NA, Wiedbusch MD, Schmorrow SG, Amon MJ, Azevedo R (2023) A complex systems approach to analyzing pedagogical agents' scaffolding of self-regulated learning within an intelligent tutoring system. Metacogn Learn 18(3):659–691

6. Dindar M, Alikhani I, Malmberg J, Järvelä S, Seppänen T (2019) Examining shared monitoring in collaborative learning: A case of a recurrence quantification analysis approach. Comput Human Behav 100:335–344
7. Dindar M, Cloude EB, Kiili K (2024) An experimental study of facial expressions in collaborative teams that quit a game-based learning task: within-team competition vs no within-team competition. In: European conference on technology enhanced learning. Springer Nature Switzerland, Cham, pp 105–119
8. Duong S, Davis TJ, Bachman HJ, Votruba-Drzal E, Libertus ME (2024) Exploring dynamic structures of dyadic conversations using categorical cross recurrence quantification analysis: A tutorial. Quant Method Psychol 20(2):121–136
9. Moulder R, Booth B, Abitino A, D'Mello S (2023) Recurrence quantification analysis of eye gaze dynamics during team collaboration. In LAK23: 13th international learning analytics and knowledge conference, pp 430–440
10. Nguyen A, Li Y, Järvelä S, Liu Y, Zhao G (2023) Emotional synchrony through regulatory triggers in collaborative learning: a facial expression recognition study. In: Proceedings of the 17th international conference of the learning sciences-ICLS 2023. International Society of the Learning Sciences, pp 1038–1041
11. Poquet O, Jovanovic J, Pardo A (2023) Student profiles of change in a university course: a complex dynamical systems perspective. In LAK23: 13th international learning analytics and knowledge conference, pp 197–207
12. Sharika KM, Thaikkandi S, Nivedita, Platt ML (2024) Interpersonal heart rate synchrony predicts effective information processing in a naturalistic group decision-making task. Proc Natl Acad Sci USA 121(21):e2313801121
13. Wallot S (2017) Recurrence quantification analysis of processes and products of discourse: a tutorial in R. Discourse Process 54(5–6):382–405
14. Wallot S (2019) Multidimensional cross-recurrence quantification analysis (MdCRQA)–a method for quantifying correlation between multivariate time-series. Multivariate Behav Res 54(2):173–191
15. Wallot S, Grabowski J (2019) A tutorial introduction to Recurrence Quantification Analysis (RQA) for key-logging data. In: Observing writing-insights from keystroke logging and handwriting, pp 163–189
16. Wallot S, Leonardi G (2018) Analyzing multivariate dynamics using cross-recurrence quantification analysis (CRQA), diagonal-cross-recurrence profiles (DCRP), and multidimensional recurrence quantification analysis (MdRQA)—a tutorial in R. Front Psychol 9:2232. https://doi.org/10.3389/fpsyg.2018.02232
17. Webber Jr CL, Zbilut JP (2005) Recurrence quantification analysis of nonlinear dynamical systems. Tutorials in contemporary nonlinear methods for the behavioral sciences, 94(2005):26–94
18. Webber CL, Zbilut JP (2005) Recurrence quantification analysis of nonlinear dynamical systems. In: Riley MA, Van Orden GC (eds) Tutorials in contemporary nonlinear methods for the behavioral sciences. National Science Foundation, Arlington, VA, pp 26–94
19. Webber CL, Marwan N, Facchini A, Giuliani A (2009) Simpler methods do it better: success of recurrence quantification analysis as a general purpose data analysis tool. Phys Lett A 373(41):3753–3756
20. Whitehead R, Nguyen A, Järvelä S (2024) Exploring the role of gaze behaviour in socially shared regulation of collaborative learning in a group task. J Comput Assist Learn
21. Yan L, Martinez-Maldonado R, Zhao L, Li X, Gašević D (2023) Physiological synchrony and arousal as indicators of stress and learning performance in embodied collaborative learning. In: International conference on artificial intelligence in education. Springer Nature Switzerland, Cham, pp 602–614

Mapping Relational Dynamics with Transition Network Analysis: A Primer and Tutorial

Mohammed Saqr, Sonsoles López-Pernas, and Santtu Tikka

Abstract This chapter presents Transition Network Analysis (TNA) that captures the full breadth of the relational dynamics of a temporal process. TNA models transitions between events as a weighted directed network. In doing so, TNA brings the wealth of network analysis to the modeled process which include graph, node and edge level metrics. TNA also enables the detection of recurring patterns such as dyads or triads, and communities and clusters. More importantly, TNA allows researchers to statistically validate the findings using bootstrapping, permutation, and case dropping techniques to verify if and when the research conclusions are correct. Furthermore, TNA allows researchers to include covariates that would explain why certain patterns emerge or examine the differences across subgroups. Such statistical rigor that brings validation and hypothesis testing at each step of the analysis offers a method for researchers to build, verify and advance existing theories and develop new ones on the basis of a robust scientific approach. This chapter offers a step-by-step tutorial using the `tna` R package, illustrating all the TNA features in a case study about group regulation.

Keywords Learning analytics · Transition network analysis · Stochastic process mining · Markov models

1 Introduction

As a concept, transition networks were proposed in the seventies by the pioneering work of Schnakenberg [1]. More recently, they have become more recognized across several fields to harness their potentials for capturing the relational and temporal

M. Saqr (✉) · S. López-Pernas
University of Eastern Finland, Joensuu, Finland
e-mail: mohammed.saqr@uef.fi

S. Tikka
University of Jyväskylä, Jyväskylä, Finland

M. Saqr, S. López-Pernas (eds.), *Advanced Learning Analytics Methods*,
https://doi.org/10.1007/978-3-031-95365-1_15

dynamic systems [2]. Nevertheless, despite the needs and potential, they have not been visible in education. In this chapter, we introduce transition network analysis (TNA) as a novel method that captures the temporal and relational aspects of an unfolding process. The main principle of TNA is representing the transition matrix between events as a graph to take full advantage of graph theory potentials and the wealth of potential of network analysis [3].

Think of it as a combination of process mining and network analysis where a process map is represented and analyzed by network analysis—but of course, TNA is far richer than this. In TNA, network analysis is applied to capture the structure, time and relationships in a holistic way. For process mining models, TNA adds network measures at the node, edge and graph level which tells us what events were important (centrality measures), what transitions were central (edge centralities), and which processes are more connected (e.g., graph density) [4]. TNA also adds community finding as well as pattern mining e.g., triads and dyads to capture patterns within the process. For network analysis, TNA brings clustering of sub-networks or different network constellations of typical temporal events which is commonly referred to as tactics. Furthermore, TNA can model covariates to explain certain patterns which is not a common method in network analysis (note that co-variates are not covered in this chapter) [4]. Furthermore, TNA offers far more than a combination of either methods and includes several innovative techniques of its own which include edge verification using bootstrapping and network comparison with permutation, as well as centrality verification through case-dropping. These statistical techniques bring rigor and validation at each edge level, so we can now be sure which edge is likely to replicate in the future and that our inferences and models are valid and not just a chance [3].

2 Why TNA?

Learning is a process that can be best conceptualized as a complex dynamic system [5–7]. Such a system is a collection of interconnected components that interact over time, where interactions can enhance, impede, amplify, or reinforce each other [8]. These dynamic interactions lead to the emergence of behaviors that cannot be fully understood by analyzing individual components in isolation. Often, the interactions result in processes that are more than the simple sum of their parts which is referred to as non-linear dynamics [5, 6] (see Fig. 1). For instance, motivation catalyzes achievement which in turn catalyzes more engagement, enjoyment and motivation. These interdependence, feedback loops and non-linear dynamics are inherently complex and therefore, require a modeling method that goes beyond the traditional linear models which assume that relationships between processes are stable and predictable [5]. TNA as a dynamic probabilistic model, offers a solution that captures the uncertainties through directional probabilities between learning

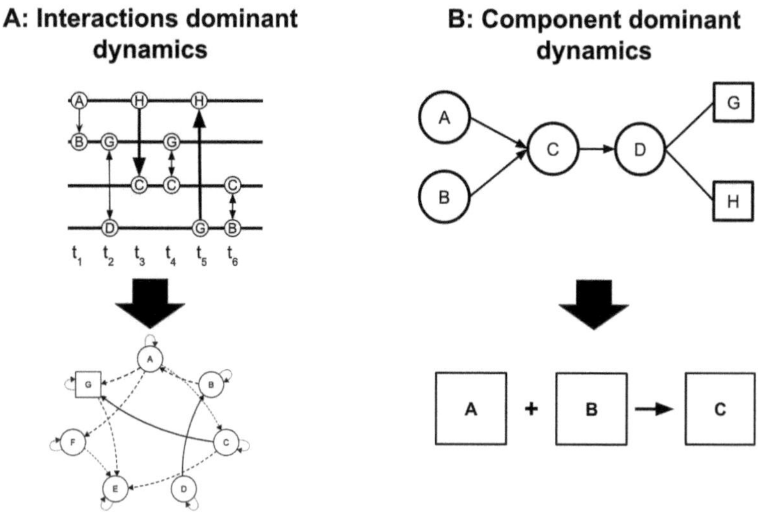

Fig. 1 Interaction dominant versus component dominant dynamics

events and accommodates the non-linear, evolving nature of learning processes. It also captures the constellations or the emergent patterns that define or shape the learning process [2].

Probabilistic processes are commonly —and indeed best— represented mathematically as matrices, where rows represent nodes and columns denote direct probabilistic interactions between them [9, 10]. Moreover, matrices are the mathematical foundation of networks [9]. Therefore, matrices and networks, particularly probabilistic networks, are a natural fit that are conceptually and mathematically aligned. As such, using network models to represent probabilistic processes has become a very common practice [10]. For instance, several probabilistic network disciplines have recently become popular e.g., psychological networks. Also, similar to TNA, similar probabilistic directed transition models have been proposed and empirically validated in various fields such as Markovian network models and dynamic network models from Markov process data [1, 2, 11].

Despite TNA demonstrated fit and alignment, an integrated framework that takes advantage of transition data, their relational and temporal interactions and the mathematical potential of matrix representations—as TNA does—has not been fully embraced or extended within educational research. TNA as an approach, not only aligns with established practices in other disciplines but also enhances the analytics repertoire for researchers by giving the power and interpretability to complex educational data and contributing to theory and rigor [3].

3 The Theoretical Grounding of TNA

TNA uses stochastic process modeling to capture the dynamics of the learning process, namely, Markov models [11]. Markov models align well with the view that the learning process is an outcome of a stochastic data generating process that led to the overall configuration of the process that we observe [12–14]. The data-generating process is what produces various network configurations or patterns based on rules, constraints, or guiding principles—for instance, self-regulation. This means that while the system operates within certain boundaries defined by the framework or principle, the specific ways in which it changes or evolves is rather random and therefore can't be strictly determined [4, 13]. That is, the transitions are governed by a stochastic process. The stochastic nature of changes introduces variability, making the system's precise future states unpredictable but statistically describable [13].

Therefore, to capture the data generating process and given that the data generating process is partly random (i.e., stochastic), it can be statistically modeled using probability theory [2, 10, 13]. That is, a method that assumes that events are probabilistically dependent on the preceding ones like Markov models. This view has been entertained by several theorists across the years who described such stochastic process e.g., [12–14]. For instance, Winne and Perry conceptualized the dynamics of self-regulation through three interconnected levels: occurrence, contingency, and patterns—that together depict how the learning process unfolds [14].

- **Occurrences** are states, events or actions, such as discussion moves or parts of learning tasks or any learning activity a student makes during learning. Occurrences are the building blocks of the learning process, the events and the elements that make up the full structure. Occurrences are represented as nodes in TNA.
- **Contingencies** represent the sequential dependency or possibility of transitions between events and how certain actions lead to or follow others. For example, after a learner reads a note on a particular topic, she reviews those notes (another occurrences). Contingency is the recording of the transition event or the transition probability when the student moves between two events. This contingency is represented as edges in TNA.
- **Patterns** are the recurring combinations of occurrences and contingencies that emerge throughout the learning process. Patterns represent consistent behaviors that typify a learner's approach to studying or learning. For instance, a pattern might show that a student repeatedly engages in self-evaluation after studying a lesson (contingency). According to Winne [15], "researchers assume mental operations generate behavior, and those cognitive and meta cognitive events are what theories seek to account for" (p. 272). In other words, patterns are the hall mark of the data generation process and the backbone of the structure. After all, a learning process is made of several units, some of which are prominent and repetitive giving it the shape and form it is.

4 The Building Blocks of TNA

The building blocks of TNA are the transitions between events that make up the transition process. A transition is a conditional relationship between an *occurrence* and another *occurrence*, from A $\rightarrow$ B (or a *contingency*) [2, 14]. TNA models these transition in sequential data to compute the transition probability between events [2, 11].The result is a transition matrix that is represented as a weighted directed network where the weights are the transition probabilities from an event to the other and the direction represents the direction of these transitions [2, 11].

Let us see an example of an imaginary transition matrix representing students learning process. We have five learning activities or *occurrences* (e.g., *Watch Video*, *Quiz*, *Read Materials*, *Submit Assignment*, and *Feedback*). The *contingencies* are the transitions between these occurrences or the probability of a student moving from one activity to another. We can see that transitioning from *Watch Video* to *Quiz* occurred 60 out of 100 times which gives us an estimated probability of 0.60 to move from *Watch Video* to *Quiz* (or contingency) (Table 1). We also see each time a student starts a quiz, they did *Submit Assignment* which represents an estimated probability of 1 (Fig. 2).

Table 1 Frequency of transitions between occurrences

From	To	Frequency
Watch video	Quiz	60
Watch video	Read materials	10
Quiz	Submit assignment	100
Read materials	Submit assignment	20
Read materials	Feedback	30
Read materials	Quiz	50
Submit assignment	Feedback	100
Watch video	Feedback	30

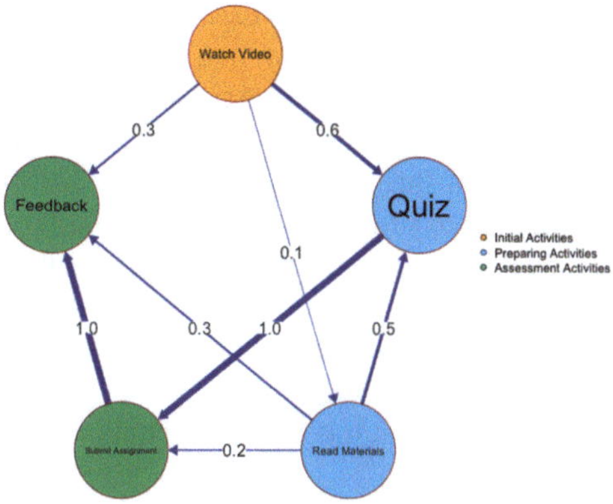

Fig. 2 A visualization of the transition network of the imaginary example in Table 1

In the example, we can see patterns as they emerge when students consistently repeat a behavior, these patterns typify their approach to studying or managing learning tasks. For instance, a repeated pattern of moving from *Read Materials* to *Quiz* reveals a strategy of preparing with taking quizzes before attempting assignments. We also see another pattern of transitioning from *Submit assignment* to *Feedback* which reflects meta-cognitive self-evaluation. It is also evident that *Feedback* receives most transitions (has high in-strength centrality) which makes it the central event in the learning process (sum of estimated transition probabilities = $1 + 0.3 + 0.3$) and that *Submit assignment* bridges connections to other events (has high betweenness centrality).

4.1 Network Representation

In TNA, the learning process is represented as a directed weighted network. This network can be described in the following way:

- **Nodes** (V): These represent the different learning events, such as watching a video, taking a quiz, or submitting an assignment. It can also be states, dialogue moves, roles in collaborative activity, motivation states or any event that can be represented as a sequence unit (occurrences).
- **Edges** (E): These represent the transitions between activities, showing the direction of transitioning from one activity to the next.
- **Weights** (W): Weights represent the probability of transitioning from an event or state to another (contingencies).

The building blocks of TNA are weighted edges that under-grid the structure and dynamics of the transition processes. The edges weight indicates the likelihood (possibility) of transition and the arrows reflects the direction of the transition. Let's take some examples from our example network.

- From *Watch Video* to *Quiz* ($v_1 \rightarrow v_2$, ($W = 0.6$): means that there is a directed edge from *Watch Video* (v_1) to *Quiz* (v_2), with a weight of 0.6. This means there is a 60% probability that students proceed to take the quiz after watching the video, making it a common next step.
- From *Quiz* to *Submit Assignment* ($v_2 \rightarrow v_4$, ($W = 1$): means that there is a directed edge from *Quiz* (v_2) to *Submit Assignment* (v_4) with a weight of 1. This indicates that all students (100%) are likely to move from completing a quiz directly to submitting an assignment. It may be because this is a mandatory step, meaning every student follows this path.
- From *Read Materials* to *Quiz* ($v_3 \rightarrow v_2$, ($W = 0.5$): means there is an edge from *Read Materials* (v_3) to *Quiz* (v_2) with a weight of 0.5. This signifies that there is 50% probability that students choose to take a quiz after reading materials, indicating that it is relatively common.

The TNA network $G = (V, E, W)$ is thus composed of the nodes (V), the edges (E), and weights (W). The resulting directed weighted network G provides a complete view of the learning process, and shows how students navigate through different activities, which transitions are most probable, and which activities are important within the overall learning process.

4.2 *Mathematical Grounding of TNA*

Given the stochastic nature of the learning process —as we assume it is the generating mechanism of the data—, TNA uses stochastic modeling (i.e., Markov models) [4, 10, 13]. Markov models mine the transition probabilities from sequences of activities by capturing the likelihood of moving from one state to the next as a probability [4, 9, 10]. For a given sequence of events, the transition probability $P(v_j \mid v_i)$ is estimated from data as:

$$\widehat{P(v_j \mid v_i)} = \frac{n(v_i \to v_j)}{\sum_{k=1}^{S} n(v_i \to v_k)}$$

where $n(v_i \to v_j)$ is the count of observed transitions from state v_i to v_j, and the denominator represents the total outgoing transitions from state v_i with S denoting the total number of states. If there are no transitions from v_i, then the probability cannot be estimated. These estimated probabilities are assembled into a **transition matrix** T, where each element T_{ij} represents the estimated probability of transitioning from v_i to v_j. For example, the matrix of our example looks like this.

$$T = \begin{pmatrix} 0 & 0.6 & 0.4 & 0 \\ 0 & 0 & 0.2 & 0.8 \\ 0.3 & 0.5 & 0 & 0.2 \\ 0 & 0 & 0 & 1 \end{pmatrix}$$

Matrices and networks together offer a seamless representation of probabilistic models, where transition matrices underpin the mathematical structure and graph theory methods provide the tools and foundations for the analysis.

4.3 *TNA Features*

4.3.1 Visualization

Visualization of the TNA models allows researchers to get a bird's-eye view of the learning process, capture the essence of the full structure, the connectivity of the events, identify important patterns, central learning events as well as how they are temporally related. Comparative plots (either by subtraction or permutation) allows researchers to understand the differences between groups of students—such as high-achieving versus low-achieving students or those in different contexts that may influence strategies or influence behavior [4, 9, 10, 16]

TNA offers powerful visualization features that rely on the `qgraph` package with several enhancements [17]. First, TNA graphs capture the full transition process in an intuitive network visualization with optimized layouts. Also, the ability to fix the layout across several plots allowing the comparison across several transition processes. Further, the visualization in TNA can accept other layouts from, e.g., the `igraph` package as well as any other custom designed layout (see `qgraph` manual for this) [18]. Furthermore, the versatility of `qgraph` allows researchers to add more information to the plot e.g., confidence intervals, thresholds and themes etc. The flexible and customization visualization options —as we will see later— offers researchers several possibilities of visualizing patterns (e.g., dyads and triads), communities and any custom configurations.

4.3.2 Network Measures

Network measures provide important insights into the overall structure of the process (graph-level measures), the importance or the roles of the events (centrality measures) in the process as well as the centrality of transitions (edge centrality) [3, 9, 19]. TNA estimates many of these measures that are appropriately situated for the nature of the transition networks (a probabilistic directed and weighted network).

4.3.2.1 Network-Level Measures

TNA offers several basic network measurements that start with the simple measures like node count which indicates the number of unique events or activities in the network and edge count which represent the number of unique directed transitions between these nodes. However, given that TNA networks may be fully connected or saturated (each node is connected to all other nodes), some measures may not be meaningful in such scenarios. For instance, network *density* reflects how completely connected the network is, or the proportion of actual connections to all possible ones. In cases of fully connected networks, density is 1 and therefore, does not offer much information. However there are several other situations (non-connected networks, pruned networks, or bootstrapped models which will be discussed later) that network density will be useful in showing the network structure [3, 9]. Similarly, measures like the mean out-strength and in-strength which refer to the sum of transition probabilities of outgoing or incoming interactions to each node, respectively may need to be understood according to the network connectivity (fully saturated or not) [9].

Centralization (both out-degree and in-degree) reflects how centralized the network is around particular nodes showing a dominant state, or event. These measures will show which events are at the center of the process, or whether an event or a state dominates the process [20]. *Reciprocity* reflects the likelihood that two nodes have mutual connections, reflecting the mutuality of interactions. It may be important in some processes to examine their reciprocal interactions such as collaborative processes where we expect some stronger transitions between, e.g., argumentation and knowledge building [9, 21].

These measures together provide a concise overview of network connectivity, distribution, and flow of dynamics in the transition process. Notable to say that no existing process mining model includes or allows or operationalizes network measures to understand the learning process and therefore, all of such measures are among the strengths of TNA.

4.3.3 Node-Level Measures

Centrality measures provide a quantification of the role or importance of a state or an event in the process [20]. With centrality measures, researchers can rank events according to their value in, for instance, bridging the interactions (betweenness centrality) or receiving most transitions (in-strength centrality) [22]. In doing so, centrality measures can reveal which behaviors or cognitive states are central to the learning process, either as destinations of frequent transitions, as starting points for various actions, or as bridges between different learning activities or as a key to a spreading phenomenon. Using centrality measures, researchers can identify important events that can be targets for intervention or improvement [3, 9].

The choice of what centrality measure to apply depends heavily on the specific research [20]. From a probabilistic perspective, certain centrality measures reflect the inherent properties of how transitions are structured, specifically, out-strength centrality is consistently equal to 1 across all nodes, it means that the transition probabilities from each event always sum 1 [4]. Therefore, it is meaningless in some situations. However, in TNA networks where we delete insignificant edges and self-transitions, out-strength becomes useful and in that case reflects stability, lack of change (being stuck in a state) or repetitive actions [10].

It is also important to mention that in TNA, the raw or absolute values of centrality measures are not inherently meaningful on their own. It is the relative values of these measures that matter, allowing researchers to rank nodes and identify their relative importance or influence within the network [9]. This comparison helps determine which events are central, which nodes serve as key bridges, and which activities are most likely to receive transitions in the learning process. Last, it is important to note that not all centrality measures are directly transferable to TNA and researchers have to think how and why a measure makes sense in their own context [23, 24].

4.3.4 Edge-Level Measures

In TNA, edge centrality measures quantify the importance of *transitions* between events, rather than the importance of the events themselves, they provide insights into how critical a particular transition is in to the flow of transitions or interactions in the process [18]. In particular, betweenness centrality of an edge, for example, reflects how often a transition bridges other transitions. In doing so, edge centrality measures help researchers understand not just which nodes are important, but which transitions are important in guiding the learning process and maintaining the integrity of the sequence of activities.

4.3.5 Patterns

Patterns represent the fundamental building blocks of the structure and dynamics of the learning process. As we discussed before, patterns provide insights into, e.g., the behavior and strategies that learners use while studying or interacting with learning materials [3, 15, 19]. Furthermore, capturing repeated consistent patterns allows us to build theories and generalizable inferences [14]. For instance, if we repeatedly found a strong pattern of transition between planning and task enactment in project work and this pattern is associated with project success, we can therefore recommend such approach [25]. We can also design to strengthen such pattern and monitor students when doing it or support them. On the functionality side, TNA allows the identification of several patterns: cliques (dyads and triads and other cliques) and communities. A network clique is a subset of nodes in a graph where every pair of nodes is directly connected to each other by an edge. In network terms, cliques represent tightly-knit communities or groups, closely related entities or systems or interdependent nodes that shape how learning unfolds over time [3] .

4.3.6 Cliques: Dyads

A dyad is the simplest pattern in TNA representing a transition between two nodes. A dyad becomes interesting when it is a mutual dyad (bidirectional) and when the edge weights are high, e.g., more than 0.15 in either direction. Strong mutual dyads (henceforth dyads for brevity) indicate the strong interdependence between states or events that they are recurrently occurring. Analyzing dyads helps to identify important sequences, recurrent or consistent transitional patterns of learning behavior or strategies. For instance, consistently moving from *reading materials* to *taking a quiz* indicating a strong self-evaluative strategy. A student commonly transitioning between disengagement and engagement indicates instability and problematic approach. On the overall picture (Fig. 3-left), dyads are the backbone of the whole process and the constellation of dyads help characterize and describe the full process.

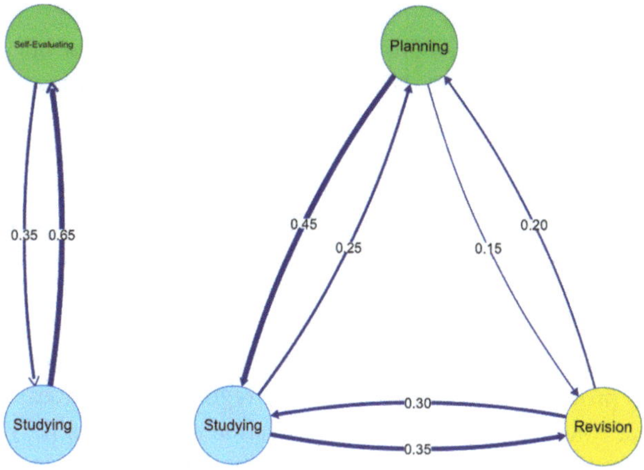

Fig. 3 Dyads (left) and triads (right)

4.3.7 Cliques: Triads

Triads represent connections between three nodes capturing more complex relationships that go beyond direct transitions. In graph theory, the constellation of three nodes carries a very central position and hence multiple names and configurations (triples, triads, triangles). In TNA, cliques of three nodes where each node is connected to the others in either direction indicates a strong interdependent subgroup of nodes that form a strong core of the process. Triads represent higher-order dependencies in learning behavior where one activity not only follows another but influence or depends on subsequent events. For instance, the triad in the example above (Fig. 3-right) shows strong temporal interdependence between studying, revising and planning reflecting a well-managed approach to learning. The analysis of triads is particularly important in TNA, as it can highlight the presence of reciprocal or reinforcing relationships, where learning activities form a cohesive structure that encourages repeated engagement [3].

4.3.8 Communities

Communities are groups of nodes that are more closely related or densely interconnected together more than to other nodes in the network. Specifically, in TNA, communities are groups of states or events that frequently transition between one-another or share similar dynamics [26]. Also, communities are cohesive sequences or succession of activities that are more likely to co-occur together, has a typical pathways or recurring behaviors. Unlike cliques, which have a fixed or predefined structure (2-clique or 3-clique), communities are data-driven based on the connectivity patterns in the data which makes them more descriptive of real-world structure [9]. In doing so, communities reveal existing structures, unravel typical configurations or constellations or actually mine them from the learning process. Communities are not strictly fully connected but show higher-than-average connectivity where the grouping indicates a higher level of association and interaction compared to the broader network [26].

In the context of TNA, communities can be groups of dialogue moves, utterances, or states, or learning events that are more densely connected with each other or follow each other. Identifying communities help uncover these — latent or hidden— clusters of related interactions or behaviors during the learning process showing e.g., how learners collaborate or self-regulate or approach their learning. For instance, in a collaborative learning scenario, communities might form around self-evaluative utterances, feedback exchanges, or task-enactment dialogue moves. Identifying these clusters provides insight into common regulatory practices or patterns of interaction that contribute to effective collaboration and learning [23].

Furthermore, identifying communities of behavior or events in TNA can contribute or advance our theory building and consequently understanding of learning. Think of identifying communities of behavior as uncovering latent variables. These communities represent underlying patterns of interaction inferred from densely connected behaviors into a simplified, meaningful structure suggesting the presence of an underlying construct or a behavioral mechanism that generated such a pattern. This data-driven approach helps provide evidence of existing constructs, concepts or validating existing ones. As such, identifying communities can help refine theoretical models or develop new ones or help discover and describe the dynamics of behavior. Using communities of inter-related events or factors to infer constructs or theoretical models has been an established tradition in several fields, e.g., factor analysis, network modeling or semantic networks [3, 20, 23, 27].

4.3.9 Clusters

Clusters represent typical transition networks that recur together across time [23, 28]. Unlike communities, clusters involve the entire network where groups of states or events are similarly interconnected and each exhibit a distinct transition pattern (latent or hidden) with its own set of transition probabilities [4]. This means that clusters capture the dynamics in learning behaviors, revealing typical relations that learners frequently adopt as units across different instances. Identifying clusters of behavior is a very common approach in learning analytics under several names or labels, e.g., tactics, clusters, or strategies. Here, clusters are similar to those in the sequence mining literature where temporal patterns are inferred from the data to reveal typical behavioral approaches [29–31]. Clusters will be discussed in a devoted chapter [23].

4.3.10 Sub-networks or Network Comparison

Oftentimes, we encounter two predefined conditions, such as high versus low achievers, or different course types (e.g., practical versus theoretical, or easy task versus demanding task). In that case, we have two different processes that are defined by their context (not clustered or inferred). Comparing such groups has been commonly performed visually, e.g., comparing two process models, sequence models, or epistemic networks [29, 32]. While visual comparison may show and inform about differences, it does not tell us about how statistically significant these differences are. Where exactly the differences are statistically significant and where they are not. TNA offers a rigorous systematic method for process comparison based on permutation. In doing so, we can compare models visually, plot the differences and estimate the significance of each and every edge. Having such a rigorous comparison opens the door for researchers to draw meaningful inferences, identify theoretically significant differences, and refine our understanding of learning processes in various contexts. This statistical rigor enhances the validity of our findings and contributes to the development of more nuanced theories and conceptual frameworks [33].

4.3.11 Bootstrapping and Model Validation

Most research on networks or process mining uses descriptive methods. The validation or the statistical significance of such models are almost absent in the literature. Having validated models allows us to assess the robustness and reproducibility of our models to ensure that the insights we get are not merely a product of chance and are therefore generalizable.

TNA offers an important methodological advancement by incorporating bootstrap methods to identify significant edges within the TNA networks [19]. Bootstrapping is a re-sampling technique that entails repeatedly —usually hundreds, if not thousands of times— drawing samples from the original dataset with replacement to estimate the model for each of these samples [34, 35]. Bootstrapping does not require strong assumptions regarding the distribution of the data which makes it suitable for analyzing process data which often do not adhere to specific distributions [36]. Given that bootstrapping entails replacement, each bootstrap sample may include multiple copies of some observations while excluding others to assess variability in the estimated parameters. When edges consistently appear across the majority of the estimated models, they are considered stable and significant [36]. In doing so, bootstrapping helps effectively filter out small, negligible or spurious edges resulting in a stable model and valid model [34, 35].

Another advantage of bootstrapping is that it can effectively prune dense networks. This because one of the challenges of most probabilistic networks like TNA is that they are commonly fully connected and lack sparsity (i.e., meaning that every possible connection between nodes (events or states) is present to some degree [10, 34]. This is a common issue in most probabilistic networks [34]. Bootstrapping can help mitigate this issue through identifying and eliminating small and uncertain edges to effectively retrieve the backbone of the network. The resulting simplified network is easier to interpret and more likely to be generalizable. The integration of bootstrap methods into TNA represents a significant advancement in the validation of process models that has the potential to improve the quality of the models but also contributes to more effective data-driven theory building in education.

The bootstrap function in TNA allows for the calculation of confidence intervals and p-values for each edge weight. This level of statistical rigor provides a quantifiable measure of uncertainty and robustness for each transition in the network. Bootstrapping also help exclude transitions that may not represent true relationships within the data to include only the most meaningful interactions. The resulting process models may offer an empirical assessment for existing constructs and concepts, as well as in validating or refining hypothesis. A systematically rigorous model also allows researchers to uncover patterns and relationships that either support current theoretical frameworks or advance others.

5 Tutorial of TNA with R

TNA can analyse any data that can be represented as a sequence which has transitions or changes across time. In other words, TNA accepts any categorically ordered event data, e.g., sequence of learning events, states, phases, roles, dialogue moves or interactions to mention a few. This data can come from time-stamped learning management system data, coded interactions data, event-log data or order event data. The data can be in `stslist` format (sequence object), which is typically created using the `seqdef()` function from the `TraMineR` package. In addition, `tna` accepts wide data format data where each row represents sequence of data and each column is a time point.

The analysis in this chapter uses the `tna` R package and other packages that are necessary for data manipulation and plotting [37]. It is assumed that the reader is familiar with the R programming language. If that is not the case, it is recommended to refer to previous tutorials on the basics of R [38], data cleaning [39], basic statistics [40], and visualization [41]. It is also recommended to have previous knowledge about Markov models [4].

In the first step of the analysis, we begin by loading the necessary packages. We will use `tidyverse` for data manipulation, wrangling, and ensuring the data is clean and organized for analysis [42]. Most importantly, we load the `tna` package which provides a wide array of functions for estimating, visualizing and manipulating `tna` models [37]. Besides loading the required libraries, the next code chunk also loads the data that we are going to use throughout the tutorial, which is a built-in dataset in the `tna` package (`group_regulation`) and represents interactions between groups of learners trying to regulate their learning based on the data in this study [3].

```
# Install `tna` if you haven't already
# install.packages("tna")
library("tna")
library("tidyverse")
data("group_regulation")
```

5.1 *Building* `tna` *Model*

TNA analysis starts by building the main TNA object which is called `model`. A model contains all the information necessary for further analysis e.g., plotting, estimation of centralities, or comparison. To estimate a TNA model we use the function `tna()` with one argument which is a `stslist` object (sequence object) usually created by the `seqdef()` function of `TraMineR`. However, the `tna` package can also process event data with the `prepare_data()` function to first convert such data into a proper format.[1] Furthermore, `tna()` also accepts standard data frames in wide format (to learn how to create a sequence object you may need to refer to this chapter [28]). The function estimates a Markov model from the data where the initial and transition probabilities are estimated directly from the observed initial state probabilities and transition frequencies. The TNA model is structured as a list containing the elements that facilitate the analysis of transition networks, these elements are described below:

- **Initial Probabilities (`inits`)**: These define the likelihood of starting in a particular state at the beginning of the process (at the first time point before any transitions happen). In an educational context, an initial probability represents the probability that a student begins in a specific state (such as "engaged" or "motivated") before any activities or interventions occur. These probabilities provide a snapshot of where students start within the process and help frame the subsequent transitions.
- **Transition Probabilities (`weights`)**: These describe the likelihood of moving from one state to another at each step in the process. Transition probabilities capture how students transitions, moves or follows between different learning states (engaged, motivated or achiever) or events (e.g., assessment, reading or submission).
- **Labels (`labels`)**: The descriptive names for each node in the network and is included to enhance the interpretability of the analysis. The labels are automatically retrieved from the alphabet of the sequence object or the categories in the data frame.

[1] Learn more about data preparation at: https://sonsoles.me/tna/articles/prepare_data.html.

- **Data (`data`)**: This is a version of the sequence data (or data frame) that contains all the necessary sequence data, converted into an internal format used by the `tna` package for further analysis (permutation, bootstrapping, etc.).

Together these elements form the basis for analysis and visualization. In the example above we see the TNA network plot with transition probabilities as directed edge weights and the initial probability as the pie (the thin ring around the nodes). In particular, we see video has an initial probability of 1 and all other events have an initial probability of 0. You can use `print(model)` to get a feel of what it has and how is it structured.

```
# Building a tna model by using the tna function with the data
model <- tna(group_regulation)
print(model)
```

```
State Labels

adapt, cohesion, consensus, coregulate, discuss, emotion, monitor, plan, synthesis

Transition Probability Matrix

             adapt cohesion consensus coregulate discuss emotion monitor  plan synthesis
adapt      0.00000    0.273     0.477      0.022   0.059   0.120   0.033 0.016    0.0000
cohesion   0.00295    0.027     0.498      0.119   0.060   0.116   0.033 0.141    0.0035
consensus  0.00474    0.015     0.082      0.188   0.188   0.073   0.047 0.396    0.0076
coregulate 0.01624    0.036     0.135      0.023   0.274   0.172   0.086 0.239    0.0188
discuss    0.07137    0.048     0.321      0.084   0.195   0.106   0.022 0.012    0.1410
 [ reached getOption("max.print") -- omitted 4 rows ]

Initial Probabilities

     adapt   cohesion  consensus coregulate    discuss    emotion    monitor       plan
     synthesis
     0.011      0.060      0.214      0.019      0.175      0.151      0.144      0.204
     0.019
```

5.2 *Plotting and Interpreting* `tna` *models*

The next step is to plot the model, and in TNA it is as easy as creating the model. We can just call the function plot and pass the model argument to it `plot(model)`. The plot is a directed weighted network where nodes take the colors and the labels retrieved from the sequence object (if available). Each node (state or event or a learning activity) is represented as a circle. For instance, we see nodes for adapt, cohesion, coregulate in the network as well as six other nodes. Around each node, there is a thin rim (pie) that represents the initial probability (`inits`) of starting in this state.

For instance, students are more likely to start in planning more than any other state (initial probability of 0.21) and showing as a pie proportional to 0.21 of the circle. We also see, that adapt has very small pie (initial probability of 0.01) which makes sense that adapt follows other actions.

The arrows from each node to the other node represent the weighted transition probability and the direction of the transition. For instance, we see that there is a strong transition from adapt to with a probability of 0.121 and from adapt to cohesion with a probability of 0.27. We can also see loops, which represent the probability of staying in the same state, or repeating the same event again in the next time point.

5.2.1 Interpretation of the Model

The model plot reveals the dynamics of self-regulated learning and co-regulation among students. We start with the initial events, where students usually begin with planning (0.20), consensus (0.21), or discussing (0.18). Students are less likely to start with some states such as coregulate (0.019), synthesis (0.02), and adapt (0.012). These smaller initial probabilities suggest that these activities are less common starting points and shows students priorities in their regulation.

Task-related processes such as planning and discussing are central in the network. The transition probabilities reveal that these processes are interconnected in various ways, reflecting the configuration of the self-regulated group interactions. Notably, transitions to consensus are particularly strong, with significant transitions from adapt (0.48) and cohesion (0.50). This indicates that adaptability and social cohesion are crucial for reaching consensus. The transition from consensus to coregulate (0.19) suggests that achieving agreement often leads to collaborative regulation.

Emotional engagement plays a significant role students' bonding and collaboration, as evidenced by the strong transition from emotion to cohesion (0.33). This highlights how emotions can facilitate social interactions and collaborative dynamics. The transition from cohesion to adapt (0.27) shows the role of social interaction in adaptability. Monitoring is another important regulatory process with notable transitions to plan (0.22) and discuss (0.38). This suggests that self-assessment prompts both further planning and discussion, essential for effective self-regulation. The transition from discussing to co-regulate (0.27) indicates that discussions often lead to collaborative regulation, further supported by the transition from co-regulate to monitor (0.09), demonstrating the cyclical nature of self-regulated learning.

Overall, the network reveals a rich interplay between cognitive and emotional processes, emphasizing task enactment and socio-emotional regulation. Understanding these transitions can help educators develop strategies to support learners.

The code below visualizes the network model (Fig. 4). Given that using `plot(model)` produces a very dense and hard to read model we will set two parameters:

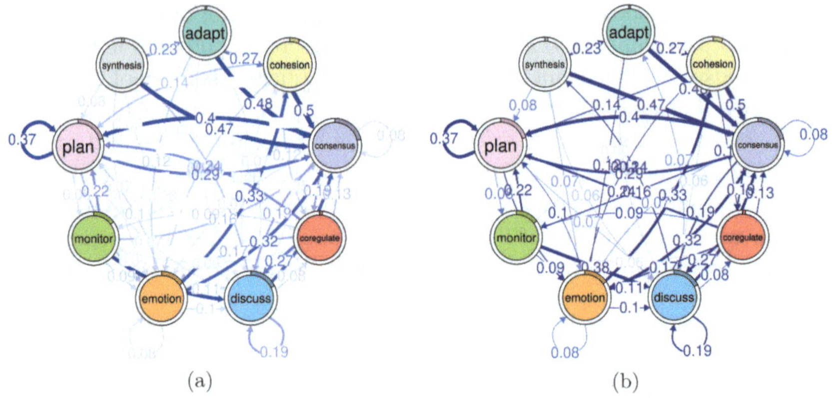

Fig. 4 TNA visualization. (**a**) Default. (**b**) Minimum = 0.05, Cut = 0.1

- **Minimum Edge Weight** (`minimum = 0.05`): Only edges with a weight of 0.05 or higher will be included in the plot.
- **Cut-Off Threshold** (`cut = 0.1`): Edges with a weight below 0.1 are entirely shown with lighter color.

```
plot(model)
plot(model, minimum = 0.05, cut = 0.1)
```

5.2.2 Pruning and Retaining Edges that "Matter"

Transition networks are commonly fully connected —or saturated— where almost all nodes have connections to all other nodes with some probability. Therefore, mechanisms are needed to retrieve the core or backbone structure of the network. In other words, to make the network sparse. Network sparsity enhances interpretability by removing the overly complex structures, making it easier to identify the important components and relationships. It also isolates the signal from the noise by removing the small noisy edges that may obscure meaningful patterns allowing researchers to focus on the important interactions. We can use the minimum argument in TNA plots to remove small edges e.g., below 0.05 (visually) for easier reading and interpretability as we did before. Worth noting that these small probabilities are retained in the model for all further computations. Researchers may be interested in removing edges that are small and negligible in weight or pruning the model. Removing these small edges keep only the strong connections or transitions that are worthy and meaningful. In a network like ours, we have 81 edges and it is hard to take into account all of them.

In TNA, small edges can be removed using the `prune()` function which removes edges using a user-specified method via the `method` argument. With the "threshold" method, edges are removed below a certain threshold value (the default is 0.05 which removes edges below 0.05). The "lowest" method deletes the lowest percentile of edges for instance, lowest = 0.1 removes all edges in the lower 10th percentile in terms of edge weights. When using these methods, the prune() function makes sure that it does not disconnect the network, if it finds that the removal of one edge disconnects the network, it retains it to keep the network connected.

Another theoretically sound method is to remove edges using a network null model. While several null models exists for non-weighted networks, only a few can be reliable in weighted networks like transition networks. TNA implements the *Disparity Filter* which is a robust network sparsification algorithm that helps extracts the backbone structure of weighted networks [43]. The disparity filter simplifies the network by discarding weak edge that are lower than a pre-defined alpha level. The values of alpha ranges from 0.05 (roughly statistically significant compared to a random model), to 0.5 (more likely to than chance). Values of 0.05 are very restrictive and removes most edges in the network, therefore, for the case of pruning TNA models, it is advisable to use the default 0.5 levels which simply means that these edges are more likely to have resulted from a learning process rather than by chance. Values of 0.05 maybe used in larger networks or in cases of theory building and could possibly be interpreted that these structure are likely to be repeated or reproducible in future scenarios. Applying the disparity filter to the TNA model is simple and can be applied using the "disparity" method in prune() with the model as argument. The code in the next chunk does three all types of pruning in TNA (Fig. 5):

1. **Threshold**: Directly removes edges with weights below 0.05, offering a straightforward cutoff approach.
2. **Lowest Percentage**: Removes the bottom 15% of edges when sorted by weight, focusing on relative edge importance.
3. **Disparity Filter**: Uses a different statistical approach (significance level $\alpha = 0.5$) to identify significant edges based on the network's topology.

Pruning with tna can also be carried out by bootstrapping, which we will demonstrate later.

```
# Pruning with different methods (using comparable parameters)
pruned_threshold <- prune(model, method = "threshold", threshold = 0.05)
pruned_lowest <- prune(model, method = "lowest", lowest = 0.15)
pruned_disparity <- prune(model, method = "disparity", level = 0.5)

# Plotting for comparison
plot(pruned_threshold)
plot(pruned_lowest)
plot(pruned_disparity)
plot(model, minimum = 0.05, cut = 0.1)
```

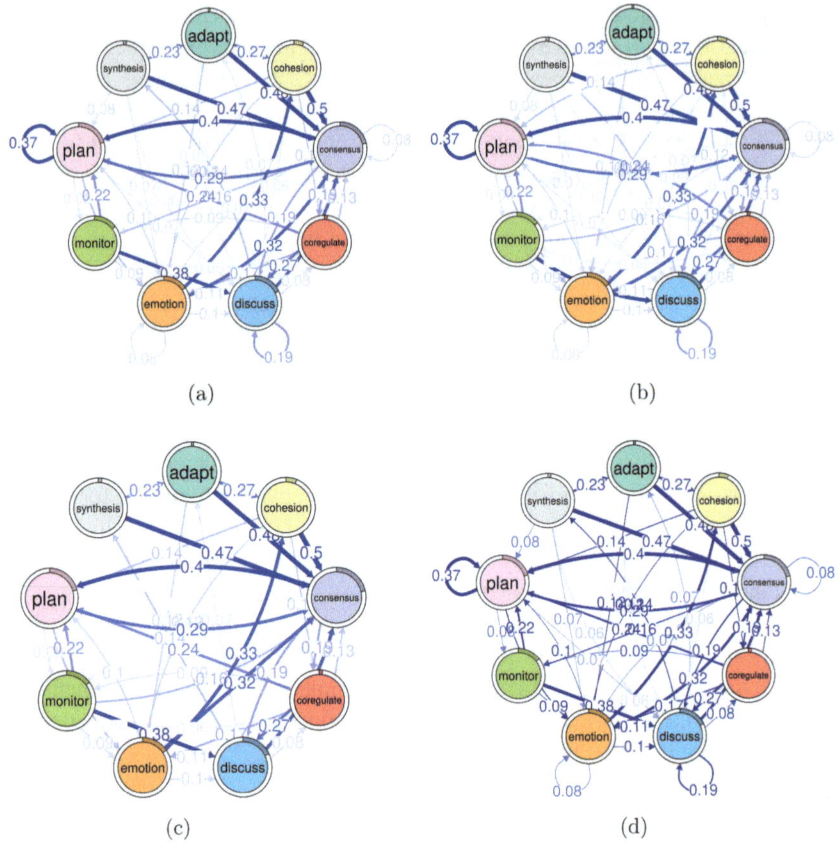

Fig. 5 Pruning with different methods. (**a**) Threshold = 0.05. (**b**) Lowest 15%. (**c**) Disparity filter. (**d**) Minimum threshold

5.3 Patterns

Patterns can be helpful to understand behavior, identify significant structures and help describe the process as detailed above. TNA supports identifying several types of patterns that can be expressed as n-clique. The `cliques()` function is designed to identify and visualize n-cliques from a TNA model. The function `cliques()` returns the number of cliques found, their respective matrices, while also allowing the user to visualize the cliques. Users can also pass custom visualization options e.g., edge, node sizes, color or layout. The arguments of `cliques` include the TNA model, `size` for specifying the size of cliques, so setting `size = 2` would find all dyads and setting `size = 3` would find all triads and so forth. The `threshold` argument sets the lowest weight for which an edge is considered part of the clique.

The `sum_weights` argument can be set to determine whether the sum of edge weights should be considered when forming cliques for instance, `threshold =`

0.1 and `sum_weights = FALSE` means that the edge has to have 0.1 weight in either direction. In contrast, `threshold = 0.1` and `sum_weights = TRUE` means, that the sum of the weights in either direction has to be 0.1 regardless of individual edge weights.

The dyads below are a list of the strong interdependent dyads. The triads show the strong well-connected structures or patterns. We do not recommend looking at 4 or 5 cliques routinely but, of course, researchers could customize based on context.

In this code, four different calls to the `cliques` function are made to identify and visualize cliques of varying sizes (with `size` values of 2, 3, 4, and 5) from a network model named `model`. Each function call focuses on a different size of cliques, using a specific weight threshold to determine which edges are considered strong enough to form part of a clique. The `show_loops` argument to `plot` controls whether self-loops (edges connecting a node to itself) should be displayed in the visual output, but loops are never included in the actual computation of cliques.

The first function call identifies cliques of size 2, known as dyads (Fig. 6), using a weight threshold of 0.1. This means only pairs of nodes that are connected by edges with transitions weights higher than 0.1 are considered. The second function identifies 3-cliques (Fig. 7), or triads, using a weight threshold of 0.05 which captures the three fully connected nodes in the network. The third call focuses on 4-node cliques, or quadruples (Fig. 8), with a lower weight threshold of 0.03. Finally, the fourth call identifies 5-cliques, or quintuples (Fig. 9), with a weight threshold of 0.1. In this case, the `sum_weights = TRUE` option means that the sum of the edge weights in both directions between nodes is considered.

```r
# Identify 2-cliques (dyads) from the TNA model, excluding loops in the visualization
# A clique of size 2 is essentially a pair of connected nodes
cliques_of_two <- cliques(
  model,
  size = 2,
  threshold = 0.1  # Only consider edges with weight > 0.1
)
print(cliques_of_two)
```

```
Number of 2-cliques: 5 (weight threshold = 0.1)
Showing 5 cliques starting from clique number 1

Clique 1:
           consensus coregulate
consensus      0.082      0.188
coregulate     0.135      0.023

Clique 2:
          consensus plan
consensus     0.082 0.40
plan          0.290 0.37
```

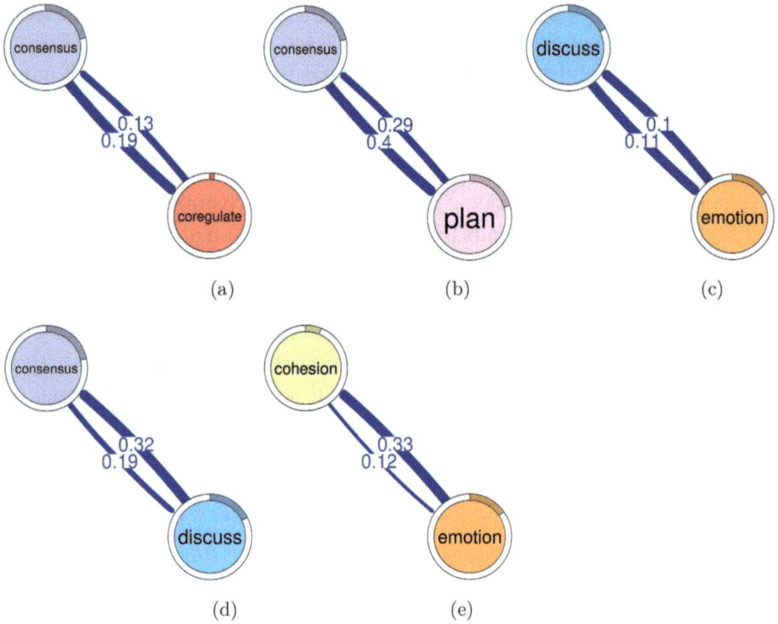

Fig. 6 Identified dyads. (**a**) Dyad 1. (**b**) Dyad 2. (**c**) Dyad 3. (**d**) Dyad 4. (**d**) Dyad 5

```
Clique 3:
        discuss emotion
discuss    0.19   0.106
emotion    0.10   0.077

Clique 4:
          consensus discuss
consensus     0.082    0.19
discuss       0.321    0.19

Clique 5:
         cohesion emotion
cohesion    0.027   0.116
emotion     0.325   0.077
```

```
plot(cliques_of_two)
```

```
# Identify 3-cliques (triads) from the TNA_Model
# A clique of size 3 means a fully connected triplet of nodes
cliques_of_three <- cliques(
```

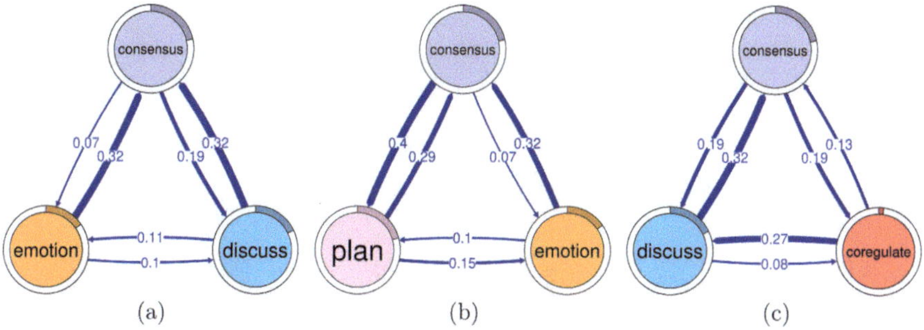

Fig. 7 Identified triads. (**a**) Triad 1. (**b**) Triad 2. (**c**) Triad 3

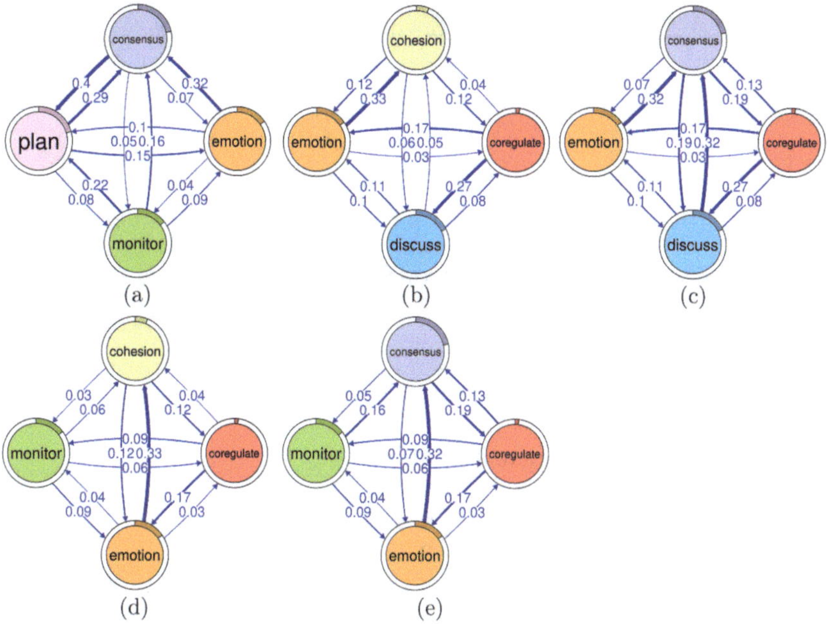

Fig. 8 Identified quadruples. (**a**) Quadruple 1. (**b**) Quadruple 2. (**c**) Quadruple 3. (**d**) Quadruple 4. (**e**) Quadruple 5

```
  model,
  size = 3,
  threshold = 0.05 # Only consider edges with weight > 0.05
)
print(cliques_of_three)
```

```
Number of 3-cliques: 3 (weight threshold = 0.05)
Showing 3 cliques starting from clique number 1
```

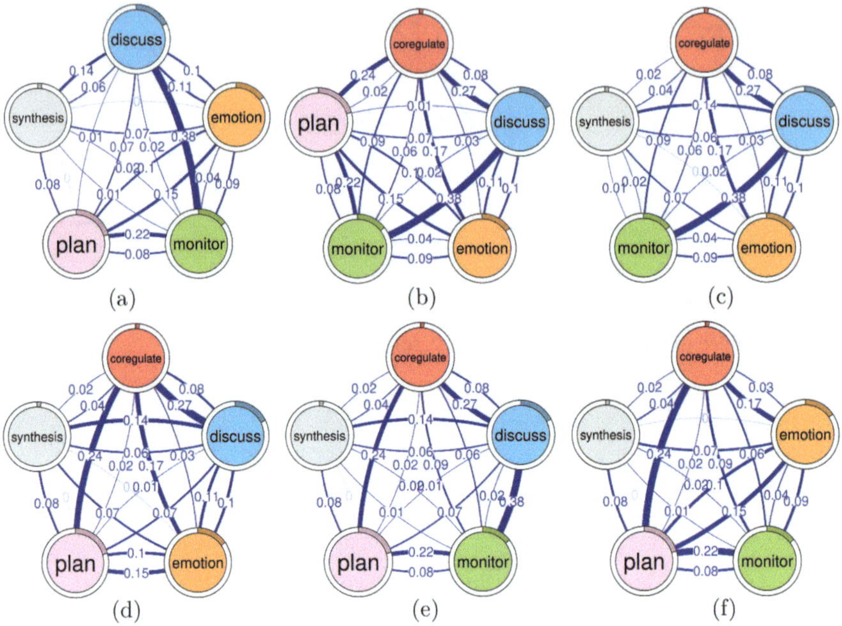

Fig. 9 Identified quintuples. (**a**) Quintuple 1. (**b**) Quintuple 2. (**c**) Quintuple 3. (**d**) Quintuple 4. (**e**) Quintuple 5. (**f**) Quintuple 6

```
Clique 1:
          consensus discuss emotion
consensus     0.082    0.19   0.073
discuss       0.321    0.19   0.106
emotion       0.320    0.10   0.077

Clique 2:
          consensus emotion plan
consensus     0.082   0.073 0.40
emotion       0.320   0.077 0.10
plan          0.290   0.147 0.37

Clique 3:
           consensus coregulate discuss
consensus      0.082      0.188    0.19
coregulate     0.135      0.023    0.27
discuss        0.321      0.084    0.19
```

```
plot(cliques_of_three)
```

```r
# Identify 4-cliques (quadruples) from the TNA_Model
# A clique of size 4 means four nodes that are all mutually connected
cliques_of_four <- cliques(
  model,
  size = 4,
  threshold = 0.03 # Only consider edges with weight > 0.03
)
print(cliques_of_four)
```

```
Number of 4-cliques: 5 (weight threshold = 0.03)
Showing 5 cliques starting from clique number 1

Clique 1:
          consensus emotion monitor plan
consensus     0.082   0.073   0.047 0.40
emotion       0.320   0.077   0.036 0.10
monitor       0.159   0.091   0.018 0.22
plan          0.290   0.147   0.076 0.37

Clique 2:
           cohesion coregulate discuss emotion
cohesion      0.027      0.119    0.06   0.116
coregulate    0.036      0.023    0.27   0.172
discuss       0.048      0.084    0.19   0.106
emotion       0.325      0.034    0.10   0.077

Clique 3:
           consensus coregulate discuss emotion
consensus      0.082      0.188    0.19   0.073
coregulate     0.135      0.023    0.27   0.172
discuss        0.321      0.084    0.19   0.106
emotion        0.320      0.034    0.10   0.077

Clique 4:
           cohesion coregulate emotion monitor
cohesion      0.027      0.119   0.116   0.033
coregulate    0.036      0.023   0.172   0.086
emotion       0.325      0.034   0.077   0.036
monitor       0.056      0.058   0.091   0.018

Clique 5:
           consensus coregulate emotion monitor
consensus      0.082      0.188   0.073   0.047
coregulate     0.135      0.023   0.172   0.086
```

```
emotion        0.320      0.034   0.077   0.036
monitor        0.159      0.058   0.091   0.018
```

```r
plot(cliques_of_four)
```

```r
# Identify 5-cliques (quintuples) from the TNA_Model, summing edge weights
# Here, the sum of both directions of an edge must meet the threshold
cliques_of_five <- cliques(
  model,
  size = 5,
  threshold = 0.015, # Consider edge weights greater than 0.1
  sum_weights = TRUE # Sum edge weights in both directions when evaluating thresholds
)
print(cliques_of_five)
```

```
Number of 5-cliques: 126 (weight threshold = 0.015)
Showing 6 cliques starting from clique number 1

Clique 1:
          discuss emotion monitor  plan synthesis
discuss     0.195   0.106   0.022 0.012    0.1410
emotion     0.102   0.077   0.036 0.100    0.0028
monitor     0.375   0.091   0.018 0.216    0.0161
plan        0.068   0.147   0.076 0.374    0.0018
synthesis   0.063   0.071   0.012 0.075    0.0000

Clique 2:
           coregulate discuss emotion monitor  plan
coregulate      0.023   0.274   0.172   0.086 0.239
discuss         0.084   0.195   0.106   0.022 0.012
emotion         0.034   0.102   0.077   0.036 0.100
monitor         0.058   0.375   0.091   0.018 0.216
plan            0.017   0.068   0.147   0.076 0.374

Clique 3:
           coregulate discuss emotion monitor synthesis
coregulate      0.023   0.274   0.172   0.086    0.0188
discuss         0.084   0.195   0.106   0.022    0.1410
emotion         0.034   0.102   0.077   0.036    0.0028
monitor         0.058   0.375   0.091   0.018    0.0161
synthesis       0.044   0.063   0.071   0.012    0.0000

Clique 4:
           coregulate discuss emotion  plan synthesis
coregulate      0.023   0.274   0.172 0.239    0.0188
discuss         0.084   0.195   0.106 0.012    0.1410
```

```
emotion          0.034    0.102    0.077 0.100    0.0028
plan             0.017    0.068    0.147 0.374    0.0018
synthesis        0.044    0.063    0.071 0.075    0.0000

Clique 5:
           coregulate discuss monitor  plan synthesis
coregulate      0.023    0.274    0.086 0.239    0.0188
discuss         0.084    0.195    0.022 0.012    0.1410
monitor         0.058    0.375    0.018 0.216    0.0161
plan            0.017    0.068    0.076 0.374    0.0018
synthesis       0.044    0.063    0.012 0.075    0.0000

Clique 6:
           coregulate emotion monitor  plan synthesis
coregulate      0.023    0.172    0.086 0.239    0.0188
emotion         0.034    0.077    0.036 0.100    0.0028
monitor         0.058    0.091    0.018 0.216    0.0161
plan            0.017    0.147    0.076 0.374    0.0018
synthesis       0.044    0.071    0.012 0.075    0.0000
```

```
plot(cliques_of_five)
```

5.3.1 Centralities

5.3.1.1 Node-Level Measures

Centrality measures are important in identifying the important events, rank their value in certain processes. The `tna` package has a built-in function `centralities()` to compute centralities using the appropriate algorithm for a directed probabilistic process. The computation of centralities is simple with just passing the model object (`model`) to the function. By default, `centralities()` removes loops from the calculations which can be changed by the user by setting `loops=TRUE`. Removing the loops would entail that all computation of centralities will be performed without considering self-transitioning or staying in the same state. In that context, out-strength will mean stability of a state, the higher out-strength centrality, the more stable the state is as it less likely to transition to other states and vice versa. In our context, for example, students use adapt (out-strength = 1), means they always follow it by other processes. Whereas, plan has the lowest out-strength = 0.63, meaning that students may repeat planning steps several times before moving on. Please note, that out-strength centrality with `loops = TRUE` will always be 1 and therefore, will be meaningless to compute in non-pruned networks. In-strength centrality reflects the sum of received transitions and indicates the node is a common pathway that states end in.

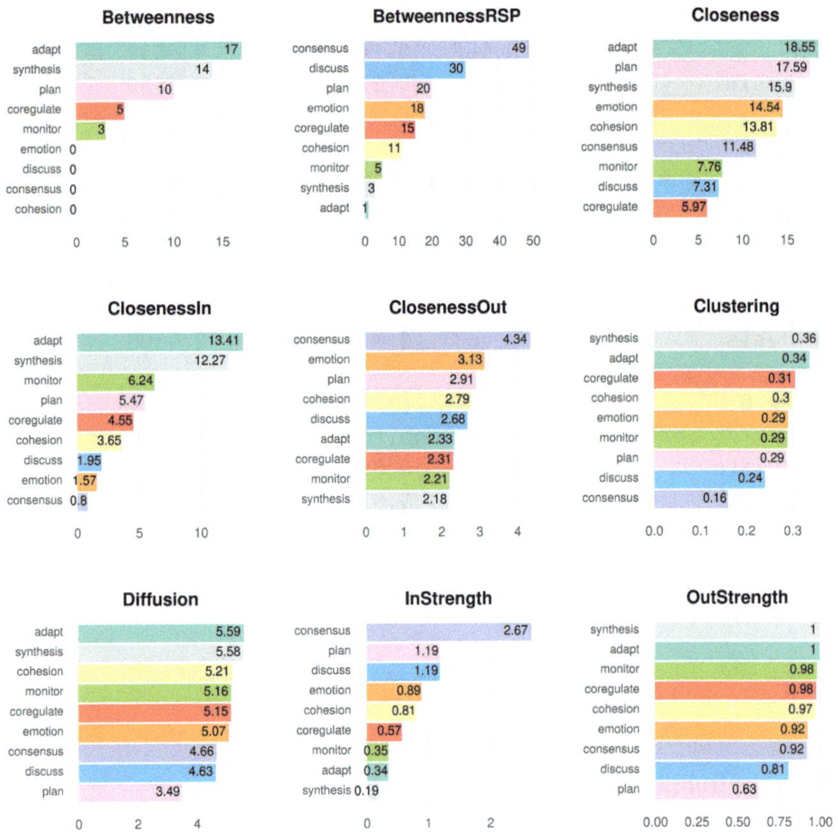

Fig. 10 Centrality measures

In our example, we see that `consensus` task received most transitions from other events (in-strength = 2.7). Betweenness centrality (based on randomized shortest paths, RSP) reflects the events that mediate or bridge other transitions. Please note, that in `tna` it is advisable to consider the Betweenness RSP as it is more appropriate for probabilistic networks. In our example, `adapt` lied between most other transitions The function also computes several other centrality measures but we won't discuss them one by one here. Researchers can use `?centralities` to read the list of possible options. The default ones are shown in Fig. 10.

```
# Compute centrality measures for the TNA model
Centralities <- centralities(model)

# Visualize the centrality measures
plot(Centralities)
```

Furthermore, given that each TNA model can be converted to an `igraph` object, researchers can compute other centralities if they so wish. In the next code, we compute the hub and authority centralities.

```
# Calculate hub scores and the authority scores for the network
hits_scores <- igraph::hits_scores(as.igraph(model))
hub_scores <- hits_scores$hubauthority_scores <- hits_scores$authority

# Print the calculated hub and authority scores for further analysis
print(hub_scores)
print(authority_scores)
```

```
    adapt   cohesion  consensus coregulate    discuss     emotion     monitor        plan  synthesis
     0.96       1.00       0.65       0.69       0.74        0.82        0.74        0.87       0.90

    adapt   cohesion  consensus coregulate    discuss     emotion     monitor        plan  synthesis
    0.122      0.301      1.000      0.195      0.439       0.333       0.122       0.511      0.059
```

5.3.1.2 Edge-Level Measures

TNA can also compute edge level measures which would rank edges according to their importance in the transition model or which transitions bridge other processes. We can do this by using the function `betweenness_network` which creates a network with betweenness centrality as edges (Fig. 11).

```
Edge_betweeness <- betweenness_network(model)
plot(Edge_betweeness)
```

5.3.2 Community Finding

Community detection identifies groups of nodes (states, events or actions) that exhibit strong interconnections within the transition process [26]. Unlike the cliques,

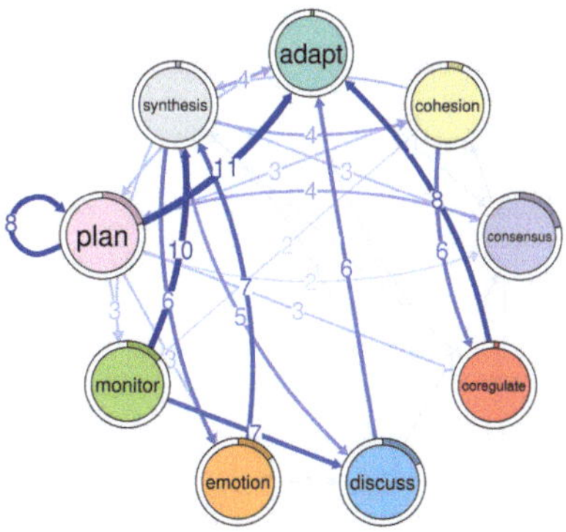

Fig. 11 Edge betweenness

communities are identified algorithmically, and can be of any size. Therefore, it could offer a more realistic way of grouping similarly inter-connected transitions. TNA offers several community detection algorithms that are suited for transition networks (usually small, weighted and directed networks). Each of these algorithms offers a unique perspective to the grouping [18].

- The *walktrap* algorithm detects tightly-knit groups of transitions based on random walks on the graph. It assumes that random walks will frequently stay within the same community before transitioning to another. finding smaller, cohesive communities within the network.
- The *Fast Greedy* algorithm detects communities by optimizing modularity where the algorithm begins with each node in its community and merges nodes based on increases in modularity until no further improvements are possible.
- *Infomap* detects communities by optimizing a flow-based model that captures how information moves through the network.
- The *Edge Betweenness* algorithm focuses on the edges (transitions) in the network rather than the nodes by measuring how many shortest paths uses this transition. Edges with high betweenness are removed iteratively to reveal communities. This method is useful in TNA for uncovering transitions that act as bridges between different learning behaviors.
- The *Spin Glass* algorithm tries to find communities with strong internal connections and weak external connections.

The code below identifies and then visualizes community structures within a `tna` network model. It begins by using the `communities()` function with the `model` argument to detect communities. The result is stored in the `communities` object (Table 2). Next, the code visualizes these communities using the `plot()` function, specifying the `leading_eigen` method for community detection (Fig. 12). Other algorithms can be specified in the same way, for example to use the spinglass algorithm we could pass the argument `method = "spinglass"`. Furthermore, the `communities` object contains the number of communities by each algorithm and the communities assignment.

Table 2 Community assignment

State	walktrap	fast_greedy	label_prop	infomap	edge_betweenness	leading_eigen	spinglass
Adapt	1	1	1	1	1	1	1
Cohesion	1	1	1	1	1	1	1
Consensus	1	1	1	1	1	2	1
Coregulate	1	2	1	1	1	3	2
Discuss	1	2	1	1	1	3	2
Emotion	1	1	1	1	1	1	1
Monitor	1	2	1	1	1	2	2
Plan	1	3	1	1	1	2	1
Synthesis	1	1	1	1	1	3	1

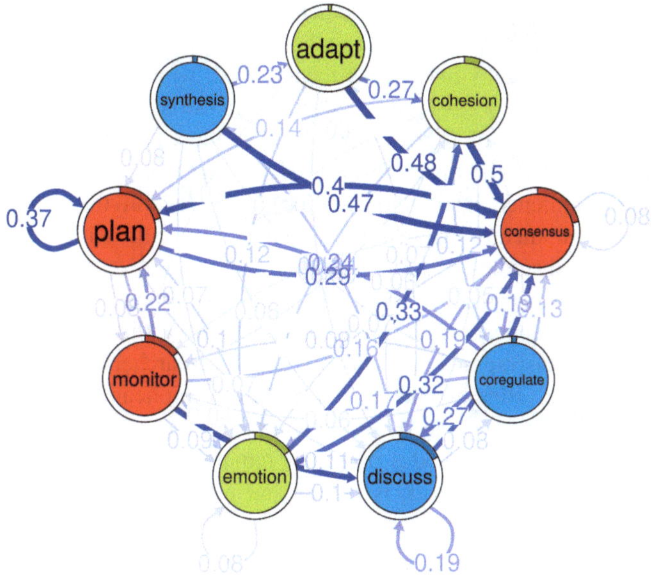

Fig. 12 Communities detected using the `leading_eigen` algorithm

```
communities <- communities(model)
print(communities)
```

```
        walktrap      fast_greedy       label_prop          infomap edge_betweenness
               1                3                1                1                1
   leading_eigen        spinglass
               3                2
```

```
communities <- communities(model)
plot(communities, method = "leading_eigen")
```

5.4 *Network Inference*

5.4.1 Bootstrapping

Bootstrapping is a robust technique for validating edge weight accuracy and stability and consequently the model as a whole. Through bootstrapping, researchers can verify each edge, determine their statistical significance, and obtain a confidence interval for the transition probabilities. Bootstrapping is performed by repeatedly resampling the original data and creating new datasets. The new datasets (bootstrap samples) are created by randomly selecting data points (individual sequences in the case of sequence data) with replacement, meaning that some data points may appear multiple times in a sample while others may not appear at all [36]. For each bootstrap sample, the TNA network is recalculated, which involves recalculating the edge weight. This process is repeated many times (typically 1000), generating an empirical distribution of edge weight for each edge. As such, we compare the original edge weight to the distribution of edge weights across bootstrap samples and compute confidence intervals (CIs) for each edge. In this context, a 95% confidence interval means that the true edge weight falls within 95% of such intervals in repeated studies.

Bootstrapping also provides a p-value for edge significance by assessing how often a given edge appears with a weight significantly different from a chosen threshold value. Bootstrapping can therefore allow researchers to verify each edge (how likely a given edge would appear in future replications of the network structure) and also, the whole TNA network robustness i.e., if the network's key features (e.g., important nodes or edges) remain consistent across re-sampled datasets, it suggests the network is robust and reliable. Therefore, we can test or verify hypothesis or advance an existing one. To carry out bootstrapping, we use the `bootstrap()` function with the `model` as an argument. The function has default value of 1000 bootstrap iterations (argument `iter`), and a higher number of iterations is of course better, but 1000 is often sufficient. We choose a threshold value of 0.05 to compare our bootstrapped transition probabilities via the argument `threshold`. This means that if we consistently observe that an edge is above this threshold in the bootstrapped samples, we deem it statistically significant.

This code below performs bootstrapping using the `bootstrap()` function, with a `seed = 265` for reproducibility. After the bootstrapping process, we can print a summary of the results showing relevant significance values for each edge in the network (CI and p-values). The result object also contains several other elements:

- `weights_orig`: The original transition matrix of the TNA model
- `weights_sig`: A matrix showing only statistically significant transitions (other weights are set to zero).
- `weights_mean`: The mean transition matrix across all bootstrap samples.
- `weights_sd`: The standard deviation matrix across all bootstrap samples.
- `ci_lower`: The lower bound matrix of the bootstrap confidence intervals for transitions.
- `ci_upper`: The upper bound matrix of the bootstrap confidence intervals for transitions.
- `p_values`: A matrix of bootstrap p-values for each transition
- `summary`: A `data.frame` containing the p-values, edge weights and the CIs for each edge in the network.

Then we print specifically those edges that we found to be non-significant by the bootstrap (p-value greater than 0.05), offering insights into the stability of the network structure.

```
# Perform bootstrapping on the TNA model with a fixed seed for reproducibility
set.seed(265)
boot <- bootstrap(model, threshold = 0.05)

# Print a summary of the bootstrap results
print(summary(boot))
```

```
   from          to  weight p_value   sig ci_lower ci_upper
2 adapt    cohesion 0.00295   0.506 FALSE  0.00047   0.0053
3 adapt   consensus 0.00474   0.163 FALSE  0.00301   0.0064
4 adapt  coregulate 0.01624   0.149 FALSE  0.01026   0.0217
5 adapt     discuss 0.07137   0.001  TRUE  0.06328   0.0790
6 adapt     emotion 0.00247   0.556 FALSE  0.00058   0.0042
7 adapt     monitor 0.01117   0.315 FALSE  0.00516   0.0161
8 adapt        plan 0.00097   0.551 FALSE  0.00014   0.0016
 [ reached 'max' / getOption("max.print") -- omitted 71 rows ]
```

```
# Show the non-significant edges (p-value >= 0.05 in this case)
# These are edges that are less likely to be stable across bootstrap samples
print(boot, type = "nonsig")
```

```
Non-significant Edges

       from          to  weight p_value ci_lower ci_upper
2     adapt    cohesion 0.00295   0.506  0.00047   0.0053
3     adapt   consensus 0.00474   0.163  0.00301   0.0064
4     adapt  coregulate 0.01624   0.149  0.01026   0.0217
6     adapt     emotion 0.00247   0.556  0.00058   0.0042
7     adapt     monitor 0.01117   0.315  0.00516   0.0161
8     adapt        plan 0.00097   0.551  0.00014   0.0016
11 cohesion    cohesion 0.02714   0.093  0.01894   0.0354
18 cohesion   synthesis 0.03374   0.232  0.01892   0.0471
 [ reached 'max' / getOption("max.print") -- omitted 19 rows ]
```

5.4.2 Permutation

TNA uses advanced statistical methods to compare models and determine if the differences between them are statistically meaningful. A straightforward but less precise way is to look at the raw differences in edge weights between models. This method can give a quick visual comparison, but it doesn't provide the statistical rigor needed to confirm that these differences aren't just due to random chance.

To tackle this issue, TNA uses a more rigorous approach through permutation. Permutation involves the generation of a distribution of differences by repeatedly shuffling and comparing the data. For each edge, the test provides a p-value, which helps researchers identify statistically significant differences. This rigorous approach ensures that the insights gained from TNA are not merely artifacts of chance but reflect true underlying differences. No other statistical method that offers such rigor in comparison of groups.

Let's look at an example using TNA to compare the transition networks of high achievers and low achievers with data from the `group_regulation` dataset. First, we create two separate TNA models—one for high achievers and one for low achievers. Then, we use the `plot_compare()` function to plot a simple comparison between these models, which just subtracts the edge weights without assessing statistical significance (Fig. 13a).

To determine the statistical significance of the differences between the two models, a permutation test is performed using the `permutation_test()` function. The `measures="Betweenness"` argument specifies that the test also should compute the differences in betweenness centrality. The results of the permutation test are then plotted using the `plot()` function, which displays the significant differences between the transition networks of high-achievers and low-achievers (Fig. 13b). Finally, we can use the `print()` function to print the results of the permutation test or just print specific parts of the results, like the significantly different edges, or centralities.

```
# Create TNA for the high-achievers subset (rows 1 to 1000)
Hi <- tna(group_regulation[1:1000, ])

# Create TNA for the low-achievers subset (rows 1001 to 2000)
Lo <- tna(group_regulation[1001:2000, ])

# Plot a comparison of the "Hi" and "Lo" models
# The 'minimum' parameter is set to 0.001, so edges with weights >= 0.001 are shown
plot_compare(Hi, Lo, minimum = 0.01)

# Run a permutation test to determine statistical significance of differences
# between "Hi" and "Lo"
# The 'iter' argument is set to 1000, meaning 1000 permutations are performed
Permutation <- permutation_test(Hi, Lo, iter = 1000, measures = "Betweenness")

# Plot the significant differences identified in the permutation test
plot(Permutation, minimum = 0.01)
```

```
print(Permutation$edges$stats)
```

```
               edge_name diff_true effect_size p_value
1         adapt -> adapt   0.00000         NaN   1.000
2      cohesion -> adapt   0.00533       1.991   0.061
3     consensus -> adapt  -0.00132      -0.763   0.412
4    coregulate -> adapt   0.01122       2.002   0.048
5       discuss -> adapt  -0.09616     -11.482   0.001
6       emotion -> adapt   0.00167       0.907   0.460
```

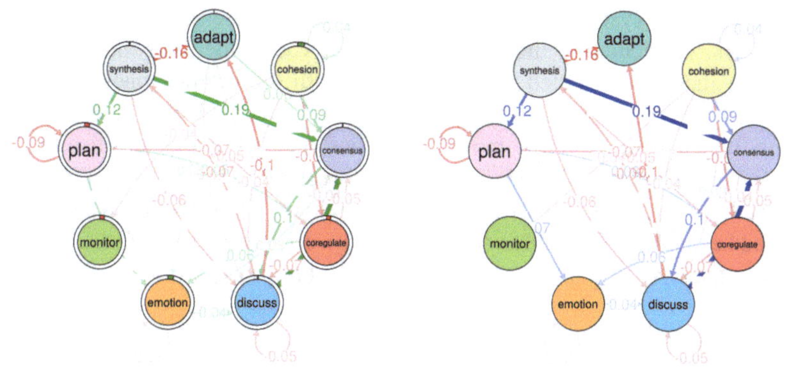

(a) (b)

Fig. 13 Network comparison. (**a**) Subtraction of high achievers minus low achievers. (**b**) Significant differences identified in the permutation test

```
7        monitor -> adapt  -0.00019       -0.034   0.943
8           plan -> adapt   0.00077        0.976   0.221
9      synthesis -> adapt  -0.15825       -4.684   0.001
10      adapt -> cohesion  -0.01476       -0.375   0.749
11   cohesion -> cohesion   0.03711        4.627   0.001
12  consensus -> cohesion   0.01056        3.452   0.001
 [ reached 'max' / getOption("max.print") -- omitted 69 rows ]
```

```
print(Permutation$centralities$stats)
```

```
       state  centrality diff_true effect_size p_value
1      adapt Betweenness        -9       -1.31   0.221
2   cohesion Betweenness         1        0.79   0.315
3  consensus Betweenness         0         NaN   1.000
4 coregulate Betweenness         3        1.02   0.323
5    discuss Betweenness         2        4.55   0.006
6    emotion Betweenness         0         NaN   1.000
7    monitor Betweenness        -6       -2.02   0.084
8       plan Betweenness       -10       -2.01   0.106
9  synthesis Betweenness        20        3.45   0.002
```

5.4.3 Estimating the Stability of Centralities

While most centrality calculation methods in traditional network analysis have no statistical method to verify their strength or reliability, TNA introduces a statistical method, the case-dropping bootstrap which is a technique to estimate the stability of centrality measures in networks. The “case-dropping”—as the name implies—refers to systematically removing a subset of observations or nodes from the network, recalculating the centrality indices each time, and then evaluating how much these recalculated values correlate with the original ones. This helps determine the correlation-stability of centrality measures. This technique allows us to assess whether the centrality indices are robust and not overly sensitive to slight variations in the network data. A high correlation-stability coefficient indicates that the centrality indices are robust, meaning they remain relatively consistent and the identified key nodes are reliably central in the network. Conversely, if the correlation coefficients vary widely or are consistently low, it indicates that the results may not be dependable.

The correlation-stability coefficients are useful for several reasons. Firstly, they emphasize the reliability of findings by ensuring that the identified influential nodes are not artifacts of a particular dataset or influenced by random variations. Secondly, they improve the generalizability of the results, as stable centrality measures are more likely to reflect underlying patterns that persist across different samples or conditions.

5.4.4 Interpreting the Results of the Case-Dropping Bootstrap for Centrality Indices

The case-dropping bootstrap results in a series of correlation coefficients that represent the relationship between the centrality measures in the original network and those in the networks with data removed. These coefficients typically range from -1 to 1, where values closer to 1 indicate a strong positive correlation, meaning the centrality rankings remain consistent even when data is omitted. A high average coefficient (e.g., above 0.7) suggests that the centrality measure is stable and reliable; the nodes identified as most central in the original network remain central even when some data are missing. This stability implies that findings about the importance of certain nodes are robust and not heavily influenced by specific data points. If the average correlation coefficient is moderate (e.g., between 0.5 and 0.7), it indicates that there is some variability in the centrality measures when data is removed, however it is still acceptable. A low average correlation coefficient (e.g., below 0.5) points to instability in the centrality measure and the results may not be reliable. This instability could stem from a small sample size or measurement errors in the data. It's also important to compare the stability of different centrality measures. For example, degree centrality often shows higher stability because it relies on direct connections, whereas betweenness centrality may be less stable due to its dependence on the shortest paths, which can change significantly with the removal of edges or nodes.

The code below assesses the centrality stability of centrality measures using the function named `estimate_centrality_stability()` (a shorthand alias `estimate_cs()` is also provided) which takes the model as the main argument. The results can be plotted in the usual way (Fig. 14).

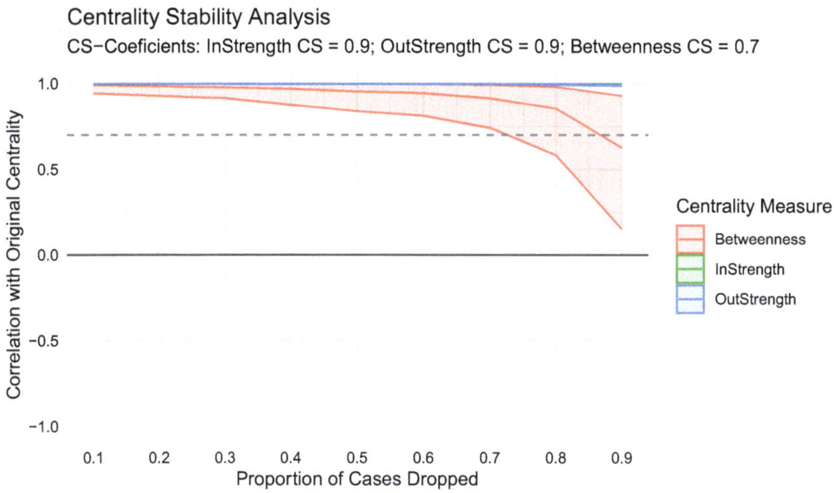

Fig. 14 Results of the case-dropping bootstrap for centrality indices

```
Centrality_stability <- estimate_centrality_stability(model, detailed = FALSE)
plot(Centrality_stability)
```

6 Conclusions

This chapter introduced Transition Network Analysis (TNA) as a novel methodological framework that models the temporal dynamics of the learning process. TNA provides a theoretically grounded approach to modeling the stochastic nature of learning processes. This aligns with established theoretical frameworks, such as Winne and Perry's conceptualization of self-regulated learning through occurrences, contingencies, and patterns. TNA operationalizes these theoretical constructs through its network representation, allowing researchers to empirically investigate theoretical propositions about learning dynamics. The method's ability to identify communities, clusters, and recurring patterns offers new opportunities for theory building in educational research. Thus, TNA can help researchers validate existing theoretical constructs through data-driven pattern identification, discover new behavioral mechanisms and learning strategies, develop more nuanced understanding of how different learning activities interconnect, and identify critical transition points that may serve as targets for intervention.

In a way, TNA can be thought of as combining the temporal aspects of process mining with the analytical power of network theory to offer researchers a robust tool for understanding the complex dynamics of learning processes. The use of graph representation enhances PM by enabling researchers to examine the role of specific events in shaping learning processes through centrality measures. For example, measures such as betweenness centrality highlight nodes that bridge transitions, while in-degree centrality identifies events that receive the most transitions. Edge-level centralities provide further insights into the underlying dynamics of transitions, offering a deeper understanding of how learning unfolds. These measures and capabilities are hardly present in process mining with some exceptions like fuzzy miner, yet they are rather few, incomplete and not easy to interpret [44]. Also, TNA employs community detection to explore events that co-occur, revealing patterns that reflect behaviors and strategies. On the other hand, PM enhances network analysis by enabling the identification of clusters representing typical transition patterns, which are widely used to capture students' tactics and strategies. However, TNA offers more than combination of either methods. The incorporation of bootstrapping provides a rigorous way to assess model stability and filter out unreliable connections. Similarly, permutation and case-dropping for assessing statistical inferences of models.

A significant contribution of TNA is its incorporation of robust statistical validation techniques. The implementation of bootstrapping methods addresses a critical gap in current process mining and network analysis approaches in education. This advancement enables identification of significant transitions through confidence intervals and p-values, helps filter out spurious connections leading to more reliable models, provides a framework for assessing the generalizability of findings, and supports more rigorous hypothesis testing and theory validation. Furthermore, TNA's ability to systematically compare different learning processes through permutation testing represents a methodological advance over simple visual comparison. Permutation allows researchers to identify statistically significant differences between learning processes across different contexts.

However, several limitations should be considered when applying TNA. The method assumes Markovian properties, which may not fully capture some aspects of learning processes (refer to [24] for a frequency-based approach). Interpretation of network measures requires careful consideration of the educational context, and the approach may be computationally intensive for very large datasets. Additionally, some network measures may not be meaningful in fully connected networks. Future research should focus on developing additional validation techniques specific to educational contexts, extending the method to handle multi-level and longitudinal data, investigating the integration of TNA with other analytical approaches, creating user-friendly tools and guidelines for educational practitioners, and exploring applications in different educational domains and contexts.

References

1. Schnakenberg J (1976) Network theory of microscopic and macroscopic behavior of master equation systems. Rev Mod Phys 48:571–85. https://doi.org/10.1103/revmodphys.48.571
2. Zou Y, Donner RV, Marwan N, Donges JF, Kurths J (2019) Complex network approaches to

nonlinear time series analysis. Phys Rep 787:1–97. https://doi.org/10.1016/j.physrep.2018.10.005

3. Saqr M, López-Pernas S, Törmänen T, Kaliisa R, Misiejuk K, Tikka S (2024) Transition network analysis: a novel framework for modeling, visualizing, and identifying the temporal patterns of learners and learning processes. In: Proceedings of the 15th international learning analytics and knowledge conference, pp 351–361
4. Helske J, Helske S, Saqr M, López-Pernas S, Murphy K (2024) A modern approach to transition analysis and process mining with markov models in education. Springer, Cham, pp 381–427
5. Hilpert JC, Marchand GC (2018) Complex systems research in educational psychology: aligning theory and method. Educ Psychol 53:185–202. https://doi.org/10.1080/00461520.2018.1469411
6. Kaplan A, Garner JK (2020) Steps for applying the complex dynamical systems approach in educational research: a guide for the perplexed scholar. J Exp Educ 88:486–502. https://doi.org/10.1080/00220973.2020.1745738
7. Saqr M, Schreuder MJ, López-Pernas S (2024) Why educational research needs a complex system revolution that embraces individual differences, heterogeneity, and uncertainty. In: Saqr M, López-Pernas S (eds) Learning analytics methods and tutorials: a practical guide using r. Springer, Cham, pp 723–34
8. Saqr M, López-Pernas S (2024) Mapping the self in self-regulation using complex dynamic systems approach. Brit J Educ Technol 55:1376–1397. https://doi.org/10.1111/bjet.13452
9. Newman M (2018) Networks, 2nd edn. Oxford University Press, London
10. Koller D, Friedman N (2009) Probabilistic graphical models: principles and techniques. MIT Press, Cambridge
11. Nicolis G, Cantú AG, Nicolis C (2005) Dynamical aspects of interaction networks. Int J Bifur Chaos Appl Sci Eng 15:3467–80. https://doi.org/10.1142/s0218127405014167
12. Reimann P (2009) Time is precious: variable- and event-centred approaches to process analysis in CSCL research. Int J Comput-Support Collaborat Learn 4:239–57. https://doi.org/10.1007/s11412-009-9070-z
13. Reimann P (2019) Methodological progress in the study of self-regulated learning enables theory advancement. Learn Instruct 72:101269. https://doi.org/10.1016/j.learninstruc.2019.101269
14. Winne PH, Perry NE (2000) Measuring self-regulated learning. In: Handbook of self-regulation, Academic, Cambridge, pp 531–66
15. Winne PH (2010) Improving measurements of self-regulated learning. Educ. Psychol 45:267–276. https://doi.org/10.1080/00461520.2010.517150
16. Borgatti SP, Mehra A, Brass DJ, Labianca G (2009) Network analysis in the social sciences. Science 323:892–95. https://doi.org/10.1126/science.1165821
17. Epskamp S, Cramer AOJ, Waldorp LJ, Schmittmann VD, Borsboom D (2012) Qgraph: network visualizations of relationships in psychometric data. J Stat Softw 48:1–18. https://doi.org/10.18637/jss.v048.i04
18. Csardi G, Nepusz T (2006) The igraph software package for complex network research. Int J Complex Syst 11:1695
19. Saqr M, Peeters W (2022) Temporal networks in collaborative learning: a case study. Brit J Educ Technol 53:1283–1303. https://doi.org/10.1111/bjet.13187
20. Saqr M, López-Pernas S, Conde-González MÁ, Hernández-García Á (2024) Social network analysis: a primer, a guide and a tutorial in R. Springer, Cham, pp 491–518
21. Saqr M, Nouri J, Vartiainen H, Tedre M (2020) Robustness and rich clubs in collaborative learning groups: a learning analytics study using network science. Sci Rep 10:14445. https://doi.org/10.1038/s41598-020-71483-z
22. Saqr M, Elmoazen R, Tedre M, López-Pernas S, Hirsto L (2022) How well centrality measures capture student achievement in computer-supported collaborative learning? – A systematic review and meta-analysis. Educ Res Rev 35:100437. https://doi.org/10.1016/j.edurev.2022.100437

23. López-Pernas S, Tikka S, Saqr M (2025) Mining patterns and clusters with transition network analysis: a heterogeneity approach. In: Saqr M, López-Pernas S (eds) Advanced learning analytics methods: AI, precision and complexity. Springer, Cham
24. Saqr M, López-Pernas S, Tikka S (2025) Capturing the breadth and dynamics of the temporal processes with frequency transition network analysis: a primer and tutorial. In: Saqr M, López-Pernas S (eds) Advanced learning analytics methods: AI, precision and complexity. Springer, Cham
25. Malmberg J, Saqr M, Järvenoja H, Järvelä S (2022) How the monitoring events of individual students are associated with phases of regulation. J Learn Anal 9:77–92. https://doi.org/10.18608/jla.2022.7429
26. Hernández-García Á, Cuenca-Enrique C, Traxler A, López-Pernas S, Conde-González MÁ, Saqr M (2024) Community detection in learning networks using R. Springer, Cham, pp 519–540
27. Epskamp S, Rhemtulla M, Borsboom D (2017) Generalized network psychometrics: combining network and latent variable models. Psychometrika 82:904–927. https://doi.org/10.1007/s11336-017-9557-x
28. Saqr M, López-Pernas S, Helske S, Durand M, Murphy K, Studer M, Ritschard G (2024) Sequence analysis in education: principles, technique, and tutorial with R. Springer, Cham, pp 321–354
29. Saint J, Gašević D, Matcha W, Uzir NA, Pardo A (2020) Combining analytic methods to unlock sequential and temporal patterns of self-regulated learning. In: Proceedings of the tenth international conference on learning analytics & knowledge, pp 402–411. https://doi.org/10.1145/3375462.3375487
30. Saqr M, López-Pernas S, Jovanović J, Gašević D (2023) Intense, turbulent, or wallowing in the mire: a longitudinal study of cross-course online tactics, strategies, and trajectories. Internet Higher Educ 57:100902. https://doi.org/10.1016/j.iheduc.2022.100902
31. Fan Y, Matcha W, Uzir NA, Wang Q, Gaševic D (2021) Learning analytics to reveal links between learning design and self-regulated learning. Int J Artif Intell Educ 31:980–1021
32. Elmoazen R, Saqr M, Tedre M, Hirsto L (2022) A systematic literature review of empirical research on epistemic network analysis in education. IEEE Access 10:17330–17348. https://doi.org/10.1109/access.2022.3149812
33. van Borkulo CD, van Bork R, Boschloo L, Kossakowski JJ, Tio P, Schoevers RA, Borsboom D, Waldorp LJ (2023) Comparing network structures on three aspects: a permutation test. Psychol Methods 28:1273–1285. https://doi.org/10.1037/met0000476
34. Epskamp S, Borsboom D, Fried EI (2017) Estimating psychological networks and their accuracy: a tutorial paper. Behav Res Methods 50:195–212. https://doi.org/10.3758/s13428-017-0862-1
35. Borsboom D, Deserno MK, Rhemtulla M, Epskamp S, Fried EI, McNally RJ, et al (2021) Network analysis of multivariate data in psychological science. Nat Rev Methods Primers 1(1):58
36. Mooney CZ, Duval RD (1994) Bootstrapping: a nonparametric approach to statistical inference. J. Am Stat Assoc 89:1150. https://doi.org/10.2307/2290969
37. Tikka S, López-Pernas S, Saqr M (2025) tna: an R package for transition network analysis. Appl Psychol Measur 01466216251348840
38. Tikka S, Kopra J, Heinäniemi M, López-Pernas S, Saqr M (2024) Getting started with R for education research. In: Saqr M, López-Pernas S (eds) Learning analytics methods and tutorials: a practical guide using R. Springer, Berlin
39. Kopra J, Tikka S, Heinäniemi M, López-Pernas S, Saqr M (2024) An R approach to data cleaning and wrangling for education research. In: Saqr M, López-Pernas S (eds) Learning analytics methods and tutorials: a practical guide using R. Springer, Berlin
40. Tikka S, Kopra J, Heinäniemi M, López-Pernas S, Saqr M (2024) Introductory statistics with R for educational researchers. In: Saqr M, López-Pernas S (eds) Learning analytics methods and tutorials: a practical guide using R. Springer, Berlin

41. López-Pernas S, Misiejuk K, Tikka S, Kopra J, Heinäniemi M, Saqr M (2024) Visualizing and reporting educational data with R. In: Learning analytics methods and tutorials. Springer, Cham, pp 151–194
42. Wickham H, Averick M, Bryan J, Chang W, McGowan LD, François R, Grolemund G, Hayes A, Henry L, Hester J, Kuhn M, Pedersen TL, Miller E, Bache SM, Müller K, Ooms J, Robinson D, Seidel DP, Spinu V, Takahashi K, Vaughan D, Wilke C, Woo K, Yutani H (2019) Welcome to the tidyverse. J Open Sour Softw 4:1686. https://doi.org/10.21105/joss.01686
43. Neal ZP (2022) Backbone: an R package to extract network backbones. PLOS ONE 17:e0269137. https://doi.org/10.1371/journal.pone.0269137
44. Pan Y, Zhang L (2021) Automated process discovery from event logs in BIM construction projects. Autom Construct 127:103713. https://doi.org/10.1016/j.autcon.2021.103713

Capturing the Breadth and Dynamics of the Temporal Processes with Frequency Transition Network Analysis: A Primer and Tutorial

Mohammed Saqr, Sonsoles López-Pernas, and Santtu Tikka

Abstract This chapter presents Frequency-Based Transition Network Analysis (FTNA), a novel method to model the relational dynamics and the transitions between states or events based on the frequency of occurrence of transitions. Compared to TNA based on Markov models, FTNA is well-suited when the research focus is on describing, summarizing, or visually analyzing the observed data without the probabilistic assumptions and constraints. Compared to process mining, FTNA leverages statistical techniques such as pruning, bootstrapping and permutation to validate and compare models. Moreover, FTNA employs networks as a lens to represent and analyze transitions, which provides a rich family of metrics and analyses such as centrality measures, communities and patterns. In this chapter, we offer an introduction to the method and its main features, along with a step-by-step tutorial in R using a case study in group collaboration.

Keywords Learning analytics · Transition network analysis · Stochastic process mining

1 Introduction

Capturing the dynamics of learning has become an increasingly important goal for educators and researchers due to the value it brings to understanding the learning process [1]. This chapter introduces a novel method for capturing such dynamics based on how events are dynamically interrelated [2]. Frequency-Based Transition Network Analysis (FTNA) is a type of transition network analysis (TNA) where the network structure is based on modelling the frequency of transitions. FTNA models

M. Saqr (✉) · S. López-Pernas
University of Eastern Finland, Joensuu, Finland
e-mail: mohammed.saqr@uef.fi

S. Tikka
University of Jyväskylä, Jyväskylä, Finland

M. Saqr, S. López-Pernas (eds.), *Advanced Learning Analytics Methods*,
https://doi.org/10.1007/978-3-031-95365-1_16

the count or number of occurrence of each transition between states and uses them as weights reflecting the magnitude of dynamic relationships [3]. This approach is a straightforward representation of the relational dynamics within the learning process, capturing how often one state follows another without conditioning on dependencies or making strong assumptions about the process.

In the previous chapter, we have studied TNA, which assumes that the learning process follows a Markov process where the transitions are probabilities and the current states depends on the previous state [2, 4]. In comparison, FTNA is well-suited when the research focus is on describing, summarizing, or visually analyzing the observed data without the probabilistic assumptions —and constraints— of Markov modeling. Additionally, there are contexts where FTNA is more suitable (see below). As a method, FTNA has a rich tool set of techniques and large array of analytical functions to harness the dynamics of the learning process and its temporality. These tools enable researchers to identify the dominant events, the notable patterns, as well as to compare different processes [4, 5]. More importantly, given that FTNA inherits most of the capabilities of TNA, it can be used to draw inferences with its statistical methods, e.g., permutation and bootstrapping [2].

2 The Basic Principles of FTNA

FTNA captures the dynamic relationship between events, i.e., how an event leads to another, follows another, or is dependent on a previous event. These events and the transitions between them are used to construct a transition (edge) between events. The aggregation of edges gives rise to the network of transitions [3, 6, 7]. This is similar to the well-known post-reply networks—yet in reverse direction- when a reply follows a post in forum discussions or chat messages [8]. However, no reply is required here, and the post-reply networks are not strictly sequential and only address conversations. FTNA extends such formulation to modeling transitions between any temporally related events, e.g., learning actions, strategies, roles, states, emotions to mention a few [2]. Also, FTNA models the transitions, or the succession of events rather than replies.

While FTNA allows the full potential of traditional network analysis, it extends these functions with several modeling and statistical techniques (e.g., permutation, scaling, bootstrapping etc.) that enhance the analysis and provides tools for understanding how the events unfold and draw inferences using statistical rigor [4, 5]. Similarly, FTNA enables the use of most TNA functions but with a different interpretation and modeling purpose. That said, it is important here to draw a comparison with TNA which is based on Markov modeling and explain when and why FTNA may be needed [2].

Markov-based models assume that the transition probability between events depends only on the current state (the "Markov property") and not on prior states, meaning that the process has no memory nor is it influenced by any events beyond the immediate events [9]. This assumption may be unrealistic in some processes, where transitions might depend on a broader sequence of previous states (e.g., reasoning over multiple steps). FTNA avoids this limitation, capturing sequences of behaviors without requiring this independence assumption which makes it more flexible for processes where this independence assumption may not hold. In some analyses, the goal may not be to predict future behavior based on current states (as in a Markov model) but rather to identify patterns and structural relationships among observed behaviors without imposing a model-based structure. FTNA models observed patterns of any length or breadth. This is because FTNA is flexible in representing longer-range dependencies or multi-step behaviors by aggregating transitions without conditioning on prior states. Lastly, FTNA is more appropriate for handling small datasets or processes where some transitions are rare. This is because Markov models often require substantial data to produce reliable probability estimates. In contrast, frequency-based TNA is often robust with smaller datasets, as it only tallies transitions rather than estimating conditional probabilities and therefore, does not inflate or conflate the modeled process [4, 5].

It should also be mentioned that FTNA holds resemblance with frequency based process mining [10], since both techniques can be used to represent the frequency of transitions between events or states. However, while process mining is a valuable exploration and visualization tool, it has been heavily criticized for its lack of statistical rigor when it comes to identifying (statistically) meaningful transitions and comparing networks, which are two limitations that FTNA overcomes. Moreover, current implementations in R do not allow for filtering out infrequent transitions (only infrequent nodes), which limit the opportunities for meaningful analysis and visualization [9, 10].

3 The Building Blocks of FTNA

The building blocks of FTNA are transitions between events where each event is a node, and the weight of the edge is the transition frequency between the edges. Rather than calculating probabilities, FTNA tracks the count of these transitions to reveal a straightforward picture of patterns in sequential data. Let us consider a hypothetical example where we model the learning transitions. Imagine we have five types of activities: *Watch Video*, *Quiz*, *Read Materials*, *Submit Assignment*, and *Feedback*. The FTNA model captures the frequency of students moving from one activity to another: 60 transitions from "Watch Video" to "Quiz" mean a frequency of 60, while "Quiz" leading to "Submit Assignment" shows a frequency of 100. This raw count provides a direct representation of how students move through the learning steps with no transformation or scaling (Table 1).

Table 1 Frequency of transitions in the network

From	To	Frequency (weight)
Watch video	Quiz	60
Watch video	Read materials	10
Quiz	Submit Assignment	100
Read materials	Submit assignment	20
Read materials	Feedback	30
Read materials	Quiz	50
Submit assignment	Feedback	100
Watch video	Feedback	30

4 The Mathematical Basis of FTNA

The FTNA network can be represented as $G = (V, E, W)$ with three basic elements:nodes (V), directed edges (E), and weights (W). *Nodes* serve as the elements, states or events while *edges* represent transitions between them. For example, if v_i is *Watch Video* and v_j is *Quiz*, the edge (v_i, v_j) captures the transition from watching a video to attempting a quiz. The weight function $W(v_i, v_j)$ represents the transition frequency—how often students move from activity v_i to v_j. A weight of $W(v_i, v_j) = 60$ signifies that students frequently transition from *Watch Video* to *Quiz* (60 times). From *Quiz* to *Submit Assignment* ($v_2 \rightarrow v_4$, $W(v_2, v_4) = 100$): showing that student transitioned from a quiz directly to submitting an assignment 100 times which is more than the previous edge.

Mathematically, this can be represented as follows: let T_{ij} denote the absolute frequency of transitions from state v_i to state v_j. If we consider a state space $V = \{v_1, v_2, \ldots, v_n\}$, the transition matrix T has entries defined as

$$T_{ij} = \text{number of transitions from } v_i \text{ to } v_j.$$

This approach does not normalize the values, allowing the raw frequencies to capture the total count of each transition observed in the data. However, researchers can build networks in different ways and scaling that makes it easier to read or interpret namely normalized or rank-based FTNA which are discussed below.

4.1 *FTNA Networks Based on Normalized Values*

Counts and raw frequencies may be conflated by several factors and are not directly comparable. That is why many network measures are best understood as ranks or scales. For instance, comparing across two classrooms where classroom A has 32 students and classroom B has 48. It becomes obvious that this is an unfair comparison, given that the larger classroom is expected to generate more data.

Table 2 Raw counts and normalized values

From	To	Frequency	Normalized weight
Watch video	Quiz	60	0.56
Watch video	Read materials	10	0.00
Watch video	Feedback	30	0.22
Quiz	Submit assignment	100	1.00
Read materials	Submit assignment	20	0.11
Read materials	Feedback	30	0.22
Read materials	Quiz	50	0.44
Submit assignment	Feedback	100	1.00

The same can be applied about tasks which differ in duration, workload or design. Therefore, scaling allows a better way of comparison not only in the same dataset but also for comparison when data comes from different studies. In TNA, the scaling transform raw counts into values between zero and one, which as we mentioned help to compare transitions more uniformly by accounting for different sample sizes making it directly comparable regardless of absolute occurrence. One option in FTNA is min-max normalization. In our example, given that the minimum frequency is 10 and the maximum frequency 100, the formula for normalization is:

$$\begin{aligned}\text{Normalized Weight} &= \frac{\text{Frequency} - \text{Minimum Frequency}}{\text{Maximum Frequency} - \text{Minimum Frequency}} \\ &= \frac{\text{Frequency} - 10}{100 - 10}\end{aligned}$$

For example, given that the transition from *Quiz* to *Submit Assignment* has the highest frequency, it has a normalized weight of 1.00, while *Watch Video* to *Read Materials* has a normalized weight of 0.00, because it is the least frequent transition in the network (Table 2) (Fig. 1).

Another option is to simply use max normalization, i.e., dividing the weights by the largest weight. This option can be useful when the smallest weight is positive and we would like keep it so after normalization, whereas the min-max normalization would make such weights zero.

4.2 *FTNA Networks Based on Ranked Frequencies*

The "ranked" option in TNA orders transitions in terms of their frequency and scales these ranks to fit within 0 to 1 for easier readability. Instead of counting or scaling the frequencies, this option assigns a rank based on how frequently each transition occurs, with the most common transition assigned the highest rank, and

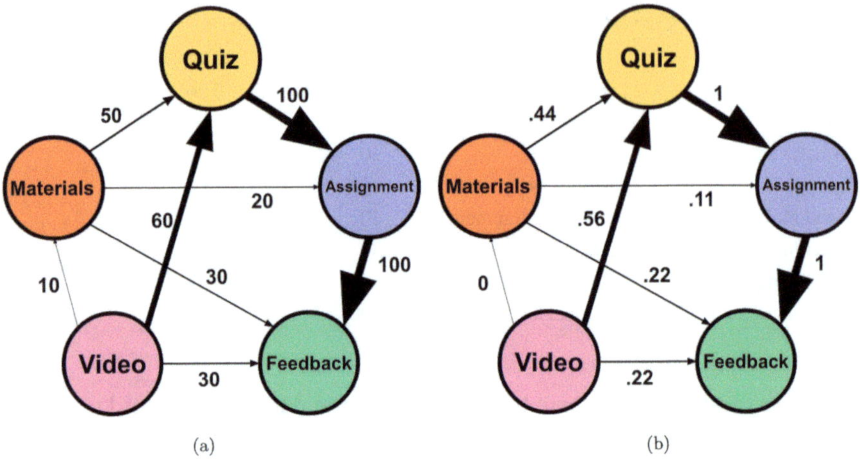

Fig. 1 Network visualizations for raw frequencies and normalized values. (**a**) Raw frequencies. (**b**) Normalized frequencies

Table 3 Final table of raw frequencies, ranks, and normalized weights

From	To	Frequency	Rank	Scaled rank
Quiz	Submit assignment	100	7.5	1.0
Submit Assignment	Feedback	100	7.5	1.0
Watch video	Quiz	60	6	0.769
Read materials	Quiz	50	5	0.615
Watch video	Feedback	30	3.5	0.385
Read materials	Feedback	30	3.5	0.385
Read materials	Submit assignment	20	2	0.154
Watch video	Read materials	10	1	0.0

then scales these ranks. In case of ties, the average rank is used. Finally, the ranks are min-max normalized. Ranking can be particularly insightful in contexts where the order of transitions —rather than their absolute counts— offers meaningful insights. Equidistant ranking is also useful when comparing two processes generated with different mechanisms, i.e., to transform the data to comparable scales.

The equidistant normalized weight can be calculated for the previous table based on the rank as follows (Table 3):

- Top 1 and 2 (100): Normalized Weight $= 1.0$
- Top 3 (60): Normalized Weight ≈ 0.769
- Top 4 (50): Normalized Weight ≈ 0.615
- Top 5 and 6 (30): Normalized Weight ≈ 0.385
- Top 7 (20): Normalized Weight ≈ 0.154
- Top 8 (10): Normalized Weight $= 0.0$

We can visualize the differences between the two networks in Fig. 2.

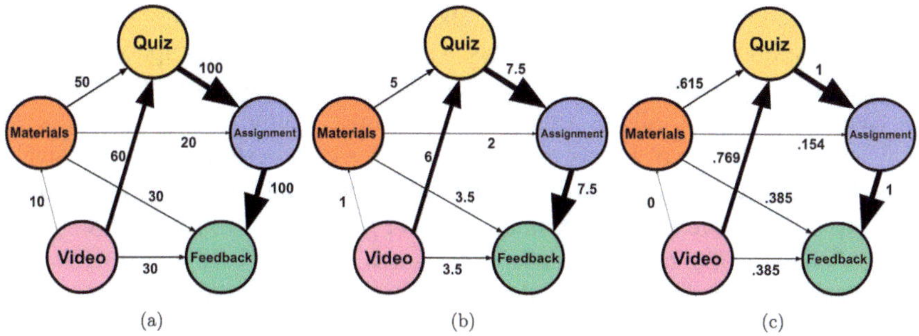

Fig. 2 Network visualizations for raw frequencies, ranks, and normalized weights. (**a**) Raw frequencies. (**b**) Rank. (**c**) Scaled rank

5 FTNA Features

FTNA inherits most of the capabilities of TNA that were discussed in the TNA chapter which will be mentioned here briefly. However, some features may not be transferable or applicable to FTNA and some features are only available in FTNA which will be discussed here [2].

5.1 *Visualization*

FTNA visualization helps researchers get a bird's eye view of the modeled process and capture patterns, central learning events, and how they are temporally related to each other as well as the strength of these connections. FTNA plotting relies on the `qgraph` framework which offers a rich and modern framework for network visualization with wide array of features [11]. These visualization features include custom and fixed layouts to enable comparison across processes as well as several customization options for node, edge shapes, forms and sizes. Comparative plots can be enhanced by having fixed layouts to allow easy comparisons. Furthermore, most FTNA output can be plotted using the plot function which includes networks, centrality objects and communities.

5.2 *Network Measures*

Similar to TNA, FTNA offers network measures at the edge, node and network levels. Network level measures provide insights about the whole network structure, e.g., network connectivity, density, and dominance of events. Node level measures —e.g., centrality measures— allow researchers to quantify or rank the importance of a state or event in the network. Researchers can find which behaviors or cognitive states are central to the learning process, which events are more important (receive more transitions), or bridges other connections (betweenness centrality) [12–14].

Given that FTNA allows multi-step modeling, centralities that capture the downstream diffusion of events or propagation of a certain state are more plausible here. For instance, diffusion centrality can help understand which events are more likely to diffuse or result in more interactions or long threads of related events, e.g., longer "chains of thoughts" or multiple steps in a learning process [15]. Given that FTNA is built with emphasis on edges, edge centralities are rather important here. Edge-level measures —e.g., edge centralities— can tell which transitions are central in the learning process and are therefore critical to the models [16].

5.3 Network Structure and Patterns

Patterns are special configurations or constellations that manifest in the data as strong connections between network nodes. They are the hallmark of the structure and the building blocks of the dynamics of the learning process. What makes patterns unique is that they tell how the data was generated, how the underlying process shaped learning or led to the emergence of certain patterns [1, 2, 17]. For instance, we expect a strong connection in a network of self-regulation between planning and task execution in well-performing students. We also expect to see strong connections between dialogue moves of argument and agreement in cohesive groups. Finding such patterns allows us to, e.g., track students, offer them support, verify existing theories about the learning processes and also test existing ones. Like TNA, patterns may include cliques: dyads (two strongly connected nodes), triads (three strongly connected nodes), or communities (of any size) which reflect underlying patterns of connections inferred from the network [2].

5.4 Network Validation

Like TNA, FTNA offers several statistical methods for verifying and establishing the significance of identified edges through bootstrapping. Bootstrapping is a re-sampling technique that in our context helps filter out small, negligible or unstable edges resulting in a rigorous model. In doing so, researchers can understand which parts of the model are weak and therefore, cannot be reliable in future analyses. Significant edges here represent transitions that are likely to appear in future iterations, and more importantly cannot be used for inference about future processes [5]. Bootstrapping is a unique feature in TNA and is not offered by other methods, e.g., process mining [10] or social network analysis [14]. The inclusion of bootstrapping in FTNA along with other statistical inference methods (permutation and case-dropping for centrality) are in fact what differentiates FTNA from traditional network analysis. This is of course in addition to the distinct data sources, the modeling choices as well as the contextual differences [2, 18].

5.5 Comparison Between Groups

Researchers and educators encounter conditions where groups may differ based on contexts, such as students who engage in collaborative versus non-collaborative learning, high versus low levels of achievement, or learning formats like problem versus project based learning. When researchers like to compare the learning process resulting from either conditions, they can do this using counts, e.g., comparison of frequencies or other techniques like process mining. Typically, such process comparisons relies on visual inspection, such as contrasting process maps or network models, which can reveal the descriptive differences but fall short of demonstrating the statistical significance or pinpointing where these differences matter the most [2, 19].

FTNA enhances this comparison by using a robust permutation-based approach to process comparison. Permutation allows not only a clear visual representation of differences but also enables researchers to estimate the statistical significance of each transition in the network and automatically quantifying the magnitude of differences in transition frequencies. This level of statistical rigor supports more meaningful inferences and contributes to theoretical development.

6 FTNA Tutorial

FTNA is capable of analyzing any data that can be represented as a temporal sequence with transitions between events or states occurring over time. FTNA is particularly well-suited for categorically ordered event data, such as sequences of learning events, phases, roles, dialogue moves, or interactions, to name a few. The source of this data can vary, including time-stamped learning management system data, coded interactions, event-log data, or ordered event data.

In this tutorial, we will use the `tna` R package [5], alongside other essential packages that facilitate data manipulation and visualization. It is assumed that the reader is familiar with the R programming language. If that is not the case, it is recommended to refer to previous tutorials on the basics of R [20], data cleaning [21], basic statistics [22], and visualization [23]. It is also recommended to have prior knowledge about Markov models [9] and to have read the previous chapter on TNA [2].

Initially, the process involves loading the necessary packages to set the groundwork for analysis. The `tna` R package is the main package that we will rely on for analysis which will build the FTNA models. The `tna` package provides a wide range of functions for estimating, visualizing, and manipulating FTNA models.

Table 4 Wide format data which can be used in FTNA

1	2	3	4	5
Submit assignment	Watch video	Read materials	Watch video	Read materials
Watch video	Read materials	Submit assignment	Read materials	Watch video
Read materials	Watch video	Watch video	Submit assignment	Read materials
Read materials	Submit assignment	Watch video	Read materials	Watch video
Watch video	Read materials	Watch video	Read materials	Submit assignment

In addition, it offers various functions for calculating network metrics, such as centrality measures, and discerning patterns like dyads and triads. The package also includes verification functions, such as bootstrapping and permutation tests. We will also use the `tidyverse` package for data manipulation, wrangling and visualization [24].

The `build_model` function from the `tna` package accept several types of data, namely, sequence data, data frames in wide format and square matrices. Sequence data must be in `stslist` format (sequence object), which is typically created using the `seqdef()` function from the `TraMineR` package [25, 26]. This sequence defines the data and its order (for guidance on creating a sequence object, please refer to [26]). Also, `tna` can accept wide data format where each row represents sequential data and each column is a time point with no extra columns. Finally, the `tna` package can also process event data with the highly flexible `prepare_data()` function. Like in Table 4, each row represents a sequence by a person, or group in a task, and column represent the order of such events.

The FTNA analysis starts by constructing the `model` which is simply performed by passing the data to the `ftna()` function or to the `build_model()` function specifying the type like `build_model(data, type = "frequency")`, or simply `ftna(data)`. The FTNA model contains all the information necessary for subsequent analyses, such as plotting, centrality estimation, or model comparison. We can also scale the frequencies using the `scaling` argument, which can be:

- **"minmax"**: This option scales the frequencies between 0 and 1 via min-max normalization such that the lowest weight becomes 0 and the largest becomes 1. Note that if all frequencies are positive, then the lowest will be converted to a weight of zero.
- **"max"**: The frequencies are divided by the largest frequency. This option also scales the frequencies such that the largest weight is 1, but the lowest will only be 0 if some transition never occurred in the original data.
- **"rank"**: This involves ranking all the frequencies from highest to lowest with the average used for ties. Ranking highlights the most significant transitions and provides a standardized way to compare processes across different contexts.

We can also perform multiple types of scaling sequentially, for example `scaling = c("rank", "minmax")` would first rank the frequencies and them scale them between 0 and 1 with the min-max normalization.

The (F)TNA model is organized as a list that has four elements to facilitate the analysis and can be accessed by the users if needed. These are:

1. **Weights** (`weights`): These are the transition weights (frequencies or scaled frequencies) of moving from one state or event to another, serving as the primary element for various network processes, analyses, and visualizations.
2. **Initial States** (`inits`): These are the probabilities of starting in each state at the first time point. Understanding these initial states helps show how events evolve in relation to their starting conditions.
3. **Labels** (`labels`): The descriptive names for each node in the network and is included to enhance the interpretability of the analysis. The labels are automatically retrieved from the alphabet of the sequence object or the categories in the data frame.
4. **Data** (`data`): This is a version of the sequence data (or data frame) that contains all the necessary sequence data, converted into an internal format used by the `tna` package for further analysis (permutation, bootstrapping etc.).

The code below loads the `tna` package. Next, it retrieves the built-in `group_regulation` dataset included in the package, which has data on group behaviors related to regulation during a collaborative project based on the paper by Saqr et al. [4]. Then we call the `ftna()` function with the `group_regulation` dataset as the data argument. The resulting model is an FTNA model with frequencies as edge weights and stored in an R object called `model`, which can be further utilized for analysis. We also use `print(model)` to view the model. The model output shows the labels, the transition frequency matrix and the initial probabilities.

```
# Install 'tna' package from CRAN if needed (uncomment if required).
# install.packages("tna")

# Load packages
library("tna")
library("tidyverse")

# Load example data provided within the 'tna' package,
# representing group regulatory interactions
data(group_regulation)

# Run FTNA on 'group_regulation' data using raw counts of
# transitions ("absolute" type) and print the result
model <- ftna(group_regulation)

# Print the output to inspect the model
print(model)
```

```
State Labels

adapt, cohesion, consensus, coregulate, discuss, emotion, monitor, plan, synthesis

Transition Frequency Matrix

          adapt cohesion consensus coregulate discuss emotion monitor plan synthesis
adapt         0      139       243         11      30      61      17    8         0
cohesion      5       46       844        202     101     196      56  239         6
consensus    30       94       519       1188    1190     460     295 2505        48
 [ reached getOption("max.print") -- omitted 6 rows ]

Initial Probabilities

    adapt  cohesion consensus coregulate   discuss   emotion   monitor      plan synthesis
    0.011     0.060     0.214      0.019     0.175     0.151     0.144     0.204     0.019
```

As we mentioned above, FTNA can be estimated with a scaled transition matrix where the highest transition frequency has the weight of 1 and the lowest has the weight of 0. The resulting transition matrix is easier to read and interpret and more importantly can be useful when comparing across datasets with different samples sizes. The following code estimates a FTNA model using the argument `scaling = "minmax"`. All of the subsequent analysis will be performed using the scaled version given its ease of use, interpretability and ease of comparison across datasets.

```
# Calculate the Transition Network Analysis (TNA) on the
↪   group_regulation
# data with scaled weights between 0 and 1
model_scaled <- ftna(group_regulation, scaling = "minmax")
print(model_scaled) # Print the FTNA model with scaled weights
```

```
State Labels

adapt, cohesion, consensus, coregulate, discuss, emotion, monitor, plan, synthesis

Transition Frequency Matrix

            adapt cohesion consensus coregulate discuss emotion monitor    plan synthesis
adapt     0.00000  0.05549     0.097    0.00439  0.0120  0.0244 0.00679 0.00319   0.00000
cohesion  0.00200  0.01836     0.337    0.08064  0.0403  0.0782 0.02236 0.09541   0.00240
consensus 0.01198  0.03752     0.207    0.47425  0.4750  0.1836 0.11776 1.00000   0.01916
 [ reached getOption("max.print") -- omitted 6 rows ]
```

```
Initial Probabilities

     adapt   cohesion  consensus coregulate    discuss    emotion    monitor       plan  synthesis
    0.0115     0.0605     0.2140     0.0190     0.1755     0.1515     0.1440     0.2045     0.0195
```

The last type of model is the ranked model which can be estimated using the argument `scaling = "rank"`.

```
# Calculate the Transition Network Analysis (TNA) on the `group_regulation`
# data with ranked weights
model_ranked <- ftna(group_regulation, scaling = "rank")
print(model_ranked) # Print the FTNA model with ranked weights
```

```
State Labels

adapt, cohesion, consensus, coregulate, discuss, emotion, monitor, plan, synthesis

Transition Frequency Matrix

           adapt cohesion consensus coregulate discuss emotion monitor plan synthesis
adapt        2.0     42.0        52       11.5    19.5    31.0      14  9.0       2.0
cohesion     4.0     25.5        72       48.0    38.0    47.0      30 51.0       5.5
consensus   19.5     36.0        67       76.0    77.0    64.0      57 81.0      28.0
 [ reached getOption("max.print") -- omitted 6 rows ]

Initial Probabilities

     adapt   cohesion  consensus coregulate    discuss    emotion    monitor       plan  synthesis
    0.0115     0.0605     0.2140     0.0190     0.1755     0.1515     0.1440     0.2045     0.0195
```

6.1 *Plotting*

Plotting the FTNA model is a simple process. We simply use the `plot()` function with the model as the argument i.e., the `model_scaled` object we estimated earlier. In fact, most FTNA objects can be visualized using the `plot()` function directly without the need to any arguments. This includes centralities, communities, permutation.

As the code below shows, the plot is overly dense and hard to read. This is because even rare and small edges are plotted. In our model, we have 9 nodes, the model includes up to 81 edges and this is far from informative due to the multidimensionality and the difficulty in picking the patterns that matter. The easiest way is to set a minimum value which will **hide (but not delete)** the edges below that threshold. In the example below, we set the minimum to 0.01 just to show all the model but you can set it to 0.05 or more if you want.

6.2 Interpretation of the Model

The model shows strong transitions from "consensus" to "plan," "discuss" to "consensus," and "emotion" to "cohesion" highlighting a network structure where agreement, dialogue, and emotional regulation play central roles in influencing and organizing the self-regulated collaboration. The transition between "consensus" and "plan" shows a high transition frequency of 1000 (simply means that it is the highest transition frequency in the network). This indicates a strong direct link where achieving consensus is usually followed by planning further steps. Another prominent transition exists from "plan" to "consensus" with a transition strength of 0.71, showing a feedback loop reinforcing a cyclic pattern of planning and consensus building within SRL.

"Discuss" also appears as a central node with strong outward transitions to "consensus" (0.51) and "emotion" (0.31), highlighting that discussions often yield mutual understanding or emotional bonding. Similarly, "emotion" shows a strong connection to "cohesion" with a transition frequency of 0.368, and to "consensus" (0.362), reflecting how emotional states may influence group cohesion and shared agreement. Other important transitions include "coregulate" to "consensus" (0.474) and "monitor" to "consensus" (0.091), revealing that collaborative regulation and monitoring are necessary for building agreement and cohesion. We also see, "adapt" and "synthesis" show lower transition frequencies overall, suggesting that these may either serve more limited roles in the learning process where synthesis or adaptation occurs with fewer preceding dependencies. In general, we see strong influence on planning and task enactment where emotions and discussion play an important role with less emphasis on adaptation or monitoring.

6.3 Pruning

In many instances, you may want to prune the model, i.e., remove small edges and continue to work with a trimmed model that contain only ***strong-enough*** edges. The `prune()` function is designed to do that by removing —pruning— edges based on specified criteria. The options of pruning are designed to remove small edges (either based on threshold or percentile) or using an algorithm that can retrieve the backbone of the network. The three options are:

- **Threshold-based pruning** allows researchers to specify a fixed numeric value, which deletes all edge weights that are less than or equal to the defined value while keeping the network fully connected (or at least weakly so).
- **Percentile-based pruning** allows to define a numeric percentile where edges with weights that fall below this specified percentile will be deleted. For instance,

if a user sets a percentile of 0.20, this will eliminate the lowest 20% of edges based on their weights and retain the rest of strong edges within the network. Both threshold-based and percentile-based pruning make sure that pruning does not disconnect the networks.
- **Disparity filter** uses a disparity filter algorithm to prune negligible edges and retain the backbone of the network. The disparity filter algorithm works by randomizing the network connections, allowing researchers to determine which edges are significantly stronger than expected by chance. Edges are retained if they are significant relative to a null model. However, disparity filter is strict and we recommend using an alpha level of around 0.5 to retain edges that are less likely to have resulted from chance. Please note that the three methods delete edges from the model.

The code below demonstrates the three approaches to pruning we just described, followed by plotting each pruned network for visual comparison (Fig. 3). First, we use the `prune()` function with the arguments `method = "threshold"` and `threshold = 0.1` to keep only the edges with weights above 0.1. The second approach, using `method = "lowest"` and `lowest = 0.15`, filters the network by retaining the top 85% of edges to remove the weakest 15%. The third method, `method = "disparity"` with `alpha = 0.5`, applies a disparity filter to keep only edges deemed statistically significant within the network structure, thereby helping to reveal meaningful links based on a significance level threshold of 0.5 or edges that are more likely than by chance.

```
# Pruning with different methods
pruned_threshold <- prune(model_scaled, method = "threshold", threshold =
↪  0.1)
pruned_lowest <- prune(model_scaled, method = "lowest", lowest = 0.15)
pruned_disparity <- prune(model_scaled, method = "disparity", alpha = 0.5)

# Plotting for comparison
plot(pruned_threshold)
plot(pruned_lowest)
plot(pruned_disparity)
plot(model_scaled, minimum = 0.05, cut = 0.1)
```

Alternatively, to plot the model without deleting any edges, you can hide small edges while retaining them for further computations by setting the `minimum` argument to a value e.g., 0.05. This will hide the edges below this value (0.05) but will retain the full model for further analysis (Fig. 4).

```
plot(model_scaled, minimum = 0.1, cut = 0.01)
```

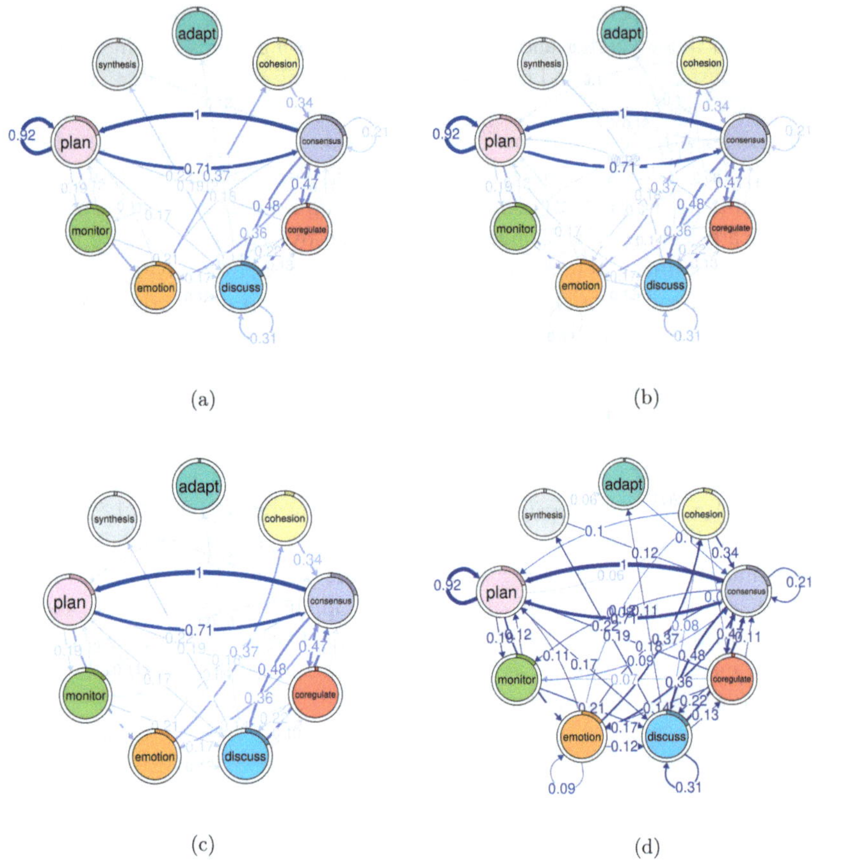

Fig. 3 FTNA scaled pruned with different methods. (**a**) Threshold = 0.15. (**b**) Lowest 15%. (**c**) Disparity filter. (**d**) Not pruned (minimum threshold)

6.4 Patterns

Patterns provide insights into the underlying mechanisms (e.g., behavior) that generated the data. These mechanisms give rise to recurring structures and connections between the learning processes. The study of such structures can help researchers interpret the underlying process, build new hypotheses or verify existing ones and identify areas that may need support (e.g., patterns indicating lack of regulation). FTNA allows users to detect and visualize n-cliques or groups of nodes that are all interconnected with transitions meeting specific criteria.

Cliques are strong, recurring patterns where nodes (representing behaviors or states) are strongly interconnected. While one can get several types of cliques, in FTNA we focus on dyads and triads. Dyads are two strongly linked nodes and triads

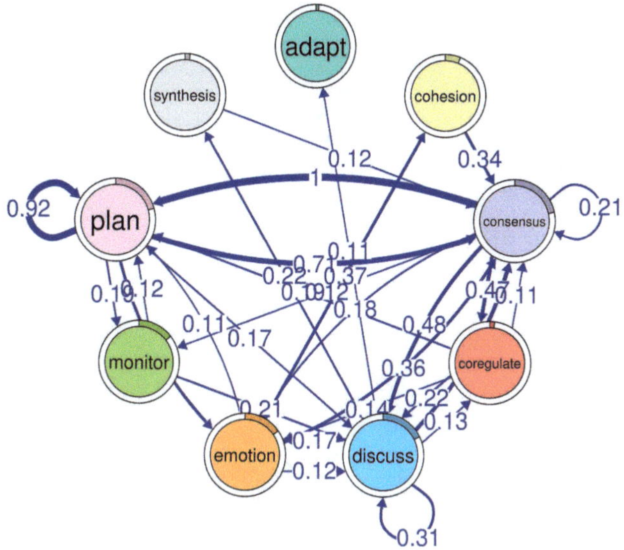

Fig. 4 FTNA scaled

are three well-connected nodes that reflect more complex dependencies. These strong structures are in a way what form or shape the entire process.

The `cliques()` function in FTNA allows users identify different cliques of different sizes by specifying the number of nodes in each clique. For instance, setting `size = 2` will identify dyads with strong pairwise mutual relationships. The function also has a `threshold` argument to get only dyads which are strongly connected above that threshold to ensure that only meaningful connections are visualized. Triads can be also identified in the same way by setting the size argument as `size = 3` which will capture patterns of three interconnected nodes above the stated threshold, in our case it is 0.05.

The clique function has a `sum_weights` argument that lets users decide if the sum of edge weights should be considered in forming the cliques. For example, setting the threshold to 0.10 and `sum_weights = FALSE` means each of the two edges should have a weight of at least 0.10 to be considered as part of the same clique. While `sum_weights = TRUE` means that only the sum of the weights of the two edges needs to be 0.1, i.e., an edge can be 0.04 and the other can be 0.06.

The code below is straightforward as it identifies and plots the cliques and shows examples of dyads, triads and 4-cliques and 5-cliques. In the first example, we set the `size = 2` and `threshold = 0.1` to get the dyads with strong transitions, while `size = 3` with a lower `threshold = 0.05` captures triads, identifying three-node interdependencies. Larger cliques, such as structures with 4 or 5 nodes (or more) can be obtained in the same way. We can also print and plot the cliques. For brevity, we only show the plot for the dyads (Fig. 5).

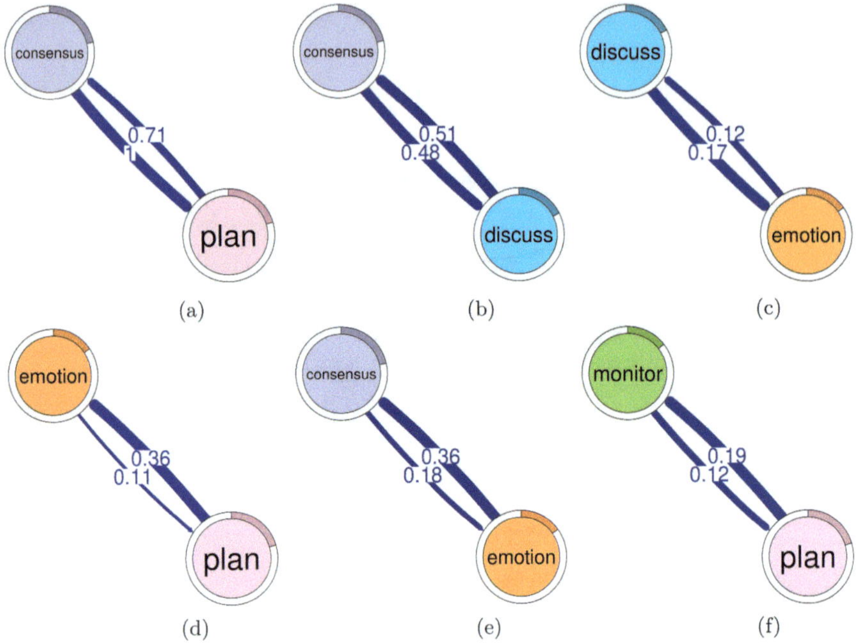

Fig. 5 Identified dyads. (**a**) Dyad 1. (**b**) Dyad 2. (**c**) Dyad 3. (**d**) Dyad 4. (**e**) Dyad 5. (**f**) Dyad 6

```
# Identify 2-cliques (dyads) from the FTNA model with a weight threshold,
# excluding loops in visualization.
# A 2-clique represents a pair of nodes that are strongly connected based on
# the specified weight threshold.
cliques_of_two <- cliques(
  model_scaled,        # The FTNA model with scaled edge weights
  size = 2,            # Looking for pairs of connected nodes (dyads)
  threshold = 0.1      # Only include edges with weights greater than 0.1
)

# Print and visualize the identified 2-cliques (dyads)
print(cliques_of_two)  # Display details of 2-cliques
```

```
Number of 2-cliques: 8 (weight threshold = 0.1)
Showing 6 cliques starting from clique number 1

Clique 1:
          consensus plan
consensus     0.207 1.00
plan          0.714 0.92
```

```
Clique 2:
          consensus discuss
consensus     0.207   0.475
discuss       0.507   0.307

Clique 3:
        discuss emotion
discuss   0.307   0.167
emotion   0.115   0.087

Clique 4:
        emotion  plan
emotion   0.087 0.113
plan      0.361 0.920

Clique 5:
          consensus emotion
consensus     0.207   0.184
emotion       0.363   0.087

Clique 6:
        monitor  plan
monitor  0.0104 0.123
plan     0.1856 0.920
```

```
plot(cliques_of_two)   # Visualize 2-cliques in the network
```

```
# Identify 3-cliques (triads) from the FTNA model.
# A 3-clique is a fully connected set of three nodes, indicating a strong
# triplet structure.
cliques_of_three <- cliques(
  model_scaled,       # The FTNA model with scaled edge weights
  size = 3,           # Looking for triplets of fully connected nodes (triads)
  threshold = 0.05    # Only include edges with weights greater than 0.05
)

# Print and visualize the identified 3-cliques (triads)
# Uncomment the code below to view the results
# print(cliques_of_three) # Display details of 3-cliques
# plot(cliques_of_three)  # Visualize 3-cliques in the network

# Identify 4-cliques (quadruples) from the FTNA model.
# A 4-clique includes four nodes where each node is connected to every other
```

```
# node in the group.
# Uncomment the code below to view the results
cliques_of_four <- cliques(
  model_scaled,       # The FTNA model with scaled edge weights
  size = 4,           # Looking for quadruples of fully connected nodes (4-cliques)
  threshold = 0.03    # Only include edges with weights greater than 0.03
)

# Print and visualize the identified 4-cliques (quadruples)
# Uncomment the code below to view the results
# print(cliques_of_four)  # Display details of 4-cliques
# plot(cliques_of_four)   # Visualize 4-cliques in the network

# Identify 5-cliques (quintuples) from the FTNA model, summing edge weights.
# Here, the sum of edge weights in both directions must meet the specified
# threshold for inclusion.
# Uncomment the code below to view the results
cliques_of_five <- cliques(
  model_scaled,       # The FTNA model with scaled edge weights
  size = 5,           # Looking for quintuples of fully connected nodes (5-cliques)
  threshold = 0.1,    # Only edges with total bidirectional weights greater than 0.1
  sum_weights = TRUE # Sum edge weight in both directions when computing  threshold
)

# Print and visualize the identified 5-cliques (quintuples)
# print(cliques_of_five)  # Display details of 5-cliques
# plot(cliques_of_five)   # Visualize 5-cliques in the network
```

6.5 *Graph Level Measures*

FTNA provides provide an easy access to all graph level measures through the `summary()` function. The summary function gives insights about nodes, the frequency and strength of transitions, and the overall connectedness of the network. The network metrics include the number of unique states (nodes) and the number of transitions (edges) which form the network's basic structure. Measures of density, mean distance, and reciprocity and other metrics which that help interpret the dynamic structure of the transition network:

- **Node Count**: The number of unique states or events being analyzed in the network.
- **Edge Count**: Total number of transitions recorded between states.

- **Network Density***: The proportion of possible transitions that are actually observed, where a density of 1 implies every possible transition between states has occurred at least once. Please note that in most cases of networks that are not pruned, density metric does not offer much value.
- **Mean Distance**: The average number of transitions (or steps) needed to move between any two states in the network reflecting the network's overall "connectivity" or flow.
- **Mean Out-Strength***: The total strength (sum of transition weights) of transitions that each state initiates.
- **Mean In-Strength***: The mean of total strength (sum of transition weights) directed towards each state.
- **Mean Out-Degree***: The mean of number of distinct states to which a particular state transitions, representing its transition "breadth."
- **Mean In-Degree***: The number of distinct states transitioning into a specific state, indicating its tendency to be a transition destination.
- **Centralization***: Measures how much the network is dominated by a central "hub" state, often a highly frequent transition point or common intermediary.
- **Reciprocity***: Reflects the tendency of transitions to be bidirectional, meaning if a transition exists from State A to B, it is likely B also transitions back to A, highlighting cyclic or reversible behavior within the network.

Please note that the measures marked with * do not offer much information in fully connected networks which is the case in most cases of FTNA and therefore should be interpreted with caution in these situations. However, the metrics are particularly useful in pruned models. See and contrast below two examples: the original model on the left side and the pruned model on the right side.

```
summary(model_scaled)
summary(pruned_disparity)
```

```
                         metric   value
1                    Node Count  9.0000
2                    Edge Count 78.0000
3               Network Density  1.0000
4                 Mean Distance  0.0240
5             Mean Out-Strength  1.1325
6               SD Out-Strength  0.8855
7              Mean In-Strength  1.1325
8                SD In-Strength  0.8778
9              Mean Out-Degree  8.6667
10               SD Out-Degree  0.7071
11 Centralization (Out-Degree)  0.0156
12  Centralization (In-Degree)  0.0156
13                  Reciprocity  0.9859
```

```
                     metric  value
                 Node Count  9.000
                 Edge Count 39.000
            Network Density  0.542
              Mean Distance  0.188
          Mean Out-Strength  0.900
            SD Out-Strength  0.700
           Mean In-Strength  0.900
             SD In-Strength  0.673
            Mean Out-Degree  4.333
              SD Out-Degree  1.225
Centralization (Out-Degree)  0.234
 Centralization (In-Degree)  0.516
                Reciprocity  0.615
```

The results of the `summary()` function of our examined networks show the network consists of 9 nodes and 78 edges, resulting in a high density of 1.0, indicating that all possible connections between nodes are present. The mean distance between nodes is minimal at 0.024, reflecting a tightly connected network where nodes are generally very close to each other in terms of path length. The average out-strength and in-strength for nodes are both approximately 1.1325, with standard deviations of 0.886 and 0.878, respectively. These values indicate some variability in the strength of connections. Similarly, the mean out-degree (the average number of connections each node has outward) is 8.67, with a low standard deviation of 0.707, showing that most nodes have nearly equal connectivity. Network centralization for both out-degree and in-degree is low 0.016. Reciprocity is 0.986 indicating that nearly all directed connections in this network are mutual, or reciprocated, which is typical of networks where bidirectional interaction is common. As mentioned above these numbers are expected in a fully connected network before pruning. Compare these values to the pruned network.

6.6 Node Level Measures

Centrality measures can help identify the influential events or states and rank their importance within the learning process [14]. The `tna` package provides the `centralities()` function designed for calculating various centralities in directed weighted networks. This function requires only the FTNA model object as input and it automatically computes a range of centralities, namely, out-strength, in-strength, closeness, betweenness, diffusion and clustering measures. By default, the function does not consider self-transitions (loops) as they do not represent changes of a state. Removing loops helps focus the analysis on transitions between distinct states rather than within-state repetitions. Of course, users can set `loops = TRUE` to count loops which may be useful in cases where researchers are estimating stability of states, e.g., remaining in the same state or other contextual reasons.

Out-strength centrality indicates how frequently a state transitions to other states which captures the change and stability within that state or that event. A higher out-strength suggests that the state often leads to different actions, while a lower out-strength implies stability, repetition or being stuck in the same state. In a learning context, a state like "plan" might have a high out-strength, indicating that students quickly transition from planning to other steps. In-strength centrality, by contrast, represents how frequently a state receives transitions and being a common destination. For instance, if "consensus" has high in-strength centrality it signifies that other learning states frequently result in consensus, reflecting group dynamics of agreement and cohesion.

Betweenness centrality reflects which states act as a bridge or mediates between other transitions. For instance, a state with high betweenness, like "synthesis", might connect various learning actions signifying its role as a juncture in the process. Given the flexible multi-step conceptualization of the network, several other centralities can be useful here —depending on the context of course— like diffusion centrality which is a good indication of diffusion of ideas or actions across the network.

The code below begins by calculating centrality measures using the `centralities()` function on `model_scaled`. After these measures are computed, we print them out to inspect the results (Table 5). Next, we visualize these centrality measures with `plot(centrality_measures)` (Fig. 6).

```
# Compute centrality measures for the FTNA model
centrality_measures <- centralities(model_scaled)

# Print the calculated centrality measures in the FTNA model
print(centrality_measures)
```

```
# Visualize the calculated centrality measures in the FTNA
↪  model
plot(centrality_measures)
```

We can also compute other centralities using the function `as.igraph(model_scaled)` which converts the FTNA model into an igraph object and thus enables all possible centralities and measures that can be computed using the `igraph` R package. For instance, the below code calculates the hub and authority scores for the FTNA network model using the `hits_scores()` function from the `igraph` package, it first converts the FTNA model (`model_scaled`) into an `igraph` object. Hub scores measure how often a node (event or state) points to other important nodes, reflecting its influence as a source. Authority scores measure how frequently a node is targeted by others, indicating its importance as a destination or key reference point. The resulting scores provide insight into which states act as influential hubs or authoritative endpoints within the learning process.

The code below exemplifies how to compute extra measures. We first convert our scaled FTNA model (`model_scaled`) into an `igraph` object. The `igraph` package supports a large array of graph analyses, including the HITS (Hyperlink-Induced Topic Search) algorithm, which calculates "hub" and "authority" scores for each node.

```
# Convert the FTNA model to an igraph object and
# calculate HITS (Hub and Authority) scores
hits_results <- igraph::hits_scores(as.igraph(model_scaled))

# Extract the hub and authority scores from the HITS results for further analysis
hub_scores <- hits_results$hub
authority_scores <- hits_results$authority
```

Table 5 Centrality measures plot for the FTNA model

State	OutStrength	InStrength	ClosenessIn	ClosenessOut	Closeness	Betweenness	BetweennessRSP	Diffusion	Clustering
Adapt	0.20	0.21	14.56	5.83	21.59	20	91	9.14	0.39
Cohesion	0.66	0.67	6.61	6.54	18.69	0	86	32.55	0.50
Consensus	2.32	2.34	1.25	5.89	7.68	0	2	87.21	0.22
Coregulate	0.77	0.82	11.28	4.02	12.98	0	83	30.27	0.63
Discuss	1.27	1.26	7.22	3.30	7.68	0	82	46.98	0.37
Emotion	1.05	1.02	4.46	6.68	16.93	0	75	43.56	0.53
Monitor	0.56	0.48	10.84	5.24	13.05	7	86	22.90	0.58
Plan	1.54	1.56	11.60	5.88	16.81	9	1	68.86	0.43
Synthesis	0.26	0.28	11.99	5.72	16.93	21	91	10.90	0.52

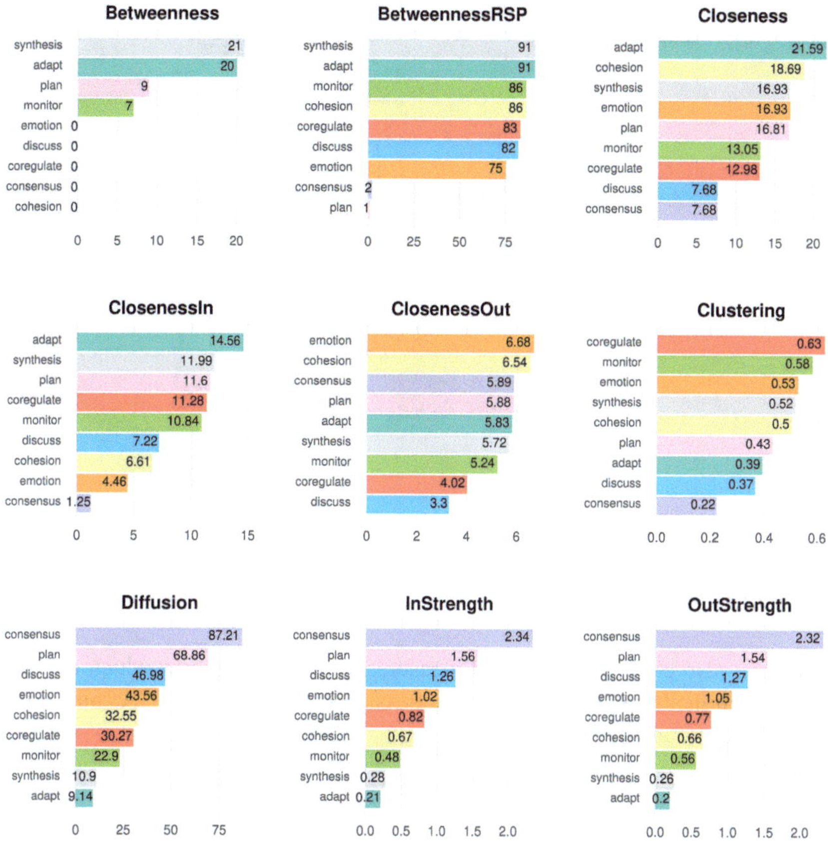

Fig. 6 Centrality measures plot for the FTNA model

```
# Print the hub and authority scores to view influential nodes
print(hub_scores)
print(authority_scores)
```

```
 adapt   cohesion  consensus coregulate   discuss   emotion   monitor      plan  synthesis
0.0556     0.2404     0.9551     0.2557    0.3768    0.3053    0.1895    1.0000     0.0738

 adapt   cohesion  consensus coregulate   discuss   emotion   monitor      plan  synthesis
0.0331     0.1292     0.6724     0.2925    0.4368    0.3437    0.1735    1.0000     0.0564
```

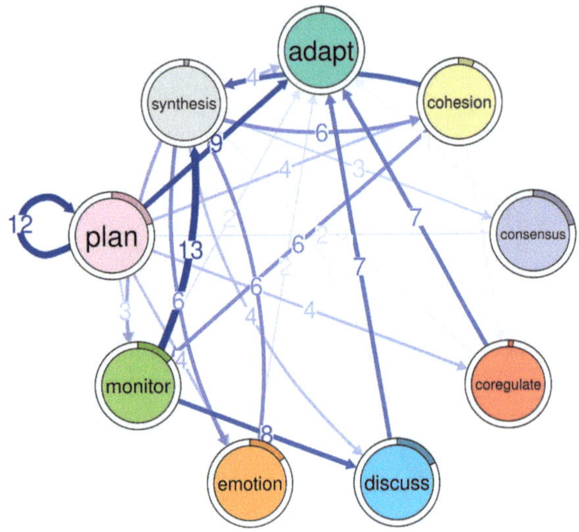

Fig. 7 Edge betweenness network

6.7 *Edge Level Measures*

FTNA can also compute edge level measures which would show which edges are important in the transition model and act as a bridge. Namely, FTNA can compute the edge betweenness (Fig. 7).

```
edge_between <- betweenness_network(model_scaled)
plot(edge_between)
```

6.8 *Community Detection*

Community detection can help identify groups of nodes —such as states, events, or actions— that exhibit strong interconnections between the nodes. Unlike cliques, which are defined by criteria of mutual connections among a fixed set of nodes and thresholds, communities are identified algorithmically and can vary in size and connection strength. This flexibility allows for a more nuanced and realistic approach to grouping transitions that share similar connectivity patterns regardless of their size, be it two nodes or five, as long as they exhibit shared interconnections.

Finding communities has been a central theme in social science for finding groups of constructs that are related (e.g., meta-cognition related nodes) to understand the structure of these constructs and their tight-knit relationships. Other examples include finding latent variables, propose new constructs and understand the semantics of discourse or behavior.

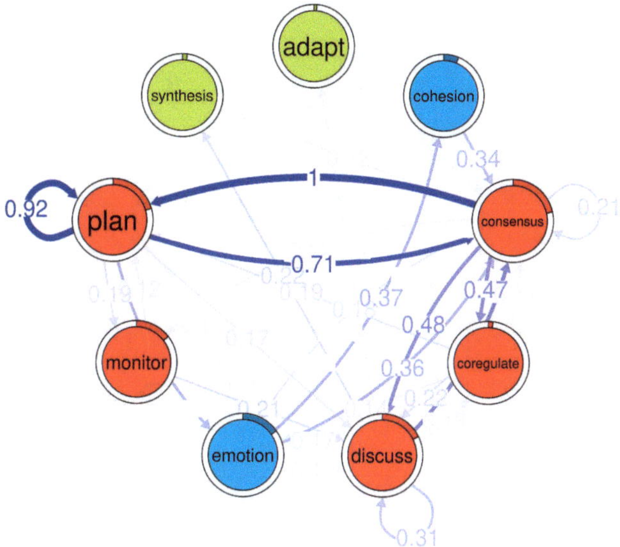

Fig. 8 Community detection using fast greedy method

TNA includes several community detection algorithms that are designed for transition networks (which are typically small, weighted, and directed). Each algorithm provides a distinct perspective on the grouping process. Some algorithms may focus on maximizing modularity, while others might emphasize flow or path length. As such the researcher can implement whatever suites their context. The code below performs community detection with the function `communities()` which takes the model argument and computes the communities, e.g., `communities(model_scaled)`. The results are stored in the `detected_communities` object which contains the results of the algorithms: Walktrap, Label Propagation, Infomap, Edge Betweenness, Leading Eigen, and Spinglass. To visualize the communities assignment, you can simply use the function plot with the name of the community that you want to use (Fig. 8). The `detected_communities` object can also be printed (Table 6).

```
detected_communities <- communities(model_scaled)
plot(detected_communities, minimum = 0.05)
```

```
print(detected_communities)
```

```
      walktrap      fast_greedy        label_prop          infomap edge_betweenness
             1                3                 1                1                1
 leading_eigen        spinglass
             3                3
```

Table 6 Community assignment

Node	walktrap	fast_greedy	label_prop	infomap	edge_betweenness	leading_eigen	spinglass
Adapt	1	1	1	1	1	1	1
Cohesion	1	2	1	1	1	3	2
Consensus	1	3	1	1	1	2	3
Coregulate	1	1	1	1	1	1	3
Discuss	1	1	1	1	1	1	3
Emotion	1	2	1	1	1	3	2
Monitor	1	1	1	1	1	1	3
Plan	1	3	1	1	1	2	3
Synthesis	1	1	1	1	1	1	3

6.9 Bootstrapping

Bootstrapping is a technique for assessing the accuracy and stability of edge weights in the network. With bootstrapping, we can estimate the likelihood that each transition would replicate or appear with similar strength in repeated sampling. In fact, bootstrapping is one of the strengths of FTNA as it allows us to verify the strength of each edge in the network and make sure that it is strong enough to be interpretable. Bootstrapping entails re-sampling the data repeatedly. Each of these sub-samples is created by randomly selecting data points from the original dataset with replacement. Then we reconstruct an FTNA network, recalculate the edge weights each time (i.e., build a new FTNA model). The process is commonly performed a large number of times (usually 1,000 iterations) to generate a distribution of edge weights for each transition in the network. Then we compare the original edge weight against the range of edge weights obtained from the 1,000 bootstrap samples and calculate confidence intervals for each edge. We can also compute p-values for each edge to estimate how different the edge is given a certain threshold value (e.g., 0.05) across the bootstrap samples. When an edge consistently exceeds this threshold, we consider it statistically significant. Bootstrapping also offers an idea about the stability and the robustness of the entire FTNA network as a whole.

To perform bootstrapping we use the `bootstrap()` function, which requires the model argument as an argument. The function typically uses a default of 1000 bootstrap iterations (`iter = 1000`), though increasing the number of iterations may improve reliability. Additionally, by setting a threshold (e.g., `threshold = 0.05`), we can directly test whether an edge's weight is consistently above this value, indicating statistical significance across samples. To view the edge data, confidence intervals and their p-values, we can print a summary output which contains the full bootstrap results.

```
# Perform bootstrapping on the FTNA model with a fixed seed for
↪  reproducibility
set.seed(265)
boot <- bootstrap(model_scaled, threshold = 0.05)

# Print the combined results data frame containing
print(summary(boot))
```

```
   from          to weight p_value   sig ci_lower ci_upper
2 adapt    cohesion 0.0020       1 FALSE 0.000397  0.00364
3 adapt   consensus 0.0120       1 FALSE 0.007822  0.01631
4 adapt coregulate 0.0128       1 FALSE 0.008502  0.01767
5 adapt     discuss 0.1126       0  TRUE 0.099843  0.12688
 [ reached 'max' / getOption("max.print") -- omitted 74 rows ]
```

```
# View non-significant edges which are less likely to be stable across
↪ bootstrap samples
print(boot, type = "nonsig")
```

```
Non-significant Edges

   from         to   weight p_value ci_lower ci_upper
2 adapt   cohesion 0.00200       1 0.000397  0.00364
3 adapt  consensus 0.01198       1 0.007822  0.01631
4 adapt coregulate 0.01277       1 0.008502  0.01767
6 adapt    emotion 0.00279       1 0.000795  0.00489
7 adapt    monitor 0.00639       1 0.003546  0.00992
 [ reached 'max' / getOption("max.print") -- omitted 34 rows ]
```

6.10 Comparing Models

Two methods can be used to compare models. The first is to compare the raw difference between the weights of the models. The second is to use a statistical technique that shows to what extent each edge differ statistically. FTNA has both methods, the first method offers a descriptive view of the differences with an intuitive visualization that shows the differences. To perform such comparison, we use the `plot_compare()` function to visualize the differences. In the plot, green edges are higher in the first model, red means lower edges and so is the case in the pie.

However, to rigorously compare models, FTNA uses a permutation test which estimates which edges are statistically significantly different and produces a p-value for each edge. The code below uses the `group_regulation` dataset which has two subsets. The first subset, `group_regulation[1:1000,]`, represents the high achievers model, while the second subset `group_regulation[1001:2000,]` represents the low achievers. For each of the subsets we create a model by applying the `tna()` function to its respective subset. Next, we perform a permutation test with `permutation_test()` to assess whether the observed differences between the "Hi" and "Lo" models are statistically significant. Setting `iter = 1000` instructs the function to run 1000 iterations, creating a distribution of differences by repeatedly shuffling and comparing the groups. Finally, we plot the significant differences identified through the permutation test using `plot(Permutation, minimum = 0.001)` (Fig. 9). This final plot visually highlights edges that have a statistically significant difference in transition weight between the "Hi" and "Lo" models.

```
# Create FTNA for the high-achievers subset (rows 1 to 1000)
Hi <- ftna(group_regulation[1:1000, ], scaling = "minmax")
```

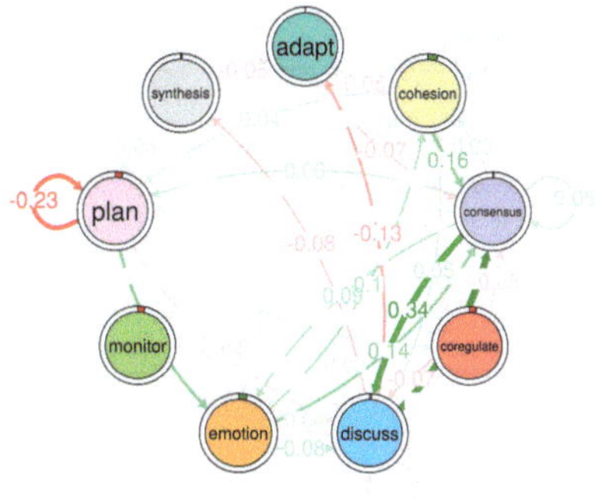

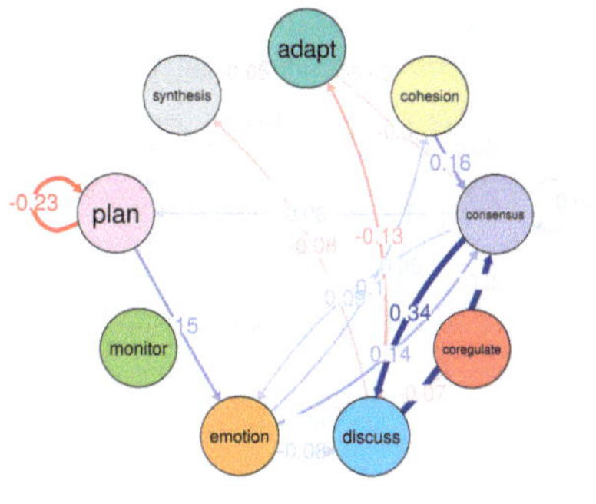

Fig. 9 Network comparison. (**a**) Subtraction of high achievers minus low achievers. (**b**) Significant differences identified in the permutation test

```
# Create FTNA for the low-achievers subset (rows 1001 to 2000)
Lo <- ftna(group_regulation[1001:2000, ], scaling = "minmax")

# Plot a comparison of the "Hi" and "Lo" models
# The 'minimum' parameter is set to 0.001, so edges with weights >= 0.001 are shown
plot_compare(Hi, Lo, minimum = 0.0001)

# Run a permutation test to determine statistical significance of
# differences between "Hi" and "Lo"
# The 'it' parameter is set to 1000, meaning 1000 permutations are performed
Permutation <- permutation_test(Hi, Lo, iter = 1000)

# Plot the significant differences identified in the permutation test
plot(Permutation, minimum = 0.01)
```

6.11 *Centrality Stability*

In contrast to most network models. FTNA has a robust mechanism to test the stability and the significance of centrality measures through case-dropping. Case-dropping involves sequentially removing data points and re-calculating correlation coefficients that compare the centrality measures in the original network with those derived from networks missing the data points. A correlation coefficient closer to 1 suggests that the centrality rankings of nodes remain stable even when data is removed which indicates that the centrality value is resilient to sample variability and more likely to generalize. A correlation coefficient above 0.7, for instance, would indicate that the centrality measure is reliable; the most central nodes in the

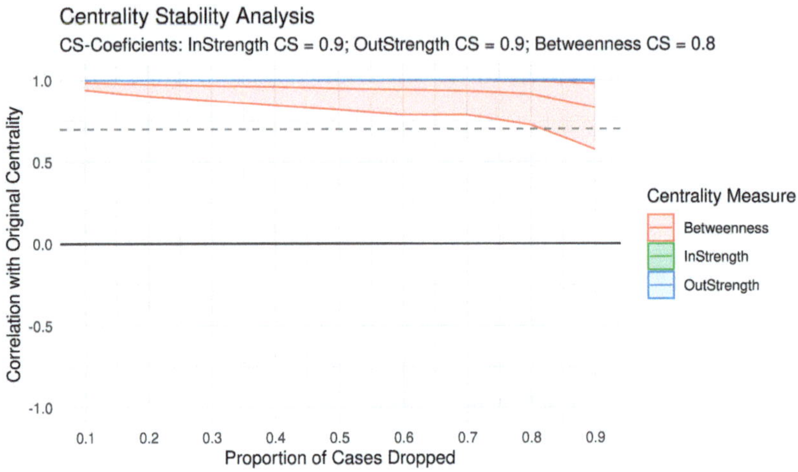

Fig. 10 Centrality stability

original FTNA network remain central even as data points are omitted. An average coefficient between 0.5 and 0.7 reflects moderate stability. However, an average coefficient below 0.5 would indicate low stability.

The function `estimate_centrality_stability()` (abbreviated as `estimate_cs()`) can be used to assess the stability of centrality measures in FTNA models. The function generates bootstrap results that can then be visualized (Fig. 10), offering a clear representation of centrality stability within the FTNA framework.

```
Centrality_stability <- estimate_centrality_stability(model_scaled)
plot(Centrality_stability)
```

7 Conclusion

This chapter presented FTNA as a flexible and robust approach to analyzing temporal data of learning processes. FTNA is a novel tool for analyzing complex learning processes. Its flexibility, scalability, and straightforward modeling of transition frequencies make it well-suited for exploring and comparing learning dynamics across varied educational contexts. Unlike traditional Transition Network Analysis (TNA) [2], which models transitions probabilistically based on Markov assumptions, FTNA focuses on the absolute frequency of transitions between states, offering a straightforward representation of learning dynamics without conditioning on prior states. This method is particularly advantageous where dependencies span multiple steps, such as collaborative learning or complex problem-solving tasks or for contexts with rare transitions, small sample sizes, or processes.

What makes FTNA stand out in comparison to existing techniques is its statistical rigor. In that, FTNA allows meaningful comparisons across different datasets or conditions with permutation. Statistical validation methods, such as bootstrapping increase the rigor of FTNA and provide a validation method for the resulting models. Additionally, FTNA inherits the network capabilities of TNA, such as centrality measures and pattern detection [18] which offer a deeper understanding of structural relationships within the learning process.

References

1. Saqr M, Peeters W (2022) Temporal networks in collaborative learning: a case study. Brit J Educ Technol 53:1283–1303. https://doi.org/10.1111/bjet.13187
2. Saqr M, López-Pernas S, Tikka S (2025) Mapping relational dynamics with transition network analysis: a primer and tutorial. In: Saqr M, López-Pernas S (eds) Advanced learning analytics methods: AI, precision and complexity. Springer, Cham
3. Zou Y, Donner RV, Marwan N, Donges JF, Kurths J (2019) Complex network approaches to nonlinear time series analysis. Phys Rep 787:1–97. https://doi.org/10.1016/j.physrep.2018.10.005
4. Saqr M, López-Pernas S, Törmänen T, Kaliisa R, Misieju, K, Tikka S (2025) Transition network analysis: a novel framework for modeling, visualizing, and identifying the temporal patterns of learners and learning processes. In: Proceedings of the 15th international learning analytics and knowledge conference, pp 351–361
5. Tikka S, López-Pernas S, Saqr M (2025) tna: an R package for transition network analysis. Appl Psychol Measur 49(6):326–328. https://doi.org/10.1177/01466216251348840 (Original work published 2025)
6. Nicolis G, Cantú AG, Nicolis C (2005) Dynamical aspects of interaction networks. Int J Bifur Chaos Appl Sci Eng 15:3467–80. https://doi.org/10.1142/s0218127405014167
7. Schnakenberg J (1976) Network theory of microscopic and macroscopic behavior of master equation systems. Rev Mod Phys 48:571–85. https://doi.org/10.1103/revmodphys.48.571
8. Poquet O, Saqr M, Chen B (2021) Recommendations for network research in learning analytics: to open a conversation. In: Poquet O, Chen B, Saqr M, Hecking T (eds) Proceedings of the NetSciLA2021 workshop "using network science in learning analytics: building bridges towards a common agenda" (NetSciLA2021). Aachen, pp 34–41
9. Helske J, Helske S, Saqr M, López-Pernas S, Murphy K (2024) A modern approach to transition analysis and process mining with markov models in education. Springer, Cham, pp 381–427
10. López-Pernas S, Saqr M (2024) The why, the how and the when of educational process mining in R. In: Learning analytics methods and tutorials. Springer, Cham, pp 467–488
11. Epskamp S, Cramer AOJ, Waldorp LJ, Schmittmann VD, Borsboom D (2012) Qgraph: network visualizations of relationships in psychometric data. J Stat Softw 48:1–18. https://doi.org/10.18637/jss.v048.i04
12. Newman M (2018) Networks, 2nd edn. Oxford University Press, London
13. Saqr M, Elmoazen R, Tedre M, López-Pernas S, Hirsto L (2022) How well centrality measures capture student achievement in computer-supported collaborative learning? – A systematic review and meta-analysis. Educ Res Rev 35:100437. https://doi.org/10.1016/j.edurev.2022.100437

14. Saqr M, López-Pernas S, Conde-González MÁ, Hernández-García Á (2024) Social network analysis: a primer, a guide and a tutorial in R. Springer, Cham, pp 491–518
15. Saqr M, López-Pernas S (2021) Modelling diffusion in computer-supported collaborative learning: a large scale learning analytics study. Int J Comput Support Collab Learn 16:441–483. https://doi.org/10.1007/s11412-021-09356-4
16. Csardi G, Nepusz T (2006) The igraph software package for complex network research. Int J Complex Syst 1695
17. Winne PH (2010) Improving measurements of self-regulated learning. Educ Psychol 45:267–276. https://doi.org/10.1080/00461520.2010.517150
18. López-Pernas S, Tikka S, Saqr M (2025) Mining patterns and clusters with transition network analysis: a heterogeneity approach. In: Saqr M, López-Pernas S (eds) Advanced learning analytics methods: AI, precision and complexity. Springer, Cham
19. van Borkulo CD, van Bork R, Boschloo L, Kossakowski JJ, Tio P, Schoevers RA, Borsboom D, Waldorp LJ (2023) Comparing network structures on three aspects: a permutation test. Psychol Methods 28:1273–1285. https://doi.org/10.1037/met0000476
20. Tikka S, Kopra J, Heinäniemi M, López-Pernas S, Saqr M (2024) Getting started with R for education research. In: Saqr M, López-Pernas S (eds) Learning analytics methods and tutorials: a practical guide using R. Springer, Berlin
21. Kopra J, Tikka S, Heinäniemi M, López-Pernas S, Saqr M (2024) An R approach to data cleaning and wrangling for education research. In: Saqr M, López-Pernas S (eds) Learning analytics methods and tutorials: a practical guide using R. Springer, Berlin
22. Tikka S, Kopra J, Heinäniemi M, López-Pernas S, Saqr M (2024) Introductory statistics with R for educational researchers. In: Saqr M, López-Pernas S (eds) Learning analytics methods and tutorials: a practical guide using R. Springer, Berlin
23. López-Pernas S, Misiejuk K, Tikka S, Kopra J, Heinäniemi M, Saqr M (2024) Visualizing and reporting educational data with R. In: Learning analytics methods and tutorials. Springer, Cham, pp 151–194
24. Wickham H, Averick M, Bryan J, Chang W, McGowan LD, François R, Grolemund G, Hayes A, Henry L, Hester J, Kuhn M, Pedersen TL, Miller E, Bache SM, Müller K, Ooms J, Robinson D, Seidel DP, Spinu V, Takahashi K, Vaughan D, Wilke C, Woo K, Yutani H (2019) Welcome to the {tidyverse}. 4:1686. https://doi.org/10.21105/joss.01686
25. Gabadinho A, Ritschard G, Müller NS, Studer M (2011) Analyzing and visualizing state sequences in R with TraMineR. J Stat Softw 40(4):1–37. https://doi.org/10.18637/jss.v040.i04
26. Saqr M, López-Pernas S, Helske S, Durand M, Murphy K, Studer M, Ritschard G (2024) Sequence analysis in education: principles, technique, and tutorial with R. Springer, Cham, pp 321–354

Mining Patterns and Clusters with Transition Network Analysis: A Heterogeneity Approach

Sonsoles López-Pernas, Santtu Tikka, and Mohammed Saqr

Abstract In this chapter, we demonstrate how we can identify and study clusters within Transition Network Analysis (TNA) to reveal the underlying heterogeneity in learners' behavioral patterns. Specifically, we rely on mixture Markov models (MMM) to identify latent subgroups characterized by unique transition probabilities, a method that can also incorporate covariates to explain the identified clusters. We employ the `tna` R package to understand the distinct transition dynamics between states or events in each cluster through the study of centrality measures, communities and cliques. Lastly, we exemplify how to implement other forms of clustering (e.g., distance based) and grouping, as well as other types of transition networks (e.g., frequency-based transition networks).

Keywords Learning analytics · Transition network analysis · Mixture Markov models · Heterogeneity · Clustering

1 Introduction

Clusters, within the context of Transition Network Analysis (TNA), represent distinct subgroups within the sequence data that exhibit unique transition patterns between states or events [1]. In other words, these clusters share similar transition dynamics that are closer to each other within the same cluster and rather different from other clusters. Identifying these clusters rests on the assumption that students' behavior varies across contexts and conditions giving rise to patterns of similar behavior that can be captured by appropriate clustering techniques. In doing so, clustering mines these unobserved patterns that may be overlooked when studying

S. López-Pernas (✉) · M. Saqr
University of Eastern Finland, Joensuu, Finland
e-mail: sonsoles.lopez@uef.fi

S. Tikka
University of Jyväskylä, Jyväskylä, Finland

M. Saqr, S. López-Pernas (eds.), *Advanced Learning Analytics Methods*,
https://doi.org/10.1007/978-3-031-95365-1_17

the data as a whole without considering the heterogeneity. Not accounting for the heterogeneity in the data may obscure individual differences [2], let alone the true underlying structure.

Clusters often arise from an underlying data generation mechanism that shapes how learners interact with tasks, contexts, and each other. The data generating mechanism reflects the interplay of cognitive, emotional, and social processes, as well as the structure and demands of the learning environment. For instance, self-regulation as a data generating mechanism shapes the transitions between states —such as planning, monitoring, and adaptation— which are driven by learners' efforts to achieve goals, respond to feedback, or enact their tasks [3]. External factors, such as the presence of collaborative tasks or deadlines, may also influence these transitions by prompting learners to emphasize specific strategies like socio-emotional regulation or task coordination. Additionally, the heterogeneity in learners' dispositions contributes to diverse patterns in how they regulate their learning as well as the facilitation and contextual variables. Together, these factors generate the distinct SRL transition patterns that are captured and represented with TNA clusters offering a window into their learning processes.

Capturing such patterns allows to accurately understand students' learning processes and their variations. Furthermore, it provides a foundation for theory building by linking these clusters to specific underlying mechanisms, e.g., self-regulation tactics and strategies. Clusters also represent episodes —sessions, epochs or times— where learners manifest certain behaviors. For instance, in a learning management system (LMS), a cluster could represent students' pattern of focus on assessment, another may represent students' engagement with their lectures, and so on. In the same vein, in self-regulated learning, clusters reveal variations in learning strategies, such as learners who adapt effectively by transitioning from monitoring to task adaptation, demonstrating advanced metacognitive skills. In contrast, some clusters may show learners stuck in cycles of monitoring and exploration without progressing, highlighting inefficiencies in self-regulation that could benefit from targeted interventions.

The learning analytics literature has a rich tradition of grouping students' similar behavior using similar methods, e.g., in sequence mining of LMS data these clusters are often referred to as tactics and aims to find clusters of online behavior [4], self-regulation [5], or engagement [6]. One can say that clusters in TNA are similar, however, the emphasis here is on clusters that share similar transition dynamics.

Clustering in TNA not only identifies subgroups, but also broadens our understanding of the strategies learners employ [7]. When we analyze the group as a whole, only the average or most common patterns are visible. Instead, clustering allows to place the focus in patterns that are unique to specific subgroups, which may have been eclipsed by the general trends [2]. Moreover, clustering allows researchers to ask questions about the conditions under which specific patterns emerge if we use covariates and explanatory variables as part of the clustering model. For instance, do learners who frequently transition between planning and monitoring perform better in problem-solving tasks, and how does this vary in collaborative versus individual learning settings? Do emotional regulation clusters correlate with sustained engagement in challenging tasks, and how can these insights guide interventions?

In TNA, clustering the sequence data is performed through identifying distinct transition patterns using mixture Markov models (MMM) [8]. Unlike simple Markov models that assume homogeneity in transition dynamics across all data, MMM accounts for the coexistence of multiple latent subgroups, each governed by its own unique transition matrix. MMM works by assigning each sequence to one of latent subgroups which share similar transition probabilities. In general it is a robust and flexible technique for clustering sequential data and more importantly it allows the inclusion of covariates that help identify or explain why certain clusters emerged.

Using covariates allows researchers to explore what factors are associated with the emergence of specific clusters [9]. For example, higher-performing groups might consistently be involved in clusters of transitions involving monitoring and adaptation, reflecting emphasis on a metacognitive approach to their tasks. Such level of analysis contextualizes the clusters and also provides actionable insights into the conditions under which specific behaviors emerge.

In this chapter, we provide a tutorial on how to identify and study clusters within TNA. We first review related works addressing the heterogeneity of transition dynamics between states or events in the education literature. We then provide a step-by-step tutorial using R on how to implement clustering in TNA using the `seqHMM` [10] and the `tna` [11] packages.

2 Related Work

Educational research has evolved from adopting a one-size-fits-all approach to acknowledging the existing heterogeneity among students and their behavior [7]. The increasing availability of data about learners has made it possible to study heterogeneity when interacting in or with learning environments, like the LMS or discussion forums. This has enabled researchers to identify distinct self-regulated learning phases, learning tactics and strategies, or collaboration patterns. For example, looking at the transitions between actions that students carried out in the LMS, Saqr et al. [4] identified four learning tactics using a mixture hidden mixture Markov model: *lecture read*, *forum read*, *forum interact*, and *diverse*. Villalobos et al. [12] clustered sequences of students' weekly interactions with the LMS using a mixture hidden Markov model and identified three distinct weekly strategies: *quiz-focused*, *material-focused*, and *diverse*.

Looking into the overall LMS engagement levels instead, López-Pernas and Saqr [13] identified three distinct trajectories of students' engagement profiles throughout each course of a study program using a mixture Markov model: *engaged*, *fluctuating* and *persistently disengaged*. They found that engagement dynamics fit the salient features of a complex dynamic system, where students that are disengaged rarely transition to higher engagement states, and engaged students —in the rare occasion they transition to lower engagement— return back to their engaged state right away. In a similar context, Saqr et al. [6] clustered students using a mixture hidden Markov model according to their patterns of LMS engagement and achievement and found three distinct trajectories of students: *engaged high-achieving starters*, *average starters*, and *disengaged starters*.

Another recent trend in the literature is the study of emotions and how they influence learning outcomes. For instance, Saqr and López-Pernas [14] used a mixture hidden Markov model to cluster students' according to the transitions between their emotions throughout two lectures. They found three trajectories: one with consistent anxiety, another one with constant happiness and alertness, and a third one with fluctuations between motivation-satisfaction and happiness-alertness. In another study, Törmänen et al. [5] used mixture hidden Markov models to cluster multimodal data: affective states and regulation were identified from video data, and activation was captured from electrodermal activity data. The authors identified four clusters of students with distinct socio-emotional interaction episodes —*frequent regulation*, *occasional regulation*, *positive*, and *negative*— which showed differences in the variability of affective states and the activation of learning regulation.

3 Tutorial: Clusters in TNA

In this section we present a step-by-step tutorial on how to identify and study clusters within TNA. Specifically, we will illustrate how to identify clusters using MMMs, visualize them, compare them, detect communities and compute centralities. We also illustrate how to implement other clustering and grouping techniques. To be able to follow the tutorial, it is assumed that the reader is familiar with the R programming language. If that is not the case, it is recommended to refer to previous tutorials on the basics of R [15], data cleaning [16], basic statistics [17], and visualization [18]. It is also recommended to have previous knowledge about Markov models [8] and to have read Chapter 15 about TNA [19].

3.1 Importing Libraries

As a first step in our tutorial, we will import the necessary libraries:

- **tidyverse**: A collection of R packages for data science, including tools for data manipulation (`dplyr`), visualization (`ggplot2`), tidying data (`tidyr`), and more, designed to work seamlessly together [20].
- **TraMineR**: A package for sequence analysis that provides methods to analyze and visualize sequential data, like event transitions or state changes [21].

- **seqHMM**: A package for sequence analysis supporting modeling and clustering of sequential data using various Markov models [10].
- **tna**: A package for implementing TNA, offering tools to estimate, visualize and investigate transitions over time using a network lens [11].
- **rio**: A package for reading data in several formats [22].

```
library(tidyverse) # install.packages("tidyverse")
library(TraMineR) # install.packages("TraMineR")
library(seqHMM) # install.packages("seqHMM")
library(tna) # install.packages("tna")
library(rio) # install.packages("rio")
```

3.2 *Preparing the Data*

We will use a simulated dataset based on a recent study by Saqr and López-Pernas [23], containing students' coded utterances in project-based learning implemented through online forum discussions. In this dataset, each row represents a team's collaboration on a problem, containing the ordered coded utterances in each column. In addition, the dataset contains the grade for each student. There are 3000 sequences in total and 10 different types of coded utterances: *agree*, *argue*, *disagree*, *discuss*, *evaluate*, *facilitate*, *organize*, *question*, *share*, and *socialize*.

```
simulated_data <- import("https://github.com/lamethods/data2/raw/main/pbl/pbl.RDS")
```

We can use the dataset to create a sequence object using the `seqdef` function from `TraMineR` and pass our coded data as an input (excluding the grades).

```
sequ <- seqdef(simulated_data[, -1])
```

3.3 *Clustering*

In our next step, we will cluster our sequence data with the aim of finding distinct groups that are governed by different transition dynamics. Specifically, we will fit an MMM to our data. A Markov model is a statistical model used to describe systems that transition between states with certain probabilities [8]. Markov models work under the assumption that the probability of moving to the next state depends only on the current state, not on the sequence of states that preceded it. A mixture Markov model extends the Markov model by allowing for multiple subpopulations or groups, each with its own set of transition probabilities (see Fig. 1). Instead of assuming a single transition structure for all data, it models the data as a combination

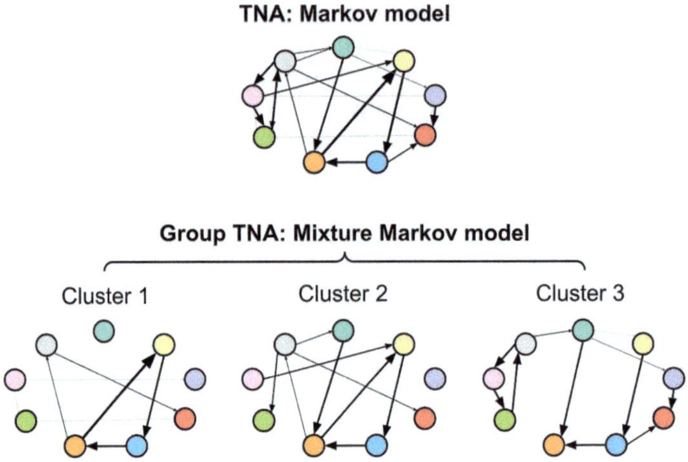

Fig. 1 Markov model vs. mixture Markov model

(or mixture) of multiple Markov models, where each component represents a different latent subgroup. For more details on these concepts, please refer to [8].

We will use the `seqHMM` library [10] to cluster our sequences of students' collaboration in groups using an MMM. When dealing with long sequences, the algorithm in charge of finding the best fitting model may take a long time. Therefore, it is recommended to provide a starting value of initial and transition probabilities that we hypothesize will be close to the real one that we expect the algorithm to find. As is the case in most clustering techniques, we do not know the optimum number of clusters. We illustrate the process for four clusters in this tutorial. In practice, we should estimate different models with different number of clusters to see which is the best fitting (lowest Bayesian Information Criterion level). Please, refer to a previous tutorial on Markov models for more details on this step [8].

To simulate the **transition probabilities**, we use the function `simulate_transition_probs` from `seqHMM`. As the first input, we provide the number of events or states in our model (ten in our case). As the second input we provide the number of clusters that we expect to find in this model (as mentioned before, we should try several combinations).

To simulate the **initial probabilities**, we extract the first coded utterance for the complete sequence object, and we use `table` on the first column to calculate the proportion of times each code appears as the first step in the collaboration. We replicate these probabilities 4 times (one for each hypothesized cluster).

```
set.seed(1234)

# Creating constants for the number of codes (10) and the number of clusters (4)
N_CODES <- length(alphabet(sequ))
N_CLUSTERS <- 4
```

```
# Transition probabilities (10 codes, 4 clusters)
trans_probs <- simulate_transition_probs(N_CODES, N_CLUSTERS, diag_c = 1)

# Initial probabilities (proportion of each code in t = 1)
init_probs <- as.numeric(prop.table(table(sequ[, 1])[1:N_CODES]))

# Replicate initial probabilities for each cluster
init_probs <- replicate(N_CLUSTERS, init_probs, simplify = FALSE)
```

We can now use `build_mmm` to define our MMM. We provide the sequence object as an input along with the number of clusters we expect to find (`n_clusters`), the transition probabilities, and the initial probabilities. In addition, an advantage of mixture Markov models —similarly, to other mixture models— is that it can take covariates. To do so, we have to provide the complete dataframe and specify the column(s) that we want to use as covariates, in our case, `Grade`. It is important to note that this dataframe should be in the same order as our sequence object. In addition, we need to provide a formula. In this case we use the formula `~ 0 + Grade`, which indicates that the cluster membership is explained by the students' grades.

```
# Build mixture Markov model for 4 clusters
mmm <- build_mmm(sequ,
                 n_clusters = N_CLUSTERS,
                 transition_probs = trans_probs,
                 data = simulated_data,
                 algorithm =
                 formula = ~ 0 + Grade,
                 initial_probs = init_probs)
```

Lastly, we can fit our model using `fit_model`. We provide the model that we have just defined (`mmm`) as an input. We can also define a series of parameters to specify the runtime of the algorithm that will find the optimum cluster solution. For example, we can specify how many CPU threads we would like to use (`threads` argument, 1 by default), that it should have 500 iterations (`restart` within `control_em` argument) as well as the optimization algorithm (`algorithm` within `control_em` argument, `"NLOPT_GD_MLSL_LDS"` by default), among others. See [8] and the `seqHMM` documentation [10] for more details. When dealing with long sequences and complex data like ours, this process can take a long time so you might need to leave your computer working for a few hours to arrive to the best solution. Once it has completed, we will have out fitted mixture Markov model in `fit_mmm`. Remember that in a real situation you would have to repeat this process for a range of clustering solutions (e.g., 2–10) to identify the optimum number of clusters.

```
fit_mmm <- fit_model(
  mmm,
  global_step = TRUE,
```

```
  local_step = TRUE,
  threads = 60,
  control_em = list(
    restart = list(times = 500,  n_optimum = 501),
    algorithm = "NLOPT_GD_MLSL_LDS"
))
```

3.4 Creating and Visualizing TNA Results

Now that we have our model fitted, we can use TNA —through the `tna` package [11]— to inspect the clusters that we have identified. We can use the function `group_tna` to create a grouped TNA model based on the clustering results. Specifically, we need to pass the MMM object (`fit_mmm$model`) as an argument.

```
tna_group <- group_tna(fit_mmm$model)
```

The functions that we have learned in Chapter 15 [19] apply also to our grouped model created from the clustering results. For instance, we can plot the model using the `plot` function, and a separate plot will be created for each cluster, each with their own initial and transition probabilities (Fig. 2). As a reminder, the pie around each node represents the initial probability of each code, whereas the arrows represent the transition probabilities between them.

We can see that Cluster 1 is characterized by discussions starting with *socialize*, with strong transitions between *share* and *discuss*, and *disagree* and *facilitate*, *argue* and *discuss*, and *agree* and *discuss*. Cluster 2 is characterized by starting with *argue*, and also a strong transition between *share* and *discuss*, but also *disagree* and *socialize*. Cluster 3 encompasses more diverse of initial probabilities —especially *discuss* and *evaluate*. This cluster has a very strong repetition of *organize*, as well as *facilitate*. Similar to the previous clusters, *share* and *discuss* is the predominant transition between different codes, followed by *agree* and *discuss*. Cluster 4 has fewer but stronger transitions, especially between *socialize* and *disagree*, *share* and *discuss*, and *argue* and *discuss*.

```
plot(tna_group)
```

We can also use the `summary` function to obtain a combined summary of the models corresponding to each cluster (Table 1). We can see differences in the overall network properties of each cluster. Cluster 3 is the most connected cluster, with the largest number of edges, and highest density value, while Cluster 4 is the least connected one. Clusters 1 and 2 lie in between, Cluster 2 being more dense and interconnected (although with higher variability), while Cluster 1 has higher reciprocity and is more stable.

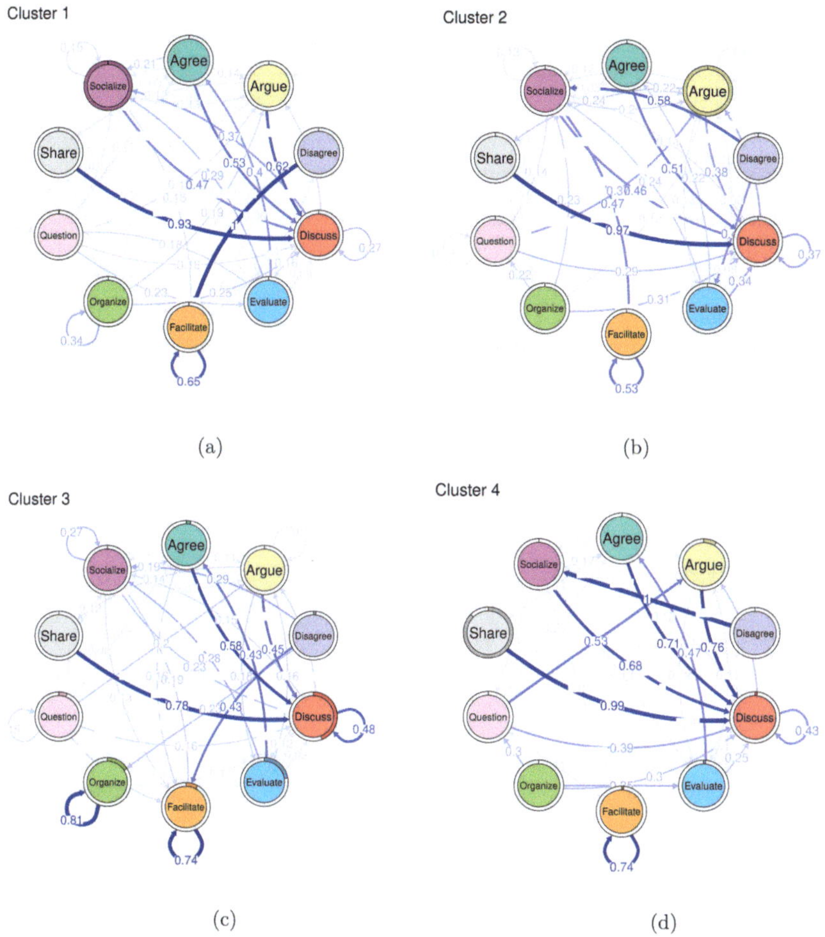

Fig. 2 Plot of each TNA model. (**a**) Socially triggered. (**b**) Argumentation triggered. (**c**) Facilitated discussions. (**d**) Information triggered

```
summary(tna_group)
```

Since we entered the grade as a covariate in the clustering process, we can study how grades are related to cluster membership. To obtain a summary of the effect of the covariates, we can use the `mmm_stats` function from `tna`. The output of this function is a data frame containing the estimate associated with each covariate (only one in our case: grade) and its confidence interval, the standard error, the t-score, and the p-value. The results are calculated taking Cluster 1 as a reference. From Table 2, we can see that grades had a small but statistically significant effect on the membership of clusters 2 and 3 (p-value < 0.05), positive in the case of Cluster 2

Table 1 Summary of the grouped tna model

Metric	Cluster 1	Cluster 2	Cluster 3	Cluster 4
Node count	10.00	10.00	10.00	10.00
Edge count	61.00	65.00	76.00	45.00
Network density	0.68	0.72	0.84	0.50
Mean distance	0.18	0.16	0.03	0.23
Mean out-strength	1.00	1.00	1.00	1.00
SD out-strength	0.97	1.12	0.86	1.31
Mean in-strength	1.00	1.00	1.00	1.00
SD in-strength	0.00	0.00	0.00	0.00
Mean out-degree	6.10	6.50	7.60	4.50
SD out-degree	2.69	3.17	1.90	1.84
Centralization (out-degree)	0.32	0.42	0.25	0.37
Centralization (in-degree)	0.32	0.30	0.25	0.37
Reciprocity	0.85	0.75	0.83	0.60

Table 2 Covariate effect

Cluster	Variable	Estimate	p_value	CI_Lower	CI_Upper	Std_Error	t_value
Cluster 2	Grade	0.004	0.00	0.003	0.006	0.001	6.31
Cluster 3	Grade	−0.007	0.00	−0.009	−0.006	0.001	−8.45
Cluster 4	Grade	0.000	0.58	−0.001	0.002	0.001	0.55

(meaning that high achievers were more likely to belong to this cluster), and negative in the case of Cluster 3. For cluster 4, the effect was not statistically significant.

```
mmm_stats(fit_mmm$model)
```

To ease the interpretation of our results henceforward, we can provide the clusters with more meaningful names. We can use the `rename_groups` function to provide the new names that are representative of each cluster's dynamics.

```
cnames <- c("Socially triggered", "Argumentation triggered",
            "Facilitated discussions", "Information triggered")
tna_group <- rename_groups(tna_group, cnames)
```

3.5 *Communities*

Similarly to simple TNA, we can study the specific patterns that characterize each cluster. For example, we can study whether we can detect distinct communities of states or events that often transition to one another to a greater extent than to the rest of the nodes, as well as compare if these communities are similar between clusters. We can obtain the community assignments using the `communities` function.

```
communities_group <- communities(tna_group)
```

We can plot the results using the `plot` function and specifying the method of community detection we want to use. Refer to [24] for more details on the different community finding algorithms. For the example below we chose the *Leading Eigen* algorithm since it clearly illustrates the existence of different communities (depicted by different colors) in each cluster (Fig. 3), each encompassing different nodes. Some similarities that can be perceived are that *argue* and *discuss* belong to the same community in each of the four clusters, and so do *disagree* and *facilitate* and *share* and *argue*. *Organize* is assigned to a community of its own in two clusters, suggesting that this action occurs rarely as a part of the collaboration process — probably more often at the beginning.

Other algorithms would have yielded different results. For example the *Infomap* [25] algorithm does not detect any communities (i.e., the whole network is a single community) in any of the clusters.

```
plot(communities_group, method = "leading_eigen")
```

3.6 *Centralities*

Centrality measures in transition networks allow to identify the role that each state or event takes in the process. For example, betweenness centrality captures how often a state acts as a bridge within the network, connecting different parts of the process. Out-degree centrality reflects the number of direct transitions originating from a state, indicating to which extent it drives the process towards other states. See [19] for a complete rationale of the role of centrality measures in TNA, and [26] for a detailed description of the most common centrality measures.

When working with clusters in TNA, since each cluster has a different transition probability matrix, the nodes also have different centrality measures. The way to compute them is the same as in regular TNA, i.e., using the `centralities` function. We can specify which specific centralities we want to compute and plot them using the `plot` function (Fig. 4). **Betweenness** centrality shows the greatest differences. In the *Facilitated discussions* cluster, *Facilitate* has the highest betweenness followed by *Disagree*. In the *Information triggered* cluster, *Argue* is the node with the highest betweenness, followed by *Socialize*. In *Argumentation triggered*, the highest betweenness is that of the *Evaluate* node, followed by *Share*. In the *Socially triggered* cluster, *Facilitate* is the node with the highest betweenness. In terms of **In-strength**, the four clusters follow similar patterns, with *Discuss* being the node with the highest value, followed by *Socialize*. Lastly, regarding **Out-strength**, *Share* and *Discuss* are the nodes with the highest overall centrality values, sharing similar patterns among the clusters. The *Facilitate* node has the lowest out-strength for all clusters, indicating that is many times a recurring action that is

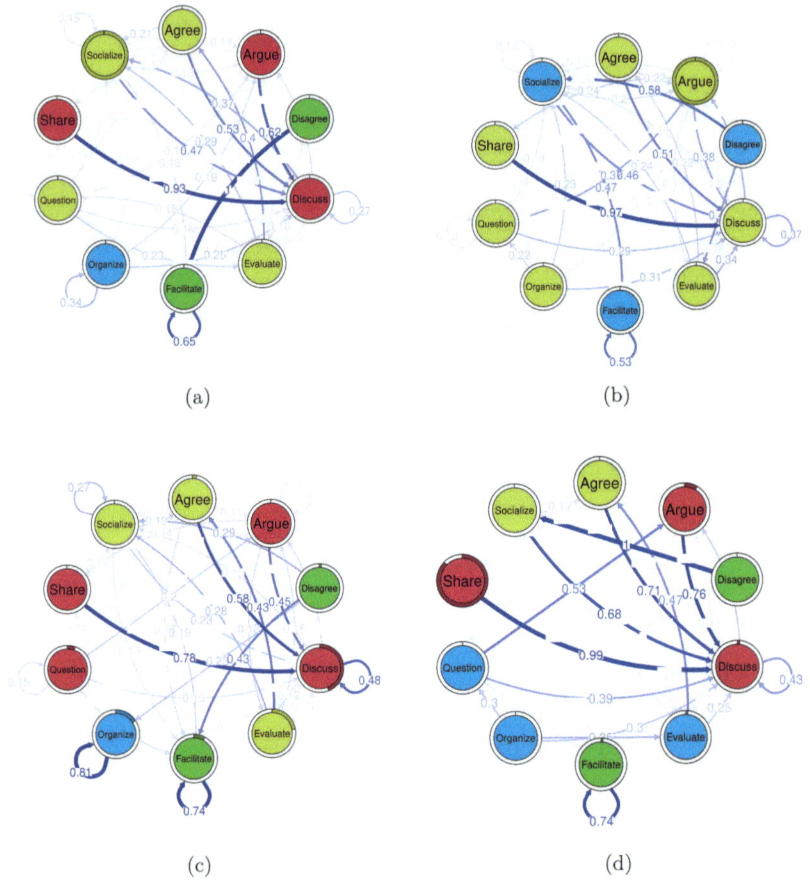

Fig. 3 Plot of each TNA model communities. (**a**) Socially triggered. (**b**) Argumentation triggered. (**c**) Facilitated discussions. (**d**) Information triggered

repeated several times in a row. Similarly, the *Facilitated discussions* cluster has a remarkably low value of out-strength centrality for *Organize*, meaning that students might be commonly stuck at that point. To better understand the meaning of each of the centrality measures, please refer to [19, 27]. We can also print the centralities for each cluster (Table 3) to find out their exact value.

```
cents <- centralities(
  tna_group,
  measures = c("Betweenness", "InStrength", "OutStrength")
)
```

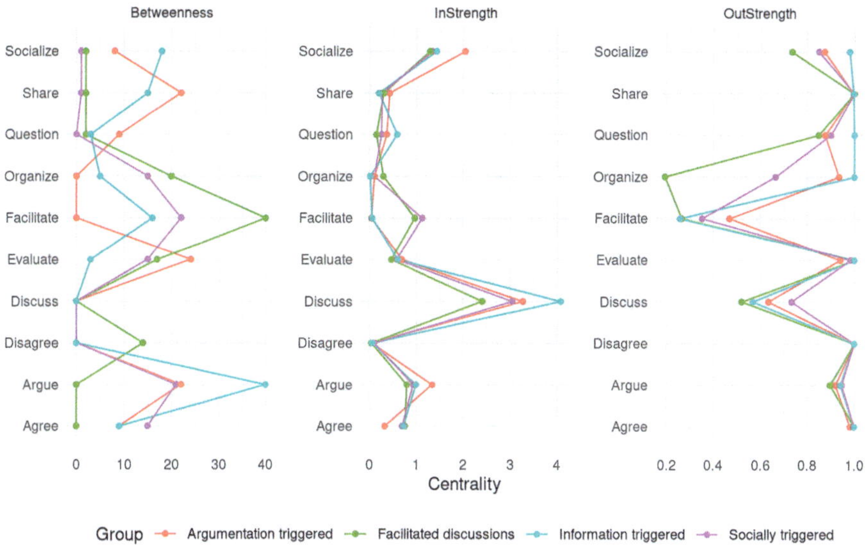

Fig. 4 Plot of each TNA model centralities

```
plot(cents, ncol = 3)
```

```
print(cents)
```

3.7 Cliques

TNA provides functions to identify groups of nodes that have a reciprocal relationship, i.e., strong transition probabilities to and from one another. These can be groups of two nodes (dyads), three (triads), and so on. When we wish to detect these patterns when working with clusters, we can use the same function as in simple TNA (`cliques`), but we need to specify for which of the clusters we want to identify these cliques. For example, if we would like to identify and plot all dyads on the *Socially triggered* cluster, we must subset the `tna_group` object by cluster name and then specify the size of the cliques (2 for dyads), as well as any additional parameters (e.g., threshold for considering strong reciprocal connections). We can plot the resulting dyads using the `plot` function. Do not forget that this function is interactive, so you will need to hit enter in the console to plot all dyads (alternatively, you may add the following argument to the `plot` function: `ask = FALSE`). Figure 5 shows that there are three dyads in the *Socially triggered* cluster: one between *Argue* and *Discuss*, *Discuss* and *Socialize*, and *Evaluate* and *Question*.

Table 3 Centrality measures for each cluster

	Betweenness				InStrength				OutStrength			
State	1. Soc. trig.	2. Arg. trig.	3. Fac. dis.	Inf. trig.	1. Soc. trig.	2. Arg. trig.	3. Fac. dis.	Inf. trig.	1. Soc. trig.	2. Arg. trig.	3. Fac. dis.	Inf. trig.
Agree	15	9	0	9	0.70	0.32	0.76	0.73	1.00	0.98	1.00	1.00
Argue	21	22	0	40	0.94	1.34	0.80	0.99	0.95	0.92	0.90	0.94
Disagree	0	0	14	0	0.10	0.04	0.04	0.06	1.00	1.00	1.00	1.00
Discuss	0	0	0	0	3.05	3.26	2.41	4.08	0.73	0.63	0.52	0.57
Evaluate	15	24	17	3	0.58	0.69	0.46	0.60	0.98	0.94	1.00	1.00
Facilitate	22	0	40	16	1.12	0.04	0.97	0.05	0.35	0.47	0.26	0.26
Organize	15	0	20	5	0.09	0.10	0.29	0.01	0.66	0.93	0.19	1.00
Question	0	9	2	3	0.25	0.36	0.14	0.59	0.90	0.87	0.85	1.00
Share	1	22	2	15	0.26	0.42	0.29	0.20	1.00	1.00	1.00	1.00
Socialize	1	8	2	18	1.34	2.04	1.29	1.43	0.85	0.87	0.73	0.98

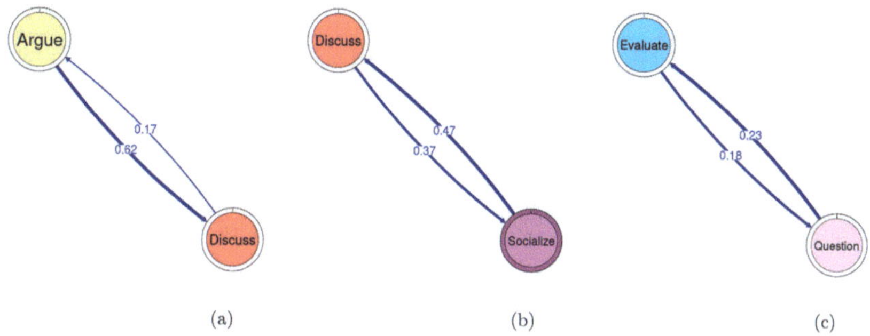

Fig. 5 Dyads for cluster 1. (**a**) Dyad 1. (**b**) Dyad 2. (**c**) Dyad 3

```
dyads1 <- cliques(tna_group[["Socially triggered"]], size = 2, threshold = 0.15)
plot(dyads1)
```

3.8 *Other Forms of Clustering and Grouping*

Throughout the tutorial we have seen how to identify and study clusters of students that are ruled by different transition patterns. However, we might want to identify groups of students through other mechanisms and only afterwards investigate their properties and transition patterns. In this section, we review different ways in which we can do this.

3.8.1 Clustering Sequences

A common way to study log data is through sequence analysis. Within this framework, several dissimilarity measures exist that allow to calculate how similar sequences of events or states are to one another, and cluster them in distinct groups. Distance measures can be broadly categorized in edit distances—such as optimal matching—, shared attributes—such as longest common subsequence (LCS), or longest common prefix (LCP), and distance based on state distributions —e.g., euclidean distance or chi-squared.

In the next code chunk, we calculate the dissimilarity between our sequences using the `seqdist` function from the `TraMineR` package, using LCP as the distance measure, which means that sequences that start in the same way will have a shorter distance between them. We then cluster the sequences using hierarchical clustering (as implemented in the `hclust` function of the `stats` package) using Ward's method minimum variance method (`"ward.D2"`). We establish the number of clusters by using the function `cutree` and retrieve the cluster assignment. We assign the result to a variable named `lcp_clusters`, to be able to operate with the

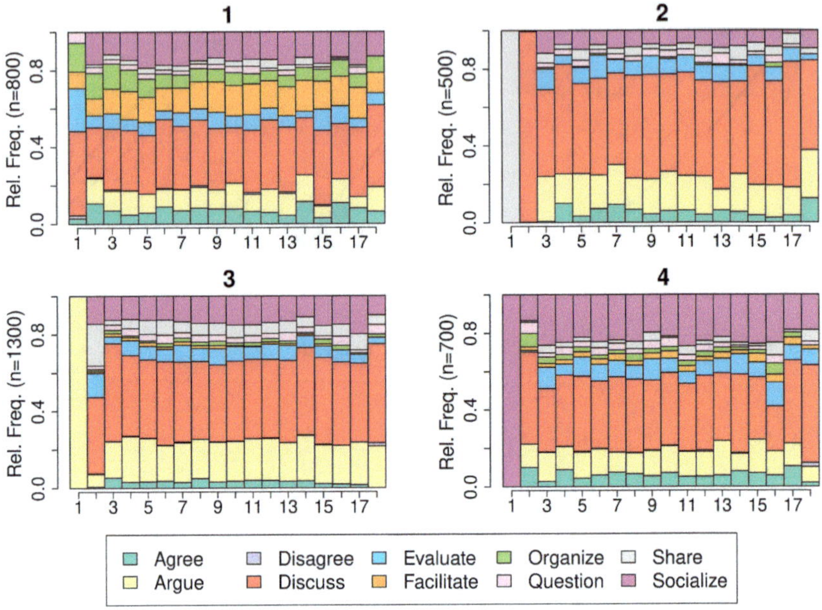

Fig. 6 Distance-based sequence clusters

cluster assignment later on. We can then visualize them using `seqdplot` (Fig. 6) in which we see the distribution of states at each time point. For more details about this process, please refer to [28]. Upon inspection, we can assign names to each of the clusters.

```
# Compute dissimilarities between sequences using the LCP method.
dissimilarities <- seqdist(sequ, method = "LCP", indel = 1)

# Perform hierarchical clustering on the computed dissimilarities
clusters_sessionsh <- hclust(as.dist(dissimilarities), method = "ward.D2")

# Cut the hierarchical clustering tree into 4 clusters
lcp_clusters <- cutree(clusters_sessionsh, k = 4)

# Plot a distribution plot (seqdplot) for the sequences
# grouped by their cluster assignments.
seqdplot(sequ, group = lcp_clusters, ncol = 5)

# Define meaningful cluster names based on interpretation of the data.
cnamesd <- c("Facilitated discussions", "Information triggered",
             "Argumentation triggered", "Socially triggered")

# Assign the corresponding cluster names to the cluster assignment vector.
lcp_clusters <- cnamesd[lcp_clusters]
```

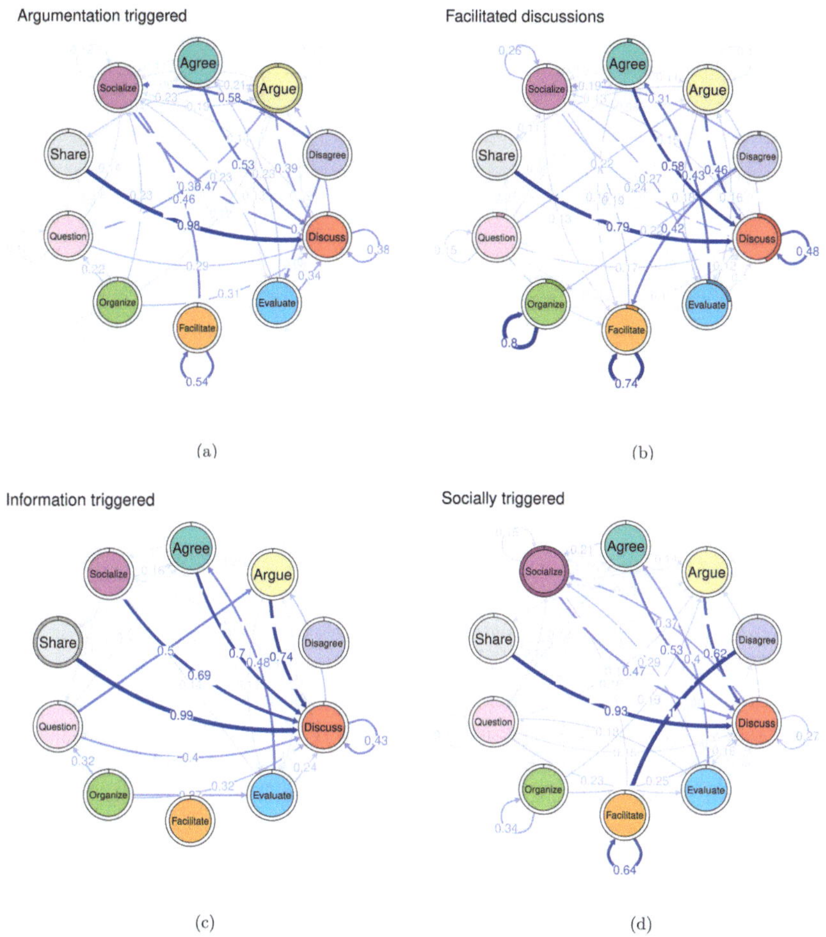

Fig. 7 TNA using distance-based clusters. (**a**) Argumentation triggered. (**b**) Facilitated discussions. (**c**) Information triggered. (**d**) Socially triggered

Once we have our cluster assignments, we can create a new clustered TNA object with the `group_tna` function, by passing the sequence object and the cluster assignments (`lcp_clusters`) as arguments. We can visualize the resulting clusters using the `plot` function as usual (Fig. 7), as well as perform all of the other operations that we have showcased throughout the chapter (centralities, communities, cliques, etc.).

```
seq_tna <- group_tna(sequ, group = lcp_clusters)
plot(seq_tna)
```

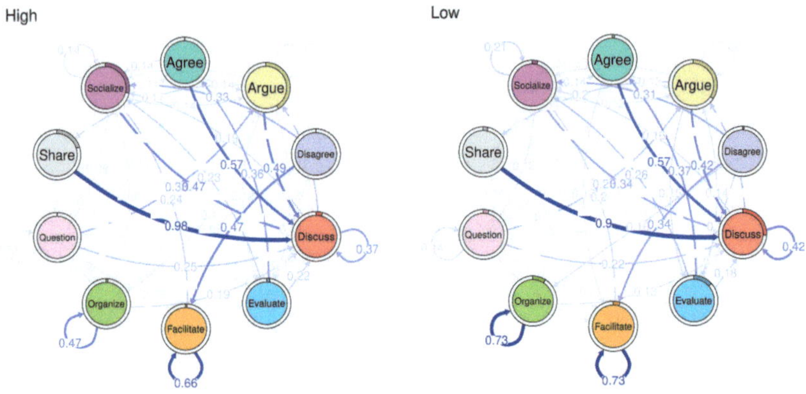

(a) (b)

Fig. 8 TNA using groups instead of clusters. (**a**) High achievers. (**b**) Low achievers

3.8.2 Fixed Groups

In addition to clustering, we can also create groups based on deterministic or known groups (e.g., gender, grades, course, etc.). In the following chunk, we create a variable named `Achievers` which is a vector containing the value `"High"` if the corresponding student has a grade above 50, and `"Low"` otherwise. We pass this vector as the `group` argument of `group_tna`, just like we did in the previous example and we can then us all the range of functions of `tna` on the two groups (Fig. 8). This helps us visualize the difference in transition probabilities between low and high achievers.

```
Achievers <- ifelse(simulated_data$Grade >
50, "High", "Low")
course_tna <- group_tna(sequ, group = Achievers)
plot(course_tna)
```

3.9 Other Forms of TNA

As demonstrated in [19], the TNA framework encompasses other forms of studying transitions, such as frequency-based transition networks. Instead of representing transition probabilities, frequency-based transition networks operate on the absolute values, i.e., the number or share of times each transition has appeared in the whole sequence object. This operationalization might be more appropriate in cases that there are not enough data to make inferences. This mode also supports cluster, allowing to inspect and compare transition frequencies among clusters.

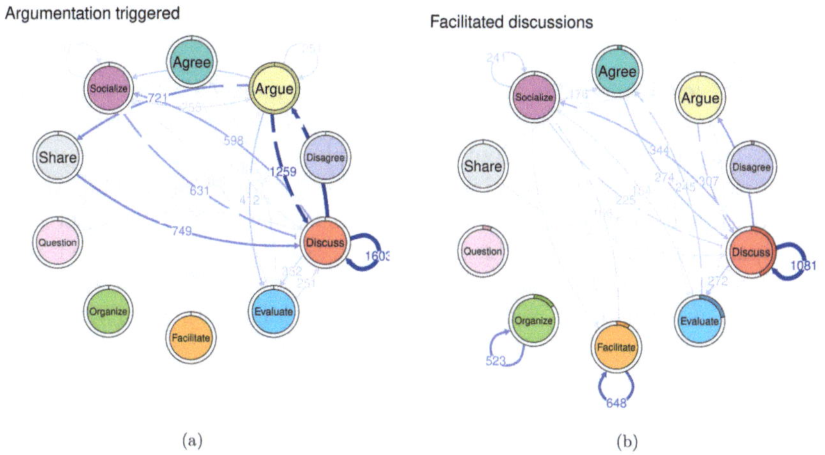

Fig. 9 FTNA of each group. (**a**) Argumentation triggered. (**b**) Facilitated discussions. (**c**) Information triggered. (**d**) Socially triggered

In the chunk below, we create a frequency-based transition network (Fig. 9), providing the results of the distance-based clustering as the `group` argument.

```
tna_group_scaled <- group_ftna(sequ, group = lcp_clusters)
plot(tna_group_scaled)
```

4 Discussion

Studying the learning process as a homogeneous phenomenon can overlook important variations that exist across different subgroups or contexts, leading to conclusions that may not fully capture the complexity of the phenomenon. Such an approach risks not only diluting the explanatory power of scientific findings but also perpetuating inequalities by failing to account for how distinct groups may experience or respond to the same processes differently. In turn, we should "capitalize on the heterogeneity that characterizes most effects in science" [2] (p. 86), so we can better understand variations, improving the relevance and applicability of our findings across diverse populations.

In this chapter, we have demonstrated how to identify clusters of transition networks, within the scope of TNA. We have illustrated this process for MMM clustering as well as for other forms of clustering. We have also showcased the features of TNA related to centralities, communities, and cliques, which provides as with a rich toolkit to study the heterogeneity of relational dynamics. Centrality measures, for instance, allow us to identify central states or transitions that shape the flow of behaviors or interactions. Influential nodes can vary significantly across subgroups, representing the diversity in how processes unfold.

Similarly, communities within the network can uncover tightly connected groups of transitions that represent cohesive patterns, dynamics, or strategies. These subnetworks might reveal distinct behavioral tendencies or shared practices that are specific to certain groups or contexts. Cliques, as smaller subsets of closely interconnected transitions, provide yet another layer of detail, capturing unique micro-patterns that might otherwise go overlooked.

Understanding the different patterns that can be identified from transition networks and how they prevail or differ among clusters, provides a comprehensive view on learning processes from a heterogeneity lens. In other words, identifying and studying different subgroups help us move beyond simplistic, one-size-fits-all interpretations to recognize the complexity and diversity inherent in behavior and interaction. Such interpretations have the potential to advance learning theories and support the design of interventions or policies that are sensitive to the diverse ways individuals and groups experience and respond to learning processes, which is one of the driving goals of learning analytics research. Embracing heterogeneity constitutes a step forward in ensuring that findings are equitable and capture diversity.

References

1. Saqr M, López-Pernas S, Törmänen T, Kaliisa R, Misiejuk K, Tikka S (2025) Transition network analysis: a novel framework for modeling, visualizing, and identifying the temporal patterns of learners and learning processes. In: Proceedings of learning analytics & knowledge (LAK'25). ACM, New York

2. Bryan CJ, Tipton E, Yeager DS (2021) Behavioural science is unlikely to change the world without a heterogeneity revolution. Nat Hum Behav 5:980–989. https://doi.org/10.1038/s41562-021-01143-3
3. Karoly P (1993) Mechanisms of self-regulation: a systems view. Annu Rev Psychol 44:23–52. https://doi.org/10.1146/annurev.ps.44.020193.000323
4. Saqr M, López-Pernas S, Jovanović J, Gašević D (2023) Intense, turbulent, or wallowing in the mire: a longitudinal study of cross-course online tactics, strategies, and trajectories. Internet High 57:100902. https://doi.org/10.1016/j.iheduc.2022.100902
5. Törmänen T, Järvenoja H, Saqr M, Malmberg J and Järvelä S (2022) A Person-Centered Approach to Study Students' Socio-Emotional Interaction Profiles and Regulation of Collaborative Learning. Front. Educ. 7:866612. https://doi.org/10.3389/feduc.2022.866612
6. Saqr M, López-Pernas S, Helske S, Hrastinski S (2023) The longitudinal association between engagement and achievement varies by time, students' profiles, and achievement state: a full program study. Comput Educ 199:104787. https://doi.org/10.1016/j.compedu.2023.104787
7. Saqr M, Schreuder MJ, López-Pernas S (2024) Why educational research needs a complex system revolution that embraces individual differences, heterogeneity, and uncertainty. In: Learning analytics methods and tutorials. Springer, Cham, pp 723–734
8. Helske J, Helske S, Saqr M, López-Pernas S, Murphy K (2024) A modern approach to transition analysis and process mining with markov models in education. In: Saqr M, López-Pernas S (eds) Learning analytics methods and tutorials: a practical guide using R. Springer, Berlin
9. Vermunt JK (2010) Latent class modeling with covariates: two improved three-step approaches. Polit Anal 18:450–469. https://doi.org/10.1093/pan/mpq025
10. Helske S, Helske J (2019) Mixture hidden markov models for sequence data: the seqHMM package in R. J Stat Softw 88:1–32. https://doi.org/10.18637/jss.v088.i03
11. Tikka S, López-Pernas S, Saqr M (2025) tna: an R package for transition network analysis. Appl Psychol Measur 49(6):326–328. https://doi.org/10.1177/01466216251348840
12. Villalobos E, Pérez-Sanagustín M, Broisin J (2024) From learning actions to dynamics: characterizing students' individual temporal behavior with sequence analysis. In: Lecture notes in computer science. Springer, Cham, pp 3–17
13. López-Pernas S, Saqr M (2024) How the dynamics of engagement explain the momentum of achievement and the inertia of disengagement: a complex systems theory approach. Comput Human Behav 153:108126. https://doi.org/10.1016/j.chb.2023.108126
14. Saqr M, López-Pernas S (2024) Momentary emotions emerge and evolve differently, yet are surprisingly stable within students. In: 2024 IEEE international conference on advanced learning technologies (ICALT). IEEE, Piscataway, pp 86–90
15. Tikka S, Kopra J, Heinäniemi M, López-Pernas S, Saqr M (2024) Getting started with R for education research. In: Saqr M, López-Pernas S (eds) Learning analytics methods and tutorials: a practical guide using R. Springer, Berlin
16. Kopra J, Tikka S, Heinäniemi M, López-Pernas S, Saqr M (2024) An R approach to data cleaning and wrangling for education research. In: Saqr M, López-Pernas S (eds) Learning analytics methods and tutorials: a practical guide using R. Springer, Berlin
17. Tikka S, Kopra J, Heinäniemi M, López-Pernas S, Saqr M (2024) Introductory statistics with R for educational researchers. In: Saqr M, López-Pernas S (eds) Learning analytics methods and tutorials: a practical guide using R. Springer, Berlin
18. López-Pernas S, Misiejuk K, Tikka S, Kopra J, Heinäniemi M, Saqr M (2024) Visualizing and reporting educational data with R. In: Learning analytics methods and tutorials. Springer, Cham, pp 151–194
19. Saqr M, López-Pernas S, Tikka S (2025) Mapping relational dynamics with transition network analysis: a primer and tutorial. In: Saqr M, López-Pernas S (eds) Advanced learning analytics methods: AI, precision and complexity. Springer, Cham
20. Wickham H, Averick M, Bryan J, Chang W, McGowan LD, François R, Grolemund G, Hayes A, Henry L, Hester J, Kuhn M, Pedersen TL, Miller E, Bache SM, Müller K, Ooms J, Robinson D, Seidel DP, Spinu V, Takahashi K, Vaughan D, Wilke C, Woo K, Yutani H (2019) Welcome to the tidyverse. J Open Sour Softw 4:1686. https://doi.org/10.21105/joss.01686

21. Gabadinho A, Ritschard G, Müller NS, Studer M (2011) Analyzing and visualizing state sequences inRwithTraMineR. J Stat Softw 40:1–37. https://doi.org/10.18637/jss.v040.i04
22. Chan C, Leeper T, Becker J, Schoch D (2023)_rio: A Swiss-army knife for data file I/O_. https://cran.r-project.org/package=rio
23. Saqr M, López-Pernas S (2023) The temporal dynamics of online problem-based learning: why and when sequence matters. Int J Comput-Support Collab Learn 18:11–37. https://doi.org/10.1007/s11412-023-09385-1
24. Hernández-García Á, Cuenca-Enrique C, Traxler A, López-Pernas S, Conde-González MÁ, Saqr M (2024) Community detection in learning networks using R. In: Learning analytics methods and tutorials. Springer, Cham, pp 519–540
25. Rosvall M, Bergstrom CT (2008) Maps of random walks on complex networks reveal community structure. Proc Natl Acad Sci USA 105:1118–1123. https://doi.org/10.1073/pnas.0706851105
26. Saqr M, López-Pernas S, Conde-González MÁ, Hernández-García Á (2024) Social network analysis: a primer, a guide and a tutorial in R. In: Learning analytics methods and tutorials. Springer, Cham, pp 491–518
27. Saqr M, López-Pernas S, Conde-González MÁ, Hernández-García Á (2024) Social network analysis: a primer, a guide and a tutorial in R. In: Learning analytics methods and tutorials. Springer, Cham, pp 491–518
28. Saqr M, López-Pernas S, Helske S, Durand M, Murphy K, Studer M, Ritschard G (2024) Sequence analysis in education: principles, technique, and tutorial with R. Springer, Berlin, pp 321–354

Part IV
Idiographic Analysis

Individualized Analytics: Within-Person and Idiographic Analysis

Mohammed Saqr, Hibiki Ito, and Sonsoles López-Pernas

Abstract Two research paradigms are commonly used: between-person (often known as nomothetic or group-based) and within-person analysis. The between-person approach captures data across different people and gathers their variations, hence the name of between-person. Within-person research captures data within the same person, gathering these variations across times or occasions and thus the name. These two approaches converge if and only if ergodicity holds. Ergodicity claims that people are similar (so that it does not matter whom to measure) and they do not change over time (so that it does not matter when to measure). However, this is often not the case in education research. Thus, the within-person approach is needed. When within-person analysis is conducted on a single person (N=1), it is known as idiographic analysis. In this chapter, we describe the main characteristics of each approach, as well as common data sources and methods used in within-person and idiographic analysis.

Keywords Learning analytics · Idiographic analysis · Within-person analysis · Nomothetic · N = 1

1 Introduction

The last decades of education research witnessed a rapid increase of data availability and advancements in quantitative methods powered by the rise in computing power and methods [1]. Such an increase has motivated the emergence of data-intensive fields, namely learning analytics and educational data mining. Learning analytics is a relatively nascent field that aims to understand and optimize learning and

M .Saqr (✉) · S. López-Pernas
University of Eastern Finland, Joensuu, Finland
e-mail: mohammed.saqr@uef.fi

H. Ito
University of Helsinki, Helsinki, Finland

M. Saqr, S. López-Pernas (eds.), *Advanced Learning Analytics Methods*,
https://doi.org/10.1007/978-3-031-95365-1_18

learning environments by utilizing the burgeoning amounts of available data through digital—and non-digital—tools [2, 3]. However, the abundance of data and the availability of methods have mostly resulted in scaling across samples, populations, and institutions rather than building highly precise person-specific solutions [4]. In other words, learning analytics tried to build algorithms that could work across a population (e.g., a course or an institution), for instance, to predict the students who may drop out using data from previous years or other classrooms [5]. The premise was that by gathering more data we will be able to use more powerful algorithms, to obtain more accurate predictions and to perform more effective interventions. However, that has not been the case. So far, few are the systems that have been implemented across institutions, and far fewer are the systems that have delivered those predictions or proved to scale up if any [6].

Several studies have proven that scaling, generalizing, and applying a one-size-fits-all strategy is ineffective, no matter how hard researchers have tried to gather more data, use fancy algorithms, or bigger samples, go with multimodal or triangulate across multiple sources of data [7, 8]. The failure of big data to deliver big returns stems from the innate fragility of the idea that the sheer size of the data will automatically lead to more accurate and effective results. The reality is that more data from a larger sample cannot solve problems of individualized learning, nor can it account for the complexity of learning or individual differences [9].

In stark contrast to the one-size-fits-all approach, person-specific methods aim to precisely model each individual, from their own data, find the distinct unique peculiarities of each, and tailor models to their own needs to deliver accurate and adaptive learning experiences [10]. By definition—or by design—the focus is narrower and the resolution is equal to a single student or what is often referred to as N=1, single subject, or idiographic. Person-specific (i.e., *idiographic*) methods have been advocated not only to address the shortcomings of other methods but also for their potential strengths and advantages [11, 12]. These methodologies originate and have been widely discussed in psychological and social sciences [13–15], however, so far under-used in education.

In this chapter, we will introduce these approaches and offer a background about idiographic and within-person analysis in general. We will first present an overview of relevant concepts such as nomothetic, idiographic, and within-person analysis. Next, we will review issues with the dominant nomothetic approach, namely, poor generalisability. Finally, we discuss the potential as well as challenges, and limitations of idiographic methods for learning analytics.

2 Background

Two types of approaches are commonly used: between-person and within-person. The first approach is widely prevalent in science at large—and in education in particular—and entails collecting data from a sample, analyzing the data, and drawing conclusions. The other approach entails collecting data several times from an individual—or multiple individuals—and analyzing these data variations across time compared to the previous self. In the next section, we will explain in detail these approaches, their rationale, and how they relate to idiographic analytics.

2.1 *Between-person (Nomothetic Approach)*

Nomothetic, between-person, inter-individual, group-level, or population-based research are mostly synonymous terms that converge to the same idea of collecting data from a group of different people to study their inter-individual variance [10, 16]. As shown in Fig. 1-A, where data are collected from different people at a point in time i.e., cross-sectional. The premise is: studying a group of individuals would allow researchers to model the overall trends or obtain summary measures e.g., mean or median. We may collect data from a group of students and we would say that the average of the group represents the summary or the average level of engagement that most students are assumed to revolve around [10, 16, 17].

Similarly, the variable-centered approach is another type of between-person research paradigm that focuses on the relationships among variables rather than the individuals themselves. All of these approaches are similar and represent the same idea of studying multiple people to capture enough variations. In all of such cases, it is assumed that the results generalize to most students. In fact, it is assumed that the findings of this approach are generalizable norms that everyone more or less conforms to. Deviations from these norms are considered outliers, noise, or

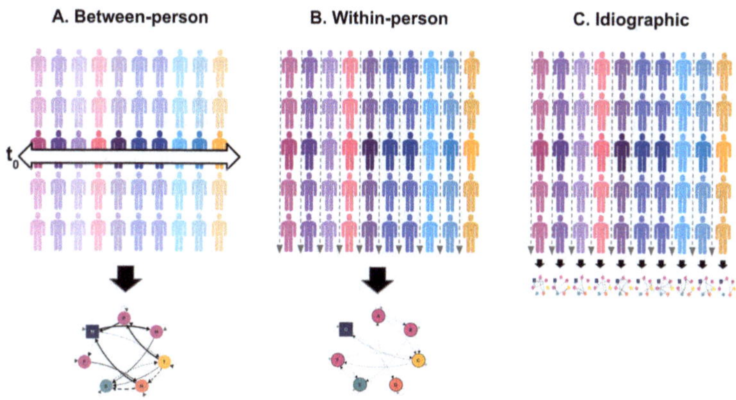

Fig. 1 The nomothetic approach (A) focuses on between-person variability where data is collected from a group of different people to create a cross-section average. The within-person approach focuses on within-person variability where data is collected from the same people several times and variance is computed compared to the same person's previous state (note the arrow directions), still we get the average of many people (B). The idiographic approach focuses on within-person variability of a single person and each is studied separately (C)

exceptions. To date, most of the research is done with this approach, and nomothetic or between-person research is dominant and can be seen everywhere [18].

2.2 Within-person

Within-person research captures the intra-individual mechanisms, variations, and changes within the same person (or sometimes multiple persons) over time or occasions. It compares individuals to their own past self to capture how individuals' experiences, behaviors, or psychological states fluctuate over time compared to their own average. As shown in Fig. 1-B, data is collected from the same persons repeatedly to measure the intra-individual variations or changes (often called the within-person variance) [10, 18].

Suppose a student scores a 9 in mathematics. In the next exam, she scores a 10. That indicates that her performance improved from her past self. If we have her average score for the last 10 exams, and it was 8.5, we can say that she is improving compared to her past self, and given that the last two exams were 9 and 10, she is on an ascending trend. If she obtained a grade of 7, it would be lower than her average. In turn, if we took another student with an average of 5, a grade of 7 would be an improvement from her own average. These are within-person variations given that we are comparing variations within the individual. This approach can be generalized to other processes, e.g., engagement, self-regulation, and well-being—to mention a few. In such a case, we will say that the engagement of a student has improved (compared to her own previous engagement). Put another way, the comparison is strictly within the same person. This is why it is called within-person.

Using a within-person approach, we can study how learning processes change by tracking the intra-individual fluctuations and the factors that affect them [19]. We can also identify the patterns and trends within the person, highlighting how different contexts or situations can impact the person's behavior. To summarize, the within-person approach collects data from a person or multiple persons over several time points, and analyzes how variables change within individuals over time.

When within-person research addresses a single person, it is referred to as idiographic (N=1 or single subject) [20–23]. One can say that it is a special case of within-person analysis and in fact, the dominant one. As shown in Fig. 1-C, idiographic research collects data from a single person and seeks to give person-specific insights about that exact person [10, 16]. Idiographic research can address many people but still would study each and every person separately as a whole unit of analysis. Therefore, findings through idiographic methods cannot be expected to generalize to other persons.

It is worth mentioning that not all within-person analysis is idiographic, some of it averages across students to create average within-person variance and therefore, it is not idiographic (because we have data from more than one student).

3 Does Group-based Research Generalize to Individuals?

The basic rationale for group-based or nomothetic research rests on the assumption that variance observed between individuals (between-person or inter-individual) mirrors the variance observed within individuals (intra-individual or within-person) [10, 19, 24]. When we gather data from many individuals to capture several situations and variability, we will be able to capture enough variations similar to what happens within the person or persons. In other words, since it may be difficult to gather data from the same person one thousand times to capture all possible variations, we can collect data from one thousand people to gather enough variations. We assume that these two variations are essentially the same. This means researchers assume that patterns of behavior that can be observed at the group level can also be generalized to the individual level. This assumption is problematic for several reasons. We will explain it in detail in the following subsection [5, 11, 13].

3.1 Ergodicity

A key theoretical framework to understand the between/within-person mechanisms is the theory of ergodicity [10, 25]. In short, a group-level variance of a variable (e.g., motivation) generalizes to an individual-level trend if and only if the variable meets the ergodicity conditions [10, 26].

1. **Homogeneity** means people are similar or homogeneous with no significant differences. The values of a variable are consistent so that if we select any individual, it would yield comparable results without impacting the study's outcomes.
2. **Stationarity** means the variable remains stable over time. So, it won't matter when we are collecting data, the variable doesn't change over time in any way.

However, we know that there is significant heterogeneity and differences among the population and many psychological constructs evolve over time. Therefore, ergodicity is neither realistic nor has been ever empirically found. In fact, the mounting evidence across all psychological processes is that they change over time, and vary across populations. Therefore, group-level analysis poorly reflects each individual or any individual in particular [10, 26, 27].

3.2 Ecological Fallacy

Ecological fallacy refers to the logical error of biased inferences about individuals based on aggregate data from groups or populations. In other words, thinking that between-person transfers as-is at an individual level. This fallacy occurs when

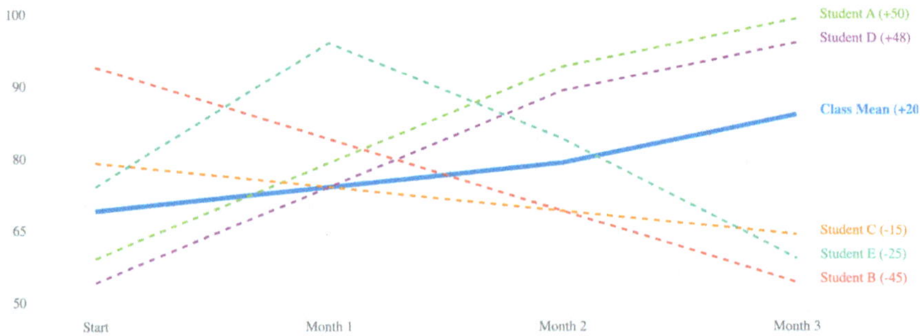

Fig. 2 An imaginary trend of 5 students. On average the blue line shows that there was an increase in the class mean. But we see that three students deteriorated and only two improved. In other words, where the mean is positive, the individual cases are not essentially so and can be even worse

researchers assume that the observed relationships at the group level hold true for all individuals within that group, leading to potentially misleading conclusions [24, 28].

Imagine a study that measures student engagement in two groups of students who are exposed to two different teaching methods (Fig. 2). Both groups have an initial engagement of 6 out of 10. After the intervention, Classroom B has an average engagement score of 9 out of 10, while Classroom B has an average score of 7 out of 10. Based on these averages, one might conclude that all students in Classroom A have improved their engagement to a greater extent than those in Classroom B. However, this is an example of ecological fallacy. In Classroom A, some students might have their engagement scores decreased (e.g., 4 or 5) and in Classroom B, some students may have their engagement improved. In other words, it is the aggregate average that improved, driven by some students in both classrooms but not each and every student. This fallacy can lead to erroneous interpretations and ineffective policy recommendations. A stark case of ecological fallacy is Simpson's paradox which we will detail next.

3.3 Simpson's Paradox

To grasp the idea of ecological fallacy even more, let us consider the following scenario, known as Simpson's paradox [29]. Simpson's paradox occurs when the direction of a relationship at the group-level is the opposite of that at the person-level, e.g., increases on average and decreases for the individual.

Figure 3 shows the simulated relationship between homework scores and hours of study of all students (N=20) over five weeks. You observe that there is a clear negative correlation between these variables on average (i.e., the more study hours, the lower the scores). Therefore, the teacher might consider reducing the amount of

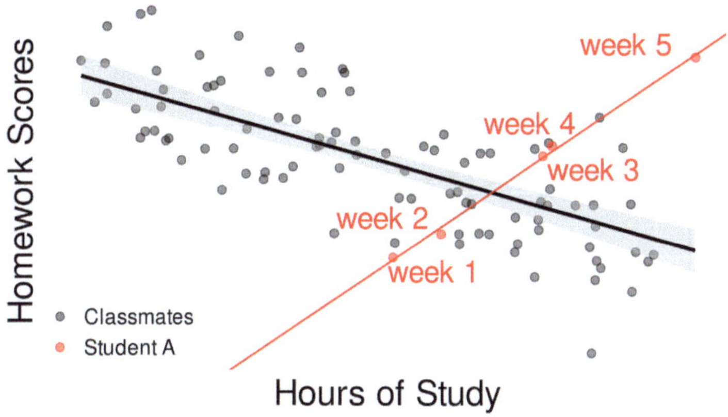

Fig. 3 Simulated data of homework scores and questionnaire records over the past five weeks for 20 students. The simple linear regression is fitted to the whole data including the data points of Student A (*black line*) and only to the data points of Student A (*red line*)

homework in the following week, or offer help to those who study a lot, but this results in relatively lower homework scores. However, if you focus on Student A, you notice that there is a totally opposite trend in her learning outcome.

In other cases, Simpson's paradox is more systematic. Let us take an example, students who are competent at mathematics are more likely to be fast in solving problems and make fewer mistakes than students who are less competent. The black line in Fig. 4 shows this average trend. One can conclude that there is a direct positive relationship between speed and maths proficiency (at between-person level). However, even within the competent group, the faster they attempt the math problems, the more likely they are to make errors. In that case, there is a direct negative relationship between the speed and accuracy of problem-solving (at the within-person level).

4 The Idiographic Approach

The previous section reviewed the theoretical rationale behind between vs. within-person approaches. In this section, we consider the potential and challenges of an idiographic, person-specific approach.

4.1 Why Idiographic Methods

Group-level insights obtained by nomothetic analyses do not necessarily apply at the individual level [4, 24, 30]. In other words, there is no *average* person

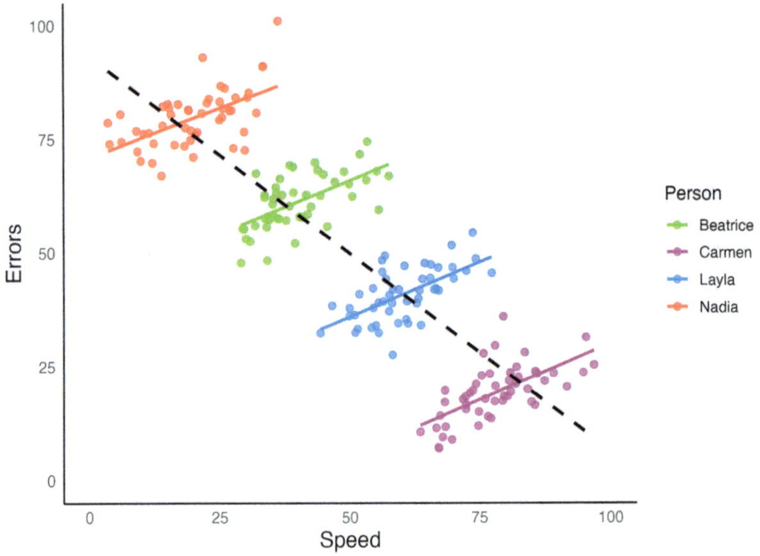

Fig. 4 Simulated data on the relationship between the number of errors and solving time of mathematics problems. On average, students with better marks solve problems faster with fewer errors. However, within each grade group of students, the faster students handle problems, the more they make mistakes

that would fit our expectations of the average. There are numerous examples in the recent literature that group-level insights do not generalize to each individual [31, 5, in psychology: 24, e.g. in education: 4, 32]. In addition, in many cases, group-level insights can be also biased toward over-represented groups of subjects [31, 32]. In learning analytics, for example, students with disabilities have been an underrepresented group in the literature [33]. Hence, the idiographic approach allows to reflect individual subjects through fine-grained person-specific analyses as well as overcomes the fairness issues encountered by nomothetic approaches.

Furthermore, idiographic methods may provide an answer to the poor generalizability of group-based analysis by offering an alternative that could provide precise personalized insights to each individual [10, 34]. In other words, individuals could rely on their own data instead of relying on analysis generalized from others. It stands to reason that if precise individualized insights or interventions are needed, we use methods that reflect the person herself.

Data collection through or with personal devices may bring privacy advantages [35–37]. Since the idiographic approach does not aggregate data across people, ideally it would be possible to complete the process of analytics, from data collection to giving feedback, within the digital devices of each individual so that the collected data are not shared by anyone else [36]. However, if idiographic analysis is used for a clinical purpose, the data is necessarily provided to the practitioner, essentially involving ethical consideration.

4.2 Challenges to Adoption and Implementation

Despite its significant potential, the adoption of idiographic methods in behavioral sciences has been remarkably slow [38], especially in education research [11]. This is primarily due to data collection as this approach necessarily requires a large amount of micro-level data from a single individual [11, 39]. However, recent technologies such as mobile phones and wearable devices have rapidly advanced data availability for idiographic analysis [11, 40]. In addition, the development of computational power and methods has also increasingly allowed for idiographic research [11, 13, 26]. Thus, more research and practice would be anticipated to employ idiographic methods in the future.

Additionally, choosing measurement intervals is another challenge in practice [13, 40]. For example, autoregressive models are a commonly used statistical technique of idiographic analysis, and the models allow to investigate the correlation between different variables over time lags [22, 26]. However, this method only reveals the correlations at a given lag. This could cause problematic evaluation or measurement unless the stationary assumption holds [40]. The stationary assumption implies that relationships between variables do not change within the measurement time, which of course, could happen. For instance, if two variables do not satisfy the stationarity condition, it is possible that their relation could be positive in the first half of the semester and negative in the second half, concluding on no relationship on average.

To address this issue, researchers can employ various methods to test for stationarity. If data is found to be non-stationary, techniques like differencing or transformation (e.g., logarithmic or seasonal adjustments) can help stabilize the mean and variance, allowing for a more accurate analysis of relationships over time. For example, Fisher et al. [41] collected idiographic (person-specific) multivariate time series data, where data were found to be non-stationary and the intervals were not uniform. So, the authors first fit an ordinary least square linear regression model so that residuals are the fluctuations from the regression line (i.e. trait), and subsequently, a cubic spline interpolation was applied to address the uneven intervals [41]. In cases where idiographic analysis may be difficult, some forms of within-person analysis could offer an approximation of the idiographic variance [42].

5 Data and Methods for Within-person and Idiographic Analysis

In this section, we provide a literature review of the existing within-person, idiographic research potentially relevant to learning analytics, focusing on common data sources and methods.

5.1 Data

Within-person analysis—and also idiographic methods—requires repeated data collection by design to be able to capture enough variations of the participants. Therefore, the hallmark of data collection for such an approach is repetitive data collection, be it through surveys, sensors, or digital data. Given the widespread adoption of quantitative methods, intensive longitudinal data collection methods have become highly common. We first delineate several data sources. Then, the key data collection methods, Ecological Momentary Assessment (EMA), and Experience Sampling Method (ESM), will be explained.

5.1.1 Data Sources

5.1.1.1 Self-reported Data

In behavioral sciences, collecting self-report data—often through some form of a questionnaire—is widely used to measure psychological constructs that are otherwise difficult to directly evaluate. The self-report approach is also used for the within-person analysis. Self-reported data collected within this context differs from the data collected when adopting a between-person approach. In between-person research, the focus is often on overall perceptions and experiences, with questionnaire items being phrased as “I often think...”, “In general, I like...”, “Overall, I do...” [27]. In within-person research, instead, the focus is on capturing in *the moment* feelings or attitudes, and therefore, questionnaire items are phrased accordingly, e.g., “Right now I feel...”, “At the moment, I am...”, etc. Also, these surveys are shorter and repeat multiple times per day to gather several time-points for each individual across time. These surveys are called experience sampling or ecological momentary surveys given that they collect experiences as they occur and therefore less prone to recall bias.

Also, well-known questionnaires that are often applied in between-person research, e.g., Motivated Strategies for Learning Questionnaire (MSLQ) [43] are not directly applicable to within-person research, as they would need to be rephrased to capture the momentary experiences of the research subjects. There are questionnaires that are already suitable for this purpose, such as the Positive and Negative Affective Scale (PANAS), which aims to capture momentary emotions [44].

A key limitation of self-report data is the potential bias in the subjective self-assessment of a participant. If everyone has different magnitudes of self-report bias, conducting between-person analysis on self-report data might introduce a significant overall bias. However, the self-report bias would potentially be reduced in within-person analysis because it focuses on the fluctuation relative to the past. Nonetheless, it may be important to complement self-report data with other data sources described below [45, 46].

5.1.1.2 Digital Log Data

Digital mobile devices, learning management systems (LMS), and applications constantly collect data for every action or click. Similarly, self-tracking apps which can log students' study habits, location, or social interactions, creating a broad picture of how these factors influence their academic performance.

Saqr [5] used the log data in an LMS from 286 students across 48 module instances. Then, mixed linear models were fit at traditional group-level and as the hybrid of between- and within-person to predict the final marks. As a result, the within-between hybrid model demonstrated better performance than the group-level nomothetic model, indicating that within-person factors or trajectories reflect academic performance [5]. Another interesting example is the analysis of self-reported data combined with digital footprints collected from smartphones by Aalbers et al. [47]. In this extensive study, 247 university students self-reported their subjective stress five times a day at random points. In addition, the log data on their smartphones such as use of specific applications were extracted. Machine learning models were trained with the survey data merged with the log data for both the group-level (nomothetic) and each individual (within-person). The findings showed that sleep hours and prolonged social network apps consistently predicted momentary subjective stress at group-level [47]. However, these variables were not sufficiently predictive for most students at within-person level, and what digital markers explain the subjective momentary stress significantly varied between persons [47].

5.1.1.3 Physiological Data

Additionally, physiological data from wearable sensors could be also used for within-person analysis. This type of data could, for instance, help track physiological responses, such as heart rate or electrodermal activity, and may offer insights into, e.g., how stress affects learning.

Dhindsa et al. [48] investigated the phenomenon of mind wandering of students during live lectures by recording electroencephalography (EEG). 15 medical students attended an orthopedic surgery lecture wearing caps to measure EEG, and students were also asked to self-report the degree of mind wandering around every four minutes during the lecture. Discriminative spatial filters and support vector machines were applied to the data to classify whether the participant was in the mind-wandering state or not. In particular, machine learning models were fit at both group (nomothetic) and individual (within-person) levels. The results verified that key EEG factors for mind wandering which had been previously reported in laboratory settings were somewhat reproduced in the live lecture setting on average, while individual patterns were highly heterogeneous [48].

5.1.1.4 Other Data Sources

Other data sources also include daily diary methods where students—or their parents and teachers—might keep a daily diary documenting their academic experiences, challenges faced, and strategies used [49]. Also, interviews with students about their learning experiences can complement quantitative data, providing a richer narrative. For example, Ordem [50] tracked a single student studying German as a second and foreign language and collected diary as well as monthly interview data for qualitative analysis of the dynamics of motivation in language learning. The result showed that the participant's motivation was negatively affected by a lack of linguistic self-confidence and the problem of social milieu [50].

5.1.2 Data Collection Methods

5.1.2.1 Ecological Momentary Assessment and Experience Sampling Methods

Ecological Momentary Assessment (EMA) involves collecting real-time data about a participant's experiences in their natural environment (hence the name ecological). The data collection can be done through observation, physiological sensors, or through self-reports. For example, in an educational setting, students might be asked to report their attitude about a specific subject or task at various points throughout the day. A student could receive a phone notification asking about their belief in their ability to do their learning tasks (self-efficacy). Typically, this is repeated several times a day and over some time (weeks or months) to record the fluctuations in self-efficacy and how they correlate with other variables e.g., performance, engagement or lesson subjects.

For instance, Respondek et al. [51] employed EMA to collect survey data from 98 undergraduate students during an exam week. In this study, participants received notifications on their mobile devices at three randomly selected times a day, and asked to report their achievement emotions. The EMA data combined with the exam grades were then analyzed using structural equation modeling. The results indicate the potential impact of previous experience, in addition to trait (within-person expected value) and situation (cross-sectional expected value at that time point or state), on the achievement emotion [51].

Experience Sampling Method (ESM) is similar to EMA—and oftentimes they are used interchangeably—but may involve scheduled prompts, encouraging students to reflect on their thoughts and experiences at preset times. As the name implies, it aims at capturing students' experiences when they happen and as close to the natural environment as it can happen, to avoid recall bias [52]. Similar to EMA, educators could use ESM to gauge students' engagement and motivation during different classes throughout the week.

Saqr and López-Pernas [53] used data from students over several weeks to study their self-regulation and reveal the idiographic variations among each student. In another study by the authors, they used EMA data to predict student studying and regulation and reported that each and every student had unique predictors [4]. Xie et al. [54] carried out ESM on 52 undergraduate students during two pre-exam weeks. In the study, participants were asked to respond through a mobile application to surveys after planned study-related events (thus experience-based sampling). Hierarchical regression analysis on the collected data revealed that situation-specific self-efficacy is highly related with engagement (i.e. how well students engaged with the planned events) [55].

Another example of the use of ESM is a more extensive study by Muenks et al. [56] who collected survey data from 902 undergraduate students after every STEM class over two weeks. The surveys asked about students' psychological experiences in the STEM (science, technology, engineering, and mathematics) classes in order to investigate the effect of their perception on professors' mindset beliefs about students' intelligence ability. The findings demonstrated that, when students perceive that their STEM professors have a fixed mindset, this perception tends to negatively affect their psychological vulnerability, in-class engagement, and even the dropout intention [56].

Given that these methods gather data at the moment, they are commonly referred to as momentary data. The analysis of momentary data either collected by ESM or EMA could reveal patterns, changes and fluctuations and most importantly, enable within-person analysis as we have enough data points with enough power to enable within-person or idiographic analysis.

5.2 *Analytical Methods*

Quantitative data collected through EMA/ESM are essentially multivariate time series, containing information about temporal processes of individuals. Here we provide several examples of approaches to handle such types of data for within-person, idiographic analysis.

5.2.1 P-technique

The most classical—yet still impactful—method for within-person analysis is the P-technique. The original P-technique developed in 1940s is essentially a factor analysis applied to data obtained from a single individual, while to date several extensions have been proposed to deal with more nuanced information of temporal changes. The factor analysis approach has an advantage over other network analysis approaches (to be subsequently presented) in that it handles latent variables, or factors. Thus, this method would be suitable for testing theory or identifying latent constructs.

For example, Musher-Eizenman et al. [57] employed the P-technique dynamic factor analysis [58] to study the relationship between children's perceived control over their work and their academic achievement taking into account concurrent and lagged effects within individuals. A total of 152 primary school pupils in Germany were included in the study, and participants were asked to self-report psychological indicators (e.g. motivation, anxiety) after each homework assignment and after the return of graded assignments for German and maths classes. First, the within-person dynamic factor analysis failed to replicate the previous findings by cross-sectional analysis that children with better control over their work achieve higher academic performance and showed that rather there are greater individual differences in the relationship between perceived control and achievement [57]. Second, the within-person analysis also demonstrated that compared to low-achieving pupils, high-achieving pupils exhibit a clearer pattern that the perceived control and performance predict further control in the subsequent time points, perhaps indicating more organized control for high-achieving children [57].

5.2.2 Graphical VAR

The recent advancement in network science has extended several psychological networks methods to within-person, idiographic analysis. This is particularly useful when an analyst is interested in the temporal changes (i.e., how the previous state affects the following state) and the contemporaneous processes (i.e., how different constructs co-occur simultaneously) of observed constructs in individual subjects [22, 42].

Graphical vector autoregression (VAR) models are especially useful for fully idiographic, person-specific analysis. They are able to capture the temporal and contemporaneous processes of psychological constructs within a single subject. For example, Saqr and López-Pernas [59] collected survey data by the EMA method from an undergraduate student (N=1) once a day for the duration of a month. The survey questions represented common dispositions of self-regulated learning such as motivation, reflection and planning. After the linear trends in variables were removed to approximately ensure the stationarity, graphical VAR was applied. The findings indicated that for the participant, the feeling of hope as well as motivation positively affect the task engagement on the following day, while working on assignment negatively correlates with the engagement on the following day [59]. Other examples can be also seen within the same authors' work on self-regulated learning where they studied the network models of 18 students and showed that each student has a different profile of self-regulation. The authors concluded that if personalization is the goal, then idiographic methods should be prioritized [53].

While the previous example utilized temporal networks, Malmberg et al. [60] employed contemporaneous networks in addition to temporal ones to investigate metacognitive monitoring in collaborative learning in the context of self-regulated learning. Video data of 12 upper secondary school students during collaborative work of an advanced physics course were collected, and monitoring events were manually coded. The within-person analysis of contemporaneous networks by graphical VAR suggested that different monitoring phases (e.g., task definition, goal setting) tend not to co-occur in a short time window, but rather monitoring events within a group lead to enactment before the group proceeds to another phase of regulation [60]. In addition, graphical VAR analysis on the within-person temporal networks showed that motivational interactions lead to monitoring of task definition and task enactment in the following step, driving the early phases of task execution [60].

5.2.3 Sequence Mining

Another approach to working with time series, especially longitudinal, data is sequence mining. While collecting self-report data through EMA/ESM from the same individuals over a long period of time (e.g. several months) could overload participants, log data or physiological data automatically collected through LMS or portable digital devices could be relatively longitudinal. Sequence mining allows for within-person analysis of such data to investigate sequential patterns of learning strategies of students or educational interventions.

Saqr et al. [5] collected LMS activity log data of 135 students for 10 successive modules in a higher education program of healthcare (corresponding to the period of three years). At the lowest level, for each student and for each module the sequences of activity log were segmented into sessions such that each session represents a non-stop learning activity. Then the series of sessions were clustered into tactics within each module for each student. Subsequently, these tactics were coded as individual module-level learning strategies for each module and each student. Finally, the series of module-level strategies were clustered over the participants to detect program-level trajectories. The results showed that at the program-level students' learning trajectories fall into three types: stable, fluctuating, and light interactive [61]. Additionally, it was demonstrated that only a small subset of students with stable, intense trajectories showed consistent strategies over different courses (i.e., transferred the strategies) [61].

5.2.4 Idiographic Machine Learning

The increasing availability of big data at the individual level allows for applying machine learning. The idiographic machine learning approach potentially exhibits enhanced predictive power while transparency and interpretability would be reduced for complex models compared to other parametric within-person analysis.

Saqr et al. [4] collected survey data from 17 secondary school students twice a day for 45 days using the EMA method. Three machine learning models are applied to predict key SRL variables (effort, metacognition, motivation, and enjoyment): RF, elastic-net regression (ENR), and the best-items scale that is cross-validated, correlation-weighted, informative, and transparent (BISCWIT). Similarly to the previous example, the results indicated that the predictability of the machine learning models differed across individual participants [4]. In addition, feature importance on prediction also varies across individuals. Moreover, the authors also applied the same machine learning techniques to the whole dataset (which is the nomothetic approach) and showed that the feature importance of the nomothetic models did not match any single individual's feature importance [4].

Aalbers et al. [47] collected survey data five times a day as well as smartphone log data for app use over a maximum of 60 days. Then three different machine learning models are applied to predict the level of subjective stress at individual and group levels: least absolute shrinkage and selection operator (LASSO) regression, support vector machine (SVM,) and random forest (RF). Features included the use of particular categories of smartphone app, sleep, and timestamp information. The results show that the predictions of nomothetic models are positively correlated with the actual stress of most participants while the models' predictions significantly differ from the actual values for a small subgroup of individuals [47]. The prediction performance was similar for the idiographic models; the idiographic modeling approach successfully predicted momentary stress, but for a minority of participants [47]. It should be noted that in this study, and often in general, the sample size is necessarily much smaller for idiographic machine learning than for the nomothetic models.

5.2.5 Group Iterative Multiple Model Estimation (GIMME)

Group Iterative Multiple Model Estimation (GIMME) integrates structural equation modeling (SEM) and dynamic modeling to capture both group-level and individual-level temporal processes [62]. It can be applied to longitudinal time series data for a group of people.

Chaku et al. [63] used a survey dataset that assessed working memory, somatic symptoms, anxiety, and intellectual interests of 26 college students over 75 days during the 2016 US presidential election and in 2017 (as a control group). Data collection employed the ESM approach, asking participants to complete the survey after 8 pm or after daily activity on electronic devices each day. The data was analyzed using one-lag GIMME, which produces group-level and person-specific networks that illustrate contemporaneous and lagged effects of the four variables on each other. The results demonstrated that on average the election group had higher network complexity, potentially reflecting the negative effect of the stressful time on learning-related processes [63]. However, the person-specific models exhibit noticeable differences from the group-level models for both the 2016 election group and the control group, which indicates remarkable heterogeneity on the relationship between negative psychological indicators and learning [63].

Thompson et al. [64] analyzed the effect of teacher feedback on students' motivation in the context of an intervention program for children who tend to exhibit challenging behavior. After screening, 58 fifth-graders with challenging behavior were included in the study. Collected data included self-monitoring surveys of daily motivation and negative teacher feedback measured as the discrepancy between self-evaluation and teacher's evaluation on the achievement of learning goal over 8 weeks. The analysis by GIMME showed that negative teacher feedback likely improves student motivation on the same day, while it negatively affects the subsequent day's motivation [64]. Nevertheless, high heterogeneity is also observed, and at the individual level, some students exhibited that higher motivation leads to negative teacher feedback, potentially reflecting overconfidence in these students [64].

6 Discussion

In this chapter, we have introduced the foundations of within-person analysis, contrasting it with the dominant nomothetic approach in education research. Specifically, we have placed our focus on a special case of within-person analysis—idiographic methods—which aim to capture unique, person-specific variations. We outline the main data sources and methods used for within-person research, providing relevant examples from the literature.

Despite the evident strengths and wealth of insights that can be gathered from adopting a within-person approach—and from idiographic analysis in particular—, the adoption of this lens in education research has been relatively limited compared to other fields like psychology [65]. In psychology, such a lens has been extensively used to examine individual differences and temporal patterns in constructs such as mood and cognition [26]. In turn, education has been slower to integrate within-person analysis, largely due to systemic challenges such as a historical reliance on nomothetic paradigms [66], a lack of expertise in temporal analyses, and the complexity of collecting repeated, high-resolution data from individuals [4].

However, although the idiographic approach has not been that common in educational research, its practical applicability has been exploited to a somewhat larger extent. For example, personalized learning analytics dashboards are heavily utilized for students to be able to visualize their own data, monitor their progress, and—ideally—obtain actionable feedback [36, 67, 68]. Students in special education have been given individualized support based on their own specific needs. Lastly, adaptive learning technologies use idiographic principles to recommend customized content and pacing [54].

Lastly, it is worth noting that idiographic methods may require a privacy and ethics framework of their own [37]. On the one hand, idiographic analysis alleviates some of the concerns related to the nomothetic approach [36], since the results yielded from this analysis are completely individualized and therefore do not undermine minorities or suffer from the generalizability issues underlined in this chapter. However, this comes at the expense of intensive data collection for each single individual, often involving sensitive data [37, 69, 70]. As such, additional safeguards must be in place to protect participants' confidentiality.

Throughout the subsequent chapters of the book, we showcase various methods suitable for within-person and idiographic analysis, accompanied by hands-on tutorials in R. For instance, we describe how to apply different forms of regression to capture between- and within-person relationships [21]. In another chapter we introduce psychological networks and demonstrate their application in modeling both contemporaneous and temporal between and within-person relationships, including the use of graphical VAR techniques [22]. Lastly, we also explore the use of automatic ML to scale the application of ML algorithms for idiographic analysis [23].

References

1. Valtonen T, López-Pernas S, Saqr M, Vartiainen H, Sointu ET, Tedre M (2022) The nature and building blocks of educational technology research. Comput Human Behav 128:107123
2. Romero C, Ventura S (2020) Educational data mining and learning analytics: an updated survey. Wiley Interdiscip Rev Data Min Knowl Discov 10:e1355
3. Siemens G (2013) Learning analytics: the emergence of a discipline. Am Behav Sci 57:1380–1400
4. Saqr M, Cheng R, López-Pernas S, Beck ED (2024) Idiographic artificial intelligence to explain students' self-regulation: toward precision education. Learn Indiv Differ 114:102499
5. Saqr M (2023) Modelling within-person idiographic variance could help explain and individualize learning. Br J Educ Technol 54:1077–1094
6. Kaliisa R, Misiejuk K, López-Pernas S, Khalil M, Saqr M (2024) Have learning analytics dashboards lived up to the hype? A systematic review of impact on students' achievement, motivation, participation and attitude. In: Proceedings of the 14th learning analytics and knowledge conference. ACM, New York, NY, USA. https://doi.org/10.1145/3636555.3636884
7. Conijn R, Snijders C, Kleingeld A, Matzat U (2017) Predicting student performance from LMS data: a comparison of 17 blended courses using moodle LMS. IEEE Trans Learn Technol 10:17–29
8. Gašević D, Dawson S, Rogers T, Gasevic D (2016) Learning analytics should not promote one size fits all: the effects of instructional conditions in predicting academic success. Internet High Educ 28:68–84. https://doi.org/10.1016/j.iheduc.2015.10.002
9. Jovanović J, Saqr M, Joksimović S, Gašević D (2021) Students matter the most in learning analytics: the effects of internal and instructional conditions in predicting academic success. Comput Educ 172:104251
10. Molenaar PCM, Campbell CG (2009) The new person-specific paradigm in psychology. Curr Dir Psychol Sci 18:112–117
11. Ito H, López-Pemas S, Saqr M (2024) A scoping review of idiographic research in education: too little, but not too late. In: 2024 IEEE international conference on advanced learning technologies (ICALT). IEEE, pp 10–12
12. Saqr M, Lopez-Pernas S (2021) Idiographic learning analytics: a definition and a case study. In: 2021 international conference on advanced learning technologies (ICALT). IEEE, pp 163–165
13. Beck ED, Jackson JJ (2021) Within-person variability. In: The handbook of personality dynamics and processes. Elsevier, Amsterdam, pp 75–100
14. Windelband W (1980) Rectorial address, strasbourg, 1894. Hist Theory 19:169
15. Windelband W, Oakes G (1980) History and natural science. Hist theory 19:165
16. Beltz AM, Wright AGC, Sprague BN, Molenaar PCM (2016) Bridging the nomothetic and idiographic approaches to the analysis of clinical data. Assessment 23:447–458
17. Saqr M, Schreuder MJ, López-Pernas S (2024) Why educational research needs a complex system revolution that embraces individual differences, heterogeneity, and uncertainty. In: Saqr M, López-Pernas S (eds) Learning analytics methods and tutorials: a practical guide using r. Springer Nature Switzerland, Cham, pp 723–734

18. Saqr M, Vogelsmeier LVDE, López-Pernas S (2024) Capturing where the learning process takes place: a person-specific and person-centered primer. Learn Indiv Differ 113:102492
19. Saqr M (2024) Group-level analysis of engagement poorly reflects individual students' processes: why we need idiographic learning analytics. Comput Human Behav 150:107991
20. Beck ED, Jackson JJ (2021) Within-person variability. Elsevier, Amsterdam, pp 75–100
21. López-Pernas S, Kayaduman H, Vogelsmeier LVDE, Saqr M (2025) The three levels of analysis: variable-centered, person-centered and person-specific analysis in education. In: Saqr M, López-Pernas S (eds) Advanced learning analytics methods: AI, precision and complexity. Springer Nature Switzerland, Cham
22. Saqr M, Dever D, López-Pernas S (2025) Idiographic networks: a tutorial on graphical vector autoregression and unified structural equation modeling. In: Saqr M, López-Pernas S (eds) Advanced learning analytics methods: AI, precision and complexity. Springer Nature Switzerland, Cham
23. Saqr M, Tlili A, López-Pernas S (2025) Automating individualized machine learning and AI prediction using AutoML: the case of idiographic predictions. In: Saqr M, López-Pernas S (eds) Advanced learning analytics methods: AI, precision and complexity. Springer Nature Switzerland, Cham
24. Fisher AJ, Medaglia JD, Jeronimus BF (2018) Lack of group-to-individual generalizability is a threat to human subjects research. Proc Natl Acad Sci USA 115:E6106–E6115
25. Molenaar PCM (2004) A manifesto on psychology as idiographic science: bringing the person back into scientific psychology, this time forever. Meas Interdiscip Res Perspectives 2:201–218
26. Piccirillo ML, Rodebaugh TL (2019) Foundations of idiographic methods in psychology and applications for psychotherapy. Clin Psychol Rev 71:90–100
27. Schuurman NK (2023) A "within/between problem" primer: About (not) separating within-person variance and between-person variance in psychology. PsyArXiv. https://doi.org/10.31234/osf.io/7zgkx
28. Sedgwick P (2015) Understanding the ecological fallacy. BMJ 351:h4773
29. Kievit RA, Frankenhuis WE, Waldorp LJ, Borsboom D (2013) Simpson's paradox in psychological science: a practical guide. Front Psychol 4:513
30. Beck ED, Jackson JJ (2022) Personalized prediction of behaviors and experiences: an idiographic person-situation test. Psychol Sci 33:1767–1782
31. Hutchinson B, Mitchell M (2019) 50 years of test (un)fairness: lessons for machine learning. In: Proceedings of the conference on fairness, accountability, and transparency. Association for computing machinery, New York, NY, USA, pp 49–58
32. Selwyn N (2019) What's the problem with learning analytics? J Learn Anal 6:11–19
33. Baek C, Aguilar SJ (2023) Past, present, and future directions of learning analytics research for students with disabilities. J Res Technol Educ 55:931–946
34. Saqr M, López-Pernas S (2021, July) Idiographic learning analytics: a definition and a case study. In: 2021 international conference on advanced learning technologies (ICALT). IEEE, pp 163–165
35. Lang C, Woo C, Sinclair J (2020) Quantifying data sensitivity: precise demonstration of care when building student prediction models. In: Proceedings of the tenth international conference on learning analytics & knowledge. ACM, New York, NY, USA. https://doi.org/10.1145/3375462.3375506
36. López-Pernas S, Saqr M (2021) Idiographic learning analytics: a within-person ethical perspective. In: Companion proceedings of the 11th international conference on learning analytics & knowledge (LAK21), pp 456–463 (Unpublished)
37. Misiejuk K, Samuelsen J, Kaliisa R, Prinsloo P (2025) Idiographic learning analytics: mapping of the ethical issues. Learn Indiv Differ 117:102599
38. Barlow DH, Nock MK (2009) Why can't we be more idiographic in our research? Perspect Psychol Sci J Assoc Psychol Sci 4:19–21
39. Hofman AD, Jansen BRJ, De Mooij SMM, Stevenson CE, Van der Maas HLJ (2018) A solution to the measurement problem in the idiographic approach using computer adaptive practicing. J Intell 6:14

40. Piccirillo ML, Beck ED, Rodebaugh TL (2019) A clinician's primer for idiographic research: considerations and recommendations. Behav Therapy 50:938–951
41. Fisher AJ, Reeves JW, Lawyer G, Medaglia JD, Rubel JA (2017) Exploring the idiographic dynamics of mood and anxiety via network analysis. J Abnormal Psychol 126:1044–1056
42. Epskamp S, Waldorp LJ, Mõttus R, Borsboom D (2018) The gaussian graphical model in cross-sectional and time-series data. Multivar Behav Res 53:453–480
43. Pintrich PR, De Groot EV (2012) Motivated strategies for learning questionnaire. PsycTESTS Dataset. https://doi.org/10.1037/t09161-000
44. Watson D, Clark LA, Tellegen A (1988) Development and validation of brief measures of positive and negative affect: the PANAS scales. J Pers Soc Psychol 54:1063–1070
45. Napa Scollon C, Prieto C-K, Diener E (2009) Experience sampling: promises and pitfalls, strength and weaknesses. In: Diener E (ed) Assessing well-being: the collected works of ed diener. Springer Netherlands, Dordrecht, pp 157–180
46. Zirkel S, Garcia JA, Murphy MC (2015) Experience-sampling research methods and their potential for education research. Educ Res (Washington, DC: 1972) 44:7–16
47. Aalbers G, Hendrickson AT, Vanden Abeele MM, Keijsers L (2023) Smartphone-tracked digital markers of momentary subjective stress in college students: idiographic machine learning analysis. JMIR mHealth uHealth 11:e37469
48. Dhindsa K, Acai A, Wagner N, Bosynak D, Kelly S, Bhandari M, Petrisor B, Sonnadara RR (2019) Individualized pattern recognition for detecting mind wandering from EEG during live lectures. PloS One 14:e0222276
49. Prieto LP, Rodríguez-Triana MJ, Odriozola-González P, Dimitriadis Y (2022) Single-case learning analytics to support social-emotional learning: the case of doctoral education. In: Social and emotional learning and complex skills assessment. Springer International Publishing, Cham, pp 251–278
50. Ordem E (2017) A longitudinal study of motivation in foreign and second language learning context. J Educ Learn 6:334
51. Respondek L, Seufert T, Nett UE (2019) Adding previous experiences to the person-situation debate of achievement emotions. Contemp Educ Psychol 58:19–32
52. Trull TJ, Ebner-Priemer UW (2009) Using experience sampling methods/ecological momentary assessment (ESM/EMA) in clinical assessment and clinical research: introduction to the special section. Psychol Assess 21:457–462
53. Saqr M, López-Pernas S (2024) Mapping the self in self-regulation using complex dynamic systems approach. Br J Educ Technol J Council Educ Technol. https://doi.org/10.1111/bjet.13452
54. Xie H, Chu H-C, Hwang G-J, Wang C-C (2019) Trends and development in technology-enhanced adaptive/personalized learning: a systematic review of journal publications from 2007 to 2017. Comput Educ 140:103599
55. Xie K, Heddy BC, Vongkulluksn VW (2019) Examining engagement in context using experience-sampling method with mobile technology. Contemp Educ Psychol 59:101788
56. Muenks K, Canning EA, LaCosse J, Green DJ, Zirkel S, Garcia JA, Murphy MC (2020) Does my professor think my ability can change? Students' perceptions of their STEM professors' mindset beliefs predict their psychological vulnerability, engagement, and performance in class. J Exp Psychol Gen 149:2119–2144
57. Musher-Eizenman DR, Nesselroade JR, Schmitz B (2002) Perceived control and academic performance: a comparison of high- and low-performing children on within-person change patterns. Int J Behav Dev 26:540–547
58. Molenaar PCM (1985) A dynamic factor model for the analysis of multivariate time series. Psychometrika 50:181–202
59. Saqr M, López-Pernas S (2021) Idiographic learning analytics: a single student (n=1) approach using psychological networks. In: Proceedings of the NetSciLA21 workshop. CEUR-WS, pp 16–22
60. Malmberg J, Saqr M, Järvenoja H, Järvelä S (2022) How the monitoring events of individual students are associated with phases of regulation: a network analysis approach. J Learn Anal 9:77–92

61. Saqr M, López-Pernas S, Jovanović J, Gašević D (2023) Intense, turbulent, or wallowing in the mire: a longitudinal study of cross-course online tactics, strategies, and trajectories. Int Higher Educ 57:100902
62. Gates KM, Molenaar PCM (2012) Group search algorithm recovers effective connectivity maps for individuals in homogeneous and heterogeneous samples. NeuroImage 63:310–319
63. Chaku N, Kelly DP, Beltz AM (2021) Individualized learning potential in stressful times: how to leverage intensive longitudinal data to inform online learning. Comput Human Behav 121:106772
64. Thompson AM, Wiedermann W, Herman KC, Reinke WM (2021) Effect of daily teacher feedback on subsequent motivation and mental health outcomes in fifth grade students: a person-centered analysis. Prev Sci Official J Soc Prev Res 22:775–785
65. Ito H, López-Pernas S, Saqr M (2024) A scoping review of idiographic research in education: too little, but not too late. In: Proceedings of the 2023 IEEE international conference on advanced learning technologies (ICALT). IEEE, pp 10–12
66. Kemmis S (1978) Nomothetic and idiographic approaches to the evaluation of learning. J Curric Stud 10:45–59
67. Gujju K, Bandi S, Moraes M (2024) Review of learning analytics dashboards for students. In: Lecture notes in networks and systems. Springer Nature Switzerland, Cham, pp 300–312
68. Kaliisa R, Misiejuk K, López-Pernas S, Khalil M, Saqr M (2024) Have learning analytics dashboards lived up to the hype? A systematic review of impact on students' achievement, motivation, participation and attitude. In: Proceedings of the 14th international learning analytics and knowledge conference (LAK'24). SoLAR
69. Beck ED, Jackson JJ (2020) Idiographic traits: a return to allportian approaches to personality. Curr Dir Psychol Sci. https://doi.org/10.1177/0963721420915860
70. Saqr M, López-Pernas S (2024) Momentary emotions emerge and evolve differently, yet are surprisingly stable within students. In: 2024 IEEE international conference on advanced learning technologies (ICALT). IEEE, pp 86–90

The Three Levels of Analysis: Variable-centered, Person-centered and Person-specific Analysis in Education

Sonsoles López-Pernas, Halil Kayaduman, Leonie V. D. E. Vogelsmeier, and Mohammed Saqr

Abstract This chapter explores three analytical approaches—variable-centered, person-centered, and person-specific—that span a spectrum of generalizability and individualization. The variable-centered approach emphasizes generalizable insights by analyzing relationships across a population, often overlooking individual heterogeneity. The person-centered approach identifies distinct subgroups within populations to reveal shared patterns and characteristics among similar individuals. The person-specific or idiographic approach shifts focus entirely to the individual, offering a detailed examination of intra-individual variability and processes over time. The chapter includes a tutorial using R, demonstrating how these methods can be applied to educational data using Experience Sampling Method (ESM) data on students' task value and happiness, we illustrate the strengths and limitations of each approach in understanding the relationship between these two variables.

Keywords Learning analytics · Nomothetic · Person-centered · Heterogeneity · Idiographic

1 Introduction

There is an ongoing scientific debate about the best approach to analyze data in order to create effective, personalized recommendations and interventions. This debate centers on two approaches: a group-based approach that seeks generalizable insights and another approach with a focus on heterogeneity and individual differences. The group-based approach assumes subjects are fundamentally similar, suggesting that group averages are representative of most individuals. In contrast, the heterogeneity

---

S. López-Pernas (✉) · H. Kayaduman · M. Saqr
University of Eastern Finland, Kuopio, Finland
e-mail: sonsoles.lopez@uef.fi

L. V. D. E. Vogelsmeier
Tilburg University, Tilburg, The Netherlands

M. Saqr, S. López-Pernas (eds.), *Advanced Learning Analytics Methods*,
https://doi.org/10.1007/978-3-031-95365-1_19

approach views each subject as unique, necessitating individualized data and methods [1–6].

One significant critique of the group-based approach is that it can obscure individual differences by averaging scores, leading to standardized recommendations that may work for some but overlook those with distinct needs [2, 3, 5, 7]. On the other hand, the idiographic approach requires intensive data collection for each individual which is "far from practical and largely exhausting" in many contexts [8] (p. 11). In this chapter, we provide an overview of these approaches and a tutorial on how to implement them using R. The tutorial implements different variations of statistical models to account for both the between- and within-person variability in repeated measures data. We base the tutorial on a recent editorial published in the journal *Learning and Individual Differences* [5].

2 Three Different Approaches

2.1 Variable-centered Approaches

In educational research, and subsequently in learning analytics research, the variable-centered approach is one of the most widely used approaches. This approach is sometimes referred to by different names, such as nomothetic, group-based, between-person, or inter-individual. In this approach, researchers typically aim to explain the relationships between variables across a population [3, 5]. To understand the population, researchers collect data from multiple individuals — commonly at the same point in time—on one or more occasions depending on their aims and research questions [3, 5, 9]. This approach relies on the homogeneity assumption, where individuals are considered more or less the same [1]. In that, the average represents the majority of individuals [2–5], and strong deviations from the average are considered outliers [5, 10]. Researchers typically conduct statistical tests to correlate or compare averages, e.g., correlations, t-test, ANOVA or regression to address research questions such as, 'What is the association between cognitive ability dimensions and job performance?' [3]. Since this approach is based on the homogeneity assumption and aims to understand the population, it typically produces results that are assumed to be generalizable to the population at large [3, 5]. However, the fact that these insights are based on a sample of the population that is indeed heterogeneous makes their generalization doubtful to each and every person due to overlooking individual differences [2, 5, 9–11]. Considering individuals' unique contexts and the shift in learning analytics toward more personalized recommendations and interventions [5, 7], complementing variable-centered approaches with other methods may be critical to ensure that each individual receives what is relevant to them especially when it comes to personalization and individualizing education.

2.2 Person-centered Approaches

The person-centered approach is another analytical method that targets individualized insights and looks at finding similarities and patterns within the data. It might also be referred to as the person-based or class-centered approach. In the person-centered approach, the researchers try to group similar individuals into separate sub-populations by considering their distinct characteristics [3–5, 12]. Similar to variable-centered approaches, the researchers collect data from multiple individuals, but the person-centered approach tries to find subgroups that are similar to each other. In other words, it tries to find hidden subgroups within the population rather than considering each individual as part of a single population [3–5]. Examples of methods include mixture regression analysis [5, 13, 14], latent class analysis [15], latent profile analysis [16], factor mixture analysis [17, 18], mixture growth curve modeling [19], latent transition analysis [20] or latent Markov factor analysis [21]. All of such methods produce different clusters—subgroups, latent groups, or classes—each with different parameters or coefficients. For instance, rather than having one regression coefficient for the whole analysis, mixture regression analysis allows all or selected coefficients to differ across (unobserved) subgroups. In that way, person-centered methods find the different heterogeneous subgroups and directly model this heterogeneity.

The potential research question might include: "What emergent subpopulations can be identified through the dimensions of cognitive ability?" [3]. With the help of this approach, the literature has revealed diverse and critical insights. For example, Kleimola et al. [15] explored nursing students' motivational profiles with 132 participants. With the help of latent class analysis, they identified three motivational profiles of vocational practical nurse students. Trautner and Schwinger [18] investigated self-concepts related to the academic ability of primary school students with mild learning challenges. They analyzed data from more than 400 students using factor mixture models and found two subgroups that differed in their levels of self-concept. In another study, Gillet et al. [20] conducted latent profile and latent transition analysis with 504 university students' data in order to examine their academic motivation profiles and the stability of these profiles. They discovered six distinct motivation profiles, which remained stable across two measurement points.

2.3 Person-specific Approaches

The person-specific approaches might be considered a novel method in learning analytics research, though they have been used previously in psychological research. This approach is sometimes referred to as idiographic, idiosyncratic, within-person, intra-individual, or single subject. In this approach, the researchers view each student as the whole unit of analysis (N = 1) [3, 6, 11, 22]. Since psychological processes such as learning or emotion can vary for each individual,

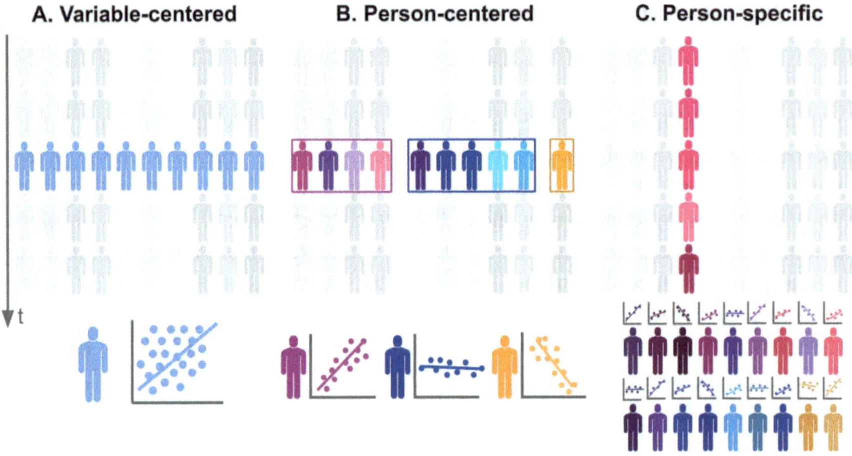

Fig. 1 Illustration of the difference between variable-centered, person-centered, and person-specific analysis

attempting to understand a person's processes using group data may overlook individual differences and variations [22]. Moreover, this can even cause ineffective recommendations and interventions for some individuals [23]. In this regard, the person-specific approaches can be valuable by conducting focused studies on individuals [3].

Unlike variable-centered and person-centered approaches, person-specific approaches do not require data to be collected from multiple individuals; even a single subject may be sufficient; however, to be able to make inferences at the person level, data must be collected at multiple time points and with enough data points for the analysis [6, 9, 11]. Figure 1 summarizes the comparison of the three approaches. In Fig. 1-A, data is collected from multiple subjects at the same time point, with all subjects considered to belong to a single group (homogeneity). In Fig. 1-B, data are again collected from multiple subjects at the same time point; however, this time, the subjects belong to different subgroups (heterogeneity and homogeneity). Conversely, in Fig. 1-C, data are collected from a single subject across multiple time points to investigate individual variations and changes in a detailed way (heterogeneity).

Researchers conduct diverse analyses using a person-specific lens such as vector autoregression [6, 10], time-series analysis [24], dynamic factor analysis [24], group iterative multiple model estimation [11], idiographic machine learning [8] or mixed linear models [9].

Compared to variable-centered and person-centered methods, person-specific approaches do not produce generalizable results since they require intensive investigation of individuals rather than attempting to understand the population, however; they yield the more individualized results [3, 5, 25]. Therefore, utilizing idiographic approaches in learning analytics research studies could be a solution to offering more personalized recommendations.

In the literature, there are a few research studies that utilize idiographic approaches. For example, Saqr and López-Pernas [25] conducted single-subject research to investigate the self-regulation strategies of a student. They gathered data daily over 30 days, and the data analysis revealed that the student needed to improve their reflection and planning strategies. Furthermore, they emphasized the potential of idiographic methods for examining students' behavior at the individual level.

In summary, each of the three approaches has its advantages and limitations. Relying on a single approach may lead to inadequate solutions in education and learning analytics. Instead, integrating and leveraging the strengths of all three approaches could provide a complete picture of the phenomenon under study and contribute to the design of more effective learning analytics interventions. Table 1 summarizes all three approaches in a detailed way.

Table 1 Summary of method comparison

	Variable-centered	Person-centered	Person-Specific
Alternative names	• Nomothetic • Group-based • Between-person • Inter-individual	• Person-based • Class-centered	• Idiographic • Idiosyncratic • Single Subject • N=1
Description	Explain how various variables are correlated or different across groups [3, 5]	Categorize similar individuals into separate sub-populations [3, 5, 12]	View each student as a whole unit of analysis [5, 6, 10, 22]
Assumption	Assumes Homogeneity [2, 3, 5]	Assumes heterogeneity[3–5]	Assumes every one is unique [5, 6, 9, 11]
Sample size	The suggested minimum sample is 30 however larger are better [3]	The suggested sample size typically is larger e.g., 200 subjects or more [3, 5], but it heavily depends on the analysis	A single subject may be sufficient but requires 30-120 or even more data-points from the same individual depending on the analytical method [3, 5, 6, 22]
Data collection	From multiple individuals [3, 5, 9]	From multiple individuals [3]	From a single individual at multiple time points (or multiple individuals however the analysis is performed individually)[6, 11]
Sample research questions	What is the association between engagement and academic performance [26]?	What different profiles of engagement can be identified and how do they change? [27]	What are the unique profiles of each student in self-regulating their own studies [10].

(continued)

Table 1 (continued)

Sample statistical analysis	• t-test [28] • ANOVA [29] • Correlation [30] • Regression [31] • Factor analysis [32] • Structural equation modeling [33]	• Mixture regression analysis [5] • Latent class analysis [15] • Latent profile analysis [16] • Factor mixture analysis [17] • Mixture growth curve modeling[34] • Latent transition analysis [20] • Latent Markov factor analysis [27] • Mixed Hidden Markov models [35]	• Vector autoregression [6] • Time-series analysis methods [24] • Dynamic factor analysis [24] • Group iterative multiple model estimation [11] • Idiographic machine learning [8] • Psychological networks [10]
Example applicatino	Studying the general relationship between two variables across a group of students to design general interventions that work for the majority.	Identifying subgroups of students with distinct profiles to tailor interventions for each subgroup's specific needs.	Analyzing how each single student fluctuates daily to create personalized interventions.

3 A Tutorial Along the Heterogeneity Spectrum

In this section we present a tutorial on variable-centered, person-centered and person-specific analysis. The tutorial assumes that the reader is familiar with the R programming language. If that is not the case, it is recommended to refer to previous tutorials on the basics of R [36], data cleaning [37], basic statistics [38], and visualization [39]. To illustrate the main principles and techniques of the three presented approaches, we will use a dataset from a previous study that used the Experience Sampling Method (ESM) to investigate the relationship between self-regulation and students' everyday experiences [40]. ESM is a method where participants report on their behaviors and feelings multiple times throughout the day, allowing for the collection of real-time data on individual experiences. Most importantly, ESM allows the collection of enough data for individual analysis.

In this study, the focus was primarily on how certain aspects such as mindfulness, self-control, and affective states influence students' daily life experiences. The data collection period involved up to 48 measurement points per participant. This intensive longitudinal design allowed the researchers to capture the changes in participants' experiences over time, with each participant providing multiple repeated measures (a requirement for person-specific analysis). Though there are many interesting variables for analysis in the dataset, in this tutorial, we will focus on the relationship between task value and happiness, in alignment with Expectancy-Value Theory (EVT). To access the data from the study to follow this tutorial, you must first fill the consent form in the dataset webpage [40] and download the data files.

3.1 Preparing the Data and Libraries

To analyze the dataset with R, we first need to install and load the necessary libraries:

- `tidyverse` is a collection of packages for data manipulation and visualization
- `flexmix` is used for finite mixture modeling
- `lme4` is for fitting linear mixed-effects models
- `modeltools` provides tools for model fitting and validation
- `jtools` helps with summarizing and visualizing regression models
- `anytime` is for converting date-time strings into proper datetime objects
- `rio` imports and exports data in several formats

```
library(tidyverse)
library(flexmix)
library(lme4)
library(modeltools)
library(jtools)
library(anytime)
library(rio)

options(scipen = 99)
```

Now we can import the data using the function `import` from `rio`. We store it in a variable named `data`.

```
# Import data downloaded from https://rdc-psychology.org/grund_fries_senker_2022
data <- import("gdal18pr18_pd.txt")
```

We will perform some cleaning and pre-processing of the data to be able to work with it. First, we convert the date and time columns of the data into a single timestamp value with a datetime format.

```
data$Timestamp = anytime(paste(data$DATE, data$TIME))
```

We then remove (using `filter`) empty responses or responses that present invalid values (which were coded with negative numbers, so we keep those above 0 only), and then we order (`arrange`) the responses within each participant (`group_by`). We tag each response with the order they occupy in the sequence of responses (`mutate`).

```
# Order responses per participant
data_seq <- data |>
  filter(KF_P06_VA >= 0) |> # Remove rows where happiness (KF_P06_VA) is invalid
  filter(KF_SDT_2 >= 0) |> # Remove entries where task value (KF_SDT_2) is invalid
  group_by(PARTICIPANT) |> # Group by participant
  arrange(Timestamp) |> # Sort by timestamp
  mutate(Sequence = seq_along(Timestamp)) |> # Create a sequential ID for responses
  ungroup()
```

We now `select` only the columns that we need, corresponding to the participant identifier, the order of the response, and the responses provided for happiness (`KF_P06_VA`) and task value (`KF_SDT_2`).

```
data_filtered <- data_seq |>
  select(PARTICIPANT, Sequence, Happiness = KF_P06_VA, TaskValue = KF_SDT_2)
```

We can preview the data to make sure we have everything we need.

```
head(data_filtered)
```

```
# A tibble: 6 x 4
  PARTICIPANT Sequence Happiness TaskValue
  <chr>          <int>     <int>     <int>
1 OeQfAU             1         7         4
2 DVxCAD             1         3         6
3 bWkgSx             1         3         4
4 DjNFOK             1         2         5
5 FeyIlD             1         5         4
6 UiTZVz             1         4         5
```

3.2 Variable-centered

The first analytical lens we are exploring is variable-centered analysis. This approach focuses on understanding the relationships between specific variables across individuals. In variable-centered analysis, the main objective is to identify general trends and patterns by analyzing how one variable (such as task value) relates to another (such as happiness) across a population. This type of analysis is typically used to study overall associations or effects, assuming that the relationships between variables are consistent across individuals, i.e., we treat all individuals as being drawn from the same population. It is a suitable approach when the goal is to make generalized conclusions about how variables relate to each other at a population level, rather than focusing on individual differences.

To illustrate the variable-centered approach, we will model the relationship between task value and happiness using a simple **linear regression**. Specifically, we will calculate how students' overall task value perception is related to their overall happiness. Since our dataset contains data collected over several time points, we can aggregate the individual observations for each participant by calculating their mean task value and mean happiness scores. This aggregation offers a more stable mean. In other words, it offers a more accurate idea about each individual given that we do not just randomly collect their data at a time-point but we rather aggregate and average their responses over many time-points. In so doing, we focus on the overall enduring attitude of task value and happiness for each participant. This allows us to look at broad, general trends across participants. The following excerpt of code calculates the mean happiness and task value for each participant and stores it in a new variable named `data_meaned`.

```
data_meaned <- data_filtered |>
  group_by(PARTICIPANT) |> # Group by participant
  summarize_at(vars(Happiness, TaskValue), mean)  # Summarize (average) columns
```

When modeling the relationship between task value and happiness, *happiness* will serve as the dependent variable—the outcome we aim to explain—while *task value* will be the independent variable—the predictor of interest. The linear regression model will provide an estimate of the relationship between task value and happiness by means of a regression coefficient. This coefficient will indicate how much we expect happiness to change for a one-unit increase in task value. A positive coefficient would suggest that higher task value is associated with higher happiness, while a negative coefficient would suggest the opposite. Furthermore, the model will provide a p-value to assess whether the relationship is statistically significant, meaning that it is unlikely to have occurred by chance. The model can be expressed mathematically as follows:

$$\text{Happiness}_i = \beta_0 + \beta_1 \cdot \text{Task Value}_i + \epsilon_i$$

Where:

- i indexes each observation.
- β_0 is the fixed intercept (overall mean happiness across all observations).
- β_1 is the fixed effect of task value (how task value affects happiness on average across observations).
- ϵ_i represents the residual error for each observation, capturing variability not explained by task value.

We can implement the linear regression using the following code. The `lm` function calculates the relationship between task value and happiness, and returns the aforementioned regression coefficient. The `summ` function is used to provide a detailed summary of the regression results, including model fit statistics, the regression coefficient, and whether the relationship is statistically significant. For a tutorial on basic statistical methods with R, please refer to [41].

```
model_lm <- lm(Happiness ~ TaskValue, data = data_meaned)
summ(model_lm) # Summarize the model
```

Observations	107
Dependent variable	Happiness
Type	OLS linear regression

F(1,105)	5.13
R^2	0.05
Adj. R^2	0.04

	Est.	S.E.	t val.	p
(Intercept)	1.97	0.92	2.15	0.03
TaskValue	0.46	0.20	2.26	0.03

Standard errors: OLS

The results show that the relationship between task value and happiness is strong (estimate = 0.46), which means that average happiness and task value are closely related. Moreover, the p-value is 0.03—which is lower than the usual significance level of 0.05—and, therefore, this relationship is statistically significant.

Following the model fitting, we can visually inspect the relationship between task value and happiness by plotting the regression results using `ggplot2`. Figure 2 shows a scatter plot (`geom_point`) of the averaged data points, along with a fitted linear regression line (`geom_smooth`) to represent the overall trend. For a tutorial on data visualization with `ggplot2`, please refer to [39].

```
label = paste0("Estimate: ", round(model_lm$coefficients["TaskValue"], 3))
# Plot the regression results
ggplot(data_meaned, aes(x = TaskValue, y = Happiness)) +
  geom_point(alpha = 0.5) + # Scatter plot with transparent points
  geom_smooth(method = "lm", fill = NA) + # Add a linear regression line
  labs(x = "Task Value", y = "Happiness") # Label the axes
```

The plot shows a small but positive relationship between task value and happiness. As task value increases, happiness tends to increase slightly. However, the

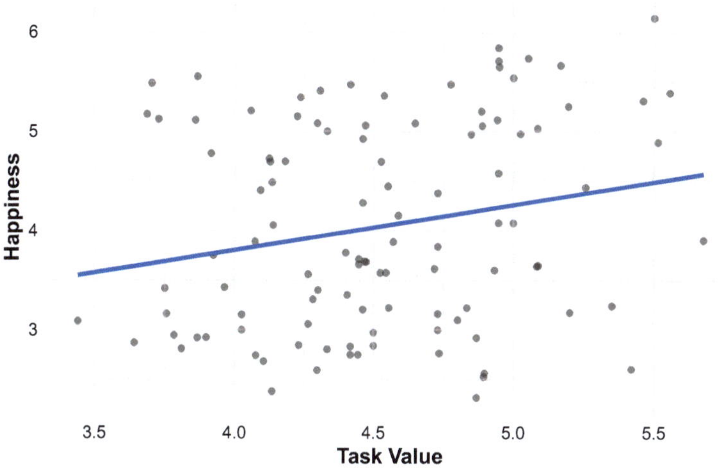

Fig. 2 Relationship between participants' mean task value and their corresponding mean happiness scores. Each point represents the average task value and happiness for a participant. The linear regression line (in *blue*) illustrates the overall trend, indicating that higher task value is associated with increased happiness

scatter points show a moderate spread around the regression line. This means that, while the general trend is upward (positive correlation), there is a lot of variability in happiness for a given level of task value, suggesting that other factors may also play a role in determining happiness.

If instead of modeling the relationship between overall task value and overall happiness we would like to model the momentary relationship—i.e., how task value and happiness are related in the same time point, we can use a **multilevel linear regression**. This approach uses the complete data (not averaged) including all individual responses per student and time point. To account for the nestedness of the data—i.e., each student has several data points—, we should add a random effect accounting for each individual person. Moreover, since we are modeling how task value and happinness are related, we need to account for the fact that different people have different baseline levels of task value and happiness. It should be noted that, while the random intercept captures individual baseline differences, the model assumes that coefficient for task value is the same for everyone. In other words, the model assumes that the way task value influences happiness is consistent across participants, even though each participant starts from a unique baseline. The model can be represented as follows:

$$Happiness_{ij} = \beta_0 + \beta_1 \cdot \text{Task Value}_{ij} + u_j + \epsilon_{ij}$$

Where:

- i indexes the repeated measures within participants.
- j indexes the participants.

- β_0 is the fixed intercept (the overall mean happiness across all participants).
- β_1 is the fixed effect of task value (how task value affects happiness on average across participants).
- u_j is the random intercept for participant j, capturing between-participant variability in baseline happiness.
- ϵ_{ij} represents the residual error for each observation, capturing within-participant variability that is not explained by task value or the random effect.

To implement the mixed-effects model described above, we will use the `lmer` function from the `lme4` package [42] to account for both the fixed effect of task value and the random intercepts for participants `(1 | PARTICIPANT)`. As explained earlier, this model allows us to estimate the overall effect of task value on happiness, while accounting for the fact that multiple observations are nested within participants, who may differ in their baseline happiness levels. As we can see in the chunk of code below, the formula is almost the same as before but we add the component representing the random intercept. Also note that we are providing the entire data, not the averaged version.

```
model_lmer <- lmer(Happiness ~ TaskValue + (1 | PARTICIPANT), data = data_filtered)
summ(model_lmer)  # Summarize the mixed-effects model
```

Observations	3866
Dependent variable	Happiness
Type	Mixed effects linear regression

AIC	12583.50
BIC	12608.54
Pseudo-R^2 (fixed effects)	0.01
Pseudo-R^2 (total)	0.43

Fixed effects

	Est.	S.E.	t val.
(Intercept)	3.61	0.12	30.14
TaskValue	0.09	0.01	6.34

Random effects

Group	Parameter	Std. Dev.
PARTICIPANT	(Intercept)	1.00
Residual		1.17

Grouping variables

Group	# groups	ICC
PARTICIPANT	107	0.42

The model shows a small but significant positive relationship between task value and happiness. However, the fixed effect of task value explains only a small portion of the variance (1%), indicating that other factors are likely influencing happiness. In turn, the Intraclass Correlation Coefficient (ICC) of 0.42 for the mixed-effects model indicates that 42% of the total variance in happiness can be attributed to differences between participants. This relatively high ICC suggests substantial variation in baseline happiness across participants, highlighting the importance of accounting for these individual differences. In the estimated model, the estimates are "informed by and shrunk toward group-level averages" [43] (p. 88), meaning that individual differences in the relationship between task value and happiness are not explicitly modeled. Instead, everyone is assumed to have the same relationship between task value and happiness. Therefore, it is still considered a variable-centered approach, as it prioritizes the average effect across all individuals rather than capturing heterogeneity in individual patterns. In contrast, a person-specific approach would allow modeling each individual's unique relationship between task value and happiness.

3.3 Person-centered

Up to this point, we have focused on the overall trend and relationship between task value and happiness across all participants. However, it is possible that different participants experience the relationship between task value and happiness in different ways. For example, some individuals may be happier with tasks they value highly (in line with the results we obtained from the variable-centered analysis), while others might experience little change in happiness regardless of task value. This heterogeneity in participant experiences suggests that subgroups or clusters of individuals may exist, each with distinct task value-happiness relationships that the overall average does not capture. When these varied patterns are averaged together in a variable-centered analysis, the heterogeneity can result in an apparent absence of a relationship if the positive and negative patterns offset each other.

In this next section of the tutorial, we will learn how to implement person-centered analysis, which shifts the focus from population-wide trends to understanding patterns across specific subgroups of individuals. Rather than modeling how one variable influences another for all participants collectively, person-centered approaches aim to identify distinct profiles, clusters, groups, or trajectories that represent unique configurations of task value and happiness within the data.

The person-centered approach relies on techniques such as cluster analysis to help us capture the heterogeneity in experiences that a variable-centered approach may overlook. The focus is on detecting subgroups of individuals that share similar patterns that differ from everyone else. As such, person-centered methods acknowledge and embrace the intrinsic heterogeneity among participants, recognizing that each subgroup has unique characteristics that may remain hidden when analyzing the population as a whole.

To illustrate the person-centered approach, we will use a **mixture regression model** to classify participants based on the relationship between task value and happiness. The goal of a mixture model is to identify subgroups (or classes) within the data where each class has a different relationship between the predictor (task value) and the outcome (happiness).

$$Happiness_i = \beta_{0k} + \beta_{1k} \cdot TaskValue_i + \epsilon_i$$

where:

- i indexes individual data points.
- k indexes the clusters (e.g., Cluster 1 and Cluster 2), where each cluster has its own parameters (β_{0k} and β_{1k}).
- β_{0k} is the intercept for cluster k, representing the baseline level of happiness specific to each cluster.
- β_{1k} is the slope for cluster k, capturing how task value influences happiness within each cluster.
- ϵ_i represents the residual error for each observation, capturing variability not explained by task value or cluster assignment.

We are using `flexmix` to fit a **mixture of linear regressions**, meaning it estimates separate regression lines for two distinct subgroups within the data (`data_meaned`). Each cluster has its own intercept (β_{0c}) and slope (β_{1c}), allowing the model to identify subgroups where the relationship between `TaskValue` and `Happiness` differ. For instance, one cluster may show a stronger positive relationship, while another may have a weaker or even negative relationship. The final model assigns each data point to one of the two clusters based on how well it fits each cluster's parameters.

To illustrate this method, we are fitting a model with two classes (k = 2), meaning that we expect to find two distinct clusters or subgroups with different relationship between task value and happiness. In practice, we would estimate the model for a range of different numbers of clusters (for example, k = 1 to k = 10) and select the optimal model. The model with the lowest BIC (Bayesian Information Criterion) indicates the best balance between model fit and complexity. For the sake of simplicity, we are using two classes here, and we rely on the `flexmix` library in R for estimating the mixture regression model. The following code—after setting a seed for reproducibility—fits a finite mixture model using `flexmix` with two clusters (k = 2)—termed "components" in the output. The `parameters` function then displays each cluster's intercepts, coefficients, and standard deviations, showing the unique relationships within each group. Finally, the code calculates the BIC for the model, which can help compare this two-cluster model with other models (e.g., with 3, 4, or more clusters) to identify the best-fitting model.

```
set.seed(022024) # Set seed for reproducibility

# Fit the mixture model with two components (k = 2)
model_k2 <- flexmix(Happiness ~ TaskValue, k = 2,
                    data = data_meaned)

# Display the parameters (intercepts, coefficients, and SDs) for each class
parameters(model_k2)
```

```
                    Comp.1 Comp.2
coef.(Intercept) -0.797 11.597
coef.TaskValue    1.069 -1.708
sigma             0.907  0.604
```

```
# Calculates the BIC to be able to compare with other models (e.g., k = 3,4,5...)
BIC(model_k2)
```

```
[1] 332
```

The results show that **Class 1** (`$Comp.1`) has a positive relationship between task value and happiness, as indicated by the coefficient of 1.07. In contrast, **Class 2** shows an even stronger negative relationship between task value and happiness, with a coefficient of -1.71. This indicates that for participants in this class, higher task value is associated with a decrease in happiness. The standard deviations (sigma values) show that Class 1 has more variability in the happiness responses to task value variations compared to Class 2. Class 2 has a smaller SD (0.60), indicating that participants in this class have less variability in their happiness responses to task value.

After identifying the two distinct classes and understanding the initial coefficients and relationships between task value and happiness, we now need to explore the model further to gain more information about each class. The next step involves refitting the mixture model using the `refit` function. This will provide a more exhaustive summary of the model for each of the identified classes, and their intercepts, coefficients, and standard deviations for each component. When refitting the model, we fine-tune the estimates for each class so that we can accurately inspect the relationships between task value and happiness within each group of participants. Thus, we ensure that we are basing our interpretations on the most reliable and detailed estimates.

```
# Refit the model to obtain more detailed summaries for each class
rm2 <- refit(model_k2)
summary(rm2)
```

```
$Comp.1
            Estimate Std. Error z value Pr(>|z|)
(Intercept)   -0.819      1.497   -0.55    0.584
TaskValue      1.074      0.327    3.28    0.001 **
---
Signif. codes:  0 '***' 0.001 '**' 0.01 '*' 0.05 '.' 0.1 ' ' 1

$Comp.2
            Estimate Std. Error z value            Pr(>|z|)
(Intercept)   11.670      1.491    7.83 0.0000000000000051 ***
TaskValue     -1.724      0.344   -5.01 0.0000005481101304 ***
---
Signif. codes:  0 '***' 0.001 '**' 0.01 '*' 0.05 '.' 0.1 ' ' 1
```

From the detailed summaries we get not only the estimate value as before, but also the standard error, z-value and p-value of both the intercept and the slope for each cluster. We see that all statistics are statistically significant except the intercept of Cluster 1.

We can extract cluster memberships (i.e., which participants belong to which class) and display the cluster sizes (i.e., the number of participants assigned to each class) to get an idea of how balanced the clusters are.

```
cluster_k2 <- clusters(model_k2)
table(cluster_k2)  # Display the number of participants in each cluster
```

```
cluster_k2
 1  2
91 16
```

We see that the first class has 91 participants assigned, and the second class has 16. It is important to take into account the distribution of participants across the classes to guide further analyses or interventions targeted at these distinct groups. To visualize the results of the mixture regression model and the differences between the two classes, we can plot the regression lines for each class alongside the participant data.

The following code creates a plot using `ggplot`. The scatter plot (`geom_point`) shows the individual data points, color-coded by class (`scale_color_manual`), while the regression lines (`geom_smooth`) represent the distinct task value-happiness relationships for each class (1 or 2). To extract the class variable—i.e., assignment of each data point to a specific cluster (component)—we use `cluster_k2` and we convert it to a factor:

```
ggplot(data_meaned, aes(x = TaskValue, y = Happiness, color = factor(cluster_k2))) +
  geom_point(alpha = 0.5) +  # Plot the points with transparency
```

```
  geom_smooth(method = "lm", se = F) +  # Plot the regression lines
  scale_color_manual(name ="Class", values = c("darkgreen","firebrick2")) # Color scale
```

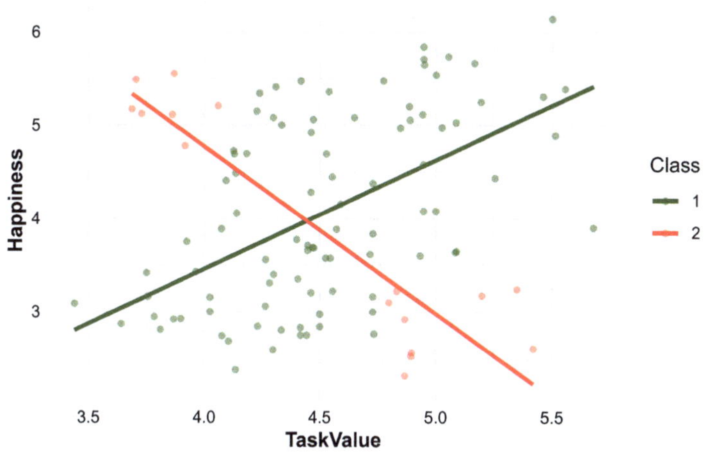

Fig. 3 Relationship between task value and happiness for two classes using a two-class mixture model- Class 1 shows a positive relationship, Class 2 shows a negative relationship

Figure 3 demonstrates the clear heterogeneity in how participants experience the relationship between task value and happiness. Some participants (Class 1) show a positive association, while others (Class 2) show a negative association. Therefore, the use of a mixture model clearly reveals that there are two distinct subgroups of participants with opposing trends, which would not have been apparent using a variable-centered approach. This is a clear example of how a person-centered approach helps uncover the diversity in how individuals experience the task value-happiness relationship.

3.4 *Person-specific*

After completing both the variable-centered and person-centered analyses, we move on to the final lens: *person-specific analysis*. The previous approaches have helped us understand general trends across the population and identify subgroups with distinct relationships between task value and happiness. In person-specific analysis, the goal is to model each individual's unique experience rather than aggregate or cluster individuals. This approach allows us to focus on within-person variability over time. Each individual is treated as their own unit of analysis, allowing us to investigate the momentary relationship between task value and happiness within

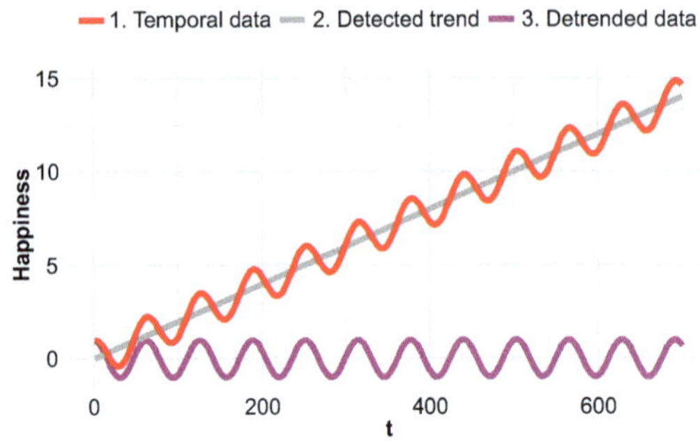

Fig. 4 Illustration of the detrending process

each person. This method is particularly important when there is a high degree of variability within individuals across different contexts and when the focus is on understanding how each person's task value relates to their happiness on a moment-to-moment basis. We will implement person-specific analysis by fitting a **linear regression** for each person including all of their individual responses.

The first step is to filter the dataset to ensure that each participant has sufficient data points for meaningful individual analysis. Specifically, we will filter out participants with fewer than 30 data points. This ensures that there are enough observations for each participant to estimate their own unique task value-happiness model. Participants with fewer data points might not provide enough information to capture meaningful within-person variability, and could lead to unreliable estimates.

```
data_filtered30 <- data_filtered |> group_by(PARTICIPANT) |> filter(n() >= 30)
```

The next step is detrending the data. Detrending is necessary when there is a temporal trend that could confound the relationship between task value and happiness. Detrending helps to remove this systematic temporal effect, allowing us to focus on the within-person relationship between task value and happiness that is not influenced by time-based trends. For example, Fig. 4 illustrates how if happiness (red line) generally increases over time for participants (e.g., due to external factors like an academic semester, weather, or life events), it could falsely inflate or mask the effect of task value on happiness. When we remove this temporal trend (grey line), we are able to isolate the effect of task value on happiness (magenta line) and ensure that the observed relationship is not simply due to changes over time.

The code below creates the `detrender` function, which takes the dataset (`data`) as an argument, the variables to detrend (`vars`), and the variables that represent time (`timevar`). The function iterates through the variables to detrend and, for each of them, it fits a linear model (`fit`) to check for a significant trend over the `timevar` variable. If the linear trend is significant ($p < 0.05$ from `anova(fit)`), the original values of the variable in the data are replaced with the residuals from `fit`, thus removing the trend component. After processing all specified variables, `to_detrend` contains the modified data where significant trends over time have been detrended for each relevant variable, and the function returns this updated dataset.

```
detrender <- function(data, vars, timevar) {
  to_detrend <- data
  for (v in seq_along(vars)) {
    # Fit linear model for each variable over time and detrend if trend is significant
    ff <- as.formula(paste0(vars[[v]], " ~ ", timevar))
    fit <- lm(ff, data = to_detrend)
    if (anova(fit)$P[1] < 0.05) {  # Check if the trend is statistically significant
      # Replace values with residuals
      to_detrend[[vars[v]]][!is.na(to_detrend[[vars[[v]]]])] <- residuals(fit)
    }
  }
  to_detrend
}
```

We now apply the `detrender` function to our dataset for the variables corresponding to happiness and task value, along the sequence variable that collects the order of the responses in time.

```
# Apply the detrending function to the happiness and task value variables
data_detrended <- detrender(data_filtered30,
                    c("Happiness","TaskValue"), "Sequence")
```

We can separate each individual student's data and create a list of all students' detrended responses so we can easily operate in each student individually.

```
# Create a list of data frames, one for each participant
data_detrended_list <- split(data_detrended, data_detrended$PARTICIPANT)
```

Now we are finally ready to conduct a person-specific analysis. We will fit a separate linear regression model for each participant to model the unique relationship between task value and happiness for each individual student. This approach helps us identify within-person differences in the relationship between these two variables for each participant. It is worth mentioning that we are estimating the contemporaneous effect of task value on happiness, meaning that we are examining how task value at a given moment is associated with happiness at the same timepoint for each individual participant. For ESM data, such as the one in this dataset, analyses typically estimate auto-regressive effects (the extent to which a variable at a given timepoint predicts itself at a subsequent timepoint) and cross-regressive effects (the extent to which one variable at a given timepoint predicts another variable at a subsequent timepoint) using lagged predictors.

The following code fits a separate linear regression model for each participant. We first create an empty list (`model_lm_person`) to store each individual model.

Then we iterate through the list of participants' data and we fit the model to each individual dataset using `lm`. We store each linear model in `model_lm_person`.

```
# Create an empty list where we will store each participant's model
model_lm_person <- list()

# Fit the models
for (k in seq_along(data_detrended_list)) {
  model_lm_person[[k]] <- lm(Happiness ~ TaskValue, data = data_detrended_list[[k]])
}
```

Now we iterate through the models and we extract the results (coefficient, p-value, etc.) for each of them using `tidy`. Then we combine (`cbind`) this information with each student's name that we extract from the data (`PARTICIPANT`). We store the model results for each student in a list (`est_person`).

```
# Create an empty list where we will store each participant's estimate
est_person <- list()

# Extract slope estimates
for (k in seq_along(model_lm_person)) {
  estimate <- tidy(model_lm_person[[k]])[2,]
  PARTICIPANT <- data_detrended_list[[k]][1, "PARTICIPANT"]
  est_person[[k]] <- cbind(PARTICIPANT, estimate)
}
```

To get an idea of how unique the relationship between task value and happiness is, we can visualize the distribution of these person-specific estimates. We first combine the individual slope estimates (`data_est_person$estimate`) into a single data frame (`data_est_person`). We do that using the `bind_rows` function, which combines a list of dataframes into a single combined dataframe and adds a new column (specified by `.id = "Person"`) to identify which list each subset of the data came from. Then, we plot a histogram (`geom_histogram`) of the slope estimates to see how the relationship between task value and happiness varies across participants (Fig. 5).

```
# Combining the individual results into one data frame
data_est_person <- bind_rows(est_person, .id = "Person")

# Histogram of the estimates to visualize the distribution across participants
ggplot(data_est_person, aes(x = estimate)) +
  # Create histogram with 20 bins
  geom_histogram(bins = 20, fill = "lightgray", color = "black", alpha = 0.7) +
  labs(x = "Estimates", y = "Frequency")  # Label the axes
```

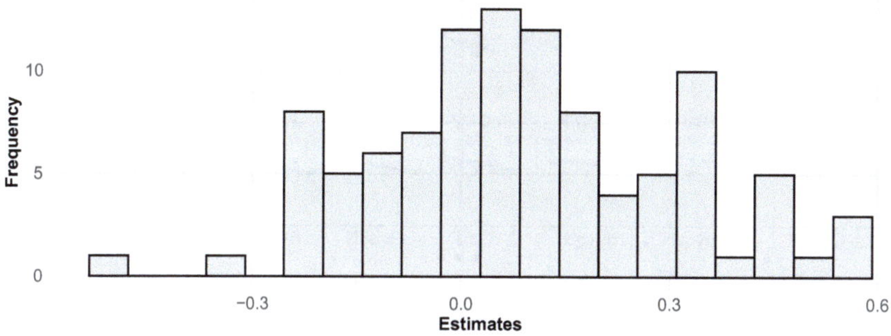

Fig. 5 Histogram of the slope estimates of each person-specific model. The mean slope is 0.09, and the standard deviation is 0.22, suggesting that there is great variability, centered around close to zero (no relationship)

To provide a diverse sample to illustrate how the person-specific method works, we will select participants for detailed analysis who display both significant and non-significant relationships between task value and happiness. This diverse sample will allow us to explore the range of individual relationships between task value and happiness, from strong predictors to negligible effects, and better understand the heterogeneity in how students experience the relationship between these variables.

The following chunk filters and selects participants based on the significance of their results. First, it identifies participants with significant results by filtering `data_est_person` (the dataframe that contains the model results for each participant) for participants whose models have a p-value < 0.05, storing these rows in the `significant` variable. Next, it selects the first 10 participants with non-significant results (p.value >= 0.05) and stores them in the variable `nonsignificant`. Finally, we obtain the list of all the selected participants for further analysis, i.e., we combine both sets of participants' IDs into a single object (`ids`), by using `rbind` to merge `significant` and `nonsignificant`, arranging the result in descending order by the `estimate` column, and extracting the `PARTICIPANT` column only which contains the identifiers.

```
# Select all participants with significant results
significant <- data_est_person |> filter(p.value < 0.05)
# Select a few (10) participants with non-significant results
nonsignificant <- data_est_person |> filter(p.value >= 0.05) |> head(10)
# Combine the IDs of the selected participants for visualization
ids <- rbind(significant, nonsignificant) |> arrange(desc(estimate)) |> pull(PARTICIPANT)
```

To visualize the regression lines for the sampled students, we will focus on plotting the relationships between task value and happiness for each of the selected participants. With this, we showcase the person-specific variability in how task value impacts happiness, showing both significant and non-significant effects from our diverse sample.

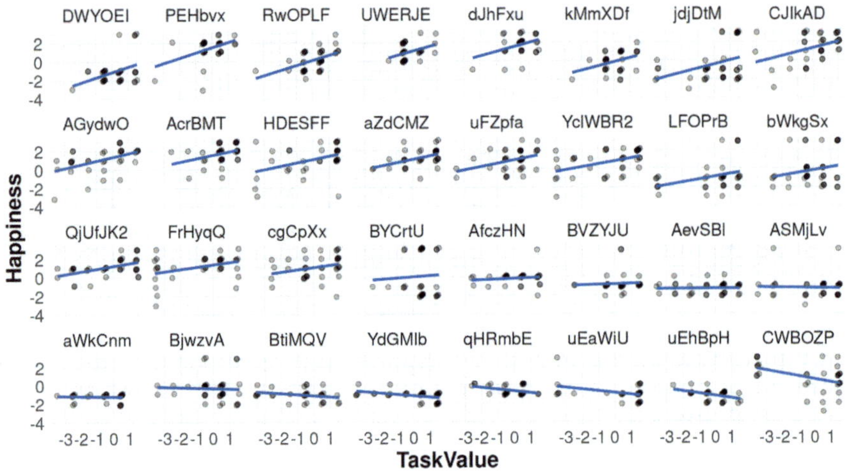

Fig. 6 Each student's individual regression

We will depart from our detrended data (`data_detrended`) and we will keep only those students that we selected in the previous step (`ids`). We will create a new column named `Person` that contains the participant identifier converted into a factor and ordered by the variable `ids`, which contains the participant identifiers ordered by the regression estimates in reverse order (from larger to smaller). This will allow us to plot the participants in this order.

```
# Get only students in the selected sample
data_in <- data_detrended |>
  filter(PARTICIPANT %in% ids) |> # Filter the selected participants
  # Convert participant ID to factor for plotting
  mutate(Person = factor(PARTICIPANT, levels = ids))
```

Now, we can proceed to plot the participants. We use `ggplot` to create the plot, `geom_jitter` to plot the data points as dots, `stat_smooth` to fit the regression line, and `facet_wrap` to create an individual subplot for each participant (Fig. 6).

```
# Plot individual regression lines for the selected participants
ggplot(data_in, aes(x = TaskValue, y = Happiness, group = Person)) +
  geom_jitter(alpha = 0.35) + # Add dots
  geom_smooth(method = "lm", fill = NA) +  # Add regression lines
  facet_wrap("Person",  ncol = 8, scales = "fixed")  # Create individual plots
```

This plot provides a clear visual representation of the individual differences in the relationship between task value and happiness. For some participants (towards the beginning of the plot), the slope is positive and significant, indicating a strong relationship between task value and happiness, while for others (towards the bottom of the plot), the slope may be close to zero or even negative, suggesting a weaker or non-existent relationship. This visualization helps capture the heterogeneity in how task value affects happiness across individuals. Chapter 22 in the book [44] presents more advanced methods for automating the generation of person-specific predictive models.

4 Discussion

In this chapter, we have described three different analytical approaches: variable-centered, person-centered, and person-specific. We have illustrated different methods to apply each of these lenses. Starting with the variable-centered approach, we used a linear model to estimate the overall relationship between task value and happiness. The results showed a positive relationship, i.e., average task value is associated with higher average happiness.

We then moved on to the person-centered approach to investigate whether there was some heterogeneity in our data that could be hidden by using a variable-centered approach. Namely, we fitted a mixture of regressions model, which aims to detect whether not one single population but rather different subpopulations exist within the data, where each subgroup can have their own (linear) relationships between task value and happiness. When setting the number of models to 2, we saw that there are two opposing groups: a larger group with a positive relationship between task value and happiness, and a smaller group with a negative relationship. Lastly, in the person-specific approach, we fitted a simple linear model to each participant's dataset. When inspecting each individual student's within-person model we saw the complete range of estimates and therefore the full range of heterogeneity of our dataset, which went from participants with a strong positive relationship between task value and happiness, to the stark opposite.

While multi-level models take into account individual differences (see the last part of Sect. 3.2), they essentially focus on group-level inferences and impose shared assumptions across individuals. The resulting coefficients are still pooled towards the group mean, which does not capture idiographic processes. The pooling towards the group mean may provide stability, but at the cost of individual differences. Pooling individual estimates towards the group mean obscures the unique patterns or processes. Thus, individuals whose data deviate significantly from the group average will see their estimates pulled towards the group mean masking their unique variability.

The differences among each of these approaches point to the need for careful consideration when analyzing students' data. It becomes apparent that making inferences at the group-level does not generalize at the individual level. Moreover, the heterogeneity of the data should be considered to make sure that all individual needs and preferences are taken into account and not hidden by the general trend that represents most students. It is precisely the ones who are not following the "default" path that need most help and support [45], so the analytical lenses that we use in research and practice need to support this principle.

In conclusion, variable-centered approaches are appropriate when we have a single measure for each student, when we do not expect this measure to fluctuate in the short term, and when we do not expect—or cannot afford to account for—great heterogeneity. We rely on variable-centered analysis when we wish to extract general trends or what works for the majority. We can use this approach when creating course-level interventions, for example, whether to extend the deadline of an assignment or introduce a certain type of learning activity. When we expect to find different trends and subgroups, we rely on person-centered methods. This allows us to create interventions for each subgroup. It is of special relevance when for one of the subgroups the intervention that works for most is not only not helpful but harmful. For example, adding gamification elements to a course might be motivating for many, irrelevant for some, and distracting or detrimental for others [46]. Lastly, when we are studying variables that change momentarily, and we have the possibility to conduct intensive data collection to capture this within-person variability, we can resort to person-specific methods to issue precise interventions tailored at each specific individual [47].

References

1. Bryan CJ, Tipton E, Yeager DS (2021) Behavioural science is unlikely to change the world without a heterogeneity revolution. Nat Hum Behav 5:980–989. https://doi.org/10.1038/s41562-021-01143-3
2. Fisher AJ, Medaglia JD, Jeronimus BF (2018) Lack of group-to-individual generalizability is a threat to human subjects research. Proc Natl Acad Sci U S A 115:E6106–E6115. https://doi.org/10.1073/pnas.1711978115
3. Howard MC, Hoffman ME (2018) Variable-centered, person-centered, and person-specific approaches: where theory meets the method. Organ Res Methods 21:846–876. https://doi.org/10.1177/1094428117744021
4. Meyer JP, Morin AJS (2016) A person-centered approach to commitment research: theory, research, and methodology. J Organ Behav 37:584–612. https://doi.org/10.1002/job.2085
5. Saqr M, Vogelsmeier LVDE, López-Pernas S (2024) Capturing where the learning process takes place: a person-specific and person-centered primer. Learn Indiv Differ 113:102492. https://doi.org/10.1016/j.lindif.2024.102492
6. Saqr M, López-Pernas S (2021) Idiographic learning analytics: a definition and a case study. In: Proceedings of the 2021 international conference on advanced learning technologies (ICALT). ieeexplore.ieee.org. https://doi.org/10.1109/ICALT52272.2021.00056
7. Saqr M, Schreuder MJ, López-Pernas S (2024) Why educational research needs a complex system revolution that embraces individual differences, heterogeneity, and uncertainty. Learn Anal Methods Tutor 723
8. Saqr M, Cheng R, López-Pernas S, Beck ED (2024) Idiographic artificial intelligence to explain students' self-regulation: toward precision education. Learn Indiv Differ 114:102499. https://doi.org/10.1016/j.lindif.2024.102499

9. Saqr M (2023) Modelling within-person idiographic variance could help explain and individualize learning. Br J Educ Technol 54:1077–1094. https://doi.org/10.1111/bjet.13309
10. Saqr M, López-Pernas S (2024) Mapping the self in self-regulation using complex dynamic systems approach. Br J Educ Technol. https://doi.org/10.1111/bjet.13452
11. Beltz AM, Wright AGC, Sprague BN, Molenaar PCM (2016) Bridging the nomothetic and idiographic approaches to the analysis of clinical data. Assessment 23:447–458. https://doi.org/10.1177/1073191116648209
12. Collie RJ, Malmberg L-E, Martin AJ, Sammons P, Morin AJS (2020) A multilevel person-centered examination of teachers' workplace demands and resources: links with work-related well-being. Front Psychol 11:626. https://doi.org/10.3389/fpsyg.2020.00626
13. Liu M, Lin T-I (2014) A skew-normal mixture regression model. Educ Psychol Meas 74:139–162. https://doi.org/10.1177/0013164413498603
14. Van Horn ML, Jaki T, Masyn K, Howe G, Feaster DJ, Lamont AE, George MRW, Kim M (2015) Evaluating differential effects using regression interactions and regression mixture models. Educ Psychol Meas 75:677–714. https://doi.org/10.1177/0013164414554931
15. Kleimola R, López-Pernas S, Väisänen S, Saqr M, Sointu E, Hirsto L (2023) Learning analytics to explore the motivational profiles of non-traditional practical nurse students: a mixed-methods approach. Empir Res Vocat Educ Train 15. https://doi.org/10.1186/s40461-023-00150-0
16. Scrucca L, Saqr M, López-Pernas S, Murphy K (2024) An introduction and R tutorial to model-based clustering in education via latent profile analysis. In: Learning analytics methods and tutorials. Springer Nature Switzerland, Cham, pp 285–317. https://doi.org/10.1007/978-3-031-54464-4/_9
17. Ning HK (2016) Examining heterogeneity in student metacognition: a factor mixture analysis. Learn Indiv Differ 49:373–377. https://doi.org/10.1016/j.lindif.2016.06.004
18. Trautner M, Schwinger M (2018) Differentiation of academic self-concept in primary school students with mild learning difficulties: a factor mixture analysis approach. Learn Indiv Differ 65:20–29. https://doi.org/10.1016/j.lindif.2018.05.004
19. Hodis FA, Meyer LH, McClure J, Weir KF, Walkey FH (2011) A longitudinal investigation of motivation and secondary school achievement using growth mixture modeling. J Educ Psychol 103:312–323. https://doi.org/10.1037/a0022547
20. Gillet N, Morin AJS, Reeve J (2017) Stability, change, and implications of students' motivation profiles: a latent transition analysis. Contemp Educ Psychol 51:222–239. https://doi.org/10.1016/j.cedpsych.2017.08.006
21. Vogelsmeier LVDE, Vermunt JK, Roekel E van, De Roover K (2019) Latent markov factor analysis for exploring measurement model changes in time-intensive longitudinal studies. Struct Equ Model 26:557–575. https://doi.org/10.1080/10705511.2018.1554445
22. Molenaar PCM, Campbell CG (2009) The new person-specific paradigm in psychology. Curr Dir Psychol Sci 18:112–117. https://doi.org/10.1111/j.1467-8721.2009.01619.x
23. Winne PH (2017) Leveraging big data to help each learner and accelerate learning science. Teach Coll Rec 119:1–24. https://doi.org/10.1177/016146811711900305
24. Molenaar PCM, Sinclair KO, Rovine MJ, Ram N, Corneal SE (2009) Analyzing developmental processes on an individual level using nonstationary time series modeling. Dev Psychol 45:260–271. https://doi.org/10.1037/a0014170
25. Saqr M, López-Pernas S (2021) Idiographic learning analytics: a single student (n= 1) approach using psychological networks. http://ceur-ws.org/Vol-2868/article_4.pdf
26. Casuso-Holgado MJ, Cuesta-Vargas AI, Moreno-Morales N, Labajos-Manzanares MT, Barón-López FJ, Vega-Cuesta M (2013) The association between academic engagement and achievement in health sciences students. BMC Med Educ 13. https://doi.org/10.1186/1472-6920-13-33
27. Saqr M, López-Pernas S, Vogelsmeier LVDE (2023) When, how and for whom changes in engagement happen: a transition analysis of instructional variables. Comput Educ 207:104934. https://doi.org/10.1016/j.compedu.2023.104934

28. Sointu E, Valtonen T, Hallberg S, Kankaanpää J, Väisänen S, Heikkinen L, Saqr M, Tuominen V, Hirsto L (2022) Learning analytics and flipped learning in online teaching for supporting preservice teachers' learning of quantitative research methods. Semin Net 18. https://doi.org/10.7577/seminar.4686
29. Ali L, Asadi M, Gašević D, Jovanović J, Hatala M (2013) Factors influencing beliefs for adoption of a learning analytics tool: an empirical study. Comput Educ 62:130–148. https://doi.org/10.1016/j.compedu.2012.10.023
30. Saqr M, Alamro A (2019) The role of social network analysis as a learning analytics tool in online problem based learning. BMC Med Educ 19:160. https://doi.org/10.1186/s12909-019-1599-6
31. Conijn R, Snijders C, Kleingeld A, Matzat U (2017) Predicting student performance from LMS data: a comparison of 17 blended courses using moodle LMS. IEEE Trans Learn Technol 10:17–29. https://doi.org/10.1109/TLT.2016.2616312
32. Hernández-García Á, Acquila-Natale E, Chaparro-Peláez J, Conde MÁ (2018) Predicting teamwork group assessment using log data-based learning analytics. Comput Human Behav 89:373–384. https://doi.org/10.1016/j.chb.2018.07.016
33. Fincham E, Whitelock-Wainwright A, Kovanović V, Joksimović S, Staalduinen J-P van, Gašević D (2019) Counting clicks is not enough: Validating a theorized model of engagement in learning analytics. In: Proceedings of the 9th international conference on learning analytics & knowledge. Association for Computing Machinery, New York, NY, USA, pp 501–510. https://doi.org/10.1145/3303772.3303775
34. López-Pernas S, Barra E, Gordillo A, Alonso Á, Quemada J (2023) Scaling student feedback in software engineering massive open online courses. IEEE Softw 40:50–57. https://doi.org/10.1109/ms.2023.3275035
35. Saqr M, López-Pernas S, Helske S, Hrastinski S (2023) The longitudinal association between engagement and achievement varies by time, students' profiles, and achievement state: a full program study. Comput Educ 199:104787. https://doi.org/10.1016/j.compedu.2023.104787
36. Tikka S, Kopra J, Heinäniemi M, López-Pernas S, Saqr M (2024) Getting started with R for education research. In: Saqr M, López-Pernas S (eds) Learning analytics methods and tutorials: a practical guide using r. Springer, Berlin
37. Kopra J, Tikka S, Heinäniemi M, López-Pernas S, Saqr M (2024) An R approach to data cleaning and wrangling for education research. In: Saqr M, López-Pernas S (eds) Learning analytics methods and tutorials: a practical guide using r. Springer, Berlin
38. Tikka S, Kopra J, Heinäniemi M, López-Pernas S, Saqr M (2024) Introductory statistics with R for educational researchers. In: Saqr M, López-Pernas S (eds) Learning analytics methods and tutorials: a practical guide using r. Springer, Berlin
39. López-Pernas S, Misiejuk K, Tikka S, Kopra J, Heinäniemi M, Saqr M (2024) Visualizing and reporting educational data with R. In: Learning analytics methods and tutorials. Springer Nature Switzerland, Cham, pp 151–194. https://doi.org/10.1007/978-3-031-54464-4/_6
40. Grund A, Fries S, Senker K (2022) Primärdaten einer experience-sampling studie zu dem zusammenhang von selbstregulation und dem alltagserleben studierender. https://doi.org/10.5160/PSYCHDATA.GDAL18PR18
41. Tikka S, Kopra J, Heinäniemi M, López-Pernas S, Saqr M (2024) Introductory statistics with R for educational researchers. In: Learning analytics methods and tutorials. Springer Nature Switzerland, Cham, pp 121–150. https://doi.org/10.1007/978-3-031-54464-4/_5
42. Bates D, Mächler M, Bolker B, Walker S (2015) Fitting linear mixed-effects models using lme4. J Stat Softw 67:1–48. https://doi.org/10.18637/jss.v067.i01
43. Beck ED, Jackson JJ (2021) Within-person variability. In: The handbook of personality dynamics and processes. Elsevier, Amsterdam, pp 75–100. https://doi.org/10.1016/b978-0-12-813995-0.00004-2
44. Saqr M, Tlili A, López-Pernas S (2025) Automating individualized machine learning and AI prediction using AutoML: the case of idiographic predictions. In: Saqr M, López-Pernas S (eds) Advanced learning analytics methods: AI, precision and complexity. Springer Nature Switzerland, Cham

45. Saqr M, López-Pernas S (2021) The dire cost of early disengagement: A four-year learning analytics study over a full program. In: Technology-enhanced learning for a free, safe, and sustainable world. Springer International Publishing, Cham, pp 122–136. https://doi.org/10.1007/978-3-030-86436-1/_10
46. Smiderle R, Rigo SJ, Marques LB, Peçanha de Miranda Coelho JA, Jaques PA (2020) The impact of gamification on students' learning, engagement and behavior based on their personality traits. Smart Learn Environ 7:1–11. https://doi.org/10.1186/s40561-019-0098-x
47. Saqr M, Ito H, López-Pernas S (2025) A comprehensive introduction to idiographic and within-person analytics. In: Saqr M, López-Pernas S (eds) Advanced learning analytics methods: AI, precision and complexity. Springer Nature Switzerland, Cham

Idiographic Networks: A Tutorial on Graphical Vector Autoregression and Unified Structural Equation Modeling

Mohammed Saqr, Daryn Dever, and Sonsoles López-Pernas

Abstract This tutorial introduces the application of advanced network analysis methods, specifically Graphical Vector Autoregression (`graphicalVAR`) and Unified Structural Equation Modeling (`uSEM`), to model learning processes as complex, dynamic systems. These approaches allow exploring both temporal and contemporaneous relationships among variables within individual learners over time. The chapter begins by conceptualizing learning as a networked system and reviewing relevant literature, discussing the advantages of probabilistic network models in education. Then, a step-by-step tutorial in the R programming language is presented so readers can learn to estimate idiographic models, visualize dynamic relationships, and interpret. As such, this tutorial aims to provide researchers with tools to analyze multivariate time-series data, which is a necessary step for truly personalized interventions in educational research

Keywords Learning analytics · Network analysis · Idiographic analysis · Complex systems · Graphical vector autoregression · Unified structural equation modeling

1 Introduction

Network analysis provides a suite of methods for examining both static and dynamic relationships within systems. These methods reveal insights into collaborations between students or student-teacher interactions, knowledge construction on a topic, social dynamics between groups, and system-wide properties, including how these connections evolve over time [1–3]. Applications of network analysis span diverse

M. Saqr (✉) · S. López-Pernas
University of Eastern Finland, Kuopio, Finland
e-mail: mohammed.saqr@uef.fi

D. Dever
University of Florida, Gainesville, FL, USA

M. Saqr, S. López-Pernas (eds.), *Advanced Learning Analytics Methods*,
https://doi.org/10.1007/978-3-031-95365-1_20

contexts, from understanding how groups of students learn on massive online platforms [4] to identifying how a single learner evolves over time [5]. Representing elements of learning processes—such as cognitive and social aspects—within a network has become a well-established research method, encompassing various analytical approaches, including social network analysis (SNA), epistemic network analysis (ENA), ordered network analysis (ONA), and temporal network analysis (TNA) [3, 6, 7]. More recently, transition network analysis (TNA) has also emerged [7].

While much of the prior work in network analysis focuses on group-level phenomena, this chapter takes an idiographic perspective, where the primary unit of analysis is the individual learner [8]. The idiographic approach diverges from traditional nomothetic methods by emphasizing the unique patterns, processes, and dynamics that characterize a single person, thereby offering a more granular understanding of learning as a personal and evolving process [9, 10]. This focus aligns with a growing recognition of the importance of personalized education, where interventions and analyses are tailored to the distinctive needs and experiences of individuals [9, 10].

This chapter focuses on a specific type of networks: probabilistic networks. These networks represent variables as nodes and the statistical relationships between them (e.g., correlations or regression) as edges. Probabilistic networks, discussed in detail in the previous book [11], have been increasingly employed to model processes as they emerge, progress, or unfold, with their elements represented as nodes [12, 13]. These networks allow researchers to understand interactions and dynamics within a process.

Here, we adopt an idiographic perspective, treating the student as the central unit of analysis. This approach examines data from a single student, continuing our exploration of individualized processes, as detailed in the chapter on idiographic machine learning [14]. In this chapter, we demonstrate how to use Graphical Vector Autoregression (GVAR) and Unified Structural Equation Modeling (uSEM) to analyze an individual's process. We begin with a background section that defines key terms and concepts and provides a concise literature review. Following this, we provide a step-by-step tutorial using R to implement the discussed techniques.

2 Background

2.1 *The Cognitive Process as a Networked System*

Representing elements of the cognitive and social processes as a network is an established research method. Such representation has afforded researchers a way to visualize the structure of these processes and to measure the magnitude of association between their elements, and to devise statistical indices that allow a precise interpretation of the resulting graphs [15]. Recent advances in network science have seen a surge in probabilistic network models, also referred to as GGMs, which are designed to examine the relationships of multivariate data or elements of a cognitive or psychological [15]. GGMs are applied within the literature to map and visualize the dynamic relationships between processes and their elements, considering them as a complex system [15–19]. GGMs commonly estimate a partial correlation between variables while controlling for the influence of all other variables and their relationships within the system. Within these networks, two nodes are connected if the covariance of the edge cannot be explained by any other variable in the network, eliminating the possibility of confounding variables [15, 16]. Resulting outputs of these models show only significant relationships between nodes, being the weight of the significant edges proportional to the strength of the relationships, as well as whether the relationship is positive or negative, and mediation pathways (i.e., indirect relationships between two variables transmitted through one or more intermediary variables within the same network) [15].

Vector autoregression (VAR) is an extension of GGM method that has enabled the capability for modeling temporal processes, i.e., how a variable predicts another in the next time window. VAR estimates a directed network (in contrast to undirected in GGM); the nodes are variables (e.g., motivation, behavior or attitude) and the link between them are temporal relationships (a variable predicts another in the next time window) [15].

The abundance of intensive time-stamped data (e.g., time-series data) has led to the existence of enough observations from individual subjects across time using e.g., data collected with experience sampling methods, observational data and physiological data, which enabled the study of an individual as a unique case (N=1) [15, 20]. The representation and estimation of the temporal network allows researchers to study the individual phenomena, the progression of behavior, predict future behavior, as well as create relevant intervention. In other words, the network allows the *idiographic* assessment of the person dispositions and their temporal dynamics. The network is commonly represented by drawing an arrow from the node that represents the variable (e.g., motivation) to the variable that it predicts in the next time window of measurement (e.g., engagement) [21].

To explain it, an example is presented in Fig. 1. We created a simulated dataset about working and achievement within an individual on a daily basis. The graph shows that motivation predicts work, as well as feeling of achievement of goals within the next day. Similarly, working predicts feeling of goal-achievement. However, engaging with work predicts slight stress the next day, and having stress negatively predicts feeling of achievement. Helping that individual could be done through offering stress management advice. Another type of network is the *contemporaneous network (associative)*, a partial correlation network that maps the correlation between the elements of the studied phenomena within the same time window. For instance, when the subject is having comfort, he/she also eats snacks at the same time. In our study this is used to study the co-temporal association between daily events, for instance, how motivation predicts working on the task within the same day.

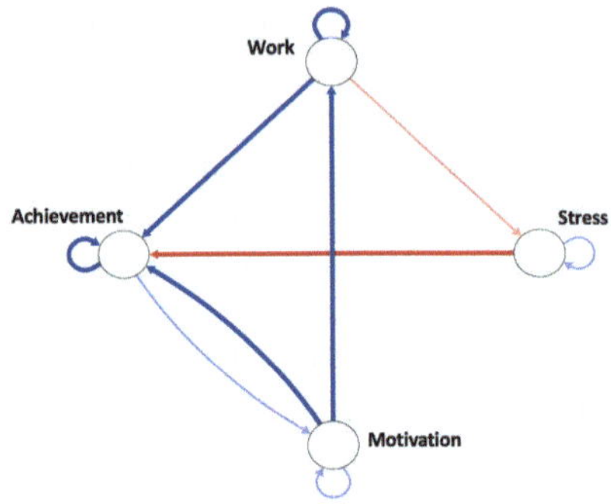

Fig. 1 A fictional temporal network of four constructs. The circles are variables. Blue lines are positive partial correlations. The thickness of the line is proportional to the magnitude of the correlation. The direction of the arrow points to the direction of the temporal correlation. Extracted from [22]

3 Review of Network Analysis Applications in Education Literature

Probabilistic network models, such as GGMs, have become an essential tool for examining how complex learning processes unfold and how the interconnectedness of psychological constructs contributes to enhanced learning outcomes. For instance, Malmberg et al. [22] explored how students monitor events and demonstrated that this monitoring facilitates functional regulatory behaviors during collaborative physics tasks. Their findings indicated that cycles of regulation during collaborative learning do not always coexist, revealing inherent limitations in group-level regulation. Moreover, they highlighted the temporal interplay between motivation and task monitoring, showing that motivation persists across different regulatory phases and significantly predicts how tasks are enacted.

Similarly, Zhou and Kang [23] employed network analysis to investigate collaborative problem-solving behaviors in teams during an astronomy simulation. Using multivariate autoregression, they identified temporal patterns in problem-solving behaviors, revealing how individual exploration within the simulation catalyzed group discussions and joint engagement. Their findings underscored the importance of individual reflections in driving group coordination and the construction of shared knowledge.

Closer to the methods discussed in this tutorial, vector autoregression (VAR) networks have been applied to examine students' self-regulation. Saqr [21] demonstrated that self-regulation profiles differ significantly across individuals, with each student exhibiting a unique regulatory pattern. Importantly, none of the students conformed to the aggregate patterns observed when data were analyzed at the group level. This underscores the critical need for idiographic approaches to capture the distinctive dynamics of each learner, challenging traditional nomothetic methodologies [24].

Additional research has used network analysis to uncover how students regulate cognitive and emotional processes, both individually and in groups, across diverse learning contexts, such as language acquisition and physics education [5, 25–31].

These studies establish network analysis as a powerful and reliable framework for identifying how learners' regulatory strategies influence motivation, engagement, and ultimately, learning outcomes.

3.1 Loading Data and Pre-processing

Before proceeding to the actual analysis, we will need to load the packages, import the data, and process it to prepare for the analysis. Below are the necessary packages:

- `tidyverse`: A renowned R package for data manipulation, wrangling and visualization [32].
- `skimr`: A package to perform summary statistics and help us get an overview of the dataset and in particular the variable distributions, any missing data, and basic descriptive statistics [33].
- `graphicalVAR`: An R package which will be used for modeling and visualizing the `graphicalVAR` models as described before [34].
- `qgraph`: An R package for network visualization [35].
- `pompom`: An R package that offers an implementation of unified structural equation modeling [36].
- `caret`: Short for Classification And REgression Training, `caret` is a set of functions that attempt to simplify the creation and evaluation of predictive models [37].
- `rio`: A package for importing data in several formats [38].

In addition, we will import an additional R file with helper functions that we will need throughout the chapter (`aux.R`).

Once we have loaded the necessary R packages and dependencies, we import the dataset `synthetic_Data_share.RDS` (step 1). This dataset contains ESM measurements of 36 students that reported how they regulate their own learning activities including aspects like planning, monitoring their progress, seeking help, managing their environment, and reflecting on feedback. We then create a vector for the variables of interest (`Vars`), which includes nine SRL processes that will be the target of our analysis (step 2).

In the next step (step 3), we create a `Day` variable for each student (by grouping by the `name` variable) using `seq_along()` to create sequential numbers for each time-point in the data for each individual. Next (step 4), given that we would like to analyze a single student, we will filter the data of a student, in our case, we chose Grace for analysis. In this step we also select only the specific SRL-related variables that we would like to analyze, defined by `Vars`, along with the new **Day** variable, preparing the filtered data for further analysis specific to Grace's behaviors.

```
# Step 1: Load the dataset
df <- import("https://github.com/lamethods/data2/raw/main/srl/srl.RDS")

# Step 2: Define variables for the graphical VAR model
Vars <- c("planning", "monitoring", "effort", "control",
          "help", "social", "organizing", "feedback", "evaluating")

# Step 3: Create a "Day" variable for each entry within a person
df <- df |>
  group_by(name) |>
  mutate(Day = seq_along(name)) |>
  ungroup()

# Step 4: Filter the data for a specific individual named 'Grace'
Grace_Data <- df |>
  filter(name == "Grace") |>
  dplyr::select(all_of(Vars), Day)
```

3.1.1 Exploring the Data

It is always a good practice to explore the data to judge the distribution of variables, missing data or any unexpected surprises in the data. We do so using the `skimr` package, which provides a detailed overview of the SRL dataset, including measures like mean, median, spread, and the presence of missing values. To do so, we simply use `skim(Grace_Data)` (Table 1).

```
Grace_descriptives <- skim(Grace_Data)
print(Grace_descriptives)
```

The next chunk of code visualizes the data and, since it is a longitudinal repeated measure dataset, we plot the data across time. For that purpose, we reshape it into a long format using `pivot_longer` to organize it so that the SRL variables (e.g., planning, monitoring) are represented in a single column (`variable`), and their corresponding values in another column (`value`). We then use `ggplot2`, where each SRL variable is plotted on the y-axis against time (`Day`) is plotted on the x-axis. We apply faceting so that each SRL construct is visualized in its own subplot (Fig. 2).

Table 1 Summary of the dataset created using `skimr`

skim_type	skim_variable	n_missing	complete_rate	numeric.mean	numeric.sd	numeric.p0	numeric.p25	numeric.p50	numeric.p75	numeric.p100	numeric.hist
Numeric	Planning	0	1	61.68	20.37	0	54.84	64.52	74.19	100	
Numeric	Monitoring	0	1	63.69	17.74	0	50.75	68.66	76.12	100	
Numeric	Effort	0	1	50.64	21.87	0	36.00	52.00	68.00	100	
Numeric	Control	0	1	43.61	25.97	0	22.58	38.71	61.29	100	
Numeric	Help	0	1	75.77	16.36	25	70.00	80.00	87.50	100	
Numeric	Social	0	1	45.38	18.29	0	31.00	48.00	56.00	100	
Numeric	Organizing	0	1	62.68	16.46	0	48.15	66.67	74.07	100	
Numeric	Feedback	0	1	60.79	24.37	0	42.14	62.86	80.71	100	
Numeric	Evaluating	0	1	46.53	18.65	0	39.13	45.65	52.72	100	
Numeric	Day	0	1	78.50	45.18	1	39.75	78.50	117.25	156	

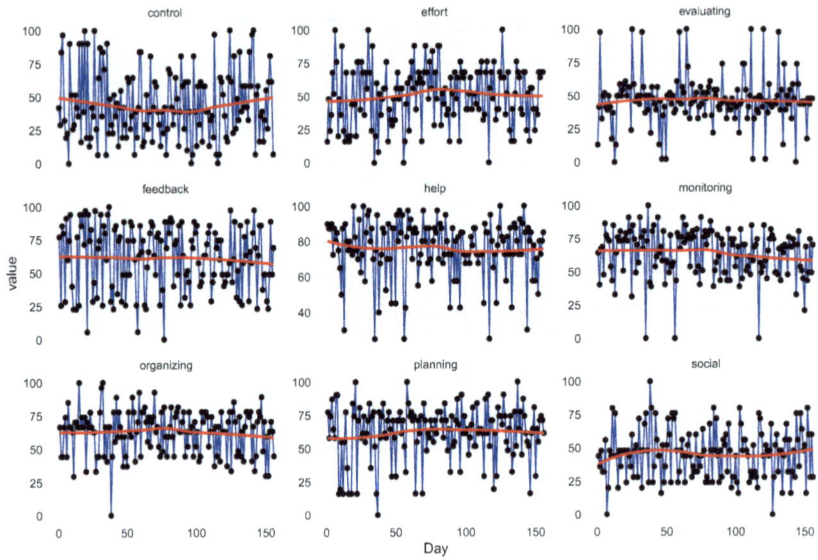

Fig. 2 Time series of variables with Smoothing

```
# Reshape the data to long format
Grace_data_long <- Grace_Data |>
  pivot_longer(cols = Vars, # Assuming 'id' and 'interaction' are the relevant columns
               names_to = "variable",
               values_to = "value")

# Plot with faceting by variable
ggplot(Grace_data_long, aes(x = Day, y = value)) +
  geom_line(color = "blue") +
  geom_point() +
  geom_smooth(method = "loess", color = "red", se = FALSE) + # Smoothing (LOESS)
  facet_wrap(~variable, scales = "free_y", ncol = 3) +
  theme_minimal()
```

3.1.2 Stationarity

Stationarity is an important pre-requisite for `graphicalVAR` models which means that the data statistical properties,—e.g., mean and variance—do not change over time. Non-stationary data, which exhibits trends or other systematic temporal patterns may lead to spurious results and misleading interpretations. Therefore, we need to examine the presence of trends, and if present, these trends need to be removed (de-trended) with appropriate techniques. De-trending improves the model's performance and the validity of its inferences and ensures that the relationships among variables are not artificially inflated or deflated by overarching trends. This is particularly important in graphicalVAR models where we focus is on understanding the network of dynamic relationships among variables. The next function detrender (defined in the file aux.R imported earlier) examines the presence of trends, and prints the output. If there are trends in the data, the functions performs detrending. As we can see the data has no trends and so we will proceed with the data without further modifications.

```
detrender(Grace_Data, vars = Vars, timevar = "Day")
```

```
No significant trend for planning - p-value: 0.305218598633039

No significant trend for monitoring - p-value: 0.0805897468195872

No significant trend for effort - p-value: 0.429465303760844

No significant trend for control - p-value: 0.972736537512727

No significant trend for help - p-value: 0.452982605134614

No significant trend for social - p-value: 0.925685121385736

No significant trend for organizing - p-value: 0.460175358168101

No significant trend for feedback - p-value: 0.456419063913471

No significant trend for evaluating - p-value: 0.993005264585157
```

3.2 *Estimation of the* graphicalVAR *Model*

Fitting the graphicalVAR model is straightforward, for that we use the graphicalVAR() function with the data argument set to the dataset of Grace data = Grace_Data. The beepvar argument specifies the time variable in the data (the temporal sequence of our observation). In our case, we use beepvar = "Day", which refers to the day-to-day sequential measurements of Grace's behaviors. We set the lambda_beta argument to 0.1 to control the regularization. A value of 0.1 is rather moderate given that we have a relatively large dataset with 156 measurements. Regularization as we mentioned before would shrink small edges, help avoid overfitting and makes the model more interpretable. Other optional arguments, like verbose, can be set to TRUE if you want to display more detailed output during the fitting process, but it defaults to FALSE to suppress extra information.

The output of the summary() function of the model of Grace provides some useful information about the data and generated networks. As the results show, the model include 9 nodes corresponding to the SRL variables (e.g., planning, monitoring, effort, etc.). The Extended Bayesian Information Criterion (EBIC) hyperparameter was at the default value of 0.5 to control the model complexity and help avoid over-fitting. The optimal EBIC score achieved was 1329.994, indicating the best trade-off between fit and complexity. The Partial Contemporaneous Correlations (PCC) network had 11 non-zero correlations which means the network was relatively sparse where about 69% of the possible edges being zero as indicated by the PCC sparsity of 0.694. The algorithm tested 50 different values for the regularization parameter to find the optimal level. The Partial Directed Correlations (PDC) or the temporal network had 9 non-zero directed edges meaning that the network was rather sparse where PDC sparsity was 0.75, meaning 75% of potential temporal connections between variables were zero. The summary tells us that both the contemporaneous and temporal networks are stored within the object inside $PCC and $PDC, objects respectively.

```
set.seed(265) # for replicability

# Fit the `graphicalVAR` model to the data from 'Grace' with 'lambda_beta'
# to 0.1 to apply moderate regularization
GraphVAR_Grace <- graphicalVAR(Grace_Data, beepvar = "Day", lambda_beta = 0.1)

# Output a summary of the fitted model to review the model's structure,
# including estimated parameters and fit indices.
summary(GraphVAR_Grace)
```

```
=== graphicalVAR results ===
Number of nodes: 9
Number of tuning parameters tested: 50
EBIC hyperparameter: 0.5
Optimal EBIC: 1329.994

Number of non-zero Partial Contemporaneous Correlations (PCC): 11
PCC Sparsity: 0.6944444
Number of PCC tuning parameters tested: 50
PCC network stored in object$PCC

Number of non-zero Partial Directed Correlations (PDC): 9
PDC Sparsity: 0.75
Number of PDC tuning parameters tested: 1
PDC network stored in object$PDC

Use plot(object) to plot the estimated networks.
```

3.3 *Visualization and Interpretation of the* `graphicalVAR` *Model*

To visualize the results of Grace, we can use the built-in `plot` function with the required network argument. In that, we use the argument `include = "PCC"` , or “`PCC"`, to get the Partial Contemporaneous Correlations network or simply the contemporaneous network. We also use the argument `include="PDC"` argument to retrieve the Partial Directed Correlations network, or simply, the temporal network.

In **graphicalVAR** contemporaneous plots, and, in fact, in most of such networks, the color of the edges reflects the direction of the correlation where blue (sometimes green) indicates a positive correlation between the two nodes and red indicates a negative correlation. The thickness of the edge indicates the strength of the relationship, with thicker edges indicating stronger associations. In this PCC network, in Fig. 3a, we see a strong positive correlation between monitoring and help suggesting that when Grace is actively monitoring her learning, she tends to seek more help. We also see a negative correlation between social and help indicating that when Grace engages in a social activity, she tends to do less monitoring of her learning. There is also a moderate association between monitoring and effort and feedback and help. These are all indicative of how Grace regulates her learning. Furthermore, in partial correlation networks, the absence of associations is also interpretable, in a way that these processes are independent of each other, e.g., we see no association between effort and control indicating that when Grace is investing good effort in her learning, she may not control her environment and distractions.

The visualization of the temporal network or the Partial Directed Correlations (PDC) network in technical terms show the temporal (lagged) relationships between self-regulated learning (SRL) behaviors. The arrows between the nodes represent

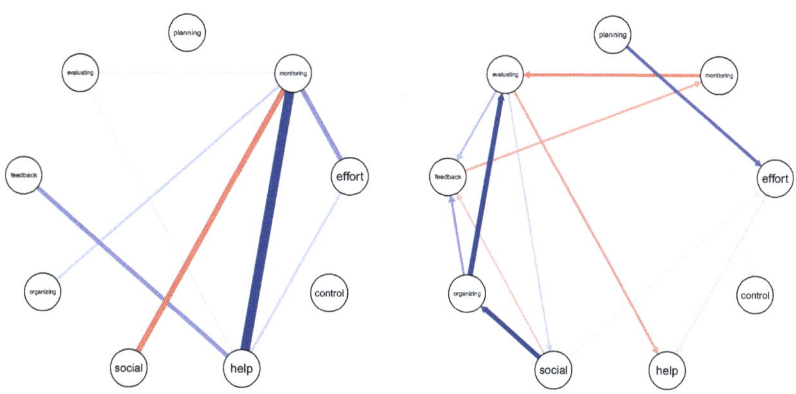

(a) (b)

Fig. 3 Graphical VAR plots. (**a**) Partial Contemporaneous Correlations (PCC). (**b**) Partial Directed Correlations (PDC) how one behavior at time t predicts another at time t+1. The color and thickness of the arrows provide information on the direction and strength of these lagged relationships in the same way as the contemporaneous network. As we can see in the figure, there is a positive relationship between planning and effort, meaning that when Grace engages in planning at one day, she is more likely to put in more effort in the next day. In the same way, we also see a strong link between organizing and evaluating indicating that time management (organizing) leads to evaluation of progress in the next day. Interestingly, we see that social is followed usually by time management (organizing) which makes sense, that when Grace socializes, she tends to catch up and organizes her duties next day. Negative links in temporal networks mean that behaviors are less likely to happen after each other. We see for instance, that evaluating is less likely to be followed by seeking help and monitoring is less likely to be followed by evaluating.

Overall, the network is informative and shows show that Grace regulates her learning. She puts her plans into effect by investing effort as well as her time management skills are a driver for evaluation and implementing feedback.

```r
# Visualize the Partial Contemporaneous Correlations (PCC) within the model
# using a circular layout.
plot(GraphVAR_Grace, include = "PCC", layout = "circle",
     theme = "colorblind", labels = Vars, titles = F)

# Similarly, visualize the Partial Directed Correlations (PDC), highlighting
# temporal directed influences between variables.
plot(GraphVAR_Grace, include = "PDC", layout = "circle",
     theme = "colorblind", labels = Vars, titles = F)
```

Given that the built-in plotting function is very basic, we can augment the plot with the `plotting` function as below. The `plotting` function creates a nicer visualization and shows the edge weights to allow judging the actual edge weights (Fig. 4).

```r
plotting <- function(Network, title = "", ...){
    qgraph(Network,
     layout = "circle",                 # Circular node arrangement
     title = title,                     # Setting a title
     loop = 0.7,                        # Curvature for self-loops
     node.width = 2,                    # Node width
     repulsion = 0.8,                   # Spacing between nodes
     label.font = 1,                    # Label font styling
     label.fill.vertical = 1,
     label.fill.horizontal = 1,
     esize = 7,                         # Edge thickness
     vsize = 6,                         # Node size
```

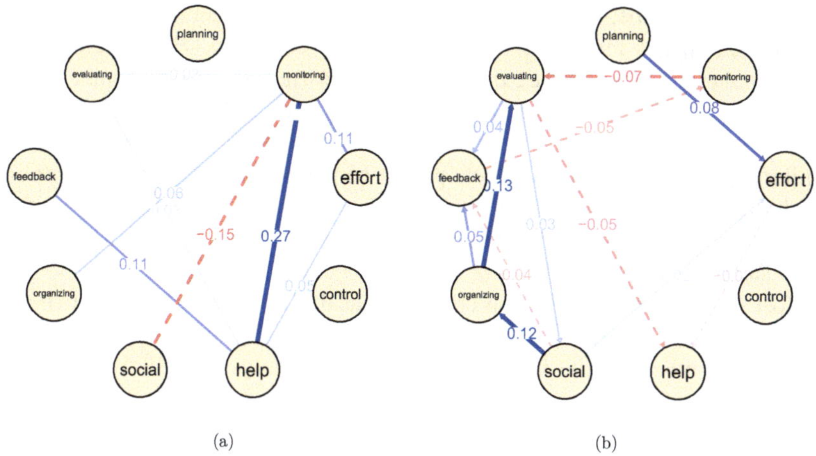

Fig. 4 Plotting the networks nicely. (**a**) Partial Contemporaneous Correlations (PCC). (**b**) Partial Directed Correlations (PDC)

```
        color = "#ffeece",          # Light pink node color
        edge.label.cex = 1.5,       # Edge label size
        edge.width = 1,             # Edge width
        border.width = 2,           # Border width for nodes
        edge.labels = TRUE,         # Show edge labels
        asize = 3,                  # Arrow size
        labels = Vars,              # Variable labels,
        negDashed = TRUE,           # Negative edges dashed
        theme = "colorblind", ...   )}

plotting(GraphVAR_Grace$PCC)
plotting(GraphVAR_Grace$PDC)
```

3.4 *Estimation of Multiple Idiographic Models n>1*

Besides estimating idiographic models, we can use `mlGraphicalVAR` to estimate multiple idiographic models for multiple people. Besides idiographic models, `mlGraphicalVAR` offers an aggregate or an average within-person and between person networks which are kind of approximation of the general group-level picture. As such, the results of `mlGraphicalVAR` contain several components: a between-person network (`betweenNet`), which captures the average stationary relationships between variables across different subjects. The `betweenNet` network is created from the mean values of each person in the data and then creates a partial correlation between the means of each person. A network of correlations between the means represents the stable average dynamics. *Please note that we did not have this betweenNet before, given that we had only one student.* Also, mlGraphicalVAR produces a fixed-effect partial directed network, fixedPDC, which captures the temporal relationships on average across the whole sample. Similarly, it produces a fixed effects partial contemporaneous network fixedPCC which captures the average contemporaneous correlations across all individuals. Think of these networks as a summary of all the networks in the data across the whole sample.

Most importantly, the function creates an analysis for each individual (like the one we did for Grace) in the data by setting the subjectNetworks argument to TRUE. In other words, it produces a specific analysis for Bob, Chen, Eve, etc and every other person in the data. To get a specific subject results, we can query the object for that person. For instance, to get the summary of the model of the first person we can use GraphVAR_all$subjecResults[[1]], to get the contemporaneous network GraphVAR_all$subjectPCC[[i]] and to get the GraphVAR_all$subjectPDC[[i]]. In that way, we can automate the analysis of a large sample and get a view of the average picture as well as detailed individual analysis of each person.

To estimate the mlGraphicalVAR model for multiple people, we use the mlGraphicalVAR function and specify the data df containing the multiple individuals' responses. Here, we need to specify the idvar = "name" and subjectNetworks = TRUE to create the individual analysis.

As we said before, the output of the mlGraphicalVAR model includes several components: The fixedPCC, fixedPDC, the betweenNet as well as the subjectPCC (the subject contemporaneous network) and subjectPDC (the subject temporal network) for each of the 36 subjects in the dataset.

The code below applies the de-trending function to assess any trends in the data for each person and in case they are present, the function de-trends these variables. Then, mlGraphicalVAR is used to estimate the models as explained above, one for each person.

```
detrended <- detrender(data = df, vars = Vars, timevar = "Day",
groupvar = "name")
# Fit the mlgraphicalVAR model to the data from 'all students
 with 'lambda_beta' to 0.1 to apply moderate regularization.
GraphVAR_all <- mlGraphicalVAR(detrended, vars = Vars,
idvar = "name", lambda_beta = 0.01,
subjectNetworks = TRUE, verbose = FALSE)
```

3.4.1 Plotting the `mlGraphicalVAR` Results

`mlGraphicalVAR` produces two types of networks: the general group-level networks, and subject specific networks. Below, we plot the group level networks with the function we created (`plotting`). The interpretation of these networks shows the SRL dynamics on average across our sample or the expected common behavior.

For instance, the between-person network (Fig. 5a) shows that students who use feedback always seek help (0.49). Similarly, we see a strong association between effort and organizing (0.33) as well as control and organizing (0.26) and feedback and evaluating (0.26). There is moderate association between organizing and planning (0.22), and effort and social (0.22), as well as help and effort (0.21), and social and planning (0.21). The rest of associations can be interpreted in the same ways.

In the contemporaneous network (Fig. 5b), we see a snapshot of how self-regulated learning constructs are associated with each other within the same day. We see a strong relationship between planning and effort (0.24), indicating that students who actively plan their work tend to regulate their effort in the same time. Another moderate association exists between effort and control (0.16), suggesting that students who manage their effort tend to exert control over their environment in the same time. We can also see other associations, like feedback and evaluating (0.17), feedback and organizing (0.13), and organizing and control (0.15). These dynamics are relatively different from the between-person given the different time-scale.

In the temporal network (Fig. 5c), the arrows represent the direction of influence between self-regulated learning (SRL) constructs over time i.e., from a day to the next. The strongest positive influence is between evaluating and control (0.02), suggesting that social interactions is followed by effort next day which could be a boost or simply that student catches up next day of socializing. There is a feedback loop between planning and social (each 0.01), indicating socially-shared regulation. There is a negative influence of organizing on monitoring (-0.02), where these two activities occur in that order.

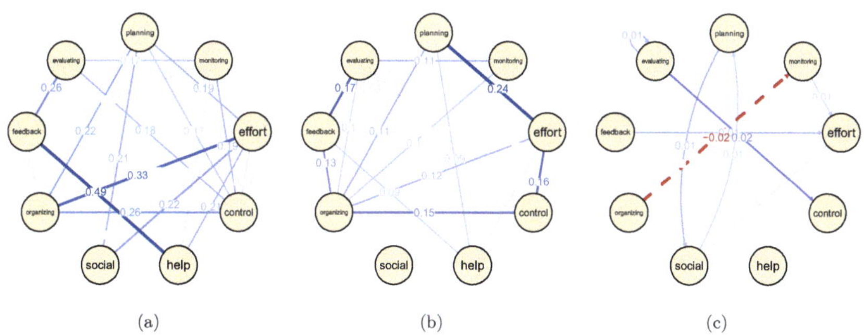

Fig. 5 Plotting the `mlGraphicalVAR` results. (**a**) Between person. (**b**) Contemporaneous network. (**c**) Temporal network

```
plotting(GraphVAR_all$betweenNet, minimum = 0.05)
plotting(GraphVAR_all$fixedPCC, minimum = 0.05)
plotting(GraphVAR_all$fixedPDC, minimum = 0.01)
```

3.4.2 Plotting and Interpreting Subject Level Networks

To plot each individual student network, we can use `GraphVAR_all$subjectPCC[[i]]` and `GraphVAR_all$subjectPCC[[i]]` to get the contemporaneous and temporal networks respectively, we can also get their names by using `GraphVAR_all$ids[[i]]` and use the plotting function to plot them. In the code below, we choose randomly five students numbered from 20 to 24 and plot their networks (Fig. 6). Of course, you can plot the whole dataset by setting the loop to iterate from 1:36 (the number of unique students in the data). The interpretation is similar, but in this case, Judy's network represents only Judy and cannot be generalized to anyone else. So is the case for Karin, Lars, Layla, and Li. Please note also the vast differences between these students. Each and every network is rather different.

```
for (i in 20:24) {
  # Apply plotting to subjectPCC and subjectPDC
  plotting(GraphVAR_all$subjectPCC[[i]], minimum = 0.05,
           title = paste("Contemporaneous network for ", GraphVAR_all$ids[[i]]))
  plotting(GraphVAR_all$subjectPDC[[i]], minimum = 0.05,
           title = paste("Temporal network for ", GraphVAR_all$ids[[i]]))
}
```

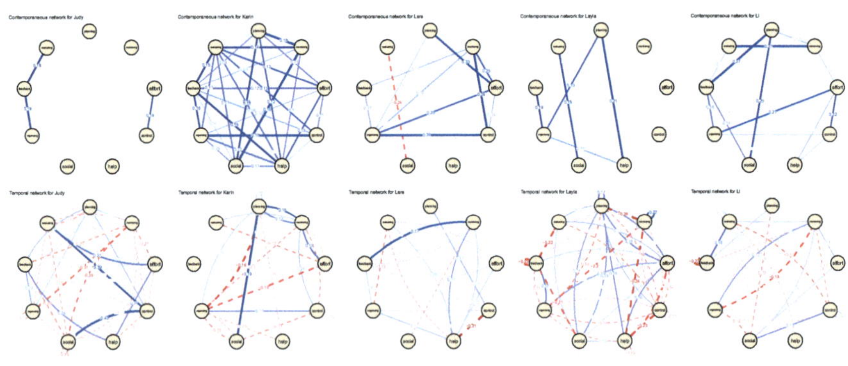

Fig. 6 Plotting the PCC and PDC individual student networks

3.5 *Unified Structural Equation Modeling (uSEM)*

Unified Structural Equation Modeling (uSEM) is another method that has the capability to offer an idiographic analysis of the individual's behavior as a complex dynamic system [39–42]. As the name implies, uSEM is built around structural equation modelling (described in [43]). The term "unified" means that the estimation method of uSEM combines both temporal (lagged relationships or t-1) and contemporaneous relationships (same time or t) simultaneously (compared to sequential estimation of the temporal and contemporaneous networks in `graphicalVAR`).

While uSEM share some similarities with `graphivalVar`, they have differences. In uSEM, the temporal relationships are estimated through a multivariate autoregressive model (MAR) and the contemporaneous relations are estimated through conventional SEM. Another difference is that uSEM estimation processes result in a single matrix that contains both the temporal and the contemporaneous relations (not two separate matrices or networks like Graphical VAR). Last, whereas graphical VAR uses regularization to prune the model, uSEM uses an iterative process where it begins with an empty model and applies Lagrange multiplier tests to assess which parameter, if added, would best improve the model's fit. In doing so, the model is refined in a step-wise process, with pathways added one at a time if they enhance overall fit. While some evidence suggests that Graphical VAR may offer better test-retest consistency in short time series, recent improvements in uSEM, such as integrating regularization may help solve these issues [44, 45].

3.5.1 uSEM Analysis

For uSEM analysis we will use the `pompom` R package which is designed for person-specific (idiographic) modeling analysis of multivariate time series data. In particular, `pompom` takes a hybrid approach that combines intraindividual variability with network analysis to model the individuals as complex dynamic systems. Also, `pompom` offers impulse response analysis metrics (iRAM) which helps quantify how variables interact dynamically, such as how one node's perturbation affects others over time (this will not be discussed here given it is beyond the scope of our chapter). In particular, `pompom` R package uSEM estimates a person-specific model, meaning each individual's data is analyzed separately [36]. If we want to analyze multiple individuals, we need to do so individually, possibly using a loop (see the section below).

uSEM requires the data to be standardized so that it captures within-person fluctuations around each person's own mean. When the data is standardized, the data can be treated as stationary, which allows us to focus on the dynamic, short-term changes in SRL without the influence of between-person differences or long-term trends. The code below standardize the data for Grace by subtracting the mean and dividing by standard deviation.

```
Grace_data_centered <- Grace_Data |>
  mutate(across(everything(), ~ scale(.x, center = TRUE, scale = TRUE))) |>
  select(-Day)
```

We then proceed to estimating the uSEM model for Grace using the same data we used before, the nine SRL variables after standardization. The estimation functionuSEM() takes four parameters: **data = Grace_data_centered**: The dataset which we have standardized to remove scale differences and convert the data to stationary and the **var.number = 9** which tells uSEM the number of variables. The **lag.order = 1** specifies a time lag of 1, meaning we are analyzing how variables affect each other from one time point to the next.The argument **verbose = FALSE** tells the model to run quietly. Finally, **trim = TRUE** tells the model to trim insignificant relationships between variables to simplify the output and show only the important interactions.

```
Usim_Grace <- uSEM(var.number = 9,
                   data = Grace_data_centered,
                   lag.order = 1,
                   verbose = FALSE,
                   trim = TRUE)
```

For the uSEM model to be reliable, we need to evaluate fitness criteria. uSEM computes four criteria: Comparative Fit Index (CFI), Tucker-Lewis Index (TLI), Root Mean Square Error of Approximation (RMSEA), and Standardized Root Mean Square Residual (SRMR). Ideally, the model should pass at least three of the evaluation criteria. The **CFI** and **TLI** assess how well the model fits the data compared to a baseline model where values closer to 1 indicates a better fit. The **RMSEA** estimates the error of approximation in the model where lower values (ideally below 0.08) indicates a good fit. The **SRMR** assesses the difference between observed and predicted correlations, with values below 0.08 being ideal.

In the code below, we use the model_summary function to evaluate the fitness of the model. We need to provide the model, tell the function that our data has nine variables, and set the time-lag order 1. The function store the evaluation criteria in Usim_Grace_fit. We can now retrieve the values of TLI, RMSEA, and SRMR from the model fit object Usim_Grace_fit and calculate the CFI by summing its values.

The code below retrieves the four criteria, and programmatically evaluates them. We see that the CFI value of 1 indicates a perfect fit, so it is evaluated as "Passed". Similarly, a TLI value higher than 0.95, an RMSEA of 0 (indicating no error in approximation), and an SRMR below 0.08 also indicate a good fit. Based on these criteria, our model passes four evaluation criteria and is therefore reliable.

```
Usim_Grace_fit <- model_summary(model.fit = Usim_Grace,
                                var.number = 9,
                                lag.order = 1)
```

```
cfi_val <- sum(Usim_Grace_fit$cfi)
tli_val <- Usim_Grace_fit$tli
rmsea_val <- Usim_Grace_fit$rmsea
srmr_val <- Usim_Grace_fit$srmr

# Print the values of each fit index
cat("CFI:", cfi_val, "->", ifelse(cfi_val > 0.95, "Passed", "Failed"), "\n")
cat("TLI:", tli_val, "->", ifelse(tli_val > 0.95, "Passed", "Failed"), "\n")
cat("RMSEA:", rmsea_val, "->", ifelse(rmsea_val < 0.08, "Passed", "Failed"), "\n")
cat("SRMR:", srmr_val, "->", ifelse(srmr_val < 0.08, "Passed", "Failed"), "\n")
```

```
CFI: 1 -> Passed

TLI: 1.001069 -> Passed

RMSEA: 0 -> Passed

SRMR: 0.04020178 -> Passed
```

The uSEM networks are a bit different because we get both temporal and contemporaneous relationships in the same plot. Here, the red edges represent negative relationships, the green edges represent positive relationships. Dashed edges indicate lag-1 temporal relationships. Solid edges indicate contemporaneous relationships. Edge width is proportional to the the strength of the relationship where thicker edges indicate stronger positive or negative relationships. As we can see in the plot, Grace has a strong temporal relationship (dashed green line) between planning and effort, suggesting that when planning increases her effort follows. We also see a negative contemporaneous relationship between monitoring and social. Compare this to the previous Graphical Var models. The default plotting function is restrictive and does not show the labels (Fig. 7a), so, we have created a slightly modified function `usemplot` to improve the plot and show the labels (Fig. 7b).

```
plot_network_graph (Usim_Grace_fit$beta, var.number = 9)

usemplot(Usim_Grace_fit$beta, var.number = 9,
         labels = colnames(Grace_data_centered))
```

3.5.2 Analysis of Multiple Individuals with uSEM

Given that `pompom` does not have a function that automates the analysis across multiple individuals, we need to iterate through each individual in our dataset and repeat the analysis. However, before doing so, we need to check the data if it has any

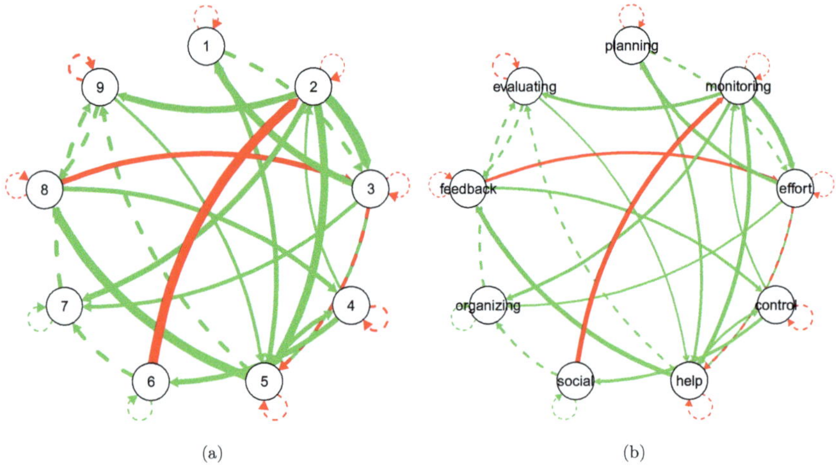

Fig. 7 Plotting the uSEM networks. (**a**) Default plotting function. (**b**) Custom plotting function

variables that are completely missing or have no variance e.g., has the same value. We do so by the function `nearZeroVar` from the `caret` package.

```
df |>
  group_by(name) |>
  reframe(across(all_of(Vars), ~ nearZeroVar(as.data.frame(.))))
```

```
# A tibble: 0 x 10
# i 10 variables: name <chr>, planning <int>, monitoring <int>, effort <int>,
#   control <int>, help <int>, social <int>, organizing <int>, feedback <int>,
#   evaluating <int>
```

Since none of the data entries present any problems, we can proceed with the analysis. The next code simply iterates through each individual in our dataset (`df`), applies a uSEM, and fit the data. First, we filter the dataset to isolate data for each individual, then define variables of interest i.e., `Vars`, and centers/scales the data. A uSEM model is run with the nine variables, and the model's fit is the estimated.

```
# Initialize a list to store the results
results_list <- list()

# Define a vector of individual names (replace with actual names in your data)
individual_names <- unique(df$name)

# Loop through each individual dataset
for (individual in individual_names) {
```

```
  # Filter the dataset for the current individual
  individual_data <- df |>
    filter(name == individual) |>
    select(all_of(Vars))

  # Center and scale the data
  centered_data <- individual_data |>
    mutate(across(everything(), ~ scale(.x, center = TRUE, scale = TRUE)))

  # Perform uSEM operations
  uSEM_result <- uSEM(var.number = 9,
                      data = centered_data,
                      lag.order = 1,
                      verbose = FALSE,
                      trim = TRUE)

  # Get the model summary
  model_fit <- model_summary(model.fit = uSEM_result,
                             var.number = 9,
                             lag.order = 1)


  # Store the results in the list
  results_list[[individual]] <- list(
    uSEM_result = uSEM_result,
    model_fit = model_fit
  )
}
```

We then plot each model by iterating through the result list. We show five as an example in Fig. 8.

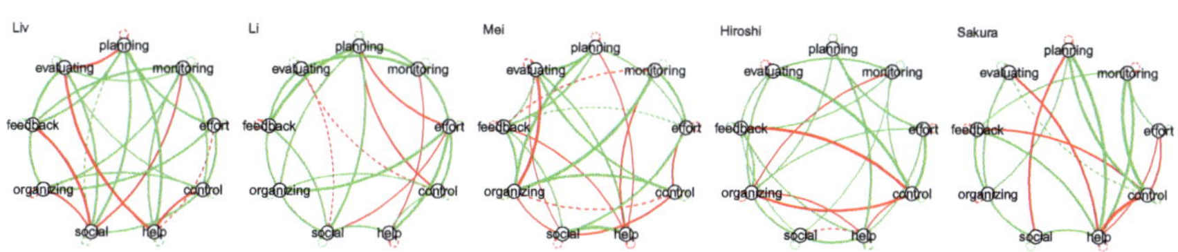

Fig. 8 Plotting multiple uSEM networks

```
for (i in 20:24) {
  usemplot(results_list[[i]]$model_fit$beta, var.number = 9, labels = Vars,
           title = individual_names[i])
}
```

4 Conclusion

Studying individual processes and the complexity of behavior is essential for understanding human dynamics, particularly in educational and psychological contexts. Behavior is inherently multifaceted, driven by the interplay of cognitive, emotional, and social processes that unfold uniquely for each person. Idiographic approaches, which focus on individual patterns rather than group averages, are crucial for capturing these complexities. They allow researchers to uncover how specific factors influence behavior over time and reveal the dynamic interactions between processes like planning, monitoring, and evaluating in self-regulated learning. This perspective respects the uniqueness of individual trajectories, providing actionable insights for personalized interventions that align with each person's needs and context. By embracing the complexity of behavior, idiographic methods offer a deeper understanding of human experiences and support more effective, tailored strategies for fostering growth and learning [10].

This chapter introduced the use of probabilistic networks, specifically graphical Vector Autoregression (graphicalVAR) and Unified Structural Equation Modeling (uSEM), to study individual learning processes. The idiographic approach provides a personalized, dynamic view of how cognitive, social, and motivational factors interrelate over time and provides insights at N=1. This approach highlights the unique profiles of students. This approach also supports more accurate personalized learning interventions. Given that the modeling captures the unique dynamics of each student, we can tailor strategies that address their specific needs. Identifying key factors that influence performance allows for targeted interventions to improve learning outcomes.

Furthermore, the integration of both temporal and contemporaneous networks enhances our ability to model the real-time interactions between variables. Temporal networks, which map how one variable predicts another in subsequent time windows, offer insight into the cause-and-effect relationships that govern student behavior over time. Contemporaneous networks, on the other hand, allow us to examine the relationships between variables within the same time window, offering a snapshot of the dynamic interdependencies at a given moment. These two complementary approaches give us a fuller picture of the cognitive, emotional, and behavioral processes that underpin learning. For instance, in our example dataset, motivation was shown to predict work behavior and achievement, while stress, in turn, negatively impacted achievement, underscoring the importance of understanding these processes as interconnected rather than isolated.

References

1. Saqr M, Poquet O, Lopez-Pernas S (2022) Networks in education: a travelogue through five decades. IEEE Access 10:32361–32380. https://doi.org/10.1109/access.2022.3159674
2. Saqr M, López-Pernas S, Conde-González MÁ, Hernández-García Á (2024) Social network analysis: A primer, a guide and a tutorial in R. In: Learning analytics methods and tutorials. Springer Nature Switzerland, Cham, pp 491–518. https://doi.org/10.1007/978-3-031-54464-4/_15
3. Cela KL, Sicilia MÁ, Sánchez S (2014) Social network analysis in E-learning environments: a preliminary systematic review. Educ Psychol Rev 27:219–246. https://doi.org/10.1007/s10648-014-9276-0
4. Fan Y, Tan Y, Raković M, Wang Y, Cai Z, Shaffer DW, Gašević D (2022) Dissecting learning tactics in MOOC using ordered network analysis. J Comput Assist Learn. https://doi.org/10.1111/jcal.12735
5. Saqr M, Lopez-Pernas S (2021) Idiographic learning analytics: a definition and a case study. 2021 international conference on advanced learning technologies (ICALT), pp 163–165. https://doi.org/10.1109/icalt52272.2021.00056
6. Elmoazen R, Saqr M, Tedre M, Hirsto L (2022) A systematic literature review of empirical research on epistemic network analysis in education. IEEE Access 10:17330–17348. https://doi.org/10.1109/access.2022.3149812
7. Saqr M, López-Pernas S, Törmänen T, Kaliisa R, Misiejuk K, Tikka S (2024) Transition network analysis: a novel framework for modeling, visualizing, and identifying the temporal patterns of learners and learning processes. https://doi.org/10.48550/ARXIV.2411.15486
8. Saqr M, López-Pernas S (2025) The three levels of analysis: Variable-centered, person-centered and person-specific analysis in education. In: Advanced learning analytics methods: AI, precision and complexity. Springer Nature Switzerland, Cham
9. Saqr M, Ito H, López-Pernas S (2025) A comprehensive introduction to idiographic and within-person analytics. In: Saqr M, López-Pernas S (eds) Advanced learning analytics methods: AI, precision and complexity. Springer Nature Switzerland, Cham
10. Saqr M, Dever D, López-Pernas S, Gernigon C, Marchand G, Kaplan A (2025) Complex dynamic systems in education: beyond the static, the linear and the causal reductionism. In: Saqr M, López-Pernas S (eds) Advanced learning analytics methods: AI, precision and complexity. Springer Nature Switzerland, Cham
11. Saqr M, Beck E, López-Pernas S (2024) Psychological Networks: a Modern Approach to Analysis of Learning and Complex Learning Processes. Springer Nature Switzerland, pp 639–671. https://doi.org/10.1007/978-3-031-54464-4_19
12. Malmberg J, Saqr M, Järvenoja H, Haataja E, Pijeira-Díaz HJ, Järvelä S (2022) Modeling the complex interplay between monitoring events for regulated learning with psychological networks. Springer International Publishing, pp 79–104. https://doi.org/10.1007/978-3-031-08076-0_4
13. Malmberg J, Saqr M, Järvenoja H, Haataja E, Pijeira-Díaz HJ, Järvelä S (2022) Modeling the complex interplay between monitoring events for regulated learning with psychological networks. Springer International Publishing, pp 79–104. https://doi.org/10.1007/978-3-031-08076-0_4
14. Saqr M, Tlili A, López-Pernas S (2025) Automating individualized machine learning and AI prediction using AutoML: The case of idiographic predictions. In: Saqr M, López-Pernas S (eds) Advanced learning analytics methods: AI, precision and complexity. Springer Nature Switzerland, Cham
15. Epskamp S, Waldorp LJ, Mõttus R, Borsboom D (2018) The gaussian graphical model in cross-sectional and time-series data. Multivar Behav Res 53:453–480. https://doi.org/10.1080/00273171.2018.1454823
16. Hevey D (2018) Network analysis: a brief overview and tutorial. Health Psychol Behav Med 6:301–328. https://doi.org/10.1080/21642850.2018.1521283

17. Artner R, Wellingerhof PP, Lafit G, Loossens T, Vanpaemel W, Tuerlinckx F (2022) The shape of partial correlation matrices. Commun Stat Theory Methods 51:4133–4150. https://doi.org/10.1080/03610926.2020.1811338
18. Hamilton M, Clarke-Midura J, Shumway JF, Lee VR (2020) An emerging technology report on computational toys in early childhood. Technol Knowl Learn 25:213–224. https://doi.org/10.1007/s10758-019-09423-8
19. Borsboom D (2017) A network theory of mental disorders. World Psychiatry 16:5–13. https://doi.org/10.1002/wps.20375
20. Molenaar PCM (2004) A manifesto on psychology as idiographic science: Bringing the person back into scientific psychology, this time forever. Meas Interdiscip Res Persp 2:201–218. https://eric.ed.gov/?id=EJ937576
21. Saqr M, López-Pernas S (2024) Mapping the self in self-regulation using complex dynamic systems approach. Br J Educ Technol 55:1376–1397. https://doi.org/10.1111/bjet.13452
22. Malmberg J, Saqr M, Järvenoja H, Järvelä S (2022) How the monitoring events of individual students are associated with phases of regulation: a network analysis approach. J Learn Anal 9:77–92. https://doi.org/10.18608/jla.2022.7429
23. Zhou Y, Kang J (2023) Enriching multimodal data: A temporal approach to contextualize joint attention in collaborative problem-solving. J Learn Anal 10:87–101. https://doi.org/10.18608/jla.2023.7989
24. Saqr M, Vogelsmeier LVDE, López-Pernas S (2024) Capturing where the learning process takes place: a person-specific and person-centered primer. Learn Indiv Differ 113:102492. https://doi.org/10.1016/j.lindif.2024.102492
25. De Neve D, Bronstein MV, Leroy A, Truyts A, Everaert J (2023) Emotion regulation in the classroom: a network approach to model relations among emotion regulation difficulties, engagement to learn, and relationships with peers and teachers. J Youth Adolesc 52:273–286. https://doi.org/10.1007/s10964-022-01678-2
26. Malmberg J, Saqr M, Järvenoja H, Haataja E, Pijeira-Díaz HJ, Järvelä S (2022) Modeling the complex interplay between monitoring events for regulated learning with psychological networks. In: Giannakos M, Spikol D, Di Mitri D, Sharma K, Ochoa X, Hammad R (eds) The multimodal learning analytics handbook. Springer International Publishing, Cham, pp 79–104. https://doi.org/10.1007/978-3-031-08076-0/_4
27. Saqr M (2024) Group-level analysis of engagement poorly reflects individual students' processes: why we need idiographic learning analytics. Comput Human Behav 150:107991. https://doi.org/10.1016/j.chb.2023.107991
28. Saqr M, Viberg O, Peteers W (2021) Using psychological networks to reveal the interplay between foreign language students' self-regulated learning tactics. In: STELLA2020 proceedings, pp 12–23
29. Qin J, Ibragim T, Zhang H (2020) E-learner's emotion regulation model with undirected gaussian graphical model. J Syst Simul 29:935–940. https://doi.org/10.16182/j.issn1004731x.joss.201705001
30. Jovanovic J, Gašević D, Yan L, Baker G, Murray A, Gasevic D (2024) Explaining trace-based learner profiles with self-reports: the added value of psychological networks. J Comput Assisted Learn 40:1481–1499. https://doi.org/10.1111/jcal.12968
31. Saqr M (2024) Group-level analysis of engagement poorly reflects individual students' processes: why we need idiographic learning analytics. Comput Human Behav 150:107991. https://doi.org/10.1016/j.chb.2023.107991
32. Wickham H, Averick M, Bryan J, Chang W, McGowan LD, François R, Grolemund G, Hayes A, Henry L, Hester J, Kuhn M, Pedersen TL, Miller E, Bache SM, Müller K, Ooms J, Robinson D, Seidel DP, Spinu V, Takahashi K, Vaughan D, Wilke C, Woo K, Yutani H (2019) Welcome to the tidyverse. J Open Source Softw 4:1686. https://doi.org/10.21105/joss.01686
33. Waring E, Quinn M, McNamara A, Arino de la Rubia E, Zhu H, Ellis S (2022) Skimr: Compact and flexible summaries of data. https://CRAN.R-project.org/package=skimr
34. Epskamp S (2024) graphicalVAR: Graphical VAR for experience sampling data. https://CRAN.R-project.org/package=graphicalVAR
35. Epskamp S, Cramer AOJ, Waldorp LJ, Schmittmann VD, Borsboom D (2012) qgraph: network visualizations of relationships in psychometric data. J Stat Softw 48:1–18

36. Yang X, Ram N, Molenaar P (2021) Pompom: Person-oriented method and perturbation on the model. https://CRAN.R-project.org/package=pompom
37. Kuhn, Max (2008) Building predictive models in r using the caret package. J Stat Softw 28:1–26. https://doi.org/10.18637/jss.v028.i05
38. Chan C, Leeper TJ, Becker J, Schoch D (2023) Rio: A swiss-army knife for data file i/o. https://cran.r-project.org/package=rio
39. Kim J, Zhu W, Chang L, Bentler PM, Ernst T (2006) Unified structural equation modeling approach for the analysis of multisubject, multivariate functional MRI data. Human Brain Mapping 28:85–93. https://doi.org/10.1002/hbm.20259
40. Beltz AM, Beekman C, Molenaar PCM, Buss KA (2013) Mapping temporal dynamics in social interactions with unified structural equation modeling: a description and demonstration revealing time-dependent sex differences in play behavior. Appl Develop Sci 17:152–168. https://doi.org/10.1080/10888691.2013.805953
41. Gates KM, Molenaar PCM, Hillary FG, Slobounov S (2011) Extended unified SEM approach for modeling event-related fMRI data. NeuroImage 54:1151–1158. https://doi.org/10.1016/j.neuroimage.2010.08.051
42. Gates KM, Molenaar PCM (2012) Group search algorithm recovers effective connectivity maps for individuals in homogeneous and heterogeneous samples. NeuroImage 63:310–319. https://doi.org/10.1016/j.neuroimage.2012.06.026
43. Jongerling J, López-Pernas S, Saqr M, Vogelsmeier LVDE (2024) Structural equation modeling with R for education scientists. Springer Nature Switzerland, pp 705–721
44. Beltz AM, Gates KM (2017) Network mapping with GIMME. Multivar Behav Res 52:789–804. https://doi.org/10.1080/00273171.2017.1373014
45. Beck ED, Jackson JJ (2021) Within-person variability. Elsevier, Amsterdam, pp 75–100. https://doi.org/10.1016/b978-0-12-813995-0.00004-2

Detecting Long-Memory Psychological Processes in Academic Settings Using Whittle's Maximum Likelihood Estimator: An Application with R

Rémi Altamore, Clément Roume, Anne Teboul, and Christophe Gernigon

Abstract Person-specific approaches are best suited to account for the complex, dynamic, and idiosyncratic processes from which cognitive, emotional, motivational, relational, and behavioral patterns emerge. Among these approaches, Whittle's approximation of the Maximum Likelihood Estimator (MLE) enables the detection of long-term memory processes in relatively short time series of data. In this chapter, we outline the principles of Whittle's MLE, illustrate its application—using R—to the motivational dynamics of approach and avoidance in an academic context, and then discuss the theoretical and practical implications of detecting long-memory processes in the field of education and learning.

Keywords Achievement motivation · Approach · Avoidance · Complex dynamical systems · Fractals · Learning analytics · Long-memory processes · Person-specific methods · Time series analysis

1 Introduction

People's cognitive, emotional, motivational, relational, and behavioral patterns emerge from complex, dynamic, and idiosyncratic processes [1]. Thus, these processes operate in an original way, at an individual level which is the most relevant scale for examining them [2]. This requires person-specific methods, particularly those based on the analysis of time series of idiographic data [3]. Some of these analysis techniques, such as the popular *Detrended Fluctuation Analysis* (DFA; [4]), can be used to detect whether a variable of interest is randomly fluctuating (white

R. Altamore (✉) · A. Teboul · C. Gernigon
EuroMov Digital Health in Motion, University Montpellier, IMT Mines Ales, Montpellier, France
e-mail: remi.altamore@umontpellier.fr

C. Roume
Institut de Recherche en Informatique, Mathématiques, Automatique et Signal (UR UHA 7499), Université de Haute-Alsace, Mulhouse, France

M. Saqr, S. López-Pernas (eds.), *Advanced Learning Analytics Methods*,
https://doi.org/10.1007/978-3-031-95365-1_21

noise[1]), or follows a movement deviating from its trajectory under the influence of random shocks (Brownian motion[2]), or obey particular auto-correlation structures that are redundant across different time scales and that reveal a kind of memory process in the form of a long-term dependency ($1/f$ noise or pink noise[3]).

In addition to its ease of implementation, DFA has the advantage of being applicable to either stationary (e.g., white noise, pink noise) or non-stationary signals (Brownian motion). This method is based on a statistical property of fractional Brownian motion [5] whereby the size of the variance to the power α is proportional to the sample size. The power value $\alpha$is the fractal exponent which provides information about the fluctuation structure of the variable of interest. When α is between 0 and 1, the time series is stationary and can be qualified as noise. From 0 to 0.5, the time series is anti-correlated to itself, which means that an increase in the value of the variable is followed by a decrease and vice versa. At 0.5 the series corresponds to white Gaussian noise. From 0.5 to 1 the time series is autocorrelated and displays long-term memory properties. For a α between 1 and 2, the series is non-stationary and is referred to as motion. A value of 1.5 corresponds to ordinary Brownian motion [6].

Despite its great popularity, DFA suffers from reliability issues when confronted with certain empirical data, particularly when the length of the time series to be analyzed is less than 500 data points. Depending on the size and target population of the studies for which this method is to be applied, it may prove very difficult to obtain such series lengths. For example, if we want to analyze walking, 500 steps represent between 10 and 12 minutes of walking, which is difficult to achieve for elderly adults with a loss of autonomy [7].

Some methods make it possible to estimate the fractal exponent of shorter time series without being affected by the estimation biases that DFA presents for such series [7–10]. This chapter presents one of these methods: Whittle's approximation of the Maximum Likelihood Estimator (MLE). Whereas the DFA output is estimated via linear regression, Whittle's estimator is a more sophisticated method that consists of solving an optimization problem [11]. As a comparison, with only 128 data points, Whittle's MLE gives results twice as accurate as the DFA on 512 data points [10], the latter offering results that were however already considered satisfactory [12].

The first part of the chapter is devoted to the presentation of the principles of Whittle's MLE, as well as their formal implementation. Next, a concrete application

---

[1] White noise is a stochastic process, that is, which is reflected by a signal that unfolds randomly over time and thus has a null autocorrelation.

[2] Originally, Brownian motion is the random movement of a particle in a liquid or gas due to successive shocks from surrounding molecules.

[3] $1/f$ noise or pink noise features distributions of temporal fluctuations of a phenomenon displaying the following properties: The frequencies of fluctuations remain invariant across time scales; the frequency of a particular magnitude of fluctuation is inversely proportional to that magnitude; lagged autocorrelations decay slowly, as an inverse power of lag (i.e., long-range temporal correlations).

of Whittle's MLE to the study of the dynamics of approach and avoidance motivation in an academic context is offered as an illustration of these principles. This illustration begins with a presentation of the motivational model referred to, continues with the application of Whittle's MLE to a time series of individual data relating to this model, and ends with a discussion of the result obtained. The chapter concludes with an attempt to reconcile variable-centered nomothetic approaches with person-specific approaches. This reconciliation, potentially capable of resolving the ergodicity issue in psychological science, requires the primacy of person-specific approaches.

2 Principles of Whittle's Maximum Likelihood Estimator (MLE)

The general principle of Whittle's estimator is to find the value of α in the interval [0,1] that maximizes the Whittle's log-likelihood function accounted for by Equation 1:

$$l_w(\alpha) = -\frac{2}{N}\sum_{j=1}^{m}\left(\ln cT'(\omega_j;\alpha) + \frac{P(\omega_j)}{cT'(\omega_j;\alpha)}\right), \tag{1}$$

where $P(\omega_j)$ is the periodogram of the time series $x(j)$ of length N; ω_j are Fourier's frequencies defined as $\omega = 2\pi j/N$; m is the integer part of $(N-1)/2$; $T'(\omega_j;\alpha)$ is the theoretical spectral density; and c is a constant used to adjust the powers of the theoretical spectrum.

In this chapter, we will use the theoretical power spectral density of an ARFIMA (Autoregressive Fractionally Integrated Moving Average[4]) model, which enables the modeling of time series with long memory. ARFIMA models are denoted ARFIMA(p,d,q) with p as the order indicating the number of time lags of the autoregressive model, q as the order of the moving-average model, and d as the fractional integration parameter ranging from -0.5 to 0.5. It corresponds to the exponent α via the conversion: $\alpha = d + ½$with α belonging to the interval [0, 1]. The theoretical power spectral density that is used here is that of the ARFIMA(0,d,0) model, which reads as follow:

[4] ARFIMA (Autoregressive Fractionally Integrated Moving Average) models are statistical tools that model time series with long memory. By allowing the differencing parameter d to take fractional values, they generalize ARIMA (Autoregressive Integrated Moving Average) models, which themselves generalize ARMA (Autoregressive Moving Average) models to non-stationary series.

$$T'(\omega_j;\alpha) = \frac{1}{2\pi}\left(2\sin\frac{\omega_j}{2}\right)^{1-2\alpha}. \tag{2}$$

An algorithm based on Whittle's estimator can be developed in three steps. The first one combines data pre-processing, via normalization of the time series and then estimation of the periodogram of this series. The second step is the estimation of α. An optimization algorithm is used to find the α value for which the Whittle's log-likelihood function, $l_w(\alpha)$, reaches its maximum value. The third step is optional and is carried out only if the α value obtained from the algorithm is that of the upper bound. In this case, the maximum of the $l_w(\alpha)$ function can be considered to belong to the interval [1, 2], which means that the time series is probably non-stationary. Indeed, the $l_w(\alpha)$ function is only applicable for stationary signals, with an α value within the interval [0,1]. Nevertheless, non-stationary series can be analyzed by using a differentiated (derivative) version of the original series and then adding 1 to the α value provided by the algorithm [13]. In summary, the third step consists of differentiating the time series, then applying the first two steps and finally adding 1 to the α value obtained from the algorithm.

3 Application of Whittle's MLE to Motivational Dynamics in an Academic Context

3.1 The Dynamical Model of Approach and Avoidance Motivation in Achievement Contexts

Motivational states regarding a goal to be achieved are properties emerging from complex dynamical systems. As such, motivational states fluctuate non-linearly, thus exhibiting various patterns such as persistence of effort or its absence, as well as oscillations and even abrupt shifts between motivated and unmotivated states [14–16]. We will illustrate how Whittle's MLE can help characterize the dynamics and the complexity of approach and avoidance motivations in an academic context, more specifically with respect to a Ph.D. student's pursuit of the goal of reaching the defense of their doctoral thesis.

Based on the complex dynamical system perspective, approach and avoidance motivational states can be conceptualized as being constrained by two competing attractors [15]. Attractors are sets of states of a system that are more or less strongly attracted and stabilized within a limited number of values, as if—metaphorically—the trajectories of these states (here, the motivational states) were constrained by the valleys of a hilly landscape. According to Tuller et al. [17], the shape of a landscape that may include one or two attractors can be accounted for by the following equation:

$$V(x) = kx - \frac{1}{2}x^2 + \frac{1}{4}x^4, \tag{3}$$

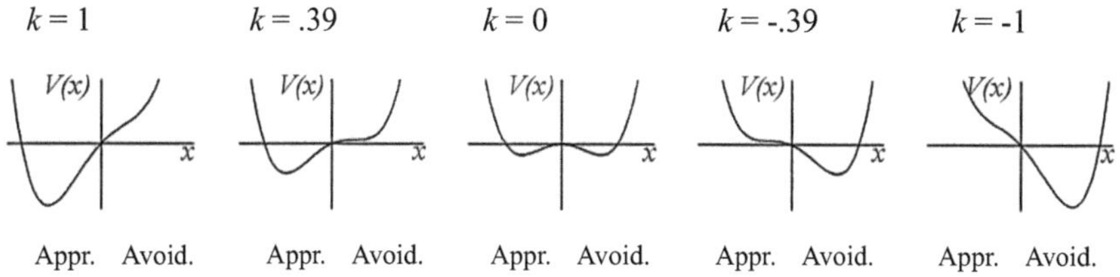

Fig. 1 Approach (Appr.) and avoidance (Avoid.) attractor landscapes as defined by Equation 3 with different values of the control parameter k (from +1 to −1)

where $V(x)$ is the potential function that constrains the different motivational states x; k is the control parameter that specifies the direction and degree of tilt of the graph of $V(x)$ towards approach (when k tends towards +1) and avoidance (when k tends towards −1) motivations (see Fig. 1).

The value of the parameter k is itself conceived as resulting from the combination of three social cognitive variables considered to be necessary and sufficient for achievement motivation: Competence expectancies with regard to a goal (e.g., [18]), as well as the benefit to the self that the prospect of achieving a goal conveys and the threat to the self associated with the prospect of failure (e.g., [19]). These three variables contribute to determining k according to Eq. (4):

$$k = (Ce \times Bs) - [Ts \times (1 - Ce)], \tag{4}$$

where Ce is the level of competence expectancies, Bs is the potential benefit to the self, and Ts is the potential threat to the self, with Ce, Bs, and $Ts \in]0;1]$.

The analysis of the time series of k values using Whittle's MLE will make it possible to test whether the motivational states of approach and avoidance do emerge from a system of elements (i.e., the internal and environmental antecedents of motivation identified in the literature) interweaving complex interactions. The time series under study comes from data collected from a Ph.D. student who responded every week—over 111 consecutive weeks—to three items related to her goal of completing her doctoral dissertation, a goal she eventually achieved. Each week, each of these three items was drawn at random from the items of each of the three dimensions Ce, Bs, and Ts of Teboul et al.'s [20] Approach-Avoidance System Questionnaire. This weekly measurement of Ce, Bs, and Ts enabled k to be calculated for each week.

3.2 *Analysis of the Time Series of k Using R*

3.2.1 Libraries Used

Whittle's MLE was originally coded using MATLAB language (see [10]). However, the MATLAB algorithm can be translated into R code using only the `stats` package, which is a basic package—automatically installed—on R. This package contains a set of useful functions for many statistical calculations. We will use `rio` to import and load data into R. We will use `dplyr`, `gt` and `ggplot2` packages to easily manipulate and visualize our data (to learn how to use these three packages, readers can refer to the previous edition of the book, see [21, 22]. These five packages can be installed using the `install.packages()` function.

```
install.packages("stats")
install.packages("rio")
install.packages("dplyr")
install.packages("gt")
install.packages("ggplot2")
```

These packages can then be loaded with the `library()` function.

```
library(stats)
library(rio)
library(dplyr)
library(gt)
library(ggplot2)
```

3.2.2 Data Visualization

First, we load the data with `import()` function from `rio` and visualize an extract of them (Table 1) with the function `gt_preview()`.

```
MotivationData <- import(paste0(
  "https://github.com/lamethods/data2/raw/main/",
  "motivation/Motivation.xlsx"
))
gt_preview(MotivationData, top_n = 8, bottom_n = 1)
```

The dataset comprises four columns of length 111: the first column is the number of weeks ('Week'), the second column is the values of competence expectancies for each week ('Ce'), the third is the values of potential benefit to the self ('Bs') and the last column the values of potential threat to the self ('Ts'). To calculate the k values according to Eq. (4), we use `mutate` from the `dplyr` package to create a fifth column and apply the equation for each week (Table 2).

```
MotivationData <- MotivationData %>%
  mutate(K = (Ce * Bs) - (Ts * (1 - Ce)))
k_values <- (MotivationData$K) # store K values in a vector
gt_preview(MotivationData, top_n = 8, bottom_n = 1)
```

Table 1 Preview of motivation data

	Week	Ce	Bs	Ts
1	1	0.5	0.5	0.5
2	2	0.5	0.5	0.5
3	3	0.6	1.0	1.0
4	4	0.7	1.0	1.0
5	5	0.5	1.0	1.0
6	6	0.5	1.0	1.0
7	7	0.3	1.0	1.0
8	8	0.5	1.0	1.0
9..110				
111	111	0.6	1.0	1.0

Table 2 Preview of motivation data with additional column

	Week	Ce	Bs	Ts	K
1	1	0.5	0.5	0.5	0.0
2	2	0.5	0.5	0.5	0.0
3	3	0.6	1.0	1.0	0.2
4	4	0.7	1.0	1.0	0.4
5	5	0.5	1.0	1.0	0.0
6	6	0.5	1.0	1.0	0.0
7	7	0.3	1.0	1.0	-0.4
8	8	0.5	1.0	1.0	0.0
9..110					
111	111	0.6	1.0	1.0	0.2

To visualize the temporal dynamics of k, we use `ggplot2` to create a line plot (see Fig. 2).

```
ggplot(MotivationData, aes(x = Week, y = K)) +
  geom_line(colour = "turquoise4", linewidth = 0.8) +
  geom_point(colour = "magenta", size = 1) +
  labs(x = "Week", y = "k") +
  scale_x_continuous(breaks = seq(0, 110, by = 10)) +
  scale_y_continuous(breaks = seq(-1, 1, by = 0.5)) +
  theme(
    panel.background = element_rect(fill = "gray95")
  )
```

The plot in Fig. 2 shows the variations in the student's motivational state with respect to her goal of defending her Ph.D. thesis. This motivational state is reflected by the value of the parameter k (vertical axis), which determines the inclination of the attractor landscape towards approach (when k tends towards +1) or avoidance (when k tends towards -1). The values of k are characterized by varied non-

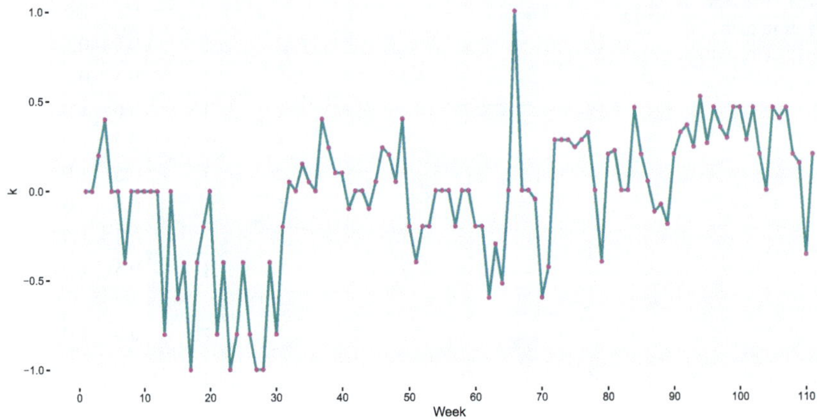

Fig. 2 Line plot of the evolution of the control parameter of approach and avoidance motivations k

linear changes over the weeks (horizontal axis). At certain periods, fluctuations in motivational state remain confined within the avoidance attractor (e.g., from week 12 to week 30). At other times, these fluctuations tend to stay within the approach attractor (e.g., from week 93 to week 102). But overall, motivation seems rather unstable, most often oscillating between the two attractors by crossing the value of $k = 0$. At first glance, motivational variability may look erratic, with no consistent trend over time. Although visual analysis of the data is essential, it is Whittle's MLE that enables the temporal structure of the variability of this signal to be detected statistically.

3.2.3 Application of Whittle's MLE Algorithm

The first step in Whittle's MLE algorithm is to estimate the power spectral density (or periodogram) of the vector containing the values of k.

```
# Power spectral density estimation of vector x
x <- (k_values)
X <- scale(x)
N <- length(X)
m <- floor((N - 1) / 2)
Y <- fft(X) # Fast Discrete Fourier Transform
P <- (1 / (pi * N)) * abs(Y[2:(m + 1)])^2 # Power spectral density estimation
w <- (2 * pi * (1:m)) / N # Fourier frequencies
```

Firstly, we give the name `x` to the `k_values` vector. Then, as a necessary step for the algorithm to work properly, the `scale()` function standardizes the vector `x` by setting its mean to 0 and its standard deviation to 1. The value `N` represents the length of the standardized vector `X` (here 111). The value `m` is half the total length of the vector `N`, minus 1 and rounded down to the nearest integer using the `floor()` function. The number `m` represents the number of frequencies used to estimate the power spectral density. The `fft()` function is an algorithm that calculates the discrete Fourier transform of the standardized vector `X`, thus adapting the signal `X` from the time domain to the frequency domain. The calculation of the vector `P` is used to estimate the periodogram (i.e., power per frequency unit) from the Fourier transform. The vector `w` represents the Fourier frequencies for `m` values.

The next step is to define the Whittle's likelihood function. To estimate the α value of the signal, Whittle's MLE compares the empirical periodogram with a theoretical spectral density, here derived from an ARFIMA(0,d,0) model which is a class of models capturing long-term memory processes in time series. The function evaluates how well an ARFIMA(0,d,0) model with a given parameter α matches the observed time series, by comparing their spectral density. This comparison is carried out via a likelihood measure. The aim is to find the α value for which the Whittle's likelihood function reaches its maximum, thus denoting the best possible fit between the theoretical model and the empirical data.

To build this function, we first set its arguments.

```
# Whittle's log-likelihood function with ARFIMA(0,d,0) theoretical PSD
WLLF <- function(a, w, P, N) {
```

The `a` parameter is the α parameter to be estimated, which controls the type of fluctuation (i.e., anti-correlated, random, long-term autocorrelated) in the ARFIMA model. The other three parameters relate to the empirical signal detailed above. The function processing aims to (a) calculate the theoretical spectral density, (b) adjust it to the empirical data, and (c) calculate Whittle's log-likelihood function between the adjusted theoretical spectral density and the empirical spectral density. The resulting value will be used to assess how well a certain parameter α fits the time series.

```
  Tp <- (1 / (2 * pi)) * (2 * sin(w / 2))^(1 - 2 * a) # ARFIMA theoretical PSD
  c <- sum(P) / sum(Tp)
  T <- c * Tp # Theoretical PSD adjusted
  lw <- -(2 / N) * sum(log(T) + (P / T)) # Whittle's log-likelihood
  return(lw)
}
```

The `Tp` vector is the theoretical spectral density for an ARFIMA(0,d,0) model. The calculation of `c` normalizes the theoretical spectral density `Tp` so that it matches the scale of the empirical spectral density `P`. The adjusted theoretical spectral density `T` is obtained by multiplying the normalization factor `c` by the theoretical spectral density `Tp`. This adjusted theoretical spectral density is the density to be compared with the empirical spectral density. Next, Whittle's log-likelihood `lw`—which measures the likelihood between the adjusted theoretical spectral density and the empirical spectral density—is calculated. The larger the resulting `lw` value, the larger is the likelihood between the theoretical spectral density and the empirical periodogram. The aim is therefore to find the value of α that maximizes this function.

The final step is to optimize Whittle's log-likelihood function to find this α value.

```
A <- optimize(function(a) WLLF(a, w, P, N), lower = 0, upper = 1, maximum = TRUE,
              tol = 0.0001)
A <- A$maximum
```

The `optimize()` function is used to find the maximum of a function over a specified interval. Here, a is the parameter α to be estimated and, as detailed above, the function `WLLF(a, w, P, N)` calculates Whittle's log-likelihood for a given α. If the series is considered stationary, the value α must lie within the interval [0,1]. Consequently, the lower bound of the interval in which α is to be sought is `lower = 0` and the upper bound is `upper = 1`. The argument `maximum = TRUE` is used to maximize the function. Finally, we specify a tolerance of 0.0001 for the convergence of the optimization algorithm, which controls the precision of the maximum search. Once `optimize()` has found the maximum of the specified function in the interval, this value is extracted and assigned to `A`.

At this stage, a resulting α value very close to 1 (i.e., $\alpha \geq 0.9999$) may suggest that the time series is non-stationary (i.e., that its mean and variance change over time). As ARFIMA(0,d,0) models are designed for stationary series, an estimate of α very close to 1 could indicate that the initial theoretical model does not represent the time series well. In this specific case, it is possible to work on a differentiated version of the original series and then add 1 to the α value yielded by the algorithm.

```
# If the time series is non-stationary
if (A >= 0.9999) {
  XDiff <- diff(X)
  YDiff <- fft(XDiff)
  mdiff <- floor((N - 2) / 2)
  Pdiff <- (1 / (pi * N)) * abs(YDiff[2:(mdiff + 1)])^2
  wdiff <- (2 * pi * (1:mdiff)) / N
  A <- optimize(function(a) WLLF(a, wdiff, Pdiff, N - 1), lower = 0, upper = 1,
                maximum = TRUE, tol = 0.0001)
  A <- A$maximum
  A <- A + 1
}
```

The `if()` condition only applies when `A` is greater than or equal to 0.9999. In this case, we apply a differentiation to the standardized vector `X` to make it stationary. Next, the remaining steps are similar to those in the previous algorithm, where the aim is to estimate an α value by maximizing Whittle's log-likelihood function, but this time on the differentiated series. The differentiation of the time series has reduced the order of integration by one degree. Therefore, after estimating α for the differentiated series, 1 is added to compensate for this differentiation.

With both algorithms implemented, we can now display the α value maximized from Whittle's log-likelihood function.

3.2.4 Result

```
A
```

```
[1] 1.000066
```

As for the time series of the control parameter k of the approach and avoidance motivations, the estimated value of α is about 1. Thus, it is the third step of the algorithm for non-stationary series that worked, after a value of $\alpha \geq 0.9999$ was delivered by the first two steps of the algorithm. The value close to 1, diverging on the lower bound, suggests that this time series may be related to $1/f$ noise. Although the length of the series analyzed is quite short and the estimated α value can therefore be considered only acceptable but not optimal, this indicative result gives us important information about the structure of the variability of the time series of k. More precisely, the variability of the series of k is neither totally random (i.e., white noise) nor rigid by its short-term deviations alone (i.e., Brownian motion), but lies between these two extremes, organizing itself between disorder and order. This type of variability is expressed by a $1/f$ noise trajectory that reflects the long-term memory of goal-directed motivation, in other words, a motivational dynamics that relies on the traces of its past history.

4 Discussion

$1/f$ noise, as an optimal compromise between the uncontrollable disorder of random series and the overly strict order of excessively predictable series, is an intrinsic signature of the dynamics of phenomena emerging from complex interactions [23–25]. Using idiographic methods, researchers have been able to find this dynamic signature in a variety of psychological phenomena or human behaviors that may result from education or training procedures, such as self-esteem [26–28], cognitive performance [29–31], language learning [32], motor learning [33], and motor performance [34–36]. By employing a person-specific method such as Whittle's MLE, the present case study offers an empirical validation of the grounding of approach and avoidance motivation as a dynamic and complex phenomenon.

As an optimal compromise between order and disorder, $1/f$ noise helps maintain the system's stability, while endowing it with the flexibility to respond to its environment. This cooperative organization between the preservation of a coherent state and adaptation to environmental influences promotes efficient operation of a system then considered healthy. For example, variability of $1/f$ noise has been found in young, healthy subjects in their walking rhythm or heart rate, while more predictable or random fluctuations in these variables have been detected in patients and the elderly [37, 38]. Similarly, mood variation has been found to show $1/f$ noise fluctuations in healthy individuals, whereas it was more organized and predictable in patients with bipolar disorder [39]. Among dyslexic children, greater trial-to-trial random variability was found in a word-naming task, thus revealing less clear examples of $1/f$ noise in their response times, compared with non-dyslexic children [40].

In the educational realm, and more specifically as regards learning, the development of new competencies has been shown to be associated with $1/f$ noise variability in performance. For example, in multilingual learning, Lowie et al. [32] conducted a repeated word-naming experiment with one participant over a period of six years. During each test session of a total of 26, the participant was asked to name 200 words in his native language (Dutch) and his second language (English). Over the course of the sessions, the improvement of skills in the second language—resulting from repeated practice—led to a series of response times in this second language that was increasingly characterized by $1/f$ distributions. In addition, 4 test sessions were conducted after a 7-day period of exclusive use of either the native language or the second language. On the one hand, this single-language immersion strengthened the $1/f$ structure of the response time distributions in the language concerned, particularly when this was the second language. On the other hand, this immersion led to whiter the time series in the language that had not been used. These findings support $1/f$ noise as an indicator of optimal coordination of a person's language subsystems.

In the domain of motor learning—in rowing [34] and skiing [33]—experts' performances have been found to follow $1/f$ dynamics, whereas more random patterns of performance variation, approaching white noise, have been evidenced among novices. Moreover, Den Hartigh et al. [35] found that higher levels of complexity, as reflected by an α value approaching 1 in the $1/f^{\alpha}$ noise expression, were related to more perturbation-resistant motor performance. In sum, learning and expertise development seem to be corollaries of a progressive installation of complexity—as reflected by $1/f$ noise distributions—within systems governing behaviors [41].

In motivational research, time series analyses designed to detect typical signatures of the complexity of motivational processes are still lacking. Yet, in the workplace context, it has been shown that medium or high levels of different motivational variables are associated with patterns of chaotic variability in these variables, while lowest levels are associated with patterns of variation in between linear and random [42, 43]. Although only one case was examined in the present chapter, the $1/f$ variation pattern that was found could reflect the typical dynamics of adaptive achievement motivation in students pursuing important goals. Thus, rather than exhibiting motivational variability that is totally random or strictly influenced by short-term events (e.g., exams, oral performances), achievement motivation in students would keep a stable and consistent state over time, while remaining capable of adapting to the constraints of the academic environment. However, the possibility of more random or rigid motivational dynamics in students who, unlike the present example, are likely to abandon their goal remains to be explored.

Although this research paradigm is still nascent in the academic literature, the analysis of long-term memory in psychological processes in students offers attractive theoretical and applied perspectives. From a theoretical point of view, the finding of signatures typical of complex dynamical systems within the time series analyzed informs about the nature of the processes governing approach and avoidance motivation. Unlike mainstream research in psychology, which views processes as structures of relationships between components [31, 44, 45], the present result supports the conceptualization of achievement motivation as a phenomenon emerging from the self-organizing complexity of interactions and nonlinear attractor dynamics [14, 15]. Indeed, whereas the component-dominant approach is currently criticized for its inability to account for the complexity, the dynamics, and the idiosyncratic character of psychological processes [1], the interaction-dominant approach inherent in the complex dynamical systems paradigm is specifically designed to account for these characteristics. Thus, this approach enables researchers to access a *process causality* of psychological phenomena [2]. Like many other psychological phenomena involved in academic achievement as well as in the harmonious development of individuals, achievement motivation will be best understood when examined through the lens of the complex dynamical systems approach [14].

In terms of practical application, the value of analyzing relatively short time series, such as that offered by Whittle's MLE, lies in the possibility of tracking the individual trajectories of psychological phenomena important to academic achievement. Today, the development of online course materials makes it possible to collect a large amount of data relating to students' logins, which can serve as behavioral indicators of their engagement in learning throughout an academic year. This is what López-Pernas and Saqr [16] did on a series of only 18 consecutive courses. Extending this monitoring on more courses would enable the collection of a number of data sufficient to implement Whittle's MLE. Then, the detection of drifts in the distribution of motivational measures towards white noise or Brownian motion could serve as a warning signal of the risk of academic failure and dropping out, and of the imminent need to deploy individualized support strategies. The implementation of such an individualized monitoring and intervention strategy would be greatly facilitated by the development of algorithms capable of automatically analyzing the time series of data relating to the successive logins and proposing exercises tailored to the needs of each student.

Like any method, Whittle's MLE has its own limitations. Although it can be used to analyze shorter series than DFA, Roume [10] has shown, for example, that with 128 data points, errors in classifying the signal as stationary or non-stationary can occur for α values close to 1, and more particularly just above 1. This is why we have considered the α value resulting from the case study—based on 111 data points—presented in this chapter as only acceptable, but not optimal. Consequently, Whittle's MLE users are encouraged, wherever possible, to analyze longer series. Otherwise, the estimated values can only be used for descriptive purposes, but hardly for statistical inference.

5 Conclusion

While variable-centered nomothetic approaches enable the finding of statistical laws concerning large populations or groups, person-specific approaches, such as the present application of Whittle's MLE, offer access to processes that are by definition intra-individual. In psychology, the laws discovered by variable-centered approaches are rarely ergodic, that is, they are hardly applicable to the person level [3, 46–48]. However, in some ways, person-specific and variable-centered approaches can go hand in hand. While typical dynamics reflecting process causality can be detected by specific time series analyses conducted at the person level, the accumulation of these individual dynamics should enable, at population level, inferential statistics that are comparable to those observed in nomothetic research. As a result, using individual cases as a starting point for generalization may lead to replicating the trends observed by variable-centered approaches, while at the same time accounting for particularities that may even contradict these trends. This way of restoring convergence between person-specific and variable-centered approaches is one promising strategy for solving the ergodicity issue in psychological science [1].

6 Suggested Readings and Example Papers

In this chapter, you have seen an introduction and tutorial on how to apply Whittle's MLE in educational research. In particular, this technique allows the use of shorter series than other more popular techniques, which is an advantage for its use in the educational field as it can be difficult to collect more than 500 data points per series. This technique has not yet been applied to the learning context, but for example to the biomedical field [49].

Readers interested in time series analysis in the field of learning may refer to the work of Lowie et al. in language learning [32] and Nourrit-Lucas et al. in motor learning [33]. In the educational field, the work of Koopmans [50, 51] also illustrates how school attendance shows dynamics in $1/f$ noise.

Finally, more broadly, the articles by Hilpert and Marchand [52] and Koopmans [53] offer a rich overview of the complex dynamical systems paradigm, its key concepts and the methodologies that can be employed for education.

References

1. Gernigon C, Den Hartigh RJR, Vallacher RR, Van Geert PLC (2024) How the complexity of psychological processes reframes the issue of reproducibility in psychological science. Persp Psychol Sci 19:952–977. https://doi.org/10.1177/17456916231187324
2. Van Geert P, De Ruiter N (2022) Toward a process approach in psychology: stepping into Heraclitus' river. Cambridge University Press, Cambridge. https://doi.org/10.1017/9781108859189
3. Saqr M, Vogelsmeier LVDE, López-Pernas S (2024) Capturing where the learning process takes place: a person-specific and person-centered primer. Learn Indiv Differ 113:102492. https://doi.org/10.1016/j.lindif.2024.102492
4. Peng C-K, Mietus J, Hausdorff JM, Havlin S, Stanley HE, Goldberger AL (1993) Long-range anticorrelations and non-Gaussian behavior of the heartbeat. Phys Rev Lett 70:1343–1346. https://doi.org/10.1103/PhysRevLett.70.1343
5. Mandelbrot BB, Van Ness JW (1968) Fractional brownian motions, fractional noises and applications. SIAM Review 10:422–437. https://www.jstor.org/stable/2027184
6. Delignières D, Marmelat V (2013) In: Richardson MJ, Riley MA, Shockley K (eds) Theoretical and methodological issues in serial correlation analysis. Springer, New York, NY, pp 127–148. https://doi.org/10.1007/978-1-4614-5465-6_7
7. Phinyomark A, Larracy R, Scheme E (2020) Fractal Analysis of human gait variability via stride interval time series. Front Physiol 11. https://doi.org/10.3389/fphys.2020.00333
8. Likens AD, Mangalam M, Wong AY, Charles AC, Mills C (2023) Better than DFA? A bayesian method for estimating the hurst exponent in behavioral sciences. ArXiv arXiv:2301.11262v1. https://www.ncbi.nlm.nih.gov/pubmed/36748008%20PMCID:%20PMC9900970
9. Mangalam M, Wilson T, Sommerfeld J, Likens AD (2023) Optimizing a bayesian method for estimating the hurst exponent in behavioral sciences. https://arxiv.org/abs/2301.12064
10. Roume C (2023) A guide to Whittle maximum likelihood estimator in MATLAB. Front Netw Physiol 3. https://doi.org/10.3389/fnetp.2023.1204757
11. Beran J, Feng Y, Ghosh S, Kulik R (2013) Long-memory processes: probabilistic properties and statistical methods. Springer, Berlin, Heidelberg. https://doi.org/10.1007/978-3-642-35512-7
12. Almurad ZMH, Delignières D (2016) Evenly spacing in detrended fluctuation analysis. Phys A Stat Mech Appl 451:63–69. https://doi.org/10.1016/j.physa.2015.12.155
13. Diebolt C, Guiraud V (2005) A note on long memory time series. Quality Quantity 39:827–836. https://doi.org/10.1007/s11135-004-0436-z
14. Gernigon C, Altamore R, Vallacher RR, van Geert PLC, Den Hartigh RJR (2025) Almost, but not quite there: research into the emergence of higher-order motivated behavior should fully embrace the dynamic systems approach. Behav Brain Sci 48:e34. https://doi.org/10.1017/S0140525X24000384
15. Gernigon C, Vallacher RR, Nowak A, Conroy DE (2015) Rethinking approach and avoidance in achievement contexts: the perspective of dynamical systems. Rev Gen Psychol 19:443–457. https://doi.org/10.1037/gpr0000055
16. López-Pernas S, Saqr M (2024) How the dynamics of engagement explain the momentum of achievement and the inertia of disengagement: a complex systems theory approach. Comput Human Behav 153:108126. https://doi.org/10.1016/j.chb.2023.108126
17. Tuller B, Case P, Ding M, Kelso JAS (1994) The nonlinear dynamics of speech categorization. J Exp Psychol Hum Percept Perform 20:3–16. https://doi.org/10.1037/0096-1523.20.1.3
18. Bandura A (1997) Self-efficacy: the exercise of control. W H Freeman/Times Books/ Henry Holt & Co, New York, NY, US
19. Crocker J, Wolfe CT (2001) Contingencies of self-worth. Psychol Rev 108:593–623. https://doi.org/10.1037/0033-295X.108.3.593
20. Teboul A, Klosek C, Montiny C, Gernigon C (2019) Development and validation of the approach-avoidance system questionnaire (AASQ). Front Psychol 10:2531. https://doi.org/10.3389/fpsyg.2019.02531

21. Kopra J, Tikka S, Heinäniemi M, López-Pernas S, Saqr M (2024) An R approach to data cleaning and wrangling for education research. In: Saqr M, López-Pernas S (eds) Springer Nature Switzerland, Cham, pp 95–119. https://doi.org/10.1007/978-3-031-54464-4_4
22. López-Pernas S, Misiejuk K, Tikka S, Kopra J, Heinäniemi M, Saqr M (2024) In: Saqr M, López-Pernas S (eds) Visualizing and reporting educational data with R. Springer Nature Switzerland, Cham, pp 151–194. https://doi.org/10.1007/978-3-031-54464-4_6
23. Delignières D, Marmelat V (2012) Fractal fluctuations and complexity: current debates and future challenges. Crit Rev Biomed Eng 40. https://doi.org/10.1615/CritRevBiomedEng.2013006727
24. Gilden DL (2001) Cognitive emissions of 1/f noise. Psychol Rev 108:33–56. https://doi.org/10.1037/0033-295x.108.1.33
25. West BJ, Shlesinger M (1990) The noise in natural phenomena. Am Sci 78:40–45. https://www.jstor.org/stable/29773862
26. Delignières D, Fortes M, Ninot G (2004) The fractal dynamics of self-esteem and physical self. Nonlinear Dyn Psychol Life Sci 8:479–510. https://www.ncbi.nlm.nih.gov/pubmed/15473949
27. De Ruiter NMP, Den Hartigh RJR, Cox RFA, Van Geert PLC, Kunnen ES (2015) The temporal structure of state self-esteem variability during parent–adolescent interactions: more than random fluctuations. Self Identity 14:314–333. https://doi.org/10.1080/15298868.2014.994026
28. Wong AE, Vallacher RR, Nowak A (2014) Fractal dynamics in self-evaluation reveal self-concept clarity. Nonlinear Dyn Psychol Life Sci 18:349–369. https://www.ncbi.nlm.nih.gov/pubmed/25196705
29. Holden JG, Van Orden GC, Turvey MT (2009) Dispersion of response times reveals cognitive dynamics. Psychol Rev 116:318–342. https://doi.org/10.1037/a0014849
30. Kello CT, Beltz BC, Holden JG, Van Orden GC (2007) The emergent coordination of cognitive function. J Exp Psychol Gen 136:551–568. https://doi.org/10.1037/0096-3445.136.4.551
31. Van Orden GC, Holden JG, Turvey MT (2003) Self-organization of cognitive performance. J Exp Psychol Gen 132:331–350. https://doi.org/10.1037/0096-3445.132.3.331
32. Lowie W, Plat R, De Bot K (2014) Pink noise in language production: a nonlinear approach to the multilingual lexicon. Ecol Psychol 26:216–228. https://doi.org/10.1080/10407413.2014.929479
33. Nourrit-Lucas D, Tossa AO, Zélic G, Delignières D (2015) Learning, motor skill, and long-range correlations. J Motor Behav 47:182–189. https://doi.org/10.1080/00222895.2014.967655
34. Den Hartigh RJR, Cox RFA, Gernigon C, Yperen NWV, Van Geert PLC (2015) Pink noise in rowing ergometer performance and the role of skill level. Motor Control 19:355–369. https://doi.org/10.1123/mc.2014-0071
35. Den Hartigh RJR, Otten S, Gruszczynska ZM, Hill Y (2021) The relation between complexity and resilient motor performance and the effects of differential learning. Front Human Neurosci 15. https://doi.org/10.3389/fnhum.2021.715375
36. Delignières D, Torre K (2009) Fractal dynamics of human gait: a reassessment of the 1996 data of Hausdorff et al. J Appl Physiol 106:1272–1279. https://doi.org/10.1152/japplphysiol.90757.2008
37. Goldberger AL, Amaral LAN, Hausdorff JM, Ivanov PCh, Peng C-K, Stanley HE (2002) Fractal dynamics in physiology: alterations with disease and aging. Proc Natl Acad Sci 99:2466–2472. https://doi.org/10.1073/pnas.012579499
38. Hausdorff JM, Mitchell SL, Firtion R, Peng C-K, Cudkowicz ME, Wei JY, Goldberger AL (1997) Altered fractal dynamics of gait: reduced stride-interval correlations with aging and Huntington's disease. J Appl Physiol (Bethesda, Md: 1985) 82:262–269. https://doi.org/10.1152/jappl.1997.82.1.262
39. Gottschalk A, Bauer MS, Whybrow PC (1995) Evidence of chaotic mood variation in bipolar disorder. Arch Gen Psych 52:947–959. https://doi.org/10.1001/archpsyc.1995.03950230061009
40. Wijnants ML, Hasselman F, Cox RFA, Bosman AMT, Van Orden G (2012) An interaction-dominant perspective on reading fluency and dyslexia. Ann Dyslexia 62:100–119. https://doi.org/10.1007/s11881-012-0067-3

41. Wijnants ML, Bosman AMT, Hasselman F, Cox RFA, Van Orden GC (2009) 1/f scaling in movement time changes with practice in precision aiming. Nonlinear Dyn Psychol Life Sci 13:79–98. https://www.ncbi.nlm.nih.gov/pubmed/19061546
42. Arrieta C, Navarro J, Vicente S (2008) Factores asociados a la emergencia de patrones diferenciales de la motivación en el trabajo [Variables related to the emergence of differential patterns in work motivation.]. Psicothema 20:745–752
43. Navarro J, Rueff-Lopes R (2015) Healthy variability in organizational behavior: empirical evidence and new steps for future research. Nonlinear Dyn Psychol Life Sci 19:529–552. https://www.ncbi.nlm.nih.gov/pubmed/26375939
44. Den Hartigh RJR, Cox RFA, Van Geert PLC (2017) In: Magnani L, Bertolotti T (eds). Complex versus complicated models of cognition. Springer International Publishing, Cham, pp 657–669. https://doi.org/10.1007/978-3-319-30526-4_30
45. Wallot S, Kelty-Stephen DG (2018) Interaction-dominant causation in mind and brain, and its implication for questions of generalization and replication. Minds Mach 28:353–374. https://doi.org/10.1007/s11023-017-9455-0
46. Fisher AJ, Medaglia JD, Jeronimus BF (2018) Lack of group-to-individual generalizability is a threat to human subjects research. Proc Natl Acad Sci 115:E6106–E6115. https://doi.org/10.1073/pnas.1711978115
47. Hamaker EL (2024) The curious case of the cross-sectional correlation. Multivar Behav Res 59:1111–1122. https://doi.org/10.1080/00273171.2022.2155930
48. Molenaar PCM, Campbell CG (2009) The new person-specific paradigm in psychology. Curr Direct Psychol Sci 18:112–117. https://doi.org/10.1111/j.1467-8721.2009.01619.x
49. Harrar K, Hamami L, Lespessailles E, Jennane R (2013) Piecewise Whittle estimator for trabecular bone radiograph characterization. Biomed Signal Process Control 8:657–666. https://doi.org/10.1016/j.bspc.2013.06.009
50. Koopmans M (2015) A dynamical view of high school attendance: an assessment of short-term and long-term dependencies in five urban schools. Nonlinear Dyn Psychol Life Sci 19:65–80. https://www.ncbi.nlm.nih.gov/pubmed/25575559
51. Koopmans M (2018) https://www.ncbi.nlm.nih.gov/pubmed/29600954On the pervasiveness of long range memory processes in daily high school attendance rates. Nonlinear Dyn Psychol Life Sci 22:243–262
52. Hilpert JC, Marchand GC (2018) Complex systems research in educational psychology: aligning theory and method. Educ Psychol 53:185–202. https://doi.org/10.1080/00461520.2018.1469411
53. Koopmans M (2020) Education is a complex dynamical system: challenges for research. J Exp Educ 88:358–374. https://doi.org/10.1080/00220973.2019.1566199

Automating Individualized Machine Learning and AI Prediction Using AutoML: The Case of Idiographic Predictions

Mohammed Saqr, Ahmed Tlili, and Sonsoles López-Pernas

Abstract This chapter introduces a tutorial on using Automated Machine Learning (AutoML) to automate and scale predictive modeling in education. In particular, we illustrate the usefulness of AutoML for idiographic analysis, where each individual student has their own particular model fitted from their own individual data. We demonstrate how AutoML simplifies the ML pipeline, enabling the creation of individually optimized models for multiple datasets. Moreover, we illustrate how to apply explainable artificial intelligence techniques to automate the interpretation of the main model predictors, offering a view of the variables that matter. The complete pipeline demonstrated in this tutorial holds potential to provide automated real-time insights based on idiographic analysis in a transparent and trustable way.

Keywords Learning analytics · Automated machine learning · Precision education · Idiographic analysis · Explainable artificial intelligence

1 Introduction

Predictive analytics is a branch of analytics that uses students' data, statistical algorithms, or machine learning (ML) techniques to forecast future outcomes [1–4]. Prediction is important for developing theories, hypothesis testing, and evaluating the relevance of the theoretical knowledge to research and practice. Therefore, prediction has been a central pillar of scientific research [5]. Moreover, predictive analytics enables organizations to make informed decisions, optimize resources, and anticipate future risks or opportunities [6]. As such, predictive analytics has been widely adopted across industries, from healthcare to marketing, due to its

M. Saqr (✉) · S. López-Pernas
University of Eastern Finland, Joensuu, Finland
e-mail: mohammed.saqr@uef.fi

A. Tlili
Beijing Normal University, Beijing, China

M. Saqr, S. López-Pernas (eds.), *Advanced Learning Analytics Methods*,
https://doi.org/10.1007/978-3-031-95365-1_22

ability to transform data into actionable insights. As data becomes increasingly diverse and voluminous, predictive analytics plays a critical role in handling big data and providing actionable insights [7]; [4]; [8], which has been a central interest of researchers in learning analytics and artificial intelligence (AI) in education [9].

In the previous chapters we studied modeling full datasets and explained them [10–13]. In this chapter we will learn how to model each student individually as a whole unit of analysis. In other words, the full dataset belongs to a single student. This is done by collecting several data points from the same student over time that is enough for building a predictive model for the student. These models are known as idiographic e.g., [8, 14, 15]. Idiographic models are novel and promise accurate insights at the resolution of single students. For a detailed reading about these models, interested readers can resort to chapter [15] or these papers [8], [16–19].

Given the need to model students individually, the process needs to be automated to process multiple datasets. Therefore, in this chapter, we will learn how to implement AutoML for a large number of idiographic datasets using `h2o`, which is a cross-platform open-source, scalable machine learning framework that supports a wide range of algorithms [20]. Among the algorithms that `h2o` supports are generalized linear models (GLM), gradient boosting machines (GBM), random forests (RF), deep learning (DL), k-means clustering, and extreme gradient boosting (XGBoost). H2O is designed to handle large datasets can efficiently use distributed computing capabilities. The `h2o` package works seamlessly with R and is also available for Python among others. It performs automation of the ML process from model selection and hyperparameter tuning and reporting of results. Also, `h2o` includes tools for model explanation and supports easy visualization and plotting of results [20].

2 Automated Machine Learning (AutoML)

The core of predictive analytics lies in building predictive models, which are often built using techniques such as regression analysis, decision trees, and neural networks [2, 4]. These models are trained to recognize complex relationships within the data, making it possible to forecast future trends with a degree of confidence [2, 7]. Constructing such predictive models entails several steps which are often repetitive which roughly includes data pre-processing and preparation for analysis, feature engineering and selection to identify the relevant variables. Next, model selection, training and evaluation of model's performance . Based on these results, the model is tuned and optimized for better accuracy [10, 11].

Since the steps to conduct the predictive analytics workflow are mostly the same each time, it is possible to automate them using AutoML [21–23]. AutoML is a relatively novel and growing area in machine learning (ML) that aims to—as the name implies—automate the process of applying ML to real-world problems

[24, 25]. In fast-paced environments, models need to be updated and redeployed regularly to adapt to new data or changing conditions. AutoML automates many steps of the ML life cycle, making it feasible to frequently refresh models as part of a continuous integration pipeline. Furthermore, when working with a large number of similar datasets, it becomes impractical to repeat the ML process for each manually. As such, AutoML allows to automate this process, optimizing ML model to each specific dataset. Besides the obvious scalability benefits, research has found that AutoML can achieve good results as traditional ML with expert tuning [8, 24, 25].

The core function of AutoML lies in optimizing ML workflows, from data preprocessing to model deployment. Recent advancements in AutoML techniques have led to substantial improvements in predictive performance [26]. For example, the framework now encompasses methodologies such as neural architecture search, which automatically designs neural networks tailored to specific tasks. This has led to models that achieve state-of-the-art performance across numerous benchmarks [27]. Additionally, the introduction of automated feature engineering techniques enables models to learn relevant features from raw data, enhancing model accuracy without extensive manual intervention [28].

Furthermore, AutoML streamlines the process of feature selection, making it easier to handle high-dimensional data [20]. The automation of hyperparameter tuning and model selection significantly reduces the time required for model development. AutoML systems may outperform traditional methods in terms of efficiency, leading to quicker time-to-insight and facilitating more rapid iterations in model improvement [29]. AutoML can also achieve competitive or superior performance compared to human experts. Through exhaustive search strategies and advanced algorithms, AutoML often yields models that may surpass human-crafted solutions [30]. In doing so, AutoML can make high-quality models accessible to non-experts while also reducing the time and computational costs traditionally associated with ML development [23]. Further, automating the time-consuming aspects of model development allows AutoML tools to lower the barriers for organizations and individuals to utilize ML, enabling a broader range of industries to integrate AI-driven solutions [31]. This is increasingly relevant, as organizations across various sectors progressively rely on data-driven decision-making to maintain competitive advantage.

There are also several challenges and opportunities within AutoML. One prominent challenge is the computational cost, particularly as the demand for larger and more complex models (such as large language models, LLMs) grows. Applying AutoML to such models, which require enormous computational resources, poses difficulties for scaling and sustainability [21]. Another issue is the environmental impact of AutoML processes, as they consume significant energy and computational resources. The integration of "Green AI" principles, which emphasize energy-efficient algorithms and workflows, has been proposed to address these concerns [32].

3 AutoML in Education Research

The use of AutoML has not been very prominent in the education scientific literature. However, an increasing number of studies recognize its potential. The study by Garmpis et al. [22] addressed the challenges posed by recent policy changes in higher education systems in Greece, which led to departmental unification and increased dropout rates. The researchers utilized AutoML to identify the best predictive models and parameters and analyzed a dataset of 23,687 students enrolled in the "Computer Networking" module, employing the open-source data science platform for model development and evaluation. The model's accuracy was assessed using established evaluation metrics, including precision, recall, and F-measure. Their findings indicated that the model accurately predicted approximately 92% of student performance, specifically identifying students at risk of dropping out or facing delays in graduation. This approach demonstrates the potential of AutoML in enhancing educational outcomes and supporting strategic interventions for student success [22]. Similarly, Zhao and Wang [33] used AutoML to predict online learning adaptability and achieved a high accuracy of 90.9%. Their empirical findings showed that AutoML yielded higher performance than traditional methods for all the evaluated metrics.

In another study, Bosch [34] conducted experiments using AutoML to streamline feature engineering for student models in the National Assessment of Educational Progress data mining competition. Bosch compared two AutoML feature engineering methods, Featuretools and TSFRESH (Time Series FeatuRe Extraction on basis of Scalable Hypothesis tests), which had rarely been applied to student interaction log data. The TSFRESH method provides a novel feature selection technique to manage over 4,000 heterogeneous features with a relatively small student sample, The study findings revealed that TSFRESH features were significantly more effective than both Featuretools and expert-engineered features [34]. However, the interpretability of these models remains limited. To overcome the limitation of interpretability in Bosch's study [34], Buñay-Guisñan and colleagues [35] developed a dashboard that combines AutoML and Explainable AI to predict student dropout, allowing non-experts to upload datasets and automatically generate predictive models, along with counterfactual explanations to help users understand the factors affecting student dropout.

In another study, Tsiakmaki et al. [36] focused on predicting students' academic performance, researchers explored the use of AutoML strategies, a critical area within Educational Data Mining and Learning Analytics. Recognizing the need for accurate learning models at a low cost, they introduced a fuzzy-based active learning method that integrates AutoML practices in a modular fashion. This approach leverages fuzzy logic to create high-performance and transparent models while reducing the time and cost of labeling data by effectively utilizing a small set of labeled data alongside a larger set of unlabeled data. Given the complexity of ML algorithms and the high-dimensional input space, selecting the right method and optimizing parameters can be challenging. To address this, they conducted numerous experiments that demonstrated the efficiency of their proposed method in accurately identifying students at risk of failure. These insights not only have the potential to enhance the learning experience but also contribute significantly to the broader science of learning.

The few existing research studies in AutoML appear to take advantage of this technique to optimize their models and to make ML accessible to non-experts. However, the potential of AutoML to automate the ML pipeline for many individual models to enable idiographic ML has barely been fulfilled. To the knowledge of the authors, only one article has come close to implementing this technique to automatically create, fit and optimize idiographic models [8], which is the aim of this tutorial. In said study, the authors automated the feature selection, as well as the model training and evaluation of the prediction of students' variables related to self-regulated learning based on previous measures.

4 Tutorial: Using `h2o` for ML for Idiographic ML

In this tutorial, we will explore how to use the `h2o` ML platform to build, evaluate, and compare idiographic models. It is assumed that the reader is familiar with the R programming language. If that is not the case, it is recommended to refer to previous tutorials on the basics of R [37], data cleaning [38], basic statistics [39], and visualization [40], as well as prediction [2, 10, 11] and explainable AI [12, 13]. When working with idiographic analysis, we treat each individual person as an independent dataset and, therefore, we need a separate ML model for each person [15]. Implementing all steps of the ML pipeline for each individual dataset is extremely time consuming for researchers, and impossible to do in real time if the ML models are to be deployed and regularly updated when new data is collected. Therefore, AutoML is more suitable for idiographic analysis.

As a first step, we need to load the essential R packages. We load the `tidyverse` package for data manipulation tasks, and the `h2o` package for AutoML. The next code loads the dataset with the `readRDS()` function. We also set a random seed that ensures the split is reproducible, helping to maintain consistency in results.

```
set.seed(265)
# Load necessary libraries
library(tidyverse)  # For data manipulation
library(h2o)     # For H2O ML platform

# Read the dataset
synthetic_data <- readRDS("https://github.com/lamethods/data2/raw/main/srl/srl.RDS")
```

The dataset used in this tutorial is a synthetic dataset based on the work by [8] which collected data using the Concise Self-Regulation Survey (CSRL) [41], as well as phone sensor and contextual data. However, we will use only the a combination of CSRL and sensor data here. In the next step, we select and define the variables that we will use as inputs for our ML models. We start by categorizing the variables into two groups: phone activity variables and self-regulated learning (SRL) variables.

The phone activity variables represent various physical and behavioral patterns related to daily activities, giving us insights into a person's physical and digital behavior, which may affect learning performance.

These include the following features:

- `"meanpace"`: average pace of movement
- `"steps"`: the number of steps taken
- `"freqscreen"`: frequency of screen use
- `"duration"`: duration of screen usage
- `"scrrenoff"`: time the screen is off
- `"nightduroff"`: duration of screen-off time during the night
- `"maxoffscreennight"`: the longest period during which the screen is off at night.

The **SRL variables** capture key aspects of cognitive and emotional self-regulation, which are crucial for understanding how learners manage and control their learning processes. These variables are derived from self-reported data reflecting various self-regulated learning strategies. They include:

- `"efficacy"`: The belief in one's ability to complete learning tasks effectively.
- `"value"`: The perceived importance or value of learning tasks.
- `"planning"`: The ability to plan and organize learning activities ahead of time.
- `"monitoring"`: Keeping track of learning progress and task completion.
- `"effort"`: The amount of effort put into learning tasks.
- `"control"`: The ability to resist distractions and stay focused on learning.
- `"help"`: Seeking help from teachers, friends, or other resources when needed.
- `"social"`: Interactions and feelings of belonging within the college or learning community.
- `"organizing"`: Organizing and structuring study time and materials.
- `"motivated"`: The level of motivation and enthusiasm towards learning and achieving better grades.
- `"feedback"`: Learning from feedback to improve performance.
- `"evaluating"`: Self-evaluating one's work to improve skills and performance.
- `"anxiety"`: The level of anxiety or stress experienced during learning or in tasks.
- `"enjoyment"`: The amount of enjoyment and satisfaction gained from completing tasks and achieving goals.

Next, we combine both the self-regulated learning (SRL) variables and the phone activity variables into a single vector of predictors. We then define the response variable as `"learning"` (how much the student learnt during this day or period or gained knowledge), which serves as the target variable that the model will attempt to predict based on the input predictors.

```
# Define the phone activity variables
phone_vars <- c("meanpace", "steps", "freqscreen", "duration",
                "scrrenoff", "nightduroff", "maxoffscreennight")

# Define the SRL variables
srl_vars <- colnames(select(synthetic_data, efficacy:enjoyment))

# Combine all variables into the Predictors list
predictors <- c(srl_vars, phone_vars)

# Set the response variable
response_variable <- "learning"
```

4.1 Building an Idiographic Model

Before working with the H2O AutoML framework, we need to initialize the `h2o` cluster. This involves allocating resources such as the number of threads and the maximum memory size for the computation. Here, initialize `h2o` with the `nthreads` parameter set to `12`, which allows `h2o` to utilize 12 CPU cores. Additionally, the `max_mem_size` parameter is set to `"12G"`, specifying that up to 12GB of memory can be used for model training and processing. You can adjust these settings based on your system's resources to ensure optimal performance during model training and evaluation. The initialization results in a "Connection successful!" message and `h2o` prints the `h2o` cluster information which include the up time, the timezone and detailed version information.

```
# Step 1: Initialize h2o with specified resources
h2o.init(nthreads = 12, max_mem_size = "12G")  # Adjust memory and threads if necessary
```

Given that idiographic models are single-subject by definition, we need to select a single student from the dataset who will be the target of our initial analysis. Using the `unique()` function, we extract a list of unique student names and choose a particular student by specifying their index (e.g., the 5th student who carries the fake name Alice). Once Alice is selected, we filter the dataset using the `filter()` function from the `dplyr` package to include only the rows that correspond to Alice's data (the 5th student). After filtering, we remove the `"name"` column, as it is no longer needed with the `select()` function. This results in a filtered dataset that contains only the relevant variables for Alice for further analysis. If we know the student already, we could of course use a simpler way by filtering only Alice data like the code below using `synthetic_data %>% filter(name == "Alice")`

```
# Step 2: Retrieve the specific name (e.g., the first unique name)
specific_name <- unique(synthetic_data$name)[5]
```

```
# Step 3: Filter the dataset for the selected person and remove the 'name' column
filtered_data <- synthetic_data %>%
  filter(name == specific_name) %>% # Filter rows where 'name' matches the selected person
  select(-name)  # Exclude the 'name' column from the data

# An alternative approach to selecting
filtered_data <- synthetic_data %>%
  filter(name == "Alice") %>% # Filter rows where 'name' matches the selected person
  select(-name)  # Exclude the 'Alice' column from the data
```

The following steps are similar to traditional ML with some minor differences. We prepare the data by first converting it into an `h2o` data frame by splitting it into training and testing sets, and defining the predictors and response variable. The data transformation into an `h2o` format is done using `as.h2o()` function. This transformation is necessary because `h2o` can only work with `h2o` data format. `h2o`'s requires this format for better optimization and efficient distributed processing and parallel computations which standard R data frames cannot provide.

```
# Disable progress bar for clarity
h2o.no_progress()
# Step 4: Convert the filtered data into an h2o frame
h2o_data <- as.h2o(filtered_data)  # Convert into an h2o format for processing
```

Next, we split the data into training and testing sets using `h2o.splitFrame()` and ratios = 0.8, which means 80% of the data and will be used to train the model, while the remaining 20% is reserved for testing. Splitting the data is standard procedure for evaluating the model's performance on unseen data and ensuring it generalizes well, also, we will use the testing set to evaluate the model. We also ensure that the seed is set using `seed = 256` for reproduciblity.

```
# Step 5: Split the h2o frame into training (80%) and testing (20%) sets
# Use seed for reproducibility
splits <- h2o.splitFrame(data = h2o_data, ratios = 0.8, seed = 256)
train_data_h2o <- splits[[1]]  # Training data (80%)
test_data_h2o <- splits[[2]]   # Testing data (20%)
```

Finally, we define the predictors and the response variable. The predictors include all the relevant features in the dataset except the target outcome, which is the `"learning"` variable. The model will use these predictors to learn and make predictions, with the response variable representing the learning outcome we aim to predict.

```
# Step 6: Define predictors and response variable for model training
predictors <- setdiff(names(filtered_data), "learning")
response_variable <- "learning"  # Specify the response variable name
```

Next, we will use `h2o`'s AutoML which will automate the complete ML pipeline for us. The function `h2o.automl()` simplifies the process of model training and evaluation by automatically testing multiple algorithms and performing hyperparameter tuning within a user-defined time limit. It stores all the models in the results, allowing us to easily compare their performance. Let's go through the parameters one by one.

- **`x = predictors`**: This specifies the **predictor variables** or features that the model will use to make predictions. The `predictors` list includes all the variables we want the model to learn from, excluding the target variable which are the SRL and phone activity data.
- **`y = response_variable`**: This is the **response variable**, also known as the target or dependent variable. In this case, it's set to `"learning"`, which represents the outcome we are trying to predict.
- **`training_frame = train_data_h2o`**: The `training_frame` is the dataset used to train the model, which must be an `h2o` Frame. In this case, it's the 80% split of the filtered data that we converted into an `h2o` Frame.
- **`nfolds = 5`**: This specifies the number of **cross-validation folds**. Cross-validation is a technique where the data is split into multiple parts, or "folds," and the model is trained on different subsets of the data while being validated on the remaining parts. By setting `nfolds = 5`, the training data is split into 5 folds, helping ensure that the model generalizes well to new data and doesn't over-fit and gives an average of the five training folds.
- **`max_runtime_secs = 900`**: This sets the **maximum time** (in seconds) that the AutoML process can run. It limits the overall time spent on training and evaluating models. In this case, we set a short time limit of 900 seconds, so the process will stop once this limit is reached, even if all possible models haven't been trained. This parameter can be adjusted based on the available computational resources and the task.
- **`seed = 256`**: When you specify a seed, the model training process can be repeated with the same results. This is important when you want to compare results or ensure consistency across different runs.

```
# Step 7: Train the model using h2o AutoML
automl_model <- h2o.automl(
  x = predictors,  # Predictor variables
  y = response_variable,  # Response variable
  training_frame = train_data_h2o,  # Training data
  nfolds = 5,  # Number of cross-validation folds
  max_runtime_secs = 900,  # Maximum runtime in seconds
  seed = 256  # Random seed for reproducibility
)
```

After `h2o` AutoML completes the job of training the models, it creates a leaderboard (a list models ranked from best to worst performing model). The leaderboard allows us to rank the resulting models that were estimated with different algorithms and hyperparameters. Given that there are multiple metrics to evaluate performance, there are several ways also to rank the algorithms. The `h2o` default methods for regression is Root Mean Squared Error (RMSE). To get the leaderboard, we use `h2o.get_leaderboard` we supply the `automl_model` given that it the object that contains the models, and set `extra_columns = "ALL"` to get all the possible evaluation parameters. To view the top 10 models in the leaderboard, we use the `head(leaderboard, 10)` function which will display a ranked list of models trained by `AutoML`.

```
# Step 8: View the leaderboard of models
leaderboard <- h2o.get_leaderboard(automl_model, extra_columns = "ALL")
head(leaderboard, 10)
```

The leaderboard has several fields which we will discuss (Table 1). The **model ID** identifies each model and specifies the algorithm used for training. For instance, the model named **GBM_grid_1_AutoML_93_20240930_00046_model_57** refers to a **Gradient Boosting Machine (GBM)** model while **StackedEnsemble_BestOfFamily_3_AutoML_93_20240930** uses the **StackedEnsemble** algorithm. The `h2o` package uses a naming convention that can provide insights about the model. In that, **StackedEnsemble_BestOfFamily_3_AutoML_93_20240930** indicates a **stacked ensemble model** that selects the best-performing models from the AutoML run, as indicated by "BestOfFamily." The number "3" shows that it is the third iteration of the model. "AutoML_93" means it is part of the 93rd AutoML run, and "20240930" is the date the AutoML process was executed, formatted as YYYYMMDD (September 30, 2024). The next field, **RMSE** (Root Mean Squared Error), is a well-known metric for evaluating regression models that measures the average prediction error. For example, the model **GBM_grid_1_AutoML_93_20240930_00046_model_57** has an **RMSE of 9.03**, which indicates that its predictions deviate by an average of 9.03 units from the actual values of the target variable (learning). Since it has the lowest RMSE on the leaderboard, this model is the best-performing model in our analysis. Please note that our variables scale ranges from 0-100 so, 9 points is still small. **MSE** (Mean Squared Error) is another evaluation metric that refers to the squared average prediction error. The **GBM_grid_1_AutoML_93_20240930_00046_model_57** model has a value of 81.53 and confirming its strong performance. **MAE** (Mean Absolute Error) measures the average magnitude of the errors in the predictions, without considering their direction. In this case, **GBM_grid_1_AutoML_93_20240930_00046_model_57** has the lowest MAE at **3.07**, indicating that its absolute prediction errors are smaller, on average, compared to other models, which have MAEs around 3.50–5.15. **RMSLE** (Root Mean Squared Logarithmic Error) is particularly useful when the target variable exhibits large variations. The **StackedEnsemble_BestOfFamily_3_AutoML_93_20240930** has an RMSLE of **0.38**, showing good performance in scenarios with varying target values. **Mean Residual Deviance** is similar to MSE and represents the average error per prediction. The **GBM_grid_1_AutoML_93_20240930_00046_model_57** has the lowest value at **81.53**, indicating strong predictive accuracy. Finally, the **training time (ms)** measures how long it took to train the model. The **StackedEnsem-**

Table 1 Leaderboard showing a list models ranked from best to worst performing model

model_id	rmse	mse	mae	rmsle	mean_res_dev	training_ms	algo
GBM_grid_1_AutoML_93_20240930_00046_model_57	9.03	81.53	3.07	0.39	81.53	384	GBM
StackedEnsemble_BestOfFamily_3_AutoML_93_20240930_00046	9.16	83.91	3.50	0.38	83.91	110	StackedEnsemble
StackedEnsemble_BestOfFamily_5_AutoML_93_20240930_00046	9.41	88.57	3.95	0.45	88.57	215	StackedEnsemble
StackedEnsemble_BestOfFamily_4_AutoML_93_20240930_00046	9.64	92.88	5.15	0.53	92.88	514	StackedEnsemble
GBM_grid_1_AutoML_93_20240930_00046_model_2	9.68	93.80	3.39	0.39	93.80	472	GBM
GBM_grid_1_AutoML_93_20240930_00046_model_54	9.76	95.30	3.24	0.40	95.30	354	GBM
GBM_grid_1_AutoML_93_20240930_00046_model_3	10.02	100.38	3.45	0.42	100.38	284	GBM
GBM_grid_1_AutoML_93_20240930_00046_model_63	10.34	106.96	3.23	0.40	106.96	462	GBM
GBM_grid_1_AutoML_93_20240930_00046_model_35	10.36	107.31	3.33	0.41	107.31	265	GBM
GBM_5_AutoML_93_20240930_00046	10.41	108.45	3.64	0.40	108.45	204	GBM

ble_BestOfFamily_3_AutoML_93_20240930 model trained in **110 milliseconds**, while the **GBM_grid_1_AutoML_93_20240930_00046_model_57** took **384 milliseconds**.

4.1.1 Best Models

To retrieve the best model based on pre-set criteria we can use the function `h2o.get_best_model`, which allow us to specify the metric that we want to use to select the best model or the algorithm (e.g., RF, GBM or GLM). To do so, we use the function `h2o.get_best_model()` with the parameter `automl_model` to obtain the best model and assign it to an R object `best_model`. In that case, `h2o` will retrieve the best model based on RMSE since it is the default ranking criteria. If you want to change that, you can use `criterion = "MAE"` for instance or other metrics e.g., RMSE.

```
# Step 9: Extract the best-performing model
best_model <- h2o.get_best_model(automl_model)
```

4.1.2 Evaluation of the Model

To evaluate how the model could perform in the wild we need to use completely new data that the model has not seen before. To test the model performance with data that the model has not seen before we can use `h2o.performance(best_model)` with the parameter `newdata = test_data_h2o`. This enables us to understand what the model can do in real-life applications. The `h2o.performance` function calculates several metrics such as MSE, RMSE, MAE, and RMSLE.

```
# Step 9.1: Evaluate the best model's performance on the test data
best_model_performance <- h2o.performance(best_model, newdata = test_data_h2o)

# Step 9.2: Print the performance metrics on test data
print(best_model_performance)

# Step 9.3: Generate predictions on the test data
predictions <- h2o.predict(best_model, test_data_h2o)
```

In our case, the metrics for the best **GBM** model were:

- **MSE**: 159.6543
- **RMSE**: 12.63544
- **MAE**: 4.978724
- **RMSLE**: 0.2364497
- **Mean Residual Deviance**: 159.6543

The **RMSE** indicates that the GBM model's predictions were **12.64 points** on average different from the values of the target variable (learning). This RMSE suggests that the model has a reasonable level of error in predicting the target variable. Similarly, the **MSE** (Mean Squared Error) is **159.65**, which reflects the squared average of the errors, while the **MAE** is **4.98**, showing the average absolute difference between predicted and actual values. The **RMSLE** is **0.24**, which is helpful for understanding the model's performance when the target variable has large variations. The **Mean Residual Deviance** is also **159.65**, representing the average error per prediction, which aligns with the MSE. Please note that these parameters are different than the above ones in the leaderboard table, since we evaluated the model with unseen data, and it is expected to see some drop in model performance.

To visualize the evaluation, we can create predictions on the test data using `h2o.predict(best_model, test_data_h2o)` to visualize the model predictions with the new data. We do so by plotting the predicted versus actual values using a scatter plot, as demonstrated below. Note that we must convert the `h2o` data frames into standard data frames to use them with traditional R packages. It may be also helpful to analyze the residuals (the differences between the actual and predicted values). Ideally, residuals should be randomly distributed around zero without any clear patterns. The code below computes the residuals, and plots them. As Fig. 1 shows, we can see that most observations are around 0 however, some extreme values are also there indicating that the model fared bad in some days. This is of course expected given the variability of daily performance.

```
#Step 2: Convert the test data and predictions to R data frames}
test_data_df <- as.data.frame(test_data_h2o) # Convert h2o test data to dataframe
predictions_df <- as.data.frame(predictions)# Convert h2o predictions to dataframe

# Step 3: Combine the predictions with the original test data
result_df <- cbind(test_data_df, Predicted_Learning = predictions_df$predict)

# Scatterplot of actual vs predicted values
ggplot(result_df, aes(x = learning, y = Predicted_Learning)) +
```

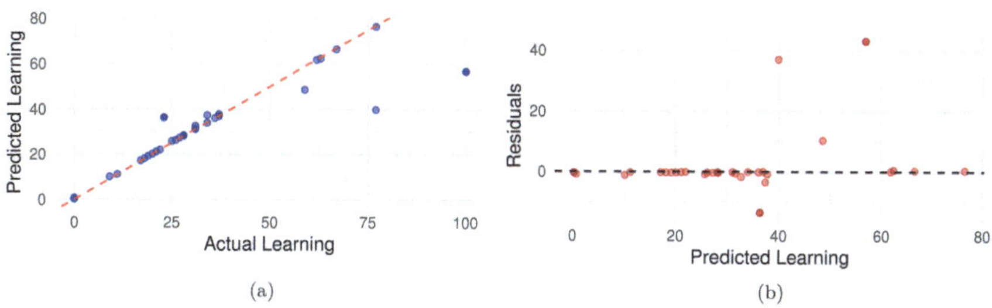

Fig. 1 Visualization of the evaluation. (**a**) Actual vs. Predicted learning. (**b**) Residuals vs. Predicted learning

```
  geom_point(color = "blue", alpha = 0.5) +  # Scatter points
  # Ideal line (y = x)
  geom_abline(slope = 1, intercept = 0, linetype = "dashed", color = "red") +
  labs(x = "Actual Learning",
       y = "Predicted Learning") +
  theme_minimal()

# Create a residual column
result_df <- result_df %>%
  mutate(Residuals = learning - Predicted_Learning)

# Scatterplot of residuals vs predicted values
ggplot(result_df, aes(x = Predicted_Learning, y = Residuals)) +
  geom_point(color = "darkred", alpha = 0.5) +
  geom_hline(yintercept = 0, linetype = "dashed") +
  labs(x = "Predicted Learning",
       y = "Residuals") +
  theme_minimal()
```

However, the best model chosen automatically by `h2o` may not be the optimal solution for every case. One has to balance accuracy, training time, and model interpretability. For instance, GBM and XGBoost are capable of handling complex relationships and often provide good predictive performance maybe computationally expensive. Random Forest (DRF) is a faster algorithm and is particularity distinguished for its inherent interpretability making it suitable when simplicity and explainability are needed. Deep Learning is adept at capturing non-linear patterns is also computationally expensive and harder to explain. Stacked ensembles combine several models to enhance performance but increases complexity and training time. In our case, we need a model that offers performance and be explainable e.g., GBM or DRF.

To select a certain model, `h2o` also allows us to extract top algorithms by type by specifying them in the function call e.g., GBM, XGBoost, DRF, DeepLearning, and StackedEnsemble. To do so, we call the function `h2o.get_best_model` with the parameter `algorithm = "algorithm_name"`. For example, to get the best GBM model, we call `h2o.get_best_model(automl_model, algorithm = "gbm")` and then we store it in the object `best_gbm`. In the same way, we can get other models like the code below.

The next code retrieves the best performing models for various ML algorithms. Specifically, it gets the best GBM, XGBoost, Distributed Random Forest, Deep Learning, and Stacked Ensemble models, and stores them in the `best_gbm`, `best_xgboost`, `best_drf`, `best_deeplearning`, and `best_stackedensemble` variables respectively.

```
best_gbm <- h2o.get_best_model(automl_model, algorithm = "gbm")
best_xgboost <- h2o.get_best_model(automl_model, algorithm = "xgboost")
```

```
best_drf <- h2o.get_best_model(automl_model, algorithm = "drf")
best_deeplearning <- h2o.get_best_model(automl_model, algorithm = "deeplearning")
best_stackedensemble <- h2o.get_best_model(automl_model, algorithm = "stackedensemble")
```

Explainability offers better transparency and helps build trust in the model's decisions. This is especially useful when making critical decisions based on ML predictions. The `h2o` package offers several well-crafted functions for evaluating the models with rich visualizations through `h2o.explain()`. Besides basic model performance metrics, it provides a wide range of diagnostic information and visualizations (Fig. 2), such as variable importance, residual analysis, and prediction explanations plots. See [12, 13] for a full explanation on explainable ML models and techniques.

The code below runs `h2o.explain()` function to analyze and explain the behavior of the model. We here choose another model, given that we already seen the GBM model. We choose the `best_drf` (a previously trained model) to be evaluated on the `test_data_h2o`. The function generates a detailed explanation of each plot and what it means, so we will not repeat them here. Of interest to our case is the Variable Importance Plot, which ranks the variables based on their contribution to the model's predictions. The variable importance plot shows that feedback, effort, and duration are the most influential variables in predicting learning outcomes. Other variables like help, mean pace, and steps also contribute, while factors such as planning, evaluating, and social engagement have comparatively less impact.

The SHAP summary plot provides detailed visualization of individual predictions by showing how much each feature contributed to pushing a prediction higher or lower for a specific observation. In SHAP plots, the x-axis represents the SHAP value (i.e., the contribution to the prediction) where positive values increasing predictions and negative values lowering them. The color scale is the normalized value of each feature (where blue is low and red is high). For instance, when effort has higher values (in red), it tends to push predictions towards the positive side of the SHAP axis, suggesting that greater effort is associated with improved learning. In contrast, lower values of effort (in blue) push the prediction towards the negative side, indicating that less effort is linked to lower predicted outcomes. Another example is the mean pace feature. Higher values of mean pace (red points) are associated with positive contributions to the prediction, while lower values (blue points) show a negative contribution. This indicates that faster pacing positively influences learning.

Partial Dependence Plots (PDPs) show how specific variables affect predictions (given that all other variables are held constant). In other words, PDPs allow us to see to which extent a change in a certain feature influences the predicted outcome if all other variables did not change. The green line indicates that **feedback** has a relatively consistent impact on predictions, with a slight increase at higher values. The histogram below highlights that most feedback values fall between 50 and 80, where the predictions remain stable, suggesting minimal change in the outcome based on feedback levels within this range.

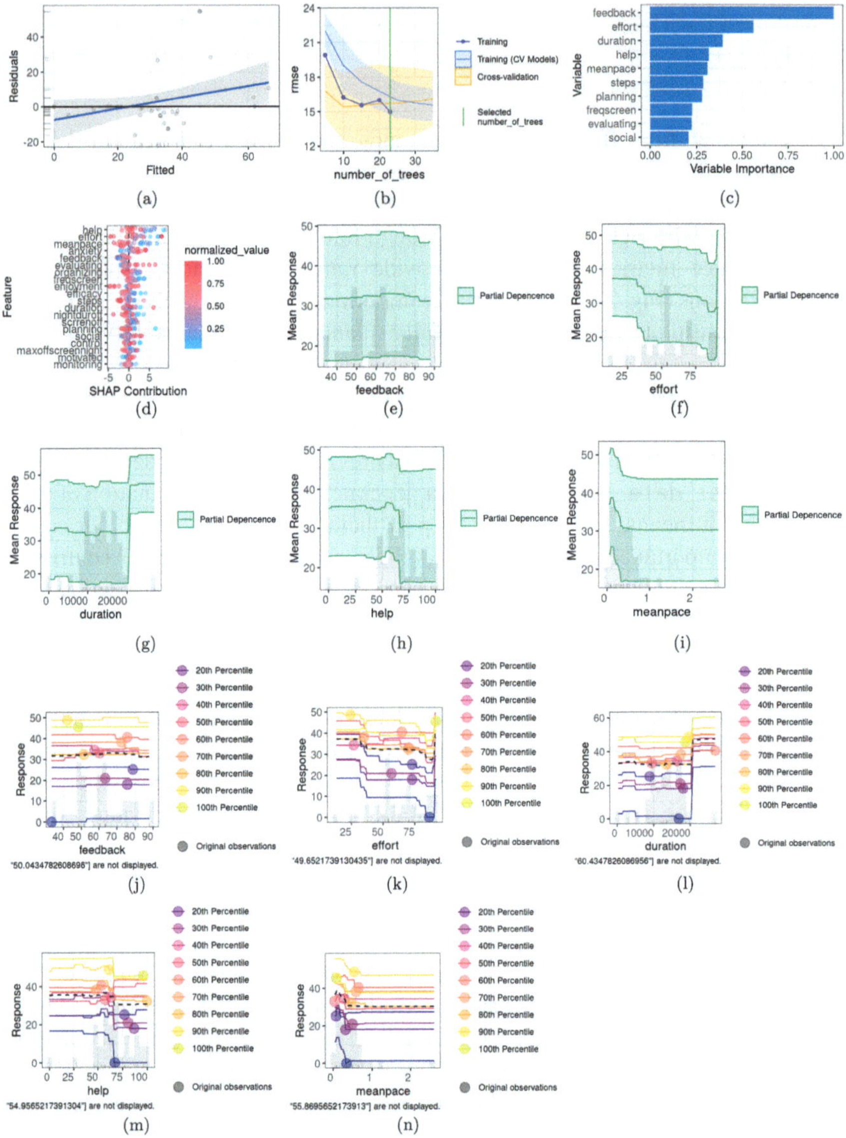

Fig. 2 Explainer. (**a**) Residual analysis. (**b**) Learning curve plot. (**c**) Variable importance. (**d**) SHAP summary. (**e**) Partial dependence plot (feedback). (**f**) Partial dependence plot (effort). (**g**) Partial dependence plot (duration). (**h**) Partial dependence plot (help). (**i**) Partial dependence plot (meanpace). (**j**) Individual conditional expectations (feedback). (**k**) Individual conditional expectations (effort). (**l**) Individual conditional expectations (duration). (**m**) Individual conditional expectations (help). (**n**) Individual conditional expectations (meanpace)

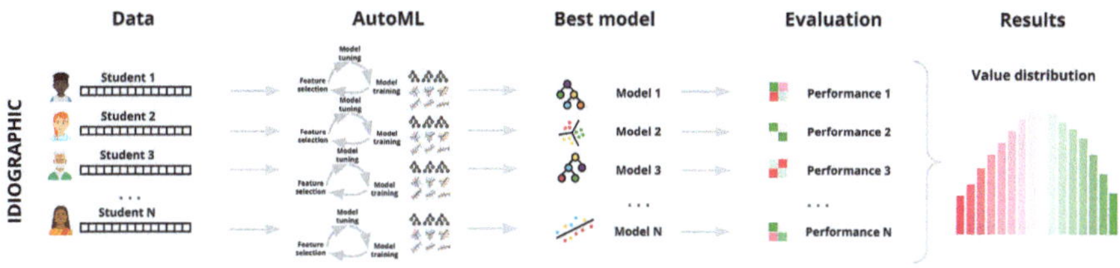

Fig. 3 AutoML as a tool for idiographic ML

The **Individual Conditional Expectation (ICE) plot** shows how a variable affects the model's predictions across different percentiles. For instance, in the duration plot, the x-axis represents **duration**, and the y-axis shows the predicted response. Each colored line represents a different percentile, with the **partial dependence** (dashed black line) indicating the average relationship between **duration** and the response. At lower **duration** values, predictions remain flat across percentiles, meaning the feature has little impact. However, around the 20,000 mark, predictions increase, especially for lower percentiles (e.g., the 0th percentile). This suggests that longer duration significantly improve predictions for these groups.

```
# Run h2o.explain to explain the model's behavior
explained_drf <- h2o.explain(best_drf, test_data_h2o)

# View the explanation output
print(explained_drf)
```

4.2 *Multiple Idiographic Models*

In the previous steps, we have automated the process of finding the best performing model for one of our students (Alice). As mentioned before, manually creating—and fine tuning—a unique model for each individual is an exhausting time-consuming process and can be impractical when done for several dozens of students as in our case. The AutoML implemented by `h2o` automates this process by handling model selection, hyperparameter tuning, evaluating and reporting the best models (Fig. 3). This is where precise `h2o` AutoML offers an efficient solution that require little interference and offers individualized ML. In fact, this is the case in most of the cases where idiographic models are estimated when we have to deal with multiple people and here where `h2o` and AutoML becomes more compelling to use.

Since each student's data patterns may be rather different, the best algorithm may vary from one to another. One student might have better results with GBM, while another might see more accurate predictions from an RF model. AutoML ensures that the best-performing model is chosen based on the characteristics of each person's data, optimizing the predictions on a per-person basis (idiographically). Also, each individual's dataset may require different hyperparameter tuning—a functionality that AutoML handles efficiently. Thus, h2o ensures that the model for each individual is not only correctly selected but also fine-tuned for best performance for each individual. The next code, goes through finding the best model for each student in our dataset. The process is technically similar to before, however, with a main difference is that we will **loop** through each person and use h2o to find the best model as we did before for the single individual.

4.2.1 Estimating Multiple Algorithms

Similar to what we did in training of the single idiographic model, we begin by loading the necessary libraries: `dplyr` for data manipulation and `h2o` for model training. We read the `synthetic_Data_share.RDS` using `readRDS()` which contains the data for the individual students. We the define the predictors, the phone activity variables that capture aspects like mean pace, steps, and screen frequency, and the SRL variables which captures students' SRL as well as self-efficacy and emotions e.g., anxiety and enjoyment. We then combine the variables into a single vector and define the response variable, `learning` as the target for our prediction.

```
# Load necessary libraries
library(dplyr)  # For data manipulation
library(h2o)    # For H2O ML platform
set.seed(265)
# Read in the dataset
synthetic_data <- readRDS("synthetic_Data_share.RDS")

# Define the phone activity variables
phone_vars <- c("meanpace", "steps", "freqscreen", "duration",
                "scrrenoff", "nightduroff", "maxoffscreennight")

# Define the SRL variables
srl_vars <- colnames(select(synthetic_data, efficacy:enjoyment))

# Combine all variables into the Predictors list
predictors <- c(srl_vars, phone_vars)

# Set the response variable
response_variable <- "learning"
```

Then, to do the estimation, we begin by initializing the `h2o` cluster with `h2o.init(nthreads = -1, max_mem_size = "12G")`. This command ensures that h2o uses all available CPU threads (`nthreads = -1`) and caps memory usage at 12GB (`max_mem_size = "12G"`). Since we are creating idiographic models for each person individually we need to allocate sufficient computational resources for the intensive tasks of fitting several hundred models running on multiple datasets.

Our dataset has a name field `name` that identifies each student dataset. Here, we use `unique(synthetic_data$name)` to extract all distinct names and later isolate their data. We will later use the list of `dataset_names` to loop through each person's data. Inside the loop, we start by filtering the current person's data. Given that the name is not a predictor, we need to remove the name column too, we do so by using `filtered_data <- synthetic_data %>% filter(name == specific_name)` to isolate the rows for the individual currently being processed, and `select(-name)` to remove the `name` column since it's no longer needed in the modeling process. After we have filtered the data for the currently modeled student, it needs to be converted to `h2o`'s data frame format. For that we use the function `as.h2o(filtered_data)` as we did before in the individual model. The next steps are familiar by now, and entail splitting the data into training and testing datasets using `h2o.splitFrame(data = h2o_data, ratios = 0.8, seed = 256)`. We split the data here into 80% for training and 20% for testing (so the the model can be evaluated on new unseen data). We also seed `seed = 256` to ensure that the split remains consistent if we run the model again.

Next, we run `h2o.automl()` for that student to automatically train multiple ML models on the individual's training data. We use the `predictors` list with all the input features (the self-regulated learning variables and phone activity data), and `response_variable` is set to `learning` the outcome we want to predict. We also set `max_runtime_secs = 300` to specify how long the AutoML process runs for each individual, please note that this number is used here for demonstration purposes and you are advised to use higher numbers.

Having trained the model, we can get the best model using `h2o.get_best_model(automl_model)` to retrieve the best-performing model for the individual. This model is chosen based on its performance (and more specifically based on RMSE which is the default criterion). The next step, we evaluate the model performance on the test data (the unseen data) using `h2o.performance(best_model, newdata = test_data_h2o)`.

In the last step, we store each student's data, the best model, and the evaluation results into a list named `results` which contains for each `specific_name`:

- **TrainingData**: The data used to train the model.
- **TestData**: The data used to evaluate the model's performance.
- **AutoMLModel**: The full AutoML object containing all the models tried during the AutoML process.
- **BestModel**: The best performing model chosen by AutoML.
- **Performance**: The performance metrics of the best model on the test data.

Storing these components helps to easily access and compare the models across individuals. For example, we can later retrieve the best model for any individual and review its performance on the test data and compare or aggregate or study the patterns.

The last two lines of code print the performance of each model for each student. As the reader can see, we created a loop that iterates through the data and performs the basic steps of analysis that we learned in the single person example for each person in our sample, and stores the results in the R object `results`.

```
# Initialize h2o with specified resources
h2o.init(nthreads = -1, max_mem_size = "60G")

# Get unique dataset names (subsets)
dataset_names <- unique(synthetic_data$name)

# Set seed for reproducibility
set.seed(2202)

# Initialize a list to store results for each subset
results <- list()

# Loop through each unique dataset (name)
for (specific_name in dataset_names) {

  # Filter the dataset for the current subset and remove the 'name' column
  filtered_data <- synthetic_data %>%
    filter(name == specific_name) %>%
    select(-name)  # Exclude the 'name' column

  # Convert the filtered data into an h2o frame
  h2o_data <- as.h2o(filtered_data)

  # Split the h2o frame into training (80%) and testing (20%) sets
  splits <- h2o.splitFrame(data = h2o_data, ratios = 0.8, seed = 256)
  train_data_h2o <- splits[[1]]  # Training data (80%)
  test_data_h2o <- splits[[2]]   # Testing data (20%)

  # Train the model using h2o AutoML
  automl_model <- h2o.automl(
    x = predictors,
    y = response_variable,
    training_frame = train_data_h2o,
    nfolds = 5,
    max_runtime_secs = 300,  # Adjust this as needed
    seed = 256
  )

  # Get the best model
  best_model <- h2o.get_best_model(automl_model,algorithm= c( "drf", "gbm",
                                                              "glm", "xgboost"))
```

```
  # Evaluate the best model on the test data
  performance <- h2o.performance(best_model, newdata = test_data_h2o)

  # Store the data, models, and performance for the current subset
  results[[specific_name]] <- list(
    TrainingData = train_data_h2o,
    TestData = test_data_h2o,
    AutoMLModel = automl_model,
    BestModel = best_model,
    Performance = performance
  )
}
```

4.2.2 Extracting and Plotting Performance Metrics for Multiple Data sets

After training all the models, now we extract the performance metrics for the multiple datasets and visualize them. We will do so by iterating through the results to extract the performance metrics, and create plots to visualize the distributions of metrics. To store the metrics, we need to create an empty data frame with columns corresponding to the metrics we want to collect: **RMSE**, **MSE**, **MAE**, and **R^2**. The `Dataset` column will store the names of each student to be able to link the results to students. We retrieve the performance object we computed before for the student: `results[[dataset_name]]$Performance` and then use the function `h2o.rmse()`, `h2o.mse()`, `h2o.mae()`, and `h2o.r2()` to we extract the values and store it `performance_data`. To plot the data, we will have to reshape it: convert it from wide to long format to make it easier plotting with **ggplot2**. In doing so, each row will represent one metric and its corresponding value for a dataset which allows us to have a single graph with multiple facets (sub-plots). Finally, we use **ggplot2** to create histograms for each metric (see last step in Fig. 3).

Figure 4 shows the distribution of the best idiographic individualized models e.g., **MAE, MSE, RMSE, R^2**. In the **MAE** plot, which measures the average absolute difference between predicted and actual values, we can see a relatively even spread across values ranging from around 2.5 to 7.5. This spread suggests that most models have a moderate error rate. Similarly, the **RMSE** plot, which places greater emphasis on larger errors, shows a wide range of values from around 4 to 16, indicating variability in prediction quality across different models. The **MSE** plot reflects squared errors, showing a heavier concentration between 0 and 150, with only a few outliers, meaning that while most models are performing relatively well, a few may struggle with larger errors. The **R^2** (Coefficient of Determination) plot indicates the proportion of variance explained by the model, shows an interesting spread where most **R^2** values are concentrated between 0.5 and 1.0, indicating that the models perform well in explaining the variance for most datasets, excluding "Diana."

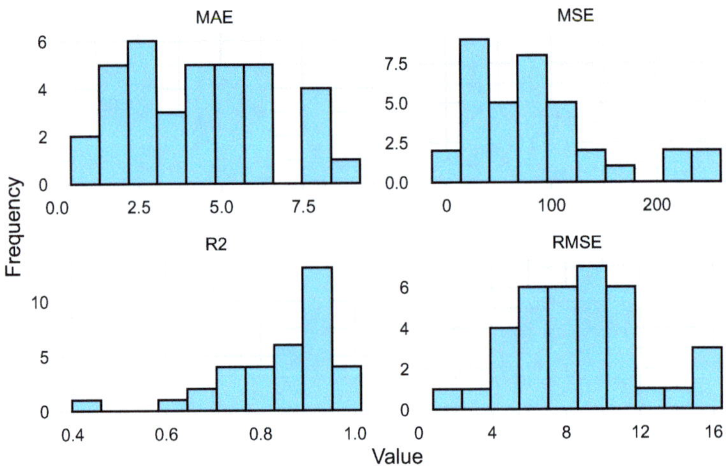

Fig. 4 Distribution of performance metrics

```
# Load necessary libraries
library(ggplot2)
library(dplyr)
library(tidyr)

# Initialize an empty dataframe to store performance metrics
performance_data <- data.frame(Dataset = character(), RMSE = numeric(),
                               MSE = numeric(), MAE = numeric(),
                               R2 = numeric(), stringsAsFactors = FALSE)

# Loop through the results list and extract performance metrics directly
# from results$Performance
for (dataset_name in names(results)) {

  # Extract performance metrics directly from the results list
  performance <- results[[dataset_name]]$Performance

  # Extract individual performance metrics (RMSE, MSE, MAE, R2) from
  # already computed data
  rmse_value <- h2o.rmse(performance)
  mse_value <- h2o.mse(performance)
  mae_value <- h2o.mae(performance)
  r2_value <- h2o.r2(performance)

  # Append the extracted metrics to the performance dataframe
  performance_data <- rbind(performance_data, data.frame(
```

```
    Dataset = dataset_name,
    RMSE = rmse_value,
    MSE = mse_value,
    MAE = mae_value,
    R2 = r2_value
  ))
}

# Reshape the data to long format for easier plotting with ggplot
performance_data_long <- performance_data %>%
  gather(key = "Metric", value = "Value", -Dataset) %>%
  filter(!(Metric == "R2" & Dataset == "Diana"))

ggplot(performance_data_long, aes(x = Value)) +
  geom_histogram(bins = 10, color = "black", fill = "skyblue") +
  facet_wrap(~ Metric, scales = "free") +
  labs(title = "Distribution of Performance Metrics",
       x = "Value", y = "Frequency") +
  theme_minimal()
```

4.2.3 Explanatory Variables

Having plotted the evaluation metrics, now we extract the top five most important variables for each student based on the best model stored in the `results` list that we created before. The code similarly loops through each student's data, retrieves the variable importance from their best model, and stores the top variables in the `top_variables` list. The code begins by creating an empty list called `top_variables`. This list will serve as a container for storing the top five variables for each student as we loop through the `results` list. As the loop iterates over the `results` list, it first extracts the name of each student and the corresponding "best" model for that student. This model is extracted using `results[[student_name]]$BestModel`. Then we extract the important variables using `h2o.varimp(best_model)` function. This function returns the variables ranked by their contribution to the model's predictive power.

From this table of variables, the script selects the top five variables by using the `head(varimp[, "variable"], 5)` function. These variables are then stored in the `top_variables` list under the student's name making it easy to analyze and compare across different models.

The process is repeated for each student in the dataset, so that every individual's top variables are extracted and stored. Once the loop has finished, the `top_variables` list contains a detailed mapping of the top five variables for every

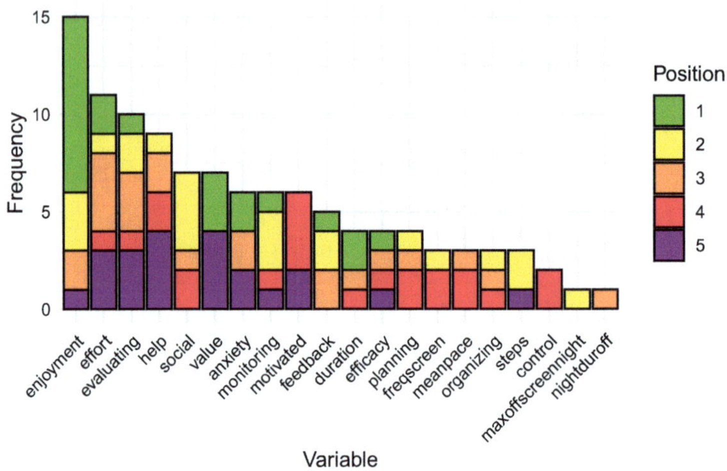

Fig. 5 Frequency of variables in top 5 positions (sorted by frequency)

student. At the end of the script, we can view or inspect the `top_variables` list to see the most important variables for each student. We offer a summary plot also for the most frequent variables and their ranking to examine which variables were most frequent, for instance, at the first position (Fig. 5).

```
```r
# Initialize an empty list to store the top 5 important
variables for each student top_variables <- list()

# Loop through each individual in the results list
for (student_name in names(results)) {

  # Extract the best model for the current student
  best_model <- results[[student_name]]$BestModel

  # Get the variable importance for the best model
  varimp <- h2o.varimp(best_model)

  # Extract the top 5 important variables
  top_5_vars <- head(varimp[, "variable"], 5)

  # Store the top 5 variables for the student in the list
  top_variables[[student_name]] <- top_5_vars
}
```
```

```
# View the top variables for each student
print(top_variables)
'''
```

```
# Initialize an empty dataframe to store variable positions for plotting
variable_positions <- data.frame(Variable = character(), Position = integer(),
                                 stringsAsFactors = FALSE)

# Loop through each student and collect variable positions
for (student_name in names(top_variables)) {

  # Get the top 5 variables for the student
  top_vars <- top_variables[[student_name]]

  # Ensure we only process students with valid top 5 variables
  if (is.null(top_vars) || length(top_vars) == 0) next

  # Assign position 1 to 5 to each variable and store in dataframe
  for (i in seq_along(top_vars)) {
    variable_positions <- rbind(variable_positions,
                                data.frame(Variable = top_vars[i], Position = i))
  }
}

# Count the total frequency of each variable and arrange them in descending order
variable_positions <- variable_positions %>%
  group_by(Variable) %>%
  mutate(Frequency = n()) %>%
  ungroup() %>%
  arrange(desc(Frequency))

# Set up a color palette using RColorBrewer
my_colors <- brewer.pal(10, "Set3")  # Choose a color palette with 5 colors

# Plot the frequency of variables in each position, arranged by overall frequency
ggplot(variable_positions,
       aes(x = reorder(Variable, -Frequency), fill = as.factor(Position))) +
  geom_bar(position = "stack", color = "black") +
  scale_fill_manual(values = my_colors) +  # Apply the selected color palette
  labs(x = "Variable",
       y = "Frequency",
       fill = "Position") +
  theme_minimal() +
  theme(axis.text.x = element_text(angle = 45, hjust = 1)) # Rotate x-axis labels
```

5 Conclusion

In this chapter, we have learned how to implement an AutoML pipeline covering several common ML algorithms. The tutorial has illustrated how to replicate this pipeline across multiple datasets belonging to different individuals, each resulting into an optimized model that most accurately predicts said individual. As exemplified, this technique is very suitable to be able to scale up idiographic analysis [15], since it requires that each individual has its very own model: a task that becomes unfeasible as the number of individuals increases. We demonstrate how AutoML simplifies the ML pipeline, enabling the creation of individually optimized models for multiple datasets efficiently. Moreover, we illustrate how to apply explainable AI techniques to automate the description of the main features and characteristics of each model. The complete pipeline demonstrated in the chapter holds potential to provide automated real-time insights based on idiographic analysis in a transparent and trustable way [42].

However, idiographic analysis is not the only use case for AutoML. AutoML can be used as an alternative to manual analysis [10, 11] in any other case since, as mentioned before, empirical results show AutoML models perform better than—or at least on par with—human-made models. AutoML is especially useful in scenarios that require continuous update of the models, for example real-time dashboards or adaptive learning systems, where it is impossible for a person to perform all of the steps of the ML pipeline in time to provide useful insights. Lastly, it should be noted that AutoML can be used for use cases beyond prediction, such as clustering [43], for example, enabling the automatic detection of students' groups or profiles.

References

1. Mishra N, Silakari S (2012) Predictive analytics: a survey, trends, applications, oppurtunities and challenges. Int J Comput Sci Inf Technol 3:4434–4438
2. Jovanovic J, López-Pernas S, Saqr M (2024) Predictive modelling in learning analytics: a machine learning approach in R. Springer Nature Switzerland, pp 197–229. https://doi.org/10.1007/978-3-031-54464-4_7
3. Brooks C, Thompson C (2017) Predictive modelling in teaching and learning. Society for Learning Analytics Research (SoLAR), Alberta, pp 61–68. https://doi.org/10.18608/hla17.005
4. Saqr M, Jovanovic J, Viberg O, Gašević D (2022) Is there order in the mess? A single paper meta-analysis approach to identification of predictors of success in learning analytics. Stud Higher Educ 47:2370–2391. https://doi.org/10.1080/03075079.2022.2061450
5. Shmueli G, Koppius OR (2011) Predictive analytics in information systems research. MIS Q 35:553–572. https://misq.umn.edu/predictive-analytics-in-information-systems-research.html
6. Adesina AA, Iyelolu TV, Paul PO (2024) Leveraging predictive analytics for strategic decision-making: enhancing business performance through data-driven insights. World J Adv Res Rev 22:1927–1934. https://doi.org/10.30574/wjarr.2024.22.3.1961

7. Olaniyi O, Shah NH, Abalaka A, Olaniyi FG (2023) Harnessing predictive analytics for strategic foresight: a comprehensive review of techniques and applications in transforming raw data to actionable insights. SSRN Electron J. https://doi.org/10.2139/ssrn.4635189
8. Saqr M, Cheng R, López-Pernas S, Beck ED (2024) Idiographic artificial intelligence to explain students' self-regulation: toward precision education. Learn Indiv Differ 114:102499. https://doi.org/10.1016/j.lindif.2024.102499
9. López-Pernas S, Oliveira E, Song Y, Saqr M (2025) AI, explainable AI and evaluative AI: an introduction to informed data-driven decision-making in education. In: Saqr M, López-Pernas S (eds) Advanced learning analytics methods: AI, precision and complexity. Springer Nature Switzerland, Cham
10. Saqr M, Misiejuk K, Tikka S, López-Pernas S (2025) Artificial intelligence: using machine learning to predict students' performance. In: Saqr M, López-Pernas S (eds) Advanced learning analytics methods: AI, precision and complexity. Springer Nature Switzerland, Cham
11. Saqr M, Misiejuk K, Tikka S, López-Pernas S (2025) Artificial intelligence: using machine learning to classify students and predict low achievers. In: Saqr M, López-Pernas S (eds) Advanced learning analytics methods: AI, precision and complexity. Springer Nature Switzerland, Cham
12. Saqr M, López-Pernas S (2025) Explainable artificial intelligence in education: a tutorial for identifying the variables that matter. In: Saqr M, López-Pernas S (eds) Advanced learning analytics methods: AI, precision and complexity. Springer Nature Switzerland, Cham
13. Saqr M, López-Pernas S (2025) Individualized explainable artificial intelligence: a tutorial for identifying local and individual predictions. In: Saqr M, López-Pernas S (eds) Advanced learning analytics methods: AI, precision and complexity. Springer Nature Switzerland, Cham
14. Saqr M, Dever D, López-Pernas S (2025) Idiographic networks: A tutorial on graphical vector autoregression and unified structural equation modeling. In: Saqr M, López-Pernas S (eds) Advanced learning analytics methods: AI, precision and complexity. Springer Nature Switzerland, Cham
15. Saqr M, Ito H, López-Pernas S (2025) A comprehensive introduction to idiographic and within-person analytics. In: Saqr M, López-Pernas S (eds) Advanced learning analytics methods: AI, precision and complexity. Springer Nature Switzerland, Cham
16. Saqr M, Lopez-Pernas S (2021) Idiographic learning analytics: a definition and a case study. 2021 international conference on advanced learning technologies (ICALT), pp 163–165. https://doi.org/10.1109/icalt52272.2021.00056
17. Ito H, López-Pemas S, Saqr M (2024) A scoping review of idiographic research in education: too little, but not too late. 2024 IEEE international conference on advanced learning technologies (ICALT), pp 10–12. https://doi.org/10.1109/icalt61570.2024.00010
18. Saqr M (2023) Modelling within-person idiographic variance could help explain and individualize learning. Br J Educ Technol 54:1077–1094. https://doi.org/10.1111/bjet.13309
19. Saqr M (2024) Group-level analysis of engagement poorly reflects individual students' processes: why we need idiographic learning analytics. Comput Human Behav 150:107991. https://doi.org/10.1016/j.chb.2023.107991
20. Fryda T, LeDell E, Gill N, Aiello S, Fu A, Candel A, Click C, Kraljevic T, Nykodym T, Aboyoun P, Kurka M, Malohlava M, Poirier S, Wong W (2024) h2o: R interface for the 'H2O' scalable machine learning platform. https://github.com/h2oai/h2o-3
21. Tornede A, Deng D, Eimer T, Giovanelli J, Mohan A, Ruhkopf T, Segel S, Theodorakopoulos D, Tornede T, Wachsmuth H, Lindauer M (2023) AutoML in the age of large language models: current challenges, future opportunities and risks. arXiv [csLG]. http://arxiv.org/abs/2306.08107
22. Garmpis S, Maragoudakis M, Garmpis A (2022) Assisting educational analytics with AutoML functionalities. Computers 11:97. https://doi.org/10.3390/computers11060097
23. Shen Z, Zhang Y, Wei L, Zhao H, Yao Q (2018) Automated machine learning: From principles to practices. arXiv [csAI]. http://arxiv.org/abs/1810.13306
24. He X, Zhao K, Chu X (2021) AutoML: a survey of the state-of-the-art. Knowl-Based Syst 212:106622. https://doi.org/10.1016/j.knosys.2020.106622

25. Fidan AS, Şimşek M, Kayhan BK (2023) A survey of the state-of-the-art AutoML tools and their comparisons. Int J Adv Natural Sci Eng Res. https://doi.org/10.59287/as-ijanser.570
26. Elsken T, Metzen JH, Hutter F (2019) Neural architecture search: a survey. J Mach Learn Res 20:1–21. http://jmlr.org/papers/v20/18-598.html
27. Barbudo R, Ventura S, Romero JR (2023) Eight years of AutoML: categorisation, review and trends. Knowl Inf Syst 65:5097–5149. https://doi.org/10.1007/s10115-023-01935-1
28. Del Valle AM, Mantovani RG, Cerri R (2023) A systematic literature review on AutoML for multi-target learning tasks. Artif Intell Rev 56:2013–2052. https://doi.org/10.1007/s10462-023-10569-2
29. Feurer M, Klein A, Eggensperger K, Springenberg J, Blum M, Hutter F (2015) Efficient and robust automated machine learning. Adv Neural Inf Process Syst 28. https://proceedings.neurips.cc/paper_files/paper/2015/file/11d0e6287202fced83f79975ec59a3a6-Paper.pdf
30. Oliveira S de, Topsakal O, Toker O (2024) Benchmarking automated machine learning (AutoML) frameworks for object detection. Information (Basel) 15:63. https://doi.org/10.3390/info15010063
31. Thirunavukarasu AJ, Elangovan K, Gutierrez L, Li Y, Tan I, Keane PA, Korot E, Ting DSW (2023) Democratizing artificial intelligence imaging analysis with automated machine learning: tutorial. J Med Int Res 25:e49949. https://doi.org/10.2196/49949
32. Castellanos-Nieves D, García-Forte L (2023) Improving automated machine-learning systems through green AI. Appl Sci (Basel) 13:11583. https://doi.org/10.3390/app132011583
33. Zhao Y, Wang P (2024) The investigation of student's online learning adaptability level prediction based on AutoML. In: 2024 IEEE 3rd international conference on electrical engineering, big data and algorithms (EEBDA). IEEE, pp 74–79. https://doi.org/10.1109/eebda60612.2024.10485809
34. Bosch N (2021) AutoML feature engineering for student modeling yields high accuracy, but limited interpretability. J Educ Data Min 13:55–79. https://doi.org/10.5281/ZENODO.5275314
35. Buñay-Guisñan P, Lara JA, Cano A, Cerezo R, Romero C (2024) Easing the prediction of student dropout for everyone integrating AutoML and explainable artificial intelligence, pp 857–861 https://doi.org/10.5281/ZENODO.12729972
36. Tsiakmaki M, Kostopoulos G, Kotsiantis S, Ragos O (2021) Fuzzy-based active learning for predicting student academic performance using autoML: a step-wise approach. J Comput High Educ 33:635–667. https://doi.org/10.1007/s12528-021-09279-x
37. Tikka S, Kopra J, Heinäniemi M, López-Pernas S, Saqr M (2024) Getting started with R for education research. In: Saqr M, López-Pernas S (eds) Learning analytics methods and tutorials: a practical guide using r. Springer, Berlin
38. Kopra J, Tikka S, Heinäniemi M, López-Pernas S, Saqr M (2024) An R approach to data cleaning and wrangling for education research. In: Saqr M, López-Pernas S (eds) Learning analytics methods and tutorials: a practical guide using r. Springer, Berlin
39. Tikka S, Kopra J, Heinäniemi M, López-Pernas S, Saqr M (2024) Introductory statistics with R for educational researchers. In: Saqr M, López-Pernas S (eds) Learning analytics methods and tutorials: a practical guide using r. Springer, Berlin
40. López-Pernas S, Misiejuk K, Tikka S, Kopra J, Heinäniemi M, Saqr M (2024) Visualizing and reporting educational data with R. In: Learning analytics methods and tutorials. Springer Nature Switzerland, Cham, pp 151–194. https://doi.org/10.1007/978-3-031-54464-4/_6
41. Saqr M, López-Pernas S (2024) Mapping the self in self-regulation using complex dynamic systems approach. Br J Educ Technol 55:1376–1397. https://doi.org/10.1111/bjet.13452
42. López-Pernas S, Saqr M (2021) Idiographic learning analytics: a within-person ethical perspective. researchgatenet. https://www.researchgate.net/profile/Sonsoles-Lopez-Pernas/publication/350886273_Idiographic_Learning_Analytics_A_Within-Person_Ethical_Perspective/links/607839718ea909241efea6f5/Idiographic-Learning-Analytics-A-Within-Person-Ethical-Perspective.pdf

43. Saqr M, López-Pernas S, Törmänen T, Vogelsmeier LVDE (2025) Stability within and heterogeneity between: an idiographic approach to the temporal dynamics of momentary self-regulated learning. In: Proceedings of the 15th international conference on learning analytics and knowledge. SoLAR; ACM (in press)

MIX
Papier aus verantwortungsvollen Quellen
Paper from responsible sources
FSC® C105338
FSC
www.fsc.org

If you have any concerns about our products,
you can contact us on
ProductSafety@springernature.com

In case Publisher is established outside the EU,
the EU authorized representative is:
Springer Nature Customer Service Center GmbH
Europaplatz 3, 69115 Heidelberg, Germany

Printed by Libri Plureos GmbH
in Hamburg, Germany